MECHANISMS OF RESISTANCE TO PLANT DISEASES

ADVANCES IN AGRICULTURAL BIOTECHNOLOGY

Akazawa T., et al., eds: The New Frontiers in Plant Biochemistry. 1983. ISBN 90-247-2829-0

Gottschalk W. and Müller H.P., eds: Seed Proteins: Biochemistry, Genetics, Nutritive Value. 1983. ISBN 90-247-2789-8

Marcelle R., Clijsters H. and Van Poucke M., eds: Effects of Stress on Photosynthesis. 1983. ISBN 90-247-2799-5

Veeger C. and Newton W.E., eds: Advances in Nitrogen Fixation Research. 1984. ISBN 90-247-2906-8

Chinoy N.J., ed: The Role of Ascorbic Acid in Growth, Differentiation and Metabolism of Plants. 1984. ISBN 90-247-2908-4

Witcombe J.R. and Erskine W., eds: Genetic Resources and Their Exploitation – Chickpeas, Faba beans and Lentils. 1984. ISBN 90-247-2939-4

Sybesma C., ed: Advances in Photosynthesis Research. Vols. I-IV. 1984. ISBN 90-247-2946-7

Sironval C., and Brouers M., eds: Protochlorophyllide Reduction and Greening. 1984. ISBN 90-247-2954-8

Fuchs Y., and Chalutz E., eds: Ethylene: Biochemical, Physiological and Applied Aspects. 1984. ISBN 90-247-2984-X

Collins G.B., and Petolino J.G., eds: Applications of Genetic Engineering to Crop Improvement. 1984. ISBN 90-247-3084-8

Chapman G.P., and Tarawali S.A., eds: Systems for Cytogenetic Analysis in *Vicia Faba* L. 1984. ISBN 90-247-3089-9

Hardarson G., and Lie T.A., eds: Breeding Legumes for Enhanced Symbiotic Nitrogen Fixation. 1985. ISBN 90-247-3123-2

Magnien E., and De Nettancourt D., eds: Genetic Engineering of Plants and Microorganisms Important for Agriculture. 1985. ISBN 90-247-3131-3

Schäfer-Menuhr A., ed: *In Vitro* Techniques – Propagation and Long Term Storage. 1985. ISBN 90-247-3186-0

Bright S.W.J., and Jones M.G.K., eds: Cerial Tissue and Cell Culture. 1985. ISBN 90-247-3190-9

Purohit S.S., ed: Hormonal Regulation of Plant Growth and Development. 1985. ISBN 90-247-3198-4

Fraser R.S.S., ed: Mechanisms of Resistance to Plant Diseases. 1985. ISBN 90-247-3204-2

Mechanisms of Resistance to Plant Diseases

edited by

R.S.S. FRASER

National Vegetable Research Station
Wellesbourne
U.K.

1985 **MARTINUS NIJHOFF/DR W. JUNK PUBLISHERS**
a member of the KLUWER ACADEMIC PUBLISHERS GROUP
DORDRECHT / BOSTON / LANCASTER

Distributors

for the United States and Canada: Kluwer Academic Publishers, 190 Old Derby Street, Hingham, MA 02043, USA
for the UK and Ireland: Kluwer Academic Publishers, MTP Press Limited, Falcon House, Queen Square, Lancaster LA1 1RN, UK
for India, Pakistan, Burma and Sri Lanka: Agro Botanical Publishers (India), Old Ginnani, Bikaner 334 001 India
for all other countries: Kluwer Academic Publishers Group, Distribution Center, P.O. Box 322, 3300 AH Dordrecht, The Netherlands

Library of Congress Cataloging in Publication Data

Main entry under title:

Mechanisms of resistance to plant diseases.

(Advances in agricultural biotechnology)
Includes index.
1. Plants--Disease and pest resistance.
I. Fraser, R. S. S. II. Series.
SB750.M43 1985 581.2'1 85-15263
ISBN 90-247-3204-2

ISBN 90-247-3204-2 (this volume)
ISBN 90-247-2790-1 (series)

Copyright

PRINTED IN THE NETHERLANDS

CONTENTS

PREFACE

Plant resistance to pathogens is one of the most important strategies of disease control. Knowledge of resistance mechanisms, and of how to exploit them, has made a significant contribution to agricultural productivity. However, the continuous evolution of new variants of pathogen, and additional control problems posed by new crops and agricultural methods, creates a need for a corresponding increase in our understanding of resistance and ability to utilize it.

The study of resistance mechanisms also has attractions from a purely academic point of view. First there is the breadth of the problem, which can be approached at the genetical, molecular, cellular, whole plant or population levels. Often there is the possibility of productive exchange of ideas between different disciplines. Then there is the fact that despite recent advances, many of the mechanisms involved have still to be fully elucidated. Finally, and compared with workers in other areas of biology, the student of resistance is twice blessed in having as his subject the interaction of two or more organisms, with the intriguing problems of recognition, specificity and co-evolution which this raises.

The number of possible approaches to the study of resistance, and the diversity of the hosts and parasites involved, inevitably dictated that this book be written by several authors. However, my policy when inviting authors to contribute was to ask them to review their own subject area in very broad terms, rather than simply to summarize their own research. This was with the aim of inferring general principles about resistance mechanisms, rather than becoming concerned with case histories of individual diseases.

On behalf of all the authors, it is a pleasure to thank many colleagues for critical readings and comments on individual Chapters. My thanks also to Linda Saw and Noreen Wood for invaluable assistance in preparation of the manuscript.

R. S. S. Fraser
May 1985

CONTRIBUTORS

Numbers indicate pages on which contributions by each author begin

I. R. CRUTE 80,197

Plant Pathology Section, National Vegetable Research Station,
Wellesbourne, Warwick CV35 9EF, U.K.

P. J. G. M. DE WIT 197,405

Department of Phytopathology, Agricultural University, 9 Binnenhaven,
6709 BD Wageningen, The Netherlands.

R. S. S. FRASER 1,13,62,143,373,425

Biochemistry Section, National Vegetable Research Station,
Wellesbourne, Warwick CV35 9EF, U.K.

J. V. GROTH 310

Department of Plant Pathology, University of Minnesota,
St. Paul, Minnesota 55108, U.S.A.

A. G. HEPBURN 425

Department of Agronomy, University of Illinois,
Turney Hall, 1102 South Goodwin Avenue
Urbana, Illinois 61801, U.S.A.

M. J. JEGER 310

Department of Plant Pathology and Microbiology, Texas A & M University,
College Station, Texas 77843, U.S.A.

J. P. RIDE 29

Department of Microbiology, University of Birmingham,
P.O. Box 363, Birmingham B15 2TT, U.K.

M. WADE 197,425

Shell Biosciences Laboratory, Shell Research Centre,
Sittingbourne, Kent ME9 8AG, U.K.

ABBREVIATIONS OF VIRUS NAMES

AMV	Alfalfa mosaic virus
BaYMV	Barley yellow mosaic virus
BBMV	Broad bean mottle virus
BCMV	Bean common mosaic virus
BeMV	Beet mosaic virus
BMV	Brome mosaic virus
BNYV	Broccoli necrotic yellows virus
BSMV	Barley stripe mosaic virus
BYDV	Barley yellow dwarf virus
BYMV	Bean yellow mosaic virus
BYV	Beet yellows virus
CaMV	Cauliflower mosaic virus
CCMV	Cowpea chlorotic mottle virus
CEV	Citrus exocortis viroid
CGMMV	Cucumber green mottle mosaic virus
CMV	Cucumber mosaic virus
CPMV	Cowpea mosaic virus
CTrV	Citrus tristeza virus
CTV	Curly top virus
FDV	Fiji disease virus
LRV	*Lycopersicon* ringspot virus
MDMV	Maize dwarf mosaic virus
MYMV	Mungbean yellow mosaic virus
PSTV	Potato spindle tuber viroid
PSV	Peanut stunt virus
PVX	Potato virus X
PVY	Potato virus Y
RMV	Ryegrass mosaic virus
RRV	Raspberry ringspot virus
SBMV	Southern bean mosaic virus
SCMV	Sugarcane mosaic virus
SEV	Severe etch virus
SMV	Soybean mosaic virus
TAV	Tomato aspermy virus
TBRV	Tomato black ring virus
TBSV	Tomato bushy stunt virus
TMV	Tobacco mosaic virus
TNDV	Tobacco necrotic dwarf virus
TNV	Tobacco necrosis virus
TRSV	Tobacco ringspot virus
TRV	Tobacco rattle virus
TSWV	Tomato spotted wilt virus
TSV	Tobacco streak virus
TuMV	Turnip mosaic virus
TVMV	Tobacco vein mottle virus
TYLCV	Tomato yellow leaf curl virus
TYMV	Turnip yellow mosaic virus
WMV-1	Watermelon mosaic virus-1
WMV-2	Watermelon mosaic virus-2
WSMV	Wheat streak mosaic virus

CHAPTER 1. SOME BASIC CONCEPTS AND DEFINITIONS IN RESISTANCE STUDIES

R. S. S. FRASER

1.1. WHY STUDY RESISTANCE?

1.1.1. Crop losses caused by disease. Crops can suffer severe losses in yield and quality as a result of many factors. In broad terms, it is estimated that about 30% of world crop production is lost every year, and that about one third of this loss is due to plant diseases (International Atomic Energy Agency, 1977). Mackenzie (1983) calculated that in the United States wheat crop, there is a 14% loss annually due to fungal, bacterial and virus diseases. There have been numerous reports of the losses caused in other crops by specific pathogens (e.g. Large, 1958; Broadbent, 1964; Kingsland, 1980).

It should be emphasized that although disease damage and crop loss may appear to be obvious and of large proportions, especially in epidemics leading to complete crop failure, the precise quantitative assessment of loss due to disease is a complex subject. It has to be viewed against the background of how yield is assessed, involving the biologically-based concepts of theoretical, primitive and attainable yield, as well as the more practically-based concepts of actual and economic yield (Zadoks & Schein, 1979). Crop losses must also be considered in the light of different methods of assessment and evaluation (Large, 1966; James, 1974; Main, 1977), and the effects of control and other treatments which lead to the concepts of potential and actual loss (Zadoks & Schein, 1979).

1.1.2. Disease agents. In this book we are concerned with mechanisms of resistance to diseases caused by those pathogenic non-living agents and micro-organisms which multiply within the host plant, that is viroids, viruses, mycoplasma-like organisms, bacteria and fungi. We are not concerned with damage caused by abiotic factors such as atmospheric pollution or water stress: these topics have recently been reviewed by Pell (1979); Laurence & Weinstein (1981) and Ayres (1984). We also exclude damage caused by angiospermic plant parasites (Knutson, 1979) and pests such as insects (Norris, 1979) and nematodes (Giebel, 1982), except in cases where these organisms act as vectors of pathogens which do fall within the scope of this book (Harris, 1981).

Obviously, the depth of knowledge on different groups of pathogenic agents varies. Viruses and pathogenic micro-organisms have been most studied, and because of their fundamentally different replicative cycles, are considered in separate Chapters. Viroids bear sufficient resemblance in many ways to viruses to consider what little is known of host resistance to them in the virus Chapters. Knowledge of the processes of pathogenesis by mycoplasmas and related organisms has lagged behind that of the other groups. Sources of resistance to several diseases caused by mycoplasma-like organisms have been identified (reviewed in Maramorosch, 1982), but exploitation in resistance breeding is at an early stage, and there appears to be little or no understanding of the mechanisms involved.

1.1.3. Disease control strategies. It quickly becomes clear that choosing a strategy for disease control is based on many factors, and not only on the extent of crop loss caused by that disease. Thus effects of natural factors such as climate and host-pathogen interactions on yield losses are only one input. Economic factors, such as cost of treatments, value of the increased yield, and whether the market can absorb the increased production are also critical (Large, 1958; Norton & Conway, 1977). Furthermore, the social and environmental side effects of any strategy deserve consideration: does creation of a cash crop economy by effective disease control in monocultures destroy a more broadly based, subsistence farming approach? Is control of a particular disease achieved only by use of chemicals with deleterious effects on beneficial organisms? Does concentration on control by resistance breeding select forms of the pathogen which eventually drive the crop species into an evolutionary cul-de-sac? The available methods of disease control must be assessed against the background of such questions. Lack of a particular method of control for a specific pathogen may stimulate research on other possible methods of control. Finally, it should be stressed that individual methods are unlikely to give fully satisfactory and durable control strategies, and that some form of integrated strategy involving various methods is more likely to succeed.

Chemotherapy. For virus diseases, chemotherapy is as yet largely an uncertain experimental tool with variable effects (Bailiss *et al.*, 1977; Fraser & Whenham, 1978). Field use in treatment of crops does not appear to have been reported, but chemical treatments have proved of value in elimination of viruses from universally-infected lines or cultivars (Simpkins et al., 1981; Cassells & Long, 1982). Treatment of mycoplasma-like diseases with tetracycline antibiotics is possible, but tends to give temporary relief rather than cure (Ishiie et al., 1967). The use of human therapeutic agents on plant diseases is also questionable (Maramorosch, 1982).

Undoubtedly, the major applications of chemical control have been with fungal, and to a lesser extent, bacterial diseases. Some of the history of chemical controls is charted in Chapter 11. While fungicides have undoubtedly given great ability to control disease, there have been problems, and some of these may increase in future. Thus the pathogen may develop resistance to the fungicide (Wade, 1982); there may be phytotoxic side effects, and environmental side effects in that beneficial organisms important to soil fertility may be damaged. Chemical screening and research, manufacture and application may all be expensive processes which reflect on the ultimate cost of treatment. The increasing statutory regulation of chemical controls in many countries is likely to make obtaining clearance for use of new chemicals, or existing chemicals on different crops and systems, an increasingly expensive process. This may tend to make future use of chemotherapy increasingly a matter for the seven major world crops, and militate against development of new disease treatments for the hundreds of minor crops.

Disease avoidance. This includes a wide variety of measures which aim to prevent infection of the crop. These are related to the method of spread or transmission of the pathogen, rather than to the intrinsic nature of the pathogen. They include use of pathogen-free seed; quarantine regulations and import restrictions; soil sterilization; direct control of vectors such as insects and nematodes by insecticides and nematocides, or their indirect control by non-toxic means such as alarm pheromones or alternative attractants. Such control measures can be effective and can have high selectivity and low environmental impact. However, once disease avoidance fails, an alternative mechanism of control operating directly within the crop or on the pathogen is required.

Biological control. Broadly speaking, this involves use of one organism or pathogenic agent to control a second, and prevent it causing disease by limiting its multiplication or spread. The mechanisms of induced resistance considered in Chapters 9 and 10, whereby prior inoculation of plants with one agent or organism confer some measure of protection against a second, challenging pathogen, are in one sense a form of biological control. The mechanism of interaction between the two inocula is likely to be indirect, involving a change in host plant metabolism.

The more direct form of biological control is where the controlling agent is directly pathogenic on, or directly inhibits spread of the plant pathogen. There are possibilities of using viruses to control insect and fungal pests of plants (Baker & Cook, 1974; Coppel & Mertins, 1977; Cook, 1985). Soils contain fungi which parasitize hyphae of fungal plant parasites. There are many indications that such interactions could be developed for disease control (reviewed in Snyder *et al*., 1976; van den Bosch *et al*., 1982). For example, a *Trichoderma* species might partly control *Sclerotium cepivorum* (onion white rot) (Entwistle & Marian, 1983). However, such control strategies generally require much further development.

Resistance. Host resistance is arguably the most important means of disease control, and has been a primary objective in the breeding programmes of numerous crops. The advantages of resistance as a control strategy are that most forms are inherited, and do not required to be applied to each individual as is necessary with chemotherapy. Resistance is generally environmentally sound, without undesirable side effects. It is normally highly selective and non-phytotoxic, although it may be associated directly or indirectly with undesirable side effects on the host. Resistance can be highly effective in disease control, and after the initial expense of the breeding programme, it is a control strategy with low running costs.

The disadvantages of resistance are that it may be ephemeral: there is a selection pressure for mutants of the pathogen which are able to overcome a particular mechanism. Thus a resistance strategy which is effective in time may require constant discovery and development of new sources of resistance, and ultimately the commitment of resources to maintain the status quo in host-pathogen balance rather than to produce net improvement. There may be no effective sources of resistance to a particular pathogen, or natural sources may become exhausted; then resistance alone is no longer viable as a control strategy. Transfer of resistance genes between cultivars, or from related wild species into crop species, can be protracted, and may by prevented by natural barriers to hybridization. There is therefore a need to accelerate gene transfers, and to bypass barriers. Breeding for resistance may be at the expense of other desirable characters, and excessive concentration on breeding for a small number of resistance genes may reduce the total range of genetic diversity of the crop. This gene erosion can affect resistance genes themselves, and also genes for other characters, if non-resistant cultivars are allowed to become extinct because they are not grown commercially.

1.1.4. <u>Approaches to the study of resistance</u>. There are many reasons for studying resistance. To varying extents, the benefits of resistance can best be maximized, and the disadvantages overcome, by an appreciation of how resistance mechanisms operate. New techniques of gene transfer and manipulation should increase our ability to utilize natural resistance genes, and understanding of how the host and pathogen interact should lead to development of novel types of control measure. These are considered in detail in Chapter 11. The great increase in mankind's ability to control animal disease stemming from knowledge of the fundamental biochemistry of the antibody-antigen interaction may be cited as a paradigm.

Resistance mechanisms also have several attractions at the purely academic level. Apart from elucidating how two organisms interact, study of how a plant reacts to a pathogen may be a useful way of making sense of how the healthy plant operates. In the case of plant viruses, which can now be completely characterized in molecular terms such as nucleic acid sequence, there may be a powerful route to probing how complex host metabolic processes, themselves relatively inaccessible, react with varying and completely definable pathogenic entities. At the population level, the host-pathogen combination allows study of processes in population dynamics and co-evolution which pose quite separate types of problem, from those of the species studied in isolation.

This book is concerned with the interactions of host and parasite at several levels. It ranges from phenomena which may be controlled by interactions at the molecular level between host and parasite factors; through effects which are the outcome of processes at the cell, tissue or whole plant level, to interactions between co-evolving populations. At each level, the objective is to examine the types of mechanism involved in the interactions. A direct consequence of this mechanistic approach is that the book does not set out to be a survey of diseases and resistance as case histories; examples are chosen for the light they may shed on mechanisms of resistance and pathogenicity rather than for reasons of, say, economic importance. However, the economic pressures on research priorities and funding inevitably ensure that there is considerable overlap between those diseases where knowledge of the mechanisms of resistance is greatest, and those of agricultural importance.

1.2. SOME WORKING DEFINITIONS

1.2.1. Why define? Resistance studies have been advanced, and hindered, by the urge of mankind to assume control of nature by defining terms. A minor but significant proportion of the phytopathological literature is devoted to lively and entertaining semantic argument, with attempts to define terms; re-define terms; define the differences between related terms, and rationalize different definitions of the same term (e.g. Federation of British Plant Pathologists, 1973; Cooper & Jones, 1983; Tavantzis, 1984; Cooper & Jones, 1984). The problem is exacerbated by the fact that virologists, mycologists, bacteriologists, entomologists and population geneticists have little unity of approach. A further problem is that many of the phenomena being described are not understood in mechanistic terms, or have not been investigated quantitatively, and such lack of knowledge must preclude precise definition. Thus the terminology should be regarded as operational and evolving, rather than as absolute and definitive. Terms may require redefinition as knowledge is accumulated, and attempts at definition may stimulation acquisition of knowledge by posing questions around the borders of the topic.

The approach adopted by the contributors to this book has been to define terms rather broadly, and to try to avoid the perils of premature over-definition. The usage represents a consensus accepted by workers covering a broad spectrum from the molecular to the population levels. Inevitably, there are differences in bias at different levels: for example, the population dynamicist makes more use of general terms such as pathogenicity, while the molecular biologist seeks more specifically defined concepts such as virulence and aggressiveness, which can be interpreted in strictly molecular terms. Figure 1.1. attempts to express graphically some of the terms defined next, and their interactions.

1.2.2. Plant factors. When a plant supports multiplication of a particular pathogen, it is a host, or in other words it is susceptible to that particular pathogen. The amount of disease that a pathogen causes in its host, expressed as host damage measured by symptom severity or loss of yield, may be at least partly distinct in mechanistic terms from the amount of multiplication of the pathogen. Plants which support pathogen multiplication without visible or severe disease symptoms are often referred to as tolerant, but the term has often been used in a loose sense and quantitative assessment of effects on yield or growth (e.g. Kooistra, 1968) are rare.

Resistance is defined very broadly in this book, as an inhibition of any

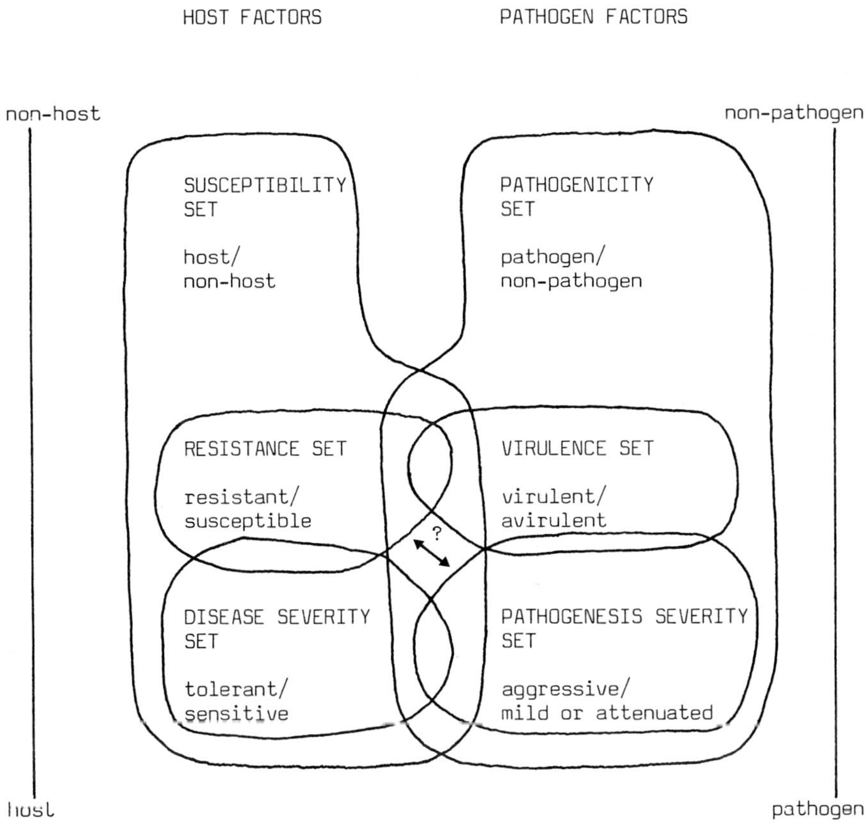

FIGURE 1.1. Some types of interactions between plants and potential pathogens, showing areas of overlap, containment and interaction between different descriptive terms. Each enclosed 'set' refers to a property of the organism, and contains two possible alternative behaviour patterns. These are described by a pair of antonyms. Where host and pathogen sets overlap, there is the possibility of interaction between host and pathogen. For hosts, the resistance and disease severity sets may overlap, and are contained within the wider context of the susceptibility set. For pathogens, the virulence and pathogenesis severity sets may interact, and are contained within the broader pathogenicity set. The pathogenesis severity set may interact with the host resistance set, either directly or through the virulence set. However, the virulence set is not thought to interact directly with the disease severity set, but only affects disease severity through the resistance set.

Note that in the host, the term 'susceptibility' is used in two separate contexts, each with its own antonym, in the susceptibility and resistance sets.

stage of the reproductive cycle of the pathogen in time or space, or of the development of pathogenic effects in the host. The term is applied at the molecular, cell, organism and population level. Thus resistance may operate against multiplication, against disease in the host, or against both.

Host resistance may be split into three basic, though probably overlapping types, based on our conception of the underlying mechanisms. Because the mechanisms are conceived as basically different, each of these types forms the subject of a major division of this book.

a) Non-host resistance or non-host immunity (Chapters 2 and 3) are where all individuals of a species are unaffected by any member of a particular pathogenic species (or virus type). For viruses, no multiplication of the pathogen, and no disease symptoms, can be detected after attempted infection. For micro-organisms, initial physical contact is made between the host and the potential pathogen, but the reproductive cycle is not completed, and pathogen development ceases at a very early stage, generally considered to be when intimate physical and chemical contact have been established. The concept of non-host resistance is therefore related to the broader question of what determines the host range of the pathogen. Although non-host resistance could involve completely different mechanisms from the second class of resistance, it could also represent the completely effective limit case of b).

b) Genetically-determined resistance (Chapters 4 - 8). This occurs when some members of a species contain a gene or genes conferring a degree of resistance to a pathogen to which the species is normally susceptible. Thus the species contains resistant and susceptible individuals, depending on their genetic make-up, and the resistance is heritable. Some authors have used the term 'immunity' to describe extremely effective, genetically-controlled resistance mechanisms, but this useage is not adopted here.

c) Induced resistance (Chapters 9 and 10) occurs when a plant which is normally susceptible to a measureable degree to a particular pathogen has a resistance to that pathogen conferred by a prior infection or treatment. This type of resistance is not heritable, but must be conferred afresh on each individual.

Note that all three types of mechanism ultimately depend on a genetic basis in the host; it is, however, only in b) that whether a plant is resistant or susceptible depends on the particular alleles at the loci involved. Note also that genetically-determined resistance mechanisms may also need to be induced or activated during the early stages of pathogenic attack, but that this induction is different from that in type c) in that the induced gene product is then directly or indirectly effective against the first, and not only against a subsequent inoculum.

1.2.3. Pathogen factors. An organism or agent which successfully establishes an infection leading to its multiplication on a particular plant species is a pathogen of that species; it is non-pathogenic on species which are non-hosts.

Within host species, there are two further terms which describe pathogen behaviour. These are qualifying attributes of the term pathogenicity, and may be to some extent overlapping. Virulence is the ability of a particular isolate of the pathogen to multiply on individuals of the host species containing genes which normally confer resistance to that pathogen. Thus virulence is specifically the ability to overcome genetically-determined resistance. Isolates which are unable to overcome such resistance are avirulent. Aggressiveness is purely a property of the pathogen, and reflects the severity of disease, and/or the amount of multiplication, in any individual of the host species, whether containing genes determining resistance or not. Aggressiveness and virulence are functional characters which need not be linked, but the mechanisms involved may well overlap (Figure 1.1).

1.2.4. Terms in plant-pathogen interactions. It will be clear from the above that many plant terms are defined in the light of interaction with different parasite isolates, and vice-versa, especially in the context of genetically determined resistance and virulence. Indeed, host and pathogen factors cannot be considered in isolation, and it is inevitable that terminologies have evolved to describe the interactions at a unified level. Vanderplank (1963) first introduced two provocative, theoretical concepts to describe host pathogen interactions: vertical and horizontal resistance. These concepts have stimulated much dissent and discussion, and their application at several levels of approach to host-pathogen interactions is discussed elsewhere in this book. At this stage and for completeness, the basic definitions and properties only are cited without qualification and without listing the many exceptions.

Vertical resistance is inherited monogenically or oligogenically, although not all such genetic mechanisms involve vertical resistance. As a consequence of its comparative genetic simplicity, vertical resistance may be readily overcome by small or limited individual evolutionary steps to virulence in the pathogen. This leads to development of a gene-for-gene relationship between host resistance genes and pathogen virulence genes, with a specific and exclusive interaction between related genes or their products. This leads to the concept of differential interactions between different host lines and pathogen isolates. Vertical resistance may, but need not be, temporary.

Horizontal resistance is inherited polygenically or oligogenically, or can be monogenic. It is more difficult for the pathogen to evolve to overcome it, because the genetic complexity and probable number of gene products involved in the more complex host genetic systems would make a resistance-breaking isolate unlikely; or from the other side, because the number of genes in the pathogen in which mutations would be required for virulence may be large, irrespective of the number of genes involved in the host. Horizontal resistance should be equally effective against all isolates of the pathogen, and differential interactions should not occur. Isolates of the pathogen differing in aggressiveness can, however, cause disease of differing degrees of severity on the host. Horizontal resistance should also be long-lasting.

The extent to which all these definitions must be questioned and qualified will, it is hoped, become apparent in subsequent Chapters.

REFERENCES

Ayres, P.G. (1984). The interaction between environmental stress injury and biotic disease physiology. Annual Review of Phytopathology 22, 53-75.

Bailiss, K.W., Cocker, F.M. & Cassells, A.C. (1977). The effect of Benlate and cytokinins on the content of tobacco mosaic virus in tomato leaf discs and cucumber mosaic virus in cucumber cotyledon discs and seedlings. Annals of Applied Biology 87, 383-392.

Baker, K.F. & Cook, R.J. (1974). Biological Control of Plant Pathogens. Freeman, San Francisco, 433 pp.

Broadbent, L. (1964). The epidemiology of tomato mosaic. VII. The effect of TMV on tomato fruit yield and quality under glass. Annals of Applied Biology 54, 209-224.

Cassells, A.C. & Long, R.D. (1982). The elimination of potato viruses X, Y, S and M in meristem and explant cultures of potato in the presence of virazole. Potato Research 25, 165-173.

Cook, R.J. (1985). Biological control of plant pathogens - theory to application. Phytopathology 75, 25-29.

Cooper, J.I. & Jones, A.T. (1983). Responses of plants to viruses; proposals for the use of terms. Phytopathology 73, 127-128.

Coppel, H.C. & Mertins, J.W. (1977). Biological Insect Pest Suppression. Springer Verlag, Berlin, 314 pp.

Federation of British Plant Pathologists (1973). A guide to the use of terms in plant pathology. Phytopathological Paper 17, Commonwealth Mycological Institute, Kew, U.K.

Entwistle, A.R, & Marian, S.E. (1983). Evaluation of Trichoderma spp. to control white rot. Annual Report of the National Vegetable Research Station for 1982, 65-66.

Fraser, R.S.S. & Whenham, R.J. (1978). Chemotherapy of plant virus disease with methyl benzimidazole-2yl-carbamate: effects on plant growth and multiplication of tobacco mosaic virus. Physiological Plant Pathology 13, 51-64.

Giebel, J. (1982). Mechanisms of resistance to plant nematodes. Annual Review of Phytopathology 20, 257-279.

Harris, K.F. (1981). Arthropod and nematode vectors of plant viruses. Annual Review of Phytopathology 19, 391-426.

International Atomic Energy Agency (1977). Induced Mutations Against Plant Diseases. IAEA, Vienna, 581 pp.

Ishiie, T., Doi, Y., Yora, K. & Asuyama, H. (1967). Suppressive effects of antibiotics of tetracycline group on symptom development of the mulberry dwarf disease (in Japanese). Annals of the Phytopathological Society of Japan, 33, 267-275.

James, W.C. (1974). Assessment of plant diseases and losses. Annual Review of Phytopathology 12, 27-48.

Jones, A.T. & Cooper, J.I., (1984). A reply to the comments of S.M. Tavantzis. Phytopathology 74, 381.

Kingsland, G.C. (1980). Effect of maize dwarf virus infection on yield and stalk strength in the field in South Carolina. Plant Disease 64, 271-273.

Knutson, D.M. (1979). How parasitic seed plants induce resistance in other plants. In 'Plant Disease; an Advanced Treatise, Vol. IV. How Pathogens induce Disease' (Horsfall, J.G. & Cowling, E.B., eds), pp. 293-312. Academic Press, New York.

Kooistra, E. (1968). Significance of the non-appearance of visible disease symptoms in cucumber (Cucumis sativus L.) after infection with Cucumis virus 2. Euphytica 17, 136-140.

Large, E.C. (1958). Losses caused by potato blight in England and Wales. Plant Pathology 7, 39-48.

Large, E.C. (1966). Measuring plant disease. Annual Review of Phytopathology 4, 9-28.

Laurence, J.A. & Weinstein, L.H. (1981). Effects of air pollutants on plant productivity. Annual Review of Phytopathology 19, 257-271.

Mackenzie, D.R. (1983). Towards the management of crop losses. In 'Challenging Problems in Plant Health' (Kommendahl, T. & Williams, P.H., eds), pp. 82-92. American Phytopathological Society, St. Paul, Minnesota.

Main, C.E. (1977). Crop destruction: the raison d'étre of plant pathology. In 'Plant Disease: an Advanced Treatise. Vol. 1. How Disease is Managed' (Horsfall, J.G. & Cowling, E.B., eds), pp. 55-78. Academic Press, New York.

Maramorosch, K. (1982). Control of vector-borne mycoplasmas. In 'Pathogens, Vectors and Plant Diseases' (Harris, K.F. & Maramorosch, K., eds), pp. 265-295. Academic Press, New York & London.

Norris, D.N. (1979). How insects induce disease. In 'Plant Disease: an Advanced Treatise, Vol. IV. How Pathogens induce Disease' (Horsfall, J.G. & Cowling, E.B., eds), pp. 239-256. Academic Press, New York.

Norton, G.A. & Conway, G.R. (1977). The economic and social context of pest, disease and weed problems. In 'The Origins of Pest, Parasite, Disease and Weed Problems' (Cherrett, J.M. & Sagar, G.R., eds), pp. 205-226. Blackwell, Oxford.

Pell E.J. (1979). How air pollutants induce disease. In 'Plant disease: an Advanced Treatise, Vol. IV. How Pathogens induce Disease' (Horsfall, J.G. & Cowling, E.B., eds), pp. 273-292. Academic Press, New York.

Simpkins, I., Walkey, D.G.A. & Neely, H. (1981). Chemical suppression of virus in plant tissue cultures. Annals of Applied Biology 99, 161-169.

Snyder, W.C., Wallis, G.W. & Smith, S.N. (1976). Biological control of plant pathogens. In 'Theory and Practice of Biological Control' (Huffaker, C.B. & Messenger, P.S., eds), pp. 521-539. Academic Press, New York.

Tavantzis, S.M. (1984). The use of terms for responses of plants to viruses: a reply to recent proposals. Phytopathology 74, 379-380.

van den Bosch, R., Messenger, P.S. & Gutierrez, A.P. (1982). An Introduction to Biological Control. Plenum Press, New York & London, 247 pp.

Vanderplank, J.E. (1963). 'Plant Disease: Epidemics and Control'. Academic Press, New York & London. 349 pp.

Wade, M. (1982). Resistance to fungicides. Span 25, 8-10.

Zadoks, J.C. & Schein, R.D. (1979). Epidemiology and Plant Disease Management. Oxford University Press, New York & Oxford. 426 pp.

CHAPTER 2. HOST-RANGE CONTROL AND NON-HOST IMMUNITY TO VIRUSES

R. S. S. FRASER

2.1. INTRODUCTION

The vast majority of plants are immune to the vast majority of plant viruses: attempts to infect produce no symptoms and no detectable virus multiplication. This effect, called non-host immunity, is one of the most intriguing problems in plant-virus interactions, and is one of the most difficult to investigate in ways that will produce meaningful results. The problem of non-host immunity is the converse of the question of what mechanisms determine host range, i.e. which species a particular virus will infect. Clearly an understanding of the mechanisms involved in determining host range might provide valuable clues to the biochemical bases of non-host immunity.

We may distinguish between two operational types of host: natural hosts, in which a particular virus occurs in the wild or in agriculture, and experimental or artificial hosts, in which infection may be induced in the laboratory, although the virus might never come in contact with the host in nature. It is reasonable to expect that a virus may have evolved a highly developed relationship with its natural hosts, while its relationship with experimental hosts would not necessarily be a product of co-evolution, and might be by chance or a result of convergent evolution. Co-evolution, however, would be likely where the experimental host was a close relative of a natural host.

Many plants are effectively non-hosts for many viruses in nature for purely passive reasons, in that they are never likely to meet with the virus because of geographical, climatic or biological factors. For example, if a virus is transmitted exclusively by a particular insect or nematode vector, then the natural host range of that virus is restricted to the host range of the vector.

However, if there is an effective mechanism for delivery of a particular virus to a number of plant species, host range may be determined by the interaction of two sets of factors: those specified by the host and those specified by the virus. This leads to the concept of a 'recognition event' between host and pathogen. An interaction may be classified as a recognition event if it involves specific interaction between two or more molecules which results in a 'go/no go' outcome. Thus interaction of pathogen molecule x with host molecule y may switch on a further stage in pathogenesis, whereas x may fail to interact with y' and thus the development of pathogenesis may be blocked. An implication of the 'go/no go' model is that recognition events are not involved in quantitative modification of the host response or pathogenesis, but are all-or-nothing switches.

Recognition events could control at any of several stages in development of pathogenesis, such as at initial attachment of the virus to the host cell surface, penetration, uncoating, transcription or translation of the viral genome. Even failure of a required recognition event at a stage as late as assembly of progeny virus could theoretically make a plant a non-host; no progeny virus would be detectable unless conditions were designed to detect uncoated viral nucleic acid. The ways by which recognition events at different stages might operate in control of host range and non-host immunity are considered in more detail later in this Chapter. Recognition events are obviously also involved in control of constitutive resistance and virulence, and these are considered in Chapter 6.

In general, it is fair to say that recognition phenomena between plants and viruses are poorly understood at the molecular level: in no case do we have an understanding of the recognition processes, and controls of outcomes, to equal our understanding of how antibodies recognise antigens, or of how hormones interact with their specific receptor sites in mammals. Paradoxically, the best understood cases of specific molecular interactions resulting in 'go/no go' decisions for plant viruses occur in insects, and control various aspects of transmission (Rochow, 1977; Sylvester, 1980). These cases are clearly outside the context of non-host immunity and resistance, but are cited as evidence that highly specific recognition systems involving plant viruses are possible.

2.2. HOST RANGE CONTROL

2.2.1. What determines the viruses to which a particular species is susceptible? Several investigators have recorded the reactions of wide ranges of taxonomically diverse species to infection by different viruses (Price,

1940; Holmes, 1946; Horvath, 1978a; 1978b; 1979a; 1979b). Much of the early data was analysed in a classical series of papers by Bald & Tinsley (1967a; 1967b; 1967c). They found that those groups of plants which were susceptible to the highest percentages of the viruses tested were relatively advanced in the phylogenetic, and thus probably in the evolutionary, sense. The most susceptible plants were not necessarily the most advanced within their own developmental groups, but in general, the more primitive families tended to be less susceptible to the overall range of viruses tested. They were, however, susceptible to viruses with natural field hosts within the group, illustrating the probable effects of co-evolution of host and virus on susceptibility.

Bald and Tinsley (1967a) developed a quasi-genetic model to explain the control of host range. The uneven distribution of susceptibility among groups of related families, or on a smaller scale among groups of related genera or species, led them to propose a multiplicity of host characters for susceptibility, distributed unevenly among species. They also postulated characters for infectivity (i.e. pathogenicity) distributed unevenly among viruses. Thus a species would only be a host for a particular virus if the pathogenicity characters of the virus matched the susceptibility characters of the plant. The implications of this model for possible mechanisms in non-host immunity are discussed in section 2.3.1.

2.2.2. What determines the host range of the virus? Some viruses such as TMV will infect several hundred species in widely separated families (Holmes, 1946; Horvath, 1978a). In contrast, other types of virus appear to be very restricted in host range. For example, several potyviruses are at present known to infect only one or a few closely related species (Hollings & Brunt, 1981).

Unrelated viruses may show a tendency to infect the same host species; an effect known as host range congruence. Thus the host range of TMV shows a degree of congruence with those of TNV, AMV and CMV (Bald and Tinsley, 1967c). In contrast, no evidence was found for host range congruence between TMV and TRV or LRV; the ability of these viruses to infect some hosts in common could be ascribed solely to chance.

An extreme case of host range congruence is the effect known as host range containment, where all tested species which are susceptible to one virus are susceptible to the second, although the second may infect further species which are immune to the first. Thus in a study of 310 species, Holmes (1946) found 83 which were susceptible to TMV and SEV; 116 which were susceptible to

TMV but not to SEV, and none which were susceptible to SEV but not to TMV. The host range of TMV therefore contains the host range of SEV. The implication is that SEV contains some but not all of the pathogenicity factors present in TMV.

An illustration of the possible influence of virus-coded factors in controlling host range is given by studying different forms within a virus group. Tobacco mosaic virus exists in a variety of forms in its different natural hosts (reviewed by Van Regenmortel, 1981). Some of these strains have highly individual characteristics: thus the legume strains are uniquely able to cause systemic infections in beans (Kassanis & Varma, 1975); the cucurbit strains (e.g. cucumber green mottle mosaic virus) are generally unable to infect tobacco and tomato hosts (Hollings et al., 1975). The implication is that these viruses have evolved a set of infectivity factors especially matched to the susceptibility factors of their natural hosts. Accepting the general principle that increased specialization reduces adaptability, this suggests that these strains may have lost the ability to infect other species during their evolution. Further evidence supporting the idea of infectivity factors comes from work with pseudorecombinants of viruses with multicomponent genomes, where ability to infect particular hosts could be shown to be located on one or more of the RNAs (Rao & Francki, 1982; Edwards et al., 1983). These determinants may be related to those controlling virulence towards resistant varieties of normally susceptible plants, and are considered in that context in Chapter 4.

To a very limited extent, the evolution of host specificity in the virus may be demonstrated by host passage experiments. If a virus isolate from one host is repeatedly passaged through a second host, some changes in the virus have been observed, although in other cases no effects were found (reviewed in Yarwood, 1979). Bawden (1956; 1958) reported transformation between the bean and tobacco forms of TMV (which are quite distinct in biochemical properties and reaction with different hosts) to be dependent on host passage, but these conclusions were challenged by Zaitlin et al. (1977). It appears that evidence for massive shifts in host range due to host passage effects is lacking, perhaps suggesting that evolution of partly-exclusive relationships between hosts and viruses is a long-term process. In contrast, evidence for evolution of virulence genes allowing the virus to overcome specific host resistance genes and thus turn resistant plants into hosts is abundant, and is considered in detail in Chapter 4. However, evolution of virulence may involve only a a single mutation in the viral genome in most cases; evolution of host-range specificity factors may be considerably more complex.

Dahl & Knight (1963) reported that one mutant prepared from a tomato isolate of TMV by nitrous acid treatment had lost the ability to infect tomato. However, this mutant was also unable to multiply systemically in tobacco, forming only small local lesions on all types of tobacco tested, including those which are normally systemic hosts of TMV. Thus the induced inability of the mutant to multiply in tomato may have been due to some general disablement rather than to any specific change in its determinants of host range.

2.3. POSSIBLE MECHANISMS IN NON-HOST IMMUNITY

2.3.1. Theoretical considerations. We may imagine three basic types of mechanism which might be involved in non-host immunity.

Positive mechanisms. In this model, all members of the non-host species contain virus-specific resistance genes which are 100% effective, and which completely prevent detectable virus multiplication. Holmes (1955) suggested that natural immunity might represent a random assemblage of individual, partially effective resistance genes giving a higher-order effect. He cited the work of Wade & Zaumeyer (1940) as a possible example: *Phaseolus vulgaris* normally reacts to inoculation with AMV by forming small necrotic lesions, in itself a resistance reaction preventing full establishment of infection. Varieties containing either of two further, independent dominant genes however appeared to be completely immune. Holmes suggested that this complete immunity might be the result of cumulative action of many individual resistance genes, perhaps as many as 20 or 40, although there was no direct evidence for this. In this model, positive non-host immunity is of course a boundary condition of constitutive resistance.

Some authors have used the term 'immunity' for varieties which show no detectable response to a particular virus normally attacking the species, or for cases where a variety is 'immune' to one isolate of the virus but not to others (e.g. Harrison *et al.*, 1974). It seems better to classify such 'immunity' as an example of extremely effective constitutive resistance, with corresponding virulence in other isolates of the virus. These cases are therefore dealt with in the appropriate Chapters, and the term 'non-host immunity' is reserved for species where no individual or variety is affected by the virus. Such definitions are inevitably somewhat arbitrary and can never be entirely satisfactory, but understanding of a mechanism is more important than classifying it.

Negative mechanisms. In this model, the non-host lacks some component or site (present in hosts) to which virus multiplication is so highly attuned that it is completely dependent upon it. In the quasi-genetic model of Bald & Tinsley (1967a), the non-host lacks suitable susceptibility genes to complement the pathogenicity genes of the virus. Possible examples of products specified by host susceptibility genes are virus-specific receptor sites, analogous to bacteriophage attachment and injection sites, or particular host enzymes such as RNA-dependent RNA polymerases to which a specific virus may be highly adapted.

Lower level explanations: physical and chemical barriers. A plant might be a non-host for rather trivial or non-specific reasons, such as physical barriers which prevent access of the virus to where it might be able to establish an infection. Experiments with protoplasts, which are considered below, may have some bearing on this question. Alternatively, non-host immunity might result from pH or ionic strength conditions within the host cell which are unfavourable for functioning of virus components and which therefore might prevent multiplication of a wide spectrum of viruses in a non-specific manner.

In the remainder of this Chapter, we consider experimental evidence which bears on these theoretical possibilities.

2.3.2. Are there virus-specific receptors? A receptor is generally considered to be a site at which the virus initially attaches to the host cell. Other functions such as penetration and uncoating may be associated with attachment. Virus-specific receptors are comparatively well understood in the cases of some bacteriophages and animal viruses. For plant viruses the picture is less clear, and there are two particular problem areas. Firstly, plant viruses are generally much smaller and of simpler structure than say the larger bacteriophages. There is thus less scope for elaboration of complex virus - host recognition mechanisms. Certainly for the rod-shaped and simpler icosahedral plant virus particles, there is no evidence of external, complex structures analogous to the tail fibres of T-even bacteriophages, which are involved in attachment site recognition. Secondly, the mode of arrival of the plant virus may militate against a requirement for specific receptor sites on the cell surface. For mechanically-transmitted viruses, the barriers of cuticle and cell wall generally require to be broken by wounding, which

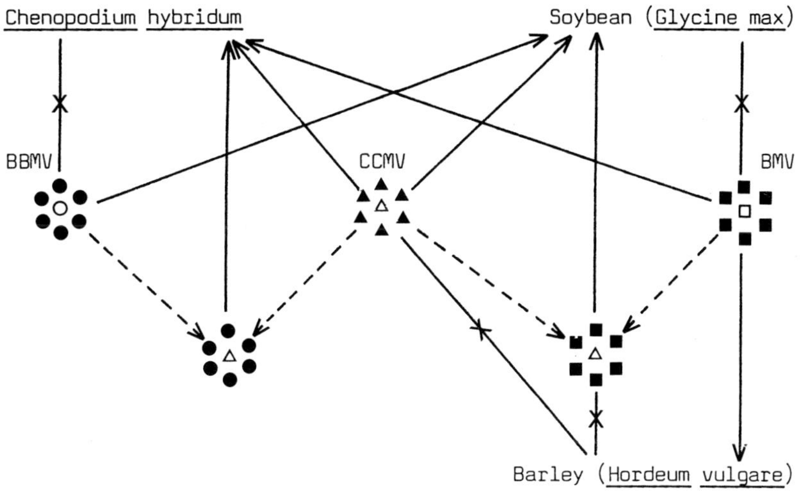

FIGURE 2.1. Host and non-host plants for three bromoviruses and mixedly reassembled particles derived from them. Open symbols represent viral RNA, and closed symbols the coat proteins. (——>) indicates successful establishment of infection on that species; (—X—) indicates that the species is a non-host for that virus. (- - ->) shows the sources of components for mixedly reconstituted particles. Drawn from data in Hiebert et al., 1968.

probably also allows penetration of the cell by the virus. Vector-transmitted viruses may be delivered directly into the host cells, for example by aphids which feed by inserting their stylets into the phloem cells. In these cases, virus-specific receptors might be present, but located inside the cell.

In plant viruses, the nucleic acid is not normally exposed on the particle surface; it is reasonable to think that the outer surface of the protein coat may be most important in recognition processes. One experimental approach has been to study the host range of mixedly-reconstituted particles, in which the RNA of one virus is re-encapsidated *in vitro* in the protein of another. In some cases such 'hybrid' or heterologously encapsidated viruses can have as high a specific infectivity as the original particles from which the components were derived (Hiebert *et al.* 1968) although in other cases heterologous encapsidation may lead to particles of low infectivity (Atabekov *et al.*, 1970). Hiebert *et al.* (1968) examined mixed reconstitution of three bromoviruses: BMV; CCMV and BBMV. Their results are summarized diagrammatically in Figure 2.1. Soybean is immune to BMV but susceptible to BBMV and CCMV. *Chenopodium hybridum* is immune to BBMV but susceptible to BMV and CCMV. CCMV RNA encapsidated in BMV protein infected soybean successfully, and CCMV RNA encapsidated in BBMV protein infected C. hybridum successfully. In each case the infectivity was as high as with the same concentration of CCMV RNA encapsidated in CCMV protein. Thus encapsidation in the coat protein of a non-hosted virus did not prevent infection by the RNA of a hosted virus. This suggests that CCMV does not rely on a CCMV-specific recognition site for infection of these two species. The results also suggest that soybean and C. hybridum were able to uncoat the CCMV RNA successfully, despite the fact that it was encapsidated in the coat proteins of non-hosted viruses. This argues against operation of a virus-specific uncoating mechanism in these cases. Atabekov et al. (1970) also reported results where encapsidation of infectious RNA in the coat protein of a non-hosted virus did not prevent establishment of infection. Shaw (1969) showed evidence for the converse case: TMV RNA was uncoated in hollyhock and corn, both non-hosts for that virus.

In contrast to these results, Atabekov (1975) reported that encapsidation of BMV RNA in TMV coat protein did prevent infection of barley. Control experiments showed that the hybrid particles were infectious to other species, and that the BMV RNA recovered from hybrid particles remained infective to barley. The barley cultivar used was stated to be immune to the TMV strain used; others have found a low level of TMV multiplication in other barley cultivars (Dodds & Hamilton, 1974). Atabekov's experiments would appear to show that mixedly reconstituted particles failed to recognize specific attachment sites, or failed to uncoat.

Another approach to the question of virus-specific recognition sites has been to encapsidate the RNA of a non-hosted virus in the protein of a hosted virus, in the hope that the protein would ensure successful recognition. Barley is immune to CCMV but systemically susceptible to BMV. Further reconstitution experiments by Hiebert et al. (1968) (Figure 2.1) showed that CCMV RNA in BMV protein did not infect barley. This is further evidence that a receptor site which recognizes specific features on the virion surface is unlikely to be the sole mechanism in non-host immunity.

A general comment about mixed-reconstitution experiments is that they must be interpreted with caution. For example, initial infection might well occur with a particle reconstituted from the RNA of a non-hosted virus and the protein of a hosted one. However, the amount of virus progeny produced directly as a result of the initial infection could be very small. Further cycles of multiplication, and virus spread, would depend on progeny RNA encapsidated in homologous coat protein. Replication of such particles would be under different constraints from the original heterologously-encapsidated inoculum. Interpretation of experiments showing failure to infect with free RNA of non-hosted viruses (Atabekov et al., 1970) requires the same caution. Experiments in which isolated protoplasts are inoculated and support one cycle of virus multiplication may be easier to interpret in this respect, and are considered in section 2.3.5.

An alternative experimental approach to the question of virus-specific receptors has been to examine competition between coat proteins and intact viruses. Novikov and Atabekov (1970) showed that proteins from BSMV or TMV would interfere with infection of Chenopodium amaranticolor by the homologous virus; when applied before or with the virus, protein reduced the numbers of lesions formed. In contrast, PVX protein did not interfere with PVX infection. For BSMV and TMV, proteins from the heterologous virus did not interfere, and protein from the homologous virus did not interfere with infection by free viral RNA. While these results may give some evidence for virus-specific receptor sites for BSMV and TMV, protein might be interfering with establishment of infection at other stages. Very high concentrations of protein were required for strong inhibition of infection; 100 to 500 times the virus concentration. This may argue against any specific receptor theory, but there could be major differences in competitive behaviour between protein subunits, and protein assembled into virions or virion-like structures.

2.3.3. Are there specific recognition events at the translational level? After uncoating, viral RNA is available for translation by host ribosomes. The products of translation in vivo have been identified for several viruses, but at present the function of many virus-coded proteins is not known. Some may be involved in replication of the viral RNA, perhaps by modification of host enzymes. Synthesis of virus coat protein, and some other virus-coded proteins, may occur in some cases by translation of monocistronic messenger RNAs rather than of complete genomic RNAs (Higgins et al., 1976; Schwinghamer & Symons, 1977).

Theoretically, host-range control and non-host immunity could operate at the translational level, by regulation of the interaction of the viral messenger RNA with host ribosomes. However, there does not yet appear to be evidence for selective controls at this level. Viral RNAs or monocistronic mRNAs are translated efficiently in vitro in cell-free systems from sources as diverse as Escherichia coli, (Glover & Wilson, 1982); wheat germ (Higgins et al., 1976) and rabbit reticulocytes (Morris-Krsinich & Hull, 1981). The products, however, often bear little relationship to those produced in vivo, so these results do not exclude the possibility of highly selective relationships between viral RNAs and the ribosomes and accessory factors of the host. However, Kiho et al. (1972) reported that TMV RNA became associated with polyribosomes equally in hosts and plants which are virtually non-hosts, in that extremely low levels of virus multiplication occurred. Wilson (1984) showed that rabbit reticulocyte ribosomes could translate not only TMV RNA, but also the RNA presented as TMV particles, a result he termed co-translational disassembly. It was only necessary to remove the few turns of coat protein sub-units at the 5' end by brief treatment at pH 8 to allow the process. As the rabbit is not currently thought to be a host of TMV, these results would suggest that the bulk of the uncoating process and initial translation are unlikely to be phases where host range are determined for this virus.

In some cases, virus proteins are produced by translation of a polycistronic messenger RNA to give a polypeptide, which is then cleaved into smaller, functional products. Although the cleavage would appear to be a possible site of action of a specific host enzyme, and thus a possible control of host range, current evidence suggests that the splitting is done by a virus-coded protein (Morris-Krsinich & Hull, 1981; Gabriel et al., 1982). In at least one case the proteolytic maturation is by internal catalysis: in TYMV the cleaving polyprotein itself contains the proteolytic activity (Morch et al., 1982).

2.3.4. Are there specific recognition events at the transcriptional level? Plant viruses have a variety of types of genome: RNA or DNA; single or double stranded; single or multicomponent; linear or circular. Not surprisingly, there is a matching diversity of mechanisms for replication. The more complex viruses, such as the rhabdoviruses, contain a virus-coded transcriptase in the particles (Toriyama & Peters, 1981). Most types, however, appear to rely, at least in part, on host enzymes for replication of the genome. Single-stranded RNA viruses utilize host RNA-dependent RNA polymerase (Fraenkel-Conrat, 1976; Romaine & Zaitlin, 1978). Viroid replication appears to involve host DNA-dependent RNA polymerase II, an enzyme normally involved in synthesis of heterogeneous nuclear RNA (Rackwitz et al., 1981). Recent schemes for replication of CaMV, a virus with a double-stranded DNA genome, suggest synthesis of a complete RNA copy of the genome then synthesis of DNA progeny by a reverse transcriptase (Hull & Covey, 1983). Whether the putative reverse transcriptase is virus or host coded is not yet clear.

Association of viral nucleic acids with host enzymes clearly offers opportunities for control and recognition events. At present, however, there is little knowledge of the enzymes involved. Modification of host enzymes by addition of virus-coded sub-units, as in the replication of the RNA of bacteriophage Qβ in *Escherichia coli* (Jockusch, 1974), clearly offers an attractive model for host-pathogen recognition. However, until recently the weight of the evidence argued against virus-coded components in RNA-dependent RNA polymerases (Romaine & Zaitlin, 1978; Gordon *et al.*, 1982). Such conclusions were drawn from studies of highly purified enzymes *in vitro*, and may not be a true reflection of the complete enzyme activity *in vivo*. At the time of writing, the latest evidence does suggest that the replicase of TYMV contains both host- and virus-coded subunits (Mouches *et al.*, 1984). Host range control and non-host immunity specified at the replicase level therefore remain largely possible.

2.3.5. *Evidence from experiments with protoplasts*. In some cases, a virus which is unable to infect whole leaves will successfully infect protoplasts isolated from them. Furusawa & Okuno (1978) showed that BMV could infect protoplasts of Japanese radish, normally a non-host for this virus. Huber *et al.* (1971; 1981) found that the tobacco strain of TMV would infect cowpea protoplasts, but infection was virtually undetectable in whole leaves. Superficially, such results might be interpreted as evidence that non-host immunity of leaves of these species acts by being a barrier to infection at the cell wall level, i.e. it prevents penetration of the virus to a stage where successful infection may be established. However, Sulzinski & Zaitlin (1982) showed that the common strain of TMV did initiate successful infection of cowpea leaves, but that the virus failed to spread from the original infection centres. Cowpea is therefore a host for this strain of TMV at the cellular level where each cell is directly inoculated - as occurs for most cells in protoplast cultures and a few cells in leaves - but is virtually a non-host at the whole-plant level.

A further example of a plant with apparent 'host' and 'non-host' regions concerns TNDV, which in tobacco plants is normally limited to phloem tissue. However, protoplasts isolated from leaf mesophyll cells can be infected successfully (Kubo & Takanami, 1979). The implication is that regions which are not normally invaded are not non-host immune because of the presence of some antiviral state in them, but do not become infected because spread of the virus from infected tissues is in some way actively prevented.

It must be stressed that the ability to infect protoplasts isolated from apparently non-host leaves is by no means a universal effect. Motoyoshi et al. (1974) found that neither tobacco leaves nor protoplasts could be infected by a wild type isolate of BMV or BMV RNA. Both however could be infected by very high concentrations of a variant form of the virus.

A report by Maekawa et al. (1981) showed that BMV would multiply to a low level in protoplasts isolated from leaves of the non-hosts radish and turnip. Virus multiplication in these protoplasts was enhanced if they were treated with actinomycin-D or ultraviolet irradiation. The latter tends to inhibit host protein synthesis without inhibiting viral protein synthesis (Sakai & Takebe, 1974). The suggestion is that BMV multiplication in the cells of these non-host plants may normally be suppressed by inhibitors, the synthesis of which requires host messenger RNA and protein synthesis. It is interesting that in protoplasts from barley, a normal host, actinomycin-D and ultraviolet light treatment actually inhibited BMV synthesis. This result would be consistent with a requirement for host-specified factors for virus multiplication.

2.3.6. Evidence from mixed infections. When plants are simultaneously infected by two viruses, there is often an interaction between them. The effects can include complementation, whereby a defective function of one virus is remedied by the effective function of the 'helper' virus. It would be interesting to ask whether multiplication of virus A could be facilitated in a non-host by the presence of an infective virus B. In the model of Bald & Tinsley (1967a), this would imply that the pathogenicity factors of virus B would be able to make good deficiencies in the pathogenicity factors of virus A. At present there seems to be only limited evidence for such effects.

Hamilton & Dodds (1970) and Hamilton & Nichols (1977) reported that TMV is normally restricted to the inoculated (first) leaf of barley. Simultaneous infection with BMV or BSMV allowed systemic multiplication of TMV to comparatively high levels. In mixed infections with BSMV, some TMV RNA was encapsidated in BSMV coat protein (Dodds & Hamilton, 1974). However, this was probably not the reason for the assisted systemic spread of TMV, because there was no evidence for encapsidation of TMV RNA in BMV coat protein when BMV was the helper virus (Hamilton & Nichols, 1977). Because TMV alone would multiply to low levels in the barley cultivar used, these experiments do not prove that a helper virus can allow a non-hosted virus to overcome complete non-host immunity. Hamilton & Nichols (1977) reported that BMV was unable to assist replication of several other viruses for which barley is not a systemic host. Cases where mixed infection may allow a virus to overcome a specific resistance gene in a normally susceptible species are considered in Chapter 6.

2.4. CONCLUSION

The biochemical and physiological mechanisms which determine host range and non-host immunity are still poorly understood. This Chapter has indicated some possible models; the frequent difficulties in interpretation of experimental results, and some directions for further study.

One longer term point deserves to be stressed. The desirability of utilizing non-host immunity in breeding for resistance is frequently mentioned. The implication is that this immunity will provide completely effective and permanent protection. Non-host immunity based on the 'positive' model outlined in section 2.3.1 could probably be exploited in this way, although there would be practical difficulties in handling large numbers of individual resistance genes by conventional breeding techniques. Non-host immunity based on a 'negative' mechanism would be much more difficult to exploit. To give an example: it might be that the complete immunity of, say, banana trees to bean common mosaic virus is because bananas lack components (present in beans) which are essential to BCMV multiplication. Clearly, under these circumstances there are no genes in banana which could be transferred to beans to confer immunity. There might be genes in bean which could be altered or deleted so conferring immunity, but that would require a completely different type of approach.

REFERENCES

Atabekov, J.G. (1975). Host specificity of plant viruses. Annual Review of Phytopathology 13, 127-145.

Atabekov, J.G., Novikov, V.K., Vishnichenko, V.K. & Javakhia, V.G. (1970). A study of the mechanisms controlling the host range of plant viruses. II. The host range of hybrid viruses reconstituted in vitro and of free viral RNA. Virology 41, 108-115.

Bald, J.G. & Tinsley, T.W. (1967a). A quasi-genetic model for plant virus host ranges. I. Group reactions within taxonomic boundaries. Virology 31, 616-624.

Bald, J.G. & Tinsley, T.W, (1967b). A quasi-genetic model for plant virus host ranges. II. Differentiation between host ranges. Virology 32, 321-327.

Bald, J.G. & Tinsley, T.W. (1967c). A quasi-genetic model for plant virus host ranges. III. Congruence and relatedness. Virology 32, 328-336.

Bawden, F.C. (1956). Reversible, host-induced, changes in a strain of tobacco mosaic virus.

Nature 177, 302-304.

Bawden, F.C. (1958). Reversible changes in strains of tobacco mosaic virus from leguminous plants. Journal of General Microbiology 18, 751-766.

Dahl, D. & Knight, C.A. (1963). Some nitrous acid-induced mutants of tomato atypical mosaic virus. Virology 21, 580-586.

Dodds, J.A. & Hamilton, R.I. (1974). Masking of the RNA genome of tobacco mosaic virus by the protein of barley stripe mosaic virus in doubly-infected barley. Virology 59, 418-426.

Edwards, M.C., Gonsalves, D. & Provvidenti, R. (1983). Genetic analysis of cucumber mosaic virus in relation to host resistance: location of determinants for pathogenicity to certain legumes and Lactuca saligna. Phytopathology 73, 269-273.

Fraenkel-Conrat, H. (1976). RNA polymerase from tobacco necrosis virus-infected and uninfected tobacco: purification of the membrane-associated enzyme. Virology 72, 23-32.

Furusawa, I. & Okuno, T. (1978). Infection with BMV of mesophyll protoplasts isolated from five plant species. Journal of General Virology 40, 489-491.

Gabriel, C.J., Derrick, K.S. & Shih, D.S. (1982). The synthesis and processing of the proteins of bean pod mottle virus in rabbit reticulocyte lysates. Virology 122, 476-490.

Glover, J.F. & Wilson, T.M.A. (1982). Efficient translation of the coat protein cistron of tobacco mosaic virus in a cell-free system from Escherichia coli. European Journal of Biochemistry 122, 485-492.

Gordon, K.H.J., Gill, D.S. & Symons, R.H. (1982). Highly purified cucumber mosaic virus-induced RNA-dependent RNA polymerase does not contain any of the full-length translation products of the genomic RNAs. Virology 123, 284-195.

Hamilton, R.I. & Dodds, J.A. (1970). Infection of barley by tobacco mosaic virus in single and mixed infection. Virology 42, 266-268.

Hamilton, R.I. & Nichols. C. (1977). The influence of bromegrass mosaic virus on the replication of tobacco mosaic virus in Hordeum vulgare. Phytopathology 67, 484-489.

Harrison, B.D., Murant, A.F. & Mayo, M.A. (1974). Distribution of determinants for symptom production, host range and nematode transmissibility between the two RNA components of raspberry ringspot virus. Journal of General Virology 22, 233-247.

Hiebert, E., Bancroft, J.B. & Bracker, C.E. (1968). The assembly in vitro of some small spherical viruses, hybrid viruses and other nucleoproteins. Virology 34, 492-508.

Higgins, T.J.V., Goodwin, P.B. & Whitfield, P.R. (1976). Occurrence of short particles in beans infected with the cowpea strain of TMV. II. Evidence that short particles contain the cistron for coat protein. Virology 71, 486-497.

Hollings, M.A. & Brunt, A.A. (1981). Potyviruses. In 'Handbook of Plant Virus Infections and Comparative Diagnosis' (Kurstak, E., ed.) pp. 732-807. Elsevier/North Holland Biomedical Press, Amsterdam.

Hollings, M., Komuro, Y. & Tochihara, H. (1975). Commonwealth Mycological Institute/Association of Applied Biologists Descriptions of Plant Viruses No. 154.

Holmes, F.O. (1946). A comparison of the experimental host ranges of tobacco etch and tobacco mosaic viruses. Phytopathology 35, 643-659.

Holmes, F.O. (1955). Additive resistances to specific viral diseases in plants. Annals of Applied Biology 42, 129-139.

Horvath, J. (1978a). New artificial hosts and non-hosts of plant viruses and their role in the identification and separation of viruses. III. Tobravirus group: tobacco rattle virus. Acta Phytopathologica Academiae Scientiarum Hungaricae 13, 51-55.

Horvath, J. (1978b). New artificial hosts and non-hosts of plant viruses and their role in the identification and separation of viruses. IV. Tobamovirus group: tobacco mosaic virus and tomato mosaic virus. Acta Phytopathologica Academiae Scientiarum Hungaricae 13, 57-73.

Horvath, J. (1979a). New artificial hosts and non-hosts of plant viruses and their role in the identification and separation of viruses. X. Cucumovirus group: cucumber mosaic virus. Acta Phytopathologica Academiae Scientiarum Hungaricae 14, 285-295.

Horvath, J. (1979b). New artificial hosts and non-hosts of plant viruses and their role in the identification and separation of viruses. XI. Tymovirus group: turnip yellow mosaic virus and belladonna mottle virus. Acta Phytopathologica Academiae Scientiarum Hungaricae 14, 207-309.

Huber, R., Hontilez, J. & Van Kammen, A. (1981). Infection of cowpea protoplasts with both the common strain and the cowpea strain of TMV. Journal of General Virology 55, 241-245.

Huber, R., Rezelman, G., Hibi, T. & Van Kammen, A. (1971). Cowpea mosaic virus infection of protoplasts from Samsun tobacco leaves. Journal of General Virology 34, 315-323.

Hull, R. & Covey, S.N. (1983). Does cauliflower mosaic virus replicate by reverse transcription? Trends in Biochemical Sciences 8, 119-121.

Ikegami, M. & Fraenkel-Conrat, H. (1980). Lack of specificity of virus-stimulated plant RNA-dependent RNA polymerases. Virology 100, 185-188.

Jockusch, H. (1974). Zellfreie Bildung eines Replikationsenzyms. Naturwissenschaften 61, 267-269.

Kassanis, B. & Varma, A. (1975). Commonwealth Mycological Institute/Association of Applied Biologists Descriptions of Plant Viruses No. 153.

Kiho, Y., Machida, H. & Oshima, N. (1972). Mechanism determining the host specificity of tobacco mosaic virus. I. Formation of polysomes containing infecting viral genome in various plants. Japanese Journal of Microbiology 16, 451-459.

Kubo, S. & Takanami, Y. (1979). Infection of tobacco mesophyll protoplasts with tobacco necrotic dwarf virus, a phloem-limited virus. Journal of General Virology 42, 387-398.

Maekawa, K., Furusawa, I & Okuno, T. (1981). Effects of actinomycin-D and ultraviolet irradiation on multiplication of brome mosaic virus in host and non-host cells. Journal of General Virology 53, 353-356.

Morch, M., Zagorski, W. & Haenni, A.L. (1982). Proteolytic maturation of the turnip yellow mosaic virus polyprotein coded in vitro occurs by internal catalysis. European Journal of Biochemistry 127, 259-265.

Morris-Krsinich, B.A.M. & Hull, R. (1981). Translation of turnip rosette virus RNA in rabbit reticulocyte lysates. Virology 114, 98-112.

Motoyoshi, F., Bancroft, J.B. & Watts, J.W. (1974). The infection of tobacco protoplasts with a variant of brome mosaic virus. Journal of General Virology 25, 31-36.

Mouches, C., Candresse, T. & Bove, J.M. (1984). Turnip yellow mosaic virus RNA-replicase contains host and virus-encoded subunits. Virology 134, 78-90.

Novikov, V.K. & Atabekov, J.G. (1970). A study of the mechanisms controlling the host range of plant viruses. I. Virus-specific receptors of Chenopodium amaranticolor. Virology 41, 101-107.

Price, W.C. (1940). Comparative host range of six plant viruses. American Journal of Botany 27, 530-541.

Rackwitz, H.R., Rhode, W. & Sänger, H.-L. (1981). DNA-dependent RNA polymerase II of plant origin transcribes viroid RNA into full length copies. Nature 291, 297-301.

Rao, L.N. & Francki, R.I.B. (1982). Distribution of determinants for symptom production and host range on the three RNA components of cucumber mosaic virus. Journal of General Virology 61, 197-205.

Rochow, W.F. (1977). Dependent virus transmission from mixed infections. In 'Aphids as Virus Vectors' (Harris, K.F. & Maramorosch, K., eds) pp. 253-276. Academic Press, New York & London.

Romaine, C. P. & Zaitlin, M. (1978). RNA-dependent RNA polymerases in uninfected and tobacco mosaic virus infected tobacco leaves: virus induced stimulation of a host polymerase activity. Virology 86, 241-253.

Sakai, F. & Takebe, I. (1974). Protein synthesis in tobacco mesophyll protoplasts induced by tobacco mosaic virus infection. Virology 62, 426-433.

Schwinghamer, M.W. & Symons, R.H. (1977). Translation of the four major RNA species of cucumber mosaic virus in plant and animal cell-free systems and in toad oocytes. Virology 79, 88-108.

Shaw, J.G. (1969). The in vivo removal of protein from tobacco mosaic virus after inoculation of tobacco leaves. II. Some characteristics of the reaction. Virology 37, 109-116.

Sulzinski, M.A. & Zaitlin, M. (1982). Tobacco mosaic virus replication in resistant and susceptible plants: in some resistant species virus is confined to a small number of initially infected cells. Virology 12, 12-19.

Sylvester, E.S. (1980). Circulative and propagative virus transmission by aphids. Annual Review of Entomology 25, 257-280.

Toriyama, S. & Peters, D. (1981). Differentiation between broccoli necrotic yellows virus and lettuce necrotic yellows virus by their transcriptase activities. Journal of General Virology 56, 59-66.

Van Regenmortel, M.H.V. (1981). Tobamoviruses. In 'Handbook of Plant Virus Infections and Comparative Diagnosis' (Kurstak, E., ed.) pp. 542-561. Elsevier/North Holland Biomedical Press. Amsterdam.

Wade, B.L. & Zaumeyer, W.J. (1940). Genetic studies of resistance to alfalfa mosaic virus and of stringiness in Phaseolus vulgaris. Journal of the American Society of Agronomy 32, 127.

Wilson, T.M.A. (1984). Co-translational disassembly of tobacco mosaic virus in vitro. Virology 137, 255-265.

Yarwood, C.E. (1979). Host passage effects with plant viruses. Advances in Virus Research 25, 169-190.

Zaitlin, M., Beachy, R. & Bruening, G. (1977). Lack of molecular hybridization between RNAs of two strains of TMV: a reconsideration of the criteria for strain relationships. Virology 82, 237-241.

CHAPTER 3. NON-HOST RESISTANCE TO FUNGI

J. P. RIDE

3.1. INTRODUCTION: DEFINITIONS AND PROBLEMS

Many fungi are pathogenic to plants, and yet plants in general continue to survive. Clearly, most species have effective means of resisting attack by the majority of potentially pathogenic fungi, the corollary being that parasitic fungi in general have a restricted range of hosts. Biotrophic fungi, such as the rusts, smuts and mildews, tend to have a much more restricted host range, frequently being limited to certain varieties of a single plant species, than do necrotrophic fungi - although even many of the latter are highly specific. The mechanistic determinants of this species specificity in host-parasite interactions are the subject of this Chapter. For example, why is a given plant species resistant to all strains of a particular fungal species (or *forma specialis*) when the fungus is known to be parasitic on other plants? Mechanisms of resistance to saprophytic fungi which have no known hosts are not considered here, although some definitions of non-host resistance would include this category.

Drawing the line between host and non-host may not always be a simple task. If a plant is only susceptible to a fungus when senescent or otherwise debilitated, is it to be called a host? Many simple factors can have a drastic influence on the ability of a fungus to be pathogenic. For instance, the growth of Botrytis cinerea in broad bean leaves is normally very restricted, yet the addition of simple nutrients (or pollen) to the inoculation droplet is sufficient to initiate spreading lesions analogous to those caused by the pathogen *B. fabae* (Harper *et al.*, 1981). However, under normal field conditions only *B. fabae* is able to spread from initial infection sites and aggressively colonize the surrounding tissues (Leach, 1955). Is broad bean a host for *B. cinerea* or not?

Despite these reservations on its circumscription, non-host resistance is usually very effective and durable, particularly in comparison to race-specific host resistance, in that minor changes in plant or fungal genotype or alterations (within reason) to environmental conditions have little effect. The possibility of transferring non-host resistance from one plant species to another is, superficially at least, attractive to crop breeders and the question of whether the mechanisms involved in non-host resistance bear any relation to those involved in host resistance is consequently both interesting and important. Unfortunately, many more physiological and biochemical studies have been devoted to host resistance than to non-host resistance. Nevertheless much can already be deduced about the defence mechanisms involved in non-host interactions and the ways in which successful pathogens overcome them. Before considering these mechanisms, however, it is appropriate to consider first the visible features of non-host interactions.

3.2. MICROSCOPY OF NON-HOST RESISTANCE TO FUNGI

A detailed structural and ultrastructural description of a resistance interaction between a fungus and a plant is the ideal preliminary to physiological and biochemical studies on the mechanisms involved. Knowledge of when and where a non-pathogen ceases its growth and any associated visible changes in the host can give valuable clues to the causes of resistance. Although detailed light- and electron-microscope studies of non-host interactions are comparatively rare there is sufficient information, particularly for the more biotrophic fungi such as the rusts, powdery mildews and Phytophthora spp., to draw some general conclusions.

In a comprehensive series of studies on non-host resistance to rusts, Heath (1972; 1974; 1977) compared the host and non-host interactions of a number of plants with a given rust, as well as those of a given plant with a number of rusts. The variation observed in the behaviour of three rusts (sunflower-, corn- and cowpea-rust) on eight different plant species (including the host for each rust) was substantial (Heath, 1977). Rusts go through complex developmental stages during successful infections and failure to grow or differentiate at any one of a number of stages could be responsible for resistance. When surface development was compared, all three rusts made fewer attempts to penetrate into the leaf on half or more of the non-hosts examined; several non-hosts supported only 20% or less of the attempted penetration sites seen in host plants. Nevertheless, in no case could surface behaviour account completely for non-host resistance, since at least some attempts at penetration occurred in all the rust/non-host combinations. When development inside leaves was compared, initial hyphal growth was less than in the respective susceptible plant only in one non-host, cabbage. This growth was less than that achieved by two of the rusts on artificial membranes, indicating inhibition of fungal growth in cabbage. The other non-hosts appeared to have little influence on the early development of the fungus within the leaf except in the development of haustorial mother cells and haustoria. The possibility that many non-hosts actively inhibit haustorium formation, perhaps via cell wall alterations, was suggested by Heath (1977).

It is clear from this work that the interactions of different non-host plant species to a particular rust can be very different, but that the interactions of a given non-host with different rusts tend to be very similar. Thus where a non-host/rust combination was characterized by a particular feature, such as the presence of host wall darkening, then a similar feature was often seen following infection of the same non-host with other rusts. Several exceptions to this rule were observed, however, indicating the dangers of generalization.

In a comparative study of the development of cucumber powdery mildew (Erysiphe cichoracearum) on its host and on non-hosts from six other angiosperm families, Johnson (1977) also demonstrated the wide varations in development that can occur on different non-hosts. In this case, however, spore germination and surface development of the parasite was nearly absent on the leaves of plants in some families, indicating that failure at this stage characterizes a few non-host interactions at least. On the other non-hosts, surface development and penetration efficiency were substantially reduced but could not completely account for resistance. It is evident from this work as well as that of Heath (1977) that non-host resistance, with rusts and mildews at least, tends not to be characterized by total cessation of all parasite growth at a single point during infection; rather it involves a gradual decline in the proportion of the parasite population which successfully completes each successive stage of development.

Comparative microscopical studies of the development of necrotrophic pathogens on non-host plants are relatively rare. Development on a susceptible host frequently involves spore germination, germ-tube growth, formation of an infection structure such as an appressorium, and penetration followed by inter- or intracellular hyphal growth. While any one of these stages may be inhibited, non-host resistance seems often to involve an inhibition of penetration or early hyphal growth. Thus in a comparative microscopical study of the host and non-host interactions of five species of Botrytis with broad bean or tulip leaves, Mansfield and Hutson (1980) observed little difference in the rates of spore germination of the fungi; however, the pathogenic B. fabae and B. tulipae were clearly distinguishable from the non-pathogens by their rapid penetration and formation of infection hyphae in their respective hosts. This faster development of the pathogens was associated with rapid and extensive host cell death. Infection hyphae of the non-pathogenic species induced some host cell death but were quickly inhibited and frequently killed during resistant reactions in non-host plants, the visible end-product being a light- or dark-brown 'hypersensitive'-like necrotic limited lesion. Microscopically, non-host reactions were also frequently characterized by the production of new wall-like deposits termed reaction material or papillae, by the plant cells at sites of attempted penetration.

In a further ultrastructural study of the interaction between four Botrytis species and broad bean leaves, Mansfield and Richardson (1981) observed that non-host interactions (B. elliptica, B. tulipae, B. cinerea) were characterized by cessation of fungal growth within the host epidermal cell wall, accompanied by death of immediately adjacent host cells and the presence of 'activated' cells (numerous Golgi bodies, mitochondria and parallel arrays of rough endoplasmic reticulum) around the dead cells. The interaction with the pathogen (B. fabae) was characterized by a more rapid and extensive cell death and the lack of activated cells. The ability to kill plant cells quickly thus appears to be an important determinant of the host range of Botrytis species.

While the inoculation of many plants with non-pathogenic fungi results in a 'hypersensitive' response where fungal penetration is followed by rapid host necrosis and a cessation of fungal growth, quite a large number of plants, particularly members of the Gramineae, show an alternative response. In this, growth of the parasite stops at the point of penetration with a new wall-like apposition called a papilla being deposited between the host plasmalemma and cell wall (Aist, 1976; Ride, 1983; Sherwood & Vance, 1980). The importance of this phenomenon is discussed later but it demonstrates that successful penetration accompanied by a hypersensitive response is not inevitably a consequence of attempted infection by 'inappropriate' necrotrophs.

Relatively few microscopical studies have directly compared non-host resistance of a plant with the resistance of varieties of the same plant to a pathogenic fungal species (or a particular race of that species). However, in those instances where a comparison has been made, non-host resistance appears more effective than varietal resistance in that fungal development ceases at an earlier stage. Thus for example the cultivar resistance of cowpea to cowpea rust involves a later cessation of fungal development than that observed in the non-host interactions of cowpea with sunflower rust or corn rust, with most infection hyphae of cowpea rust forming at least one haustorium (Heath 1977). Similarly, the non-host resistance of alfalfa to Phytophthora megasperma f.sp. glycinea, a pathogen of soybean, was expressed more rapidly and appeared to be more effective in limiting hyphal growth than host resistance to the alfalfa pathogen P. megasperma f.sp. medicaginis (Miller & Maxwell, 1984). Interestingly, however, the ultrastructural changes observed in host cells, e.g. production of wall appositions and a characteristic cell necrosis, were similar in both cases, although expressed more rapidly in response to P. megasperma f.sp. glycinea.

3.3. MECHANISMS OF NON-HOST RESISTANCE: GENERAL CONSIDERATIONS

Resistance mechanisms operative in non-hosts may be somewhat artificially subdivided into 'negative' mechanisms which depend on an absence of factors in the plant which are essential for the pathogen to grow, develop and produce disease, and 'positive' mechanisms which rely on the presence of factors which inhibit disease, either by direct inhibition of fungal growth or by inhibition of 'pathogenicity factors' produced by the pathogen. Resistance mechanisms can also be separated, again somewhat simplistically, into 'passive' or constitutive defences which are present in uninoculated plants (probably all 'negative' mechanisms and some 'positive' mechanisms are in this group) and 'active' mechanisms which are induced by microbial challenge (many 'positive' mechanisms).

Even without any knowledge of the nature of the mechanisms involved, some indication as to whether defences are constitutive or induced can be obtained by observing the effect on non-host resistance of pretreatments designed to suppress active defences. Hence, in a study of the effects of actinomycin D (an RNA synthesis inhibitor), cycloheximide, blasticidin S (protein synthesis inhibitors), and heat-shock on the development of the parasite in four non-host/rust combinations, Heath (1979a) observed increased fungal development in three out of the four combinations with at least one of the treatments. This suggested some type of inhibition of hyphal growth and haustorium formation in untreated non-host leaves. Similarly, blasticidin S induced increases in lengths of infection hyphae in three of the seven non-host interactions examined by Tani et al. (1976), and chloroform treatment of Zea mays leaves increased the development and formation of haustoria by the non-pathogen Puccinia graminis (Leath & Rowell, 1970). Cycloheximide was also effective in destroying the resistance of reed canarygrass to non-infecting fungi (Vance & Sherwood, 1976). Such experiments strongly suggest that non-host resistance frequently involves positive mechanisms of resistance, most probably induced by the challenge of the potential parasite; thus when the defensive activity of the non-host is disabled in some way, the fungus is capable of invasion and growth. The induction of defence mechanisms during the expression of non-host resistance is also supported by observations that prior inoculation with non-pathogens frequently protects a plant against subsequent attack by pathogens (e.g. Tani et al., 1980). Nevertheless, the possibility that some forms of non-host resistance may be determined by an absence of factors essential for pathogen development is worthy of consideration and is discussed next.

3.4. MECHANISMS OF RESISTANCE DEPENDENT ON AN ABSENCE OF FACTORS IN THE POTENTIAL HOST

3.4.1. Absence of factors essential for fungal differentiation or growth. Many plant parasitic fungi, particularly the more specialized biotrophs, undergo relatively complex differentiation processes or other tropic responses during the course of successful infections. Some of these developmental stages can be very difficult to reproduce in vitro and the possibility exists that they are specific responses to the chemical or physical attributes of the host. Non-host resistance could then be determined by a lack of suitable 'triggers' for differentiation, directional growth or motility.

Many fungal parasites begin the infection process by germination of conidia, chlamydospores, oospores, sclerotia or other propagules or resting structures. There is evidence that in some cases such germination is specifically stimulated by the proximity of the host plant or its exudates or extracts. Thus sclerotia of Sclerotium cepivorum, a parasite confined to members of the genus Allium, are normally dormant in unsterilized soil but will respond by mass germination to the presence of Allium spp. in the soil (Coley-Smith, 1960). Out of 63 plants from a wide range of families, only members of the genus Allium elicited a germination response, and there was no evidence of inhibition of germination by plants which failed to elicit a response (Coley-Smith & Holt, 1966). The stimulatory activity appears to be at least partly due to a number of n-propyl and allyl sulphides (Coley-Smith & King, 1969) although the production of these compounds may depend on the action of the soil microflora (King & Coley-Smith, 1969).

Germination of teliospores of the rust *Puccinia carthami* is also stimulated more by extracts and volatiles from its host, safflower, than from six other non-hosts (Klisiewicz, 1972; 1973), the stimulation apparently being due to a number of polyacetylenes (Binder *et al.*, 1977). Similarly, teliospore germination in *Uromyces appendiculatus* is activated by volatiles from its bean host (Gold & Mendgen, 1983). The volatiles do not activate six other rusts and preliminary data suggested that volatiles from the non-hosts maize and wheat do not activate bean rust spores.

Despite these examples of specific stimulation of pathogen germination by hosts, it is unlikely that this mechanism is a general determinant of host-parasite specificity. Many fungal spores germinate readily without additional nutrients or stimulants and while the germination of many pathogens is stimulated by the proximity of plants this is usually a non-specific effect. Thus although germination of chlamydospores of the bean root-rot fungus *Fusarium solani* f.sp. *phaseoli* in soil is stimulated by the presence of bean seeds, roots or root exudates (Schroth & Snyder, 1961), the effect is non-specific since 16 non-susceptible plants have the same effect (Schroth & Hendrix, 1962). Similarly, germination of oospores of the downy mildew pathogen *Peronosclerospora sorghi* is stimulated by the presence of roots of non-hosts (oats, wheat, soybean, cotton) as well as those of hosts (sorghum, maize) (Pratt, 1978). These results, coupled with the observation that fungal spores frequently germinate on non-hosts, indicate that absence of suitable stimulants of spore germination is not a common form of non-host defence. There is also little reason to believe that chemotaxis of zoospores in the soil plays any significant part in host-parasite specificity (Mitchell, 1976).

Germ-tube elongation and formation of an infection structure, such as an appressorium, typically follow the germination of a propagule of a fungal pathogen. The germ-tube may be orientated in a particular direction on its host, and the appressoria may form at particular sites (e.g. over stomata for most rusts). Further identified 'tropisms' include adherence of the germ-tubes to the host plant surface; directional emergence of penetration pegs from appressoria; and the formation and adherence of haustorial mother cells. These tropisms could possibly be specifically induced by contact or chemical stimuli present only in the host; absence of the correct stimulus could thus be the basis of non-host resistance. Certainly, microscopical observations indicate that tropic mistakes, e.g. rust appressoria forming away from stomata, are frequently more common in non-host interactions than in susceptible combinations (Heath, 1977; Wynn, 1976), at least for the more biotrophic fungi.

What is known, then, of the nature of the stimuli for these important tropic responses? Numerous studies using isolated cuticles, artificial membranes or inert leaf replicas indicate that the physical features of plant surfaces are the major factors controlling fungal differentiation (Wynn & Staples, 1981). Thus, for example, by controlling the size and frequency of surface lines and granules on vinyl plastic and nitrocellulose membranes, Dickinson (1970; 1971; 1972) was able to control the formation of appressoria and the differentiation of Puccinia recondita and P. coronata. In some cases the behaviour of a pathogen on host and non-host surfaces has been compared with its behaviour on inert replicas of those surfaces. Thus, interestingly, Wynn (1976) demonstrated that the behaviour of the bean rust fungus, Uromyces phaseoli, on two hosts (bean, lima bean) and four non-hosts (cowpea, soybean, wheat, oats) was very similar to its behaviour on polystyrene replicas of the leaf surfaces. On leaves, both the proportion of germ-tubes forming appressoria and the proportion of appressoria forming over stomates were higher on the hosts than on the non-hosts, the numbers declining with those non-hosts more distantly related to the hosts. Inoculation of leaf replicas produced almost exactly the same results as inoculation of leaves. Resistance which is characterized (at least in part) by tropic mistakes on the part of the pathogen is therefore likely to be due to the absence of a suitable physical stimulus on the surface of the non-host.

Wynn and Staples (1981) have also stressed the importance of close adherence as a critical prerequisite for a proper response to contact stimuli. Directional germ-tube growth, appressorium formation and haustorium formation in rusts are normally entirely dependent on adhesion to the appropriate plant surface. Thus, for example, failure of cucumber powdery mildew to penetrate barley leaves has been attributed to lack of close contact between germ-tubes and the cuticular surface, possibly due to an inability of the fungus to dissolve the surface wax (Staub et al., 1974). The adherence of germ-tube walls of Uromyces phaseoli to plant tissue is known to be specific in that adherence only occurs to cells of those plants in which the fungus is capable of forming haustoria (Mendgen, 1978). The causes of close adherence of fungi to plant surfaces are largely unknown but lack of a suitable binding site or surface could clearly be a cause of poor adherence, and hence a factor contributing to non-host resistance.

Puccinia graminis f.sp. tritici is an unusual rust in that infection structures can be induced by chemicals alone, as well as by physical contact of suitable surfaces (Wynn & Staples, 1981). Hence, for example, Grambow & Riedel (1977) have shown that certain volatile and phenolic fractions from the host plant wheat will elicit the formation of infection structures. Similar extracts of distant non-hosts were not so effective, indicating that a lack of suitable chemical triggers may also contribute to non-host resistance in some cases. It must also be remembered however that direct chemical inhibition of infection structure formation might occur in some non-hosts.

Despite convincing evidence that tropic mistakes do occur in non-host interactions, and that absence of suitable physical triggers is probably the major cause, it must be remembered that such mistakes rarely stop all members of the pathogen population, and that although often delayed or reduced, penetration of non-hosts frequently occurs. Further defence mechanisms are therefore presumably in operation at the post-penetration stage. Nevertheless such errors obviously reduce the challenge mounted against the plant, effectively reducing the inoculum load, such that the chances of other defences being successful are increased.

Tropic mistakes are perhaps more likely to contribute to resistance to biotrophs, where the degree of differentiation and host specificity is usually high, than to necrotrophs. Some of the latter differentiate very little, if at all, during the infection process. With those that do produce infection structures, there is little evidence for any host-specificity in their production, appressoria for example usually being formed on hosts and non-hosts alike, or even on any hard surface. It has also often been observed that wound-inoculation of non-hosts with necrotrophs, whereby the wounding circumvents the need for the parasite to produce infection structures, still results in resistant interactions (e.g. Ride, 1975).

In addition to the requirement of some pathogens for specific differentiation triggers, it is also possible that some have a specific requirement for some nutrient or stimulant of hyphal growth which is present in hosts but not in non-hosts. Such a requirement might not be surprising since the constant association of parasites with their hosts could make redundant the genetic information for the synthesis of those nutrients and growth factors which are readily available from the host. Examples of this type of specific nutrient requirement are rare, but choline and betaine have been identified as the specific stimulants determining the infection of flowers of wheat by Fusarium graminearum (Strange et al., 1974). The compounds also stimulate the growth of F. avenaceum, F. culmorum, Aspergillus niger and Diplodia maydis but not several other plant parasites (Strange & Smith, 1978). F. avenaceum and F. culmorum cause a headblight in wheat similar to that caused by F. graminearum, and both Diplodia maydis and F. graminearum are parasites of maize, a plant known to contain high choline concentrations. These specific stimulants of growth appear therefore to play some part in host-parasite specificity.

3.4.2. Absence of substances important for the production or action of 'pathogenicity factors'. In addition to mechanisms which directly affect fungal growth and development it is possible that non-hosts possess properties that counteract fungal 'pathogenicity factors', that is factors important in the causation of disease in the host species. Many necrotrophic fungal pathogens are known to secrete toxins or wall-degrading enzymes which may be responsible for some or all of the symptoms of disease. They are also thought to be frequently important for continued growth of the parasite in the host, possibly by suppressing active defence mechanisms, by releasing nutrients and (for the enzymes) removing the physical barrier of the plant cell wall. Other specific suppressors of active defences may perhaps also be produced by some biotrophic or necrotrophic parasites (e.g. Garas et al., 1979). Non-hosts, therefore, may lack the ability to induce the production of enzymes/toxins/suppressors, or lack the substrates or binding sites/targets essential for their action.

The possible importance of cell wall-degrading enzymes as determinants of specificity has been relatively little studied. However, a comparative study by Cooper et al. (1981) on the in vitro production of pectic enzymes suggested that these enzymes at least were unlikely to be universally important. The abilities of six fungal species to produce polygalacturonase and pectin lyase when cultured on the cell walls of six host or non-host plants showed wide variation, but no consistent relationship with susceptibility or resistance was apparent. Occasionally, however, enzyme levels were especially low during growth on non-host walls and the possibility, therefore, that low enzyme induction is a contributory factor in some non-host interactions cannot be ignored. Evidence of selective binding of fungal enzymes to cell walls of hosts was not apparent, indicating that lack of binding sites is unlikely to be a common resistance mechanism. The possibility exists that the in vivo production, binding or activity of wall-degrading enzymes differs substantially from that observed in vitro, but there is no firm evidence as yet that lack of factors important for enzyme production or activity is crucial to non-host resistance.

More convincing evidence of a role in the determination of species specificity is available for certain fungal toxins. Many necrotrophic fungal pathogens, but notably Helminthosporium and Alternaria species, are known to produce host-selective toxins which have the same host range as the pathogen, and appear to be major determinants of disease (Scheffer, 1983). Hence, for example, the necrosis-inducing toxins of Alternaria alternata f.sp. lycopersici, causal agent of stem canker of tomato, were shown to affect only tomato in tests on a range of species from nine plant families (Gilchrist & Grogan, 1976). Many other 'pathotypes' (formae speciales) of A. alternata are known, each one being a pathogen of only one or two plant species. In each case specific pathogenicity appears to be controlled by production of a host-selective toxin which is crucial to the establishment of the pathogen in its host in addition to causing symptoms (Nishimura & Kohmoto, 1983).

Other toxins produced by plant pathogenic fungi may be important in disease development but be less selective in action than the pathogens producing them. Thus tentoxin, a cyclic tetrapeptide produced by A. alternata, induces chlorosis in many (but not all) plant species (Durbin & Uchytil, 1977), whereas the fungus causes a chlorotic seedling blight of relatively few species. However, even though the 'host' range of a toxin may be wider than that of the fungus, resistance to that toxin could still be an important factor in the resistance of some of the non-hosts.

Fungal toxins vary widely in their structure, although they are nearly all low molecular weight secondary metabolites. For relatively few of the toxins is the site of action known for certain: for even fewer is the mechanism(s) of resistance to toxin understood. However, the A. alternata f.sp. lycopersici toxins are known to interfere with the activity of aspartate carbamoyl-transferase (ACTase), an important regulatory enzyme in pyrimidine biosynthesis. The enzyme from mung bean (a non-host) is less affected than that from tomato (Gilchrist, 1983). Binding of toxin to the enzyme from susceptible tomato plants enhances inhibition of activity by UMP, a natural feedback regulator. This synergism between UMP and toxin could have drastic effects on pyrimidine biosynthesis, although how this ultimately leads to the visible cell necrosis is not yet known. Similarly, the exact basis of the resistance of non-hosts to the toxin remains to be discovered, although the lack of a suitable receptor site on the ACTase seems most probable. Details of the molecular sites of action of the other fungal, host-selective toxins have not been completely elucidated; hence the importance in resistance of the absence of a suitable target has yet to be confirmed. However, of all the possible ways in which a plant may be resistant to a host-selective toxin, the lack of a suitable receptor site would seem the most probable (Durbin, 1983).

Of the 'non-selective' toxins, the mode of action of tentoxin seems the best understood. This chlorosis-inducing toxin interacts with the chloroplast coupling factor 1 (CF_1) of sensitive plants. Binding of toxin inhibits the associated ATPase activity and hence photophosphorylation (Steele _et al._, 1976). CF_1 from a sensitive species, such as lettuce, was shown to have a single high-affinity binding site for the toxin. Binding to CF_1 from an insensitive species, e.g. radish, was of a much lower affinity and ATPase activity was not inhibited. CF_1, therefore, seems to play a crucial role in the species specificity of the toxin, and absence of a high affinity receptor for tentoxin on CF_1 may be one reason why some plants are resistant to _A. alternata_. However, the fact that some plants are resistant to disease and yet sensitive to toxin implies that other factors are involved.

3.4.3. _Absence of 'common antigens'_. The concept that, in order to be able successfully to parasitize a plant, a fungus must possess certain key macromolecules (or regions of macromolecules) in common with its host is intriguing. That a parasite must 'mimic' its host in certain key areas to avoid 'recognition' could be one explanation of such a requirement, although many others are possible. Resistance might then be explained by the absence of matching proteins or carbohydrates, for example, in the non-host.

Attempts to search for these hypothesized 'matching' molecules have centred on the use of antisera raised in animals to protein preparations from fungi and plants, followed by a search for cross-reactive antigens by methods such as immunodiffusion. It is now clear that cross-reactive antigens do exist for many fungus-plant combinations (De Vay, 1976) but questions remain as to the specificity of the cross-reaction and its relevance to virulence or pathogenicity. Thus in a study of the interactions of cotton with three pathogens and one non-pathogen, cross-reactive antigens were detected with all the pathogens, but also with one of the non-pathogens (Charudattan & De Vay, 1972). Similarly, during a comparison of the soluble antigens of Phytophthora infestans with those of several plants, Palmerley & Callow (1978) found cross-reactions between the fungus and potato, tomato (both hosts), tobacco and two other members of the Solanaceae (all non-hosts). Common antigens were not detected between P. infestans and non-host representatives of six other higher plant families, nor between potato and two non-pathogens. The presence of common antigens in some incompatible interactions (e.g. P. infestans/tobacco) suggests that they are not determinants of specificity. Nonetheless, it has been suggested that they may be a prerequisite for (although not a guarantee of) 'basic compatibility' between host and parasite at the species level. (De Vay et al., 1981). The absence of common antigens might then be one cause of non-host resistance although the presence of such antigens would not alone determine susceptibility. However, until some function in a fungus-plant interaction can be ascribed to these antigens, the possibility that cross-reaction is simply a fortuitous consequence of a basic similarity of some cell components would seem a reasonable conclusion.

3.5. MECHANISMS OF RESISTANCE DEPENDENT ON THE PRESENCE OR PRODUCTION OF FACTORS BY THE POTENTIAL HOST

3.5.1. Preformed inhibitors of fungal growth. The inhibition of fungal growth by low molecular weight antifungal chemicals appears to be a common form of defence in plants, and at least a few of these inhibitors are known to be preformed constituents of plants. Detailed analyses of the roles of the compounds in non-host resistance are rare but correlations between insensitivity to inhibitor and pathogenicity do suggest a contribution to resistance in some cases. Thus when 30 fungal species were tested for sensitivity to the glycoalkaloid α-tomatine, an inhibitor found within tomato plants, all 14 tomato pathogens tested were among the 16 most insensitive species (Arneson & Durbin, 1968), suggesting that tolerance of tomatine is a prerequisite for successful infection of tomato. Mutants of Fusarium solani which are more tolerant of tomatine than the wild type are also more virulent (Défago & Kern, 1983; Défago et al., 1983), and although F. solani is not a non-pathogen of tomato, this evidence still indicates the potential role of α-tomatine in non-host resistance. Similarly, tolerance of the saponin avenacin, an inhibitor found within oat roots, may be an important attribute of pathogens of oats. Turner (1961) demonstrated that *Gaeumannomyces graminis* var. *avenae* produces a glycosidase enzyme (avenacinase) which detoxifies the inhibitor, similar enzyme activity being undetectable in cultures of the type species, *G. graminis* and five other non-pathogens of oats.

Inhibition of fungal growth may also occur on the surfaces of plants prior to penetration. For example, the poor germination of *Botrytis cinerea* and *B. elliptica* on broad bean leaves has been attributed in part to the epicuticular wax and in part to the epiphytic microflora, particularly bacteria (Rossall & Mansfield, 1980). The pathogen *B. fabae* germinates much more rapidly although the cause of this difference remains uncertain. In general, however, evidence for a crucial role of surface inhibitors in determining non-host resistance is lacking.

3.5.2. *Inhibitors released from immediate precursors on cellular damage*. Many so-called 'preformed' inhibitors found in higher plants are in fact released from less-toxic precursors, usually by enzyme action, following homogenization or other cellular damage. In many cases the precursor exists as a glycoside which is frequently thought to be restricted to the vacuole in intact tissue, being stored away from the activating, presumably cytoplasmic, glycosidases. Since necrotrophic fungi frequently cause cell disruption, for example via the action of pectic enzymes, it is possible that non-host resistance to this type of pathogen might involve the action of these rapidly released inhibitors. Unfortunately, much of the currently available evidence simply rests on observations that aqueous homogenates of some plants contain compounds which are fungitoxic *in vitro*. Whether the compounds are actually released in resistant interactions and whether the local levels attained at the site of penetration would be likely to be sufficiently inhibitory and occur at the right time to explain the observed cessation of fungal growth has rarely been addressed. Mansfield (1983) has also pointed out that experiments designed to observe the effects of antimetabolites on resistance would provide useful information if conducted on plants known to contain constitutive inhibitors. Treatments which are known to suppress some induced defence mechanisms, have frequently been found to make plants more susceptible to non-pathogens (see section 3.3). However, plants with constitutive inhibitors (or inhibitors released by simple conversions by pre-existing enzymes) as a principal defence mechanism would be unlikely to be affected.

Notwithstanding the above reservations, the evidence for some inhibitors is clearly suggestive of a role in non-host resistance. Hence, aqueous extracts of garlic and other Allium species contain an antibacterial and antifungal compound named allicin (diallyldisulphide oxide) which is produced by action of the enzyme alliin-lyase on the precursor alliin (S-allyl-L-cysteine-sulphoxide), the conversion taking only a few minutes (Stoll & Seebeck, 1951). Growth of isolates of Penicillium corymbiferum and P. cyclopium which are pathogenic for garlic have been shown to be much less inhibited by garlic extracts or allicin than other non-pathogenic Penicillium spp. (Smalley & Hansen, 1962; Durbin & Uchytil, 1971). Because fresh garlic is capable of releasing much more allicin than that required completely to inhibit growth of the non-pathogens in vitro, and because of its specificity for non-pathogens, it seems likely that the compound plays some part in non-host resistance (provided that fungal challenge does cause its release).

A similar role in resistance has been postulated for some of the cyanogenic glycosides known to be present in many plants. On wounding, the glycosides are converted, normally by a series of enzymic steps, to highly toxic hydrogen cyanide. The inhibitor may also be released during attack by necrotrophic fungi. For example, Millar & Higgins (1970) observed that infection of birdsfoot trefoil by Stemphylium loti resulted in cyanide release from the glucosides linamarin and lotaustralin, with enzymes of the pathogen probably assisting in the process. S. loti was markedly more tolerant of HCN than eight other non-pathogenic fungal species. Similarly Gloeocercospora sorghi, a pathogen of the cyanogenic plant Sorghum vulgare, is relatively insensitive to HCN, tolerance being associated with the induction of the enzyme formamide hydro-lyase, which converts HCN to formamide (Fry & Munch, 1975). In a further examination of the association of formamide hydro-lyase with pathogenicity to cyanogenic plants, Fry and Evans (1977) observed that all 11 fungal pathogens of cyanogenic plants tested produced relatively large amounts of HCN-induced enzyme activity. By comparison only 9 out of 14 pathogens of other plants produced detectable activity with only 4 of them producing moderate to high levels. Of 6 saprophytes tested only one showed (trace) activity. The mean specific activities for pathogens, non-pathogens and saprophytes were 30.1, 7.4 and 0.1 μmoles/min/mg protein respectively. These results suggest that ability to detoxify HCN is a prerequisite for successful pathogenicity on cyanogenic plants and indicate the potential role of the inhibitor in non-host resistance. Clearly, however, other defence mechanisms must exist in cyanogenic plants to explain why every fungus producing formamide hydro-lyase is not pathogenic for every cyanogenic species.

3.5.3. Post-infectional antifungal compounds: phytoalexins. Many plants are known to be capable of actively responding to fungal challenge by the synthesis of low molecular weight antibiotics from remote precursors, production probably depending on *de novo* synthesis of enzymes (Mansfield & Bailey, 1982). These compounds are the 'phytoalexins', a term originally proposed by Müller and Börger (1940) following observations on induced resistance to *Phytophthora infestans* in potato. Much of the early work on the induction and identification of phytoalexins utilized non-host interactions in legumes. Thus droplets containing spores of *Monilinia fructicola*, a stone-fruit pathogen, accumulated antifungal activity when placed in the seed cavities of opened bean (*Phaseolus vulgaris*) pods (Müller, 1958). Within 24 hours of inoculation, the fungus had induced death and browning of host cells around the points of attempted infection, and fungal growth had ceased. This type of 'hypersensitive' reaction involving limited host cell necrosis is typical of many resistant reactions of plants to fungi, including reactions to both biotrophic and necrotrophic parasites, and interactions of non-host, race-specific and race-non-specific types. In bean pods, the hypersensitive reaction to *M. fructicola* was accompanied by the diffusion of inhibitors into the infection droplets, such that droplets taken after 24h completely inhibited the germination of fresh *M. fructicola* spores. In contrast, control droplets of pure water became highly stimulatory when left in contact with the bean tissue. A similar hypersensitive response coupled to phytoalexin accumulation was subsequently observed in pea pods inoculated with the same non-pathogen (Cruickshank & Perrin, 1960), the antifungal principle being characterized as a pterocarpan and named pisatin (Perrin & Bottomley, 1962). The main phytoalexin from bean, a related pterocarpan named phaseollin, was isolated and characterized shortly afterwards (Perrin, 1964).

Phytoalexins are now known to be produced by many angiosperm families (Bailey & Mansfield, 1982) and a number of generalizations can be made: they are usually low molecular weight secondary metabolites with a broad spectrum of antibiotic activity; most plants so far examined appear capable of a phytoalexin response, each plant species usually producing a range of biosynthetically related phytoalexins; and the induction of phytoalexins is not specific to fungi, being usually associated with a wide range of injurious biological, physical and chemical stimuli.

Proof of the importance of a phytoalexin in resistance depends very largely on evidence that the compound accumulates to sufficiently toxic levels immediately adjacent to the pathogen at the time at which fungal growth is observed to cease. Coordinated biochemical and microscopical studies are thus essential (Mansfield, 1982), although not without problems. Difficulties here relate to determination of local *in vivo* concentrations, and estimation by *in vitro* tests of whether these concentrations would be sufficient to inhibit the fungus *in vivo*. *In vitro* bioassays are notoriously variable, the results frequently depending on the medium selected, the nature of the inoculum, the incubation time etc., and *in vivo*/*in vitro* comparisons are hence frequently difficult.

Much of the more precise work on phytoalexin accumulation in relation to fungal development has been devoted to examples of host resistance (e.g. Bailey *et al*., 1980; Yoshikawa *et al*., 1978), and these aspects are considered in Chapter 7. Detailed studies of their role in non-host interactions are relatively rare. An exception is the study by Mansfield and coworkers of the interaction of broad bean with pathogenic and non-pathogenic species of *Botrytis*. As noted previously (section 3.2) microscopy of the interactions indicates that the non-pathogens (*B. cinerea*, *B. elliptica* and *B. tulipae*) cease growth in the epidermal cell walls shortly after penetration, cessation of growth being acompanied by a limited hypersensitive-like plant cell necrosis (Mansfield & Hutson, 1980; Mansfield & Richardson, 1981). The pathogen *B. fabae* (at higher inoculum doses at least) causes more rapid and extensive host cell death than the non-pathogens and quickly forms a spreading lesion. Measurements on leaf discs cut from beneath the inoculum droplets demonstrated rapid accumulation of phytoalexins, notably wyerone acid, following inoculation with *B. cinerea*, and a smaller increase with a subsequent decline followed inoculation with *B. fabae* (Hargreaves *et al*., 1977). More detailed observations on the growth of infection hyphae in relation to the accumulation of phytoalexins within the inoculated epidermis indicated that the onset of phytoalexin accumulation in response to *B. cinerea* immediately preceded the restriction of fungal growth (Mansfield, 1982). Although the toxicity of wyerone acid is known to be dependent on pH (Hargreaves *et al*., 1977), the relatively large amounts of phytoalexin which accumulate would appear more than sufficient to explain the inhibition of fungal growth. This assumes, however, that the infection hyphae within the epidermal cell walls are actually in contact with the phytoalexins. Fluorescence studies indicate that the broad bean furanoacetylenic phytoalexins are synthesized by live cells adjacent to the necrotic lesion cells (Mansfield et al., 1974). However, Mansfield (1982) has suggested that wyerone acid at least is likely to diffuse freely throughout the lesion and thus contact the invading hyphae. Evidence in favour of this is the rapid accumulation of the acid in infection droplets containing non-pathogens.

Coordinated microscopical/biochemical studies also suggest an involvement of the phytoalexin capsidiol in the non-host resistance of pepper fruit to Phytophthora infestans (Jones et al., 1975a; 1975b). The non-pathogen rapidly penetrates the fruit and within 4 hours some of the invaded cells are necrotic. The fungus continues growing into the second or third cell layers and eventually ceases growth between 24 and 36 hours after inoculation. Measurement of capsidiol concentrations in the first few cell layers indicated trace amounts at 6 hours followed by a rapid increase between 12 and 24 hours. The concentrations at 24 hours were totally inhibitory to P. infestans in an in vitro bioassay. The question remains as to whether the fungus is actually in contact with the phytoalexin and whether the in vivo toxicity of capsidiol is sufficiently close to that suggested by the in vitro tests. Nevertheless the results support a role for the phytoalexin in this interaction.

A similar role for capsidiol in the hypersensitive resistance of leaves of pepper to P. infestans is suggested by the work of Ward (1976). Although the level of capsidiol in infection droplets was not high, the calculated concentration in hypersensitively reacting cells (calculated from the ratio of hypersensitive cell volume to infection droplet volume assuming all the phytoalexin originated from the hypersensitive cells alone) exceeded the in vitro inhibitory doses by several orders of magnitude. Similarly, assumption of a localized distribution of pisatin or phaseollin around sites of attempted infection of pea or bean by the non-pathogen Stemphylium botryosum leads to the conclusion that sufficient phytoalexin is present to account, theoretically at least, for the observed cessation of fungal growth (Heath & Higgins, 1973). The observation that conversion products of the two phytoalexins can be detected following infection as well as in artificial culture of S. botryosum supports the idea that the inhibitors are in contact with fungal hyphae in vivo.

The question of local concentrations of phytoalexins around invading hyphae is crucial. Until the exact position of phytoalexins in challenged tissues is ascertained (e.g. by fluorescent antibody techniques), calculations on probable concentrations make assumptions which are difficult to verify, e.g.

that phytoalexins are restricted to hypersensitive cells. Similarly, suggestions that phytoalexin levels (in gross tissue extracts) are too low to account for resistance may be misleading if intense localization of the inhibitors around hyphae is occurring. Interpretation of much data is thus very difficult at present. For example, Teasdale et al. (1974) suggested that the accumulation of pisatin (around 11 ug/g by 6 hours) in pea pod tissue in response to the bean pathogen Fusarium solani f.sp. phaseoli is unlikely to be solely responsible for inhibition of fungal growth (germ-tube growth ceases around 5.5 to 6 hours after inoculation) since even 100 to 200 ug/ml pisatin can be insufficient to prevent hyphal growth in vitro. However, calculations which assume localization in hypersensitive cells (e.g. Bailey & Deverall, 1971; Ward, 1976) indicate that measurements of phytoalexin concentrations in droplets or gross tissue extracts may well be several orders of magnitude lower than local concentrations. It is therefore very difficult to draw any firm conclusions for the F. solani f.sp. phaseoli/pea system from the evidence currently available.

If phytoalexins are, as seems probable, important in the non-host resistance of at least some plants to some fungi, then it naturally follows that pathogens of these plants must either (i) cause a later, slower or smaller accumulation of phytoalexins in their host, either by failing to induce the normal response or by actively suppressing induction, or (ii) be able to tolerate the levels of phytoalexins induced, either by being inherently less sensitive to their action or by enzymically degrading the compounds to less toxic products. Strategies (i) and (ii) may not necessarily be unrelated in that fungal degradation of phytoalexins may result in an observed slower rate of accumulation in the plant. Any evidence to indicate that pathogens alone are able to circumvent the potential threat of phytoalexin production obviously supports the notion that the compounds are important in non-host resistance; such evidence is now considered.

A slower or delayed accumulation of phytoalexins in response to pathogens, when compared to non-pathogens, has been observed many times, e.g. for Botrytis spp. on broad bean leaves and pods (Hargreaves et al., 1977; Mansfield, 1982), Botrytis spp. on narcissus bulbs (O'Neill & Mansfield, 1982), Erysiphe spp. on pea stems (Shiraishi et al., 1977), Fusarium spp. on pea pods (Teasdale et al., 1974) and Phytophthora spp. on pepper fruit (Jones et al., 1975b), although the observation is by no means universal (e.g. Cruickshank & Perrin (1963; 1971), but note that the concentrations here refer to the infection droplets). In many instances too, pathogens of a plant appear to be more tolerant of the appropriate phytoalexin(s), e.g. Botrytis spp. and broad bean (Hargreaves et al., 1977), Phytophthora spp. and pepper (Jones et al., 1975b), Ceratocystis fimbriata strains and sweet potato (Kojima & Uritani, 1976), and various fungi and pea (Cruickshank & Perrin, 1963; Van Etten, 1973) or bean (Cruickshank & Perrin, 1971; Van Etten, 1973). Exceptions to the pattern certainly exist (O'Neill & Mansfield, 1982; Van Etten, 1973), although to some extent exceptions would be expected even in cases where phytoalexins have a role in resistance. For example the occasional pathogen might be relatively sensitive to its host phytoalexin but still be pathogenic because of a low rate of accumulation; also the occasional non-pathogen might be tolerant of a phytoalexin but unable for other reasons to invade the host and invoke its accumulation. More disturbing perhaps are cases where a successful pathogen is known to be sensitive to a phytoalexin in vitro and in addition induces large quantities of the compound in vivo (e.g. Aphanomyces enteiches/pea; Pueppke & Van Etten, 1974; 1976), although the failure of in vitro bioassays to mimic in vivo conditions could be a simple explanation of the apparent paradox.

Insensitivity of pathogens to phytoalexins has sometimes been linked to an ability to metabolize the compounds to less toxic products (Arinze & Smith, 1980; Higgins & Millar, 1969; Macfoy & Smith, 1979; Van Etten et al., 1982). Not all pathogens of a plant however appear to be capable of phytoalexin metabolism; nor is the ability normally restricted to pathogens (Smith et al., 1982; Heuvel & Glazener, 1975). Nonetheless the ability may be crucial to pathogenicity in some cases. Thus a survey of naturally occurring isolates of Nectria haematococca (perfect stage of F. solani f.sp. pisi) revealed that all those isolates that could not demethylate pisatin were more sensitive to the phytoalexin and were non-pathogenic, or of low virulence, on pea (Van Etten et al., 1980). Crosses between isolates confirmed the close association between demethylation, tolerance of pisatin and virulence (Tegtmeier & Van Etten, 1982) which supports the view that pisatin accumulation is an important defence mechanism which must be overcome if a fungus is to be pathogenic on pea.

An interesting example of how the ability to metabolize a phytoalexin, coupled with a relatively low basic sensitivity to the inhibitor, together with the capacity to suppress phytoalexin accumulation can all contribute to pathogenicity is provided by the studies of Botrytis fabae infections of broad bean (Mansfield, 1983). B. fabae is less sensitive than other Botrytis spp. to the wyerone-type of phytoalexins produced by broad bean and is capable of metabolizing them to less toxic products. A very important additional factor however is the ability of B. fabae to kill bean cells sufficiently quickly to prevent a large accumulation of phytoalexins (Hutson & Mansfield, 1980). Cell-killing ability is possibly related to the high levels of pectic enzymes released by the pathogen (Cooper et al., 1981). This extremely necrotrophic strategy of B. fabae is also evident in Botrytis infections of other plants (O'Neill & Mansfield, 1982; Stewart & Mansfield, 1984).

The strategies employed by the more biotropic fungi must by definition be quite different, presumably involving either an active suppression of phytoalexin accumulation and hypersensitivity, or a simple failure to elicit these responses. Failure perhaps could be due to an absence or masking of elicitors, or to an absence of damage to host cells. Unfortunately, direct evidence to support any of these hypotheses is currently rather thin. Studies on the involvement of phytoalexins in resistance to many biotrophic parasites are also complicated by the fact that the hypersensitive reaction which usually accompanies phytoalexin accumulation, involving as it does host cell disorganization and death, may in itself be sufficient to prevent the development of a fungus dependent on living host cells. For necrotrophs, host cell death by itself is obviously not a problem, but the evidence clearly suggests that those fungi causing a hypersensitive reaction in plants capable of producing phytoalexins face a serious problem.

3.5.4. Structural barriers: preformed and induced. The role of the physical features of plant surfaces in controlling the differentiation and tropic responses of potential pathogens has been discussed in an earlier section (3.4.1.). In addition to this type of 'structural' defence mechanism it is also possible that the cell walls and outer layers of plants contribute directly to preventing penetration or colonization by non-pathogens. Variations in the polysaccharide and glycoprotein structure of plant primary cell walls could have some controlling effect on specificity, although studies on the induction of fungal enzymes by cell walls from various plant species make it seem unlikely that variations in wall structure have any great effect (Cooper et al., 1981). The basic cell wall may however be naturally strengthened and made considerably more resistant to fungal enzymes by overlayering or impregnation with materials such as cutin, suberin, waxes, lignin, silicon and calcium. Unfortunately there has been little investigation of the possible importance of pre-infectional levels of these compounds in non-host interactions; there is therefore as yet no convincing evidence for a role in this type of resistance (Martin, 1964; Royle, 1976). Since the preformed 'defence' barriers of most plants show similar structural features it seems unlikely that the induction or action of the fungal enzymes that degrade them would show much species specificity.

Induced structural defences appear to play a more significant role in non-host resistance. Numerous reports in the literature, many based on microscopic observations, suggest that alterations to cell walls occur in many plant species in response to attempted infection by fungi. Alterations may involve changes to existing walls or the deposition of new wall-like material. Unfortunately, knowledge of the chemical composition of the modifications and their importance in determining resistance and specificity is relatively slight. The problem here is not just one of knowing whether an alteration occurs at the right time and place to explain the observed cessation of fungal growth, but also in deciding whether the alteration is in itself sufficient to account for that inhibition. Wall alterations may have effects other than simply presenting a structural barrier to physical penetration by the invading fungus: protection against fungal enzymes and the restriction of 'molecular interchange' between host and parasite may be equally important (Ride, 1983).

Despite these difficulties, induced structural barriers would appear to play at least a part in the non-host resistance of a number of plants. For instance the deposition of new wall-like material on the interior of the host cell wall has been frequently noted in relation to the resistance of epidermal cells to fungal ingress. The attempted penetration by fungi of many plants, but particularly members of the Gramineae (Sherwood & Vance, 1980) results in the formation of such 'papillae' (or 'appositions', or 'reaction material') between the host plasmalemma and cell wall at points directly opposite the fungal appressoria and penetration pegs. Often the adjacent epidermal wall is also altered to produce a disc-shaped area frequently termed a 'halo'. The size, appearance and constituents of papillae vary enormously but callose, a β1,3-glucan, is a common component with lignin, suberin, silicon, cellulose and protein having also been identified in some cases (Aist, 1976).

Although much of the work on the role of papillae in resistance has been concerned with cases of host resistance, many correlations exist between the production of papillae and resistance to fungal penetration of non-hosts. Thus, for example, lignified papillae and haloes are produced in reed canarygrass leaves at sites of attempted penetration by non-pathogenic fungi (Sherwood & Vance, 1976). Attempted penetration by the pathogen Helminthosporium catenarium also frequently fails, with most of the appressoria being opposed by well developed lignified papillae; penetration only occurs when papillae are absent or poorly developed. Treatment of leaves with cycloheximide, to inhibit host protein synthesis, blocks papilla formation and enhances penetration (Vance & Sherwood, 1976), an effect which also occurs with many other members of the Gramineae (Sherwood and Vance, 1980).

Lignified papillae and epidermal cell walls also occur in wheat (Young, 1926; Ride & Pearce, 1979) and many cucurbits (Hammerschmidt et al., 1983) in response to attempted penetration by non-pathogenic fungi. In wheat leaves, papillae and haloes develop shortly after the formation of appressoria and are highly resistant to wall-degrading enzymes, a resistance which is largely due to the presence of lignin (Ride & Pearce, 1979). In vitro tests also indicate that the structures are extremely resistant to fungal degradation, whether the fungi are pathogens of wheat or not; even papillae and haloes taken from wheat leaves as early as 12 hours after inoculation with a non-pathogen have significant resistance to degradation, indicating the potential importance of the structures in resistance to penetration (Ride, 1980). The possible inclusion of silicon in such structures would also presumably increase their effectiveness (Kunoh & Ishizaki, 1975).

Studies such as these indicate that, in some cases at least, papillae form at the right place, probably at the right time and possibly are of the right composition to explain the inhibition of fungal penetration. However, the possibility that other defence mechanisms, operating prior to or during papilla development, are equally or more important cannot easily be ruled out. The use of general inhibitors such as cycloheximide does not solve this problem since many different defence mechanisms may be simultaneously inhibited. In some cases a search for induced antifungal compounds has been made and has proved negative (Ride & Pearce, 1979; Vance & Sherwood, 1976). However, in other instances it is clear that phytoalexins and lignified papillae may be operating in parallel. Thus inoculation of narcissus bulb tissue with non-pathogenic Botrytis spp. may result in the production of at least 12 phytoalexins in addition to lignified papillae (O'Neill & Mansfield, 1982). The relative importance of the two mechanisms remains unknown but the co-ordinate operation of such chemical and structural defences could obviously produce an extremely effective barrier to invasion and may be a strategy common to many plants.

In addition to the deposition of localized papillae in the epidermis, the cell walls of other tissues may be altered to more resistant forms in some plants following challenge by non-pathogenic fungi. Hammerschmidt (1984) demonstrated that the cell walls at the surface of sliced potato tubers undergo rapid lignification following inoculation with pathogens of other plants (Cladosporium cucumerinum and Fusarium roseum). Lignification in response to the pathogen F. roseum f.sp. sambucinum occurs several hours later, after penetration of the tuber tissue has already taken place. Similarly, the inoculation of wounded wheat leaves with non-pathogenic fungi results in the rapid deposition of a ring of lignin around the wounds, the rate of lignification being slower in response to pathogenic Septoria spp. (Ride, 1975). Wounding alone does not induce lignification in wheat leaves although lignification is a common wound-response in other plants. In fact the induction of lignification in wounded wheat leaves appears to be highly specific for filamentous fungi (Pearce & Ride, 1980). As noted with potato tuber tissue (Hammerschmidt, 1984) the induced polymer in wheat has a different structure and staining properties from the polymer in uninoculated tissue (Ride, 1975). Perhaps the production of novel polymers is just as common a response of plants to attempted infection as is the production of novel inhibitors such as phytoalexins? Certainly the deposition of the unusual lignin in wounded wheat leaves makes the tissue surrounding the wounds extremely resistant to enzymic degradation (Ride, 1975), and although the polymer may be functioning in many ways other than as a simple physical barrier (Ride, 1983), the fact that this resistant material is rapidly deposited as a specific response to filamentous fungi indicates a role in resistance.

Wall alterations other than lignification may also play a part in the non-host resistance of some plants. Silicon, for example, is deposited on and within mesophyll cell walls of bean leaves after inoculation with the incompatible cowpea rust (Heath, 1979b). Injection of exudates of the compatible bean rust, or extracts of bean rust-infected leaves, into healthy bean leaves suppresses the formation of these silicon-containing deposits on subsequent inoculation with cowpea rust and increases the frequency of haustoria (Heath, 1981). This suggests that ability to suppress formation of these deposits is important to the success of bean rust as a bean parasite, although obviously silicon deposition may not be the only induced defence mechanism.

Structural responses which result in blockage of xylem elements may be important in defence against micro-organisms which penetrate the vascular system. Occlusion of vessels by tyloses or plugs is a common response of plants to the presence of micro-organisms in the vascular system (Beckman & Taeboys, 1981; Van der Molen *et al.*, 1977) and may prevent the upward movement of spores in the transpiration stream. Tyloses are formed by a bulging of adjacent parenchyma cells through pit membranes into vessels, whereas the more amorphous gels or plugs may be composed of pectin and hemicellulose (Beckman & Zaroogian, 1967); both structures may be strengthened by the infusion and polymerization of phenolic compounds. In some instances a correlation with non-host resistance has been demonstrated. Thus extensive tyloses and gels are formed rapidly in banana roots in response to various non-pathogenic *Fusarium* spp. (Beckman & Halmos, 1962); only with the pathogenic *F. oxysporum* f.sp. *cubense* does early disappearance of gels coupled with delayed tylose formation apparently permit successive generations of spores to become systemically distributed. However, the precise importance of such physical barriers, particularly in relation to other defence mechanisms, remains unknown at present.

3.5.5. *Inhibition of pathogenicity factors.* In addition to direct inhibition of fungal growth it is possible that non-hosts may be capable of preventing pathogen development and disease by counteracting substances produced by the fungus which are essential for pathogenicity. The principle substances identified so far are toxins and wall-degrading enzymes. Resistance to toxins based on the absence of target sites has been discussed earlier (section 3.4.2.), but it is also possible that resistant plants either (i) metabolize toxins more rapidly than susceptible plants, (ii) have 'self-repair' mechanisms activated by the toxin, or (iii) in some way prevent the toxin from reaching its potential target. Experimental evidence for these possibilities is weak, although the observation that maize chloroplasts are sensitive to tentoxin (Arntzen, 1972) whereas intact maize plants are not (Fulton *et al.*, 1965) supports the view that mechanisms such as these may be important in some cases.

There is similarly no firm evidence that inhibitors of cell wall degrading enzymes play a crucial role in determining host-specificity. However, the known presence or production of inhibitors of pectic enzymes in plants, such as oxidized phenols (Byrde, 1957), glycoproteins (Albersheim & Anderson, 1971) and phytoalexins (Bull & Smith, 1981) suggest that inhibition of these important macerating enzymes may have some part to play in resistance. The precise importance of these effects in non-host resistance is not clear.

3.6. CONCLUSIONS

While much remains to be discovered about the mechanisms involved in non-host resistance to fungi, the available evidence suggests that there are numerous 'hurdles' involved, which a fungal species must overcome if it is to be a successful pathogen of a plant. Even in the same plant different hurdles may be operative against different pathogens and even against different individuals from the same pathogen population. This complex array of hurdles has also been envisaged as a succession of on/off switches with the responses of host or pathogen at each 'switching point' determining the subsequent progress of infection (Heath, 1974).

To be parasitic a fungus must possess the ability to surmount the hurdles presented by the plant, i.e. to grow on and adhere to the plant surface, to 'recognize' that surface and produce infection structures in suitable places, to penetrate the surface layers and cell walls of the plant, to cope with the nutrient/preformed inhibitor status of the tissues and to avoid or nullify active defences such as phytoalexin accumulation, lignification etc. Such a state of adaptation of a fungal species to a particular host plant has been called 'basic compatibility' (Bushnell, 1979), a term which perhaps seems more appropriate for the highly specialized biotrophic parasites than for the toxin/enzyme secreting necrotrophs.

Non-pathogenic fungal species may lack many of the properties required for basic compatibility and thus be far removed from the pathogenic status. Hence, probably, the general durability of non-host resistance. In some cases, however, the difference between a pathogen and a non-pathogen of a plant may be the production of only one 'pathogenicity factor'. Such a factor may demolish more than one hurdle at once. For example the ability rapidly to kill host cells by production of an enzyme or toxin may be sufficient to nullify all active defence mechanisms; hence perhaps the transformation of saprophytic forms of *Alternaria alternata* into host-specific pathogens by the acquisition of a host-selective toxin (Nishimura & Kohmoto, 1983). Similarly it is interesting to note the importance to *Botrytis fabae*, *B. tulipae* and *B. squamosa* of the ability to kill cells of their hosts rapidly when compared to non-pathogenic Botrytis spp. (Mansfield & Hutson, 1980; Stewart & Mansfield, 1984), an ability which may be related to a capacity to produce high levels of wall-degrading enzymes when in contact with the host. However, the more highly adapted to one particular plant species a parasite becomes, the less likely it is to be parasitic on other plants. If, for example, a fungus possesses the ability to recognize and form appressoria over the stomata of bean leaves it is likely to make 'mistakes' when meeting the physically and chemically quite different surface of a wheat leaf. Hence the high degree of host specificity exhibited by the extreme biotrophs, and the durability of non-host resistance to them.

It is interesting to compare the mechanisms involved in non-host resistance with those observed in host resistance. While many mechanisms appear to be common to both types of resistance there is a greater tendency in non-host resistance for the parasite to cease development at an early stage of the interaction due to 'negative' or preformed defence mechanisms. This is not to say that post-infectional defence mechanisms such as phytoalexin accumulation, hypersensitivity or lignification are not important; clearly they are in many non-host interactions. But it is obvious that mechanisms such as absence of suitable physical features on the plant surface or presence of constitutive antifungal compounds do have a role to play in resistance to non-pathogens whereas varietal resistance to pathogens rarely involves these factors.

A fungus facing a non-host is thus confronted by a stack of defensive 'barriers', many but not all of these barriers also being operative in varietal resistance. The general durability of non-host resistance presumably derives from the 'depth' of the stack, but the possibility that some forms of non-host resistance are only one barrier (or one pathogenicity factor) away from susceptibility should not be forgotten.

REFERENCES

Aist, J.R. (1976). Papillae and related wound plugs of plant cells. Annual Review of Phytopathology 14, 145-163.

Albersheim, P. & Anderson, A.J. (1971). Proteins from plant cell walls inhibit polygalacturonases secreted by plant pathogens. Proceedings of the National Academy of Sciences U.S.A. 68, 1815-1819.

Arinze, A.E. & Smith, I.M. (1980). Antifungal furanoterpenoids of sweet potato in relation to pathogenic and non-pathogenic fungi. Physiological Plant Pathology 17, 145-155.

Arneson, P.A. & Durbin, R.D. (1968). The sensitivity of fungi to α-tomatine. Phytopathology 58, 536-537.

Arntzen, C.J. (1972). Inhibition of photophosphorylation by tentoxin, a cyclic tetrapeptide. Biochimica et Biophysica Acta 283, 539-542.

Bailey, J.A. & Deverall, B.J. (1971). Formation and activity of phaseollin in the interaction between bean hypocotyls (Phaseolus vulgaris) and physiological races of Colletotrichum lindemuthianum. Physiological Plant Pathology 1, 435-449.

Bailey, J.A. & Mansfield, J.W. (eds) (1982). Phytoalexins. Blackie, Glasgow. 334 pp.

Bailey, J.A., Rowell, P.M. & Arnold, G.M. (1980). The temporal relationship between infected cell death, phytoalexin accumulation and the inhibition of hyphal development during resistance of Phaseolus vulgaris to Colletotrichum lindemuthianum. Physiological Plant Pathology 17, 329-339.

Beckman, C.H. & Halmos, S. (1962). Relation of vascular occluding reactions in banana roots to pathogenicity of root-invading fungi. Phytopathology 52, 893-897.

Beckman, C.H. & Talboys, P.W. (1981). Anatomy of resistance. In 'Fungal Wilt Diseases of Plants' (Mace, M.E., Bell, A.A. & Beckman, C.H., eds), pp. 487-521. Academic Press, London.

Beckman, C.H. & Zaroogian, G.E. (1967). Origin and composition of vascular gel in infected banana roots. Phytopathology 57, 11-13.

Binder, R.G., Klisiewicz, J.M. & Waiss, A.C. (1977). Stimulation of germination of Puccinia carthami teliospores by polyacetylenes from safflower. Phytopathology 67, 472-474.

Bull, C.A. & Smith, D.A. (1981). Pectic enzyme inhibition by the phytoalexin kievitone. Phytopathology 71, 206.

Bushnell, W.R. (1979). The nature of basic compatibility: comparisons between pistil-pollen and host-parasite interaction. In 'Recognition and Specificity in Plant Host-Parasite Interactions' (Daly, J.M. & Uritani, I., eds), pp. 211-227. Japan Scientific Societies Press, Tokyo.

Byrde, R.J.W. (1957). The varietal resistance of fruits to brown rot II. The nature of resistance in some varieties of cider apple. Journal of Horticultural Science 32, 227-238.

Charudattan, R. & DeVay, J.E. (1972). Common antigens among varieties of Gossypium hirsutum and isolates of Fusarium and Verticillium species. Phytopathology 62, 230-234.

Coley-Smith, J.R. (1960). Studies of the biology of Sclerotium cepivorum Berk IV. Germination of sclerotia. Annals of Applied Biology 48, 8-18.

Coley-Smith, J.R. & Holt, R.W. (1966). The effect of species of Allium on germination in soil of sclerotia of Sclerotium cepivorum Berk. Annals of Applied Biology 58, 273-278.

Coley-Smith, J.R. & King, J.E. (1969). The production by species of Allium of alkyl sulphides and their effect on the germination of sclerotia of Sclerotium cepivorum Berk. Annals of Applied Biology 64, 289-301.

Cooper, R.M., Wardman, P.A. & Skelton, J.E.M. (1981). The influence of cell walls from host and non-host plants on the production and activity of polygalacturonide-degrading enzymes from fungal pathogens. Physiological Plant Pathology 18, 239-255.

Cruickshank, I. & Perrin, D.R. (1960). Isolation of a phytoalexin from Pisum sativum L. Nature 187, 799-800.

Cruickshank, I. & Perrin, D.R. (1963). Studies on phytoalexins VI. The effect of some factors on its formation in Pisum sativum L. and the significance of pisatin in disease resistance. Australian Journal of Biological Science 16, 111-128.

Cruickshank, I & Perrin, D.R. (1971). Studies on phytoalexins XI. The induction, antimicrobial spectrum and chemical assay of phaseollin. Phytopathologische Zeitschrift 70, 209-229.

Défago, G. & Kern, H. (1983). Induction of Fusarium solani mutants insensitive to tomatine, their pathogenicity and aggressiveness to tomato fruits and pea plants. Physiological Plant Pathology 22, 29-37.

Défago, G., Kern, H. & Sedlar, L. (1983). Genetic analysis of tomatine insensitivity, sterol content and pathogenicity for green tomato fruits in mutants of Fusarium solani. Physiological Plant Pathology 22, 39-43.

DeVay, J.E. (1976). Protein specificity in plant disease development: protein sharing between host and parasite. In 'Specificity in Plant Diseases'. (Wood, R.K.S. & Graniti, A., eds), pp. 199-212. Plenum Press, New York.

DeVay, J.E., Wakeman, R.J., Kavanagh, J.A. & Charudattan, R. (1981). The tissue and cellular location of a major cross-reactive antigen shared by cotton and soil-borne fungal parasites. Physiological Plant Pathology 18, 59-66.

Dickinson, S. (1970). Studies in the physiology of obligate parasitism. VII. The effect of a curved thigmotropic stimulus. Phytopathologische Zeitschrift 69, 115-124.

Dickinson, S. (1971). Studies in the physiology of obligate parasitism. VIII. An analysis of fungal responses to thigmotropic stimuli. Phytopathologische Zeitschrift 70, 62-70.

Dickinson, S. (1972). Studies in the physiology of obligate parasitism. IX. The measurement of thigmotropic stimulus. Phytopathologische Zeitschrift 73, 347-358.

Durbin, R.D. (1983). The biochemistry of fungal and bacterial toxins and their modes of action. In 'Biochemical Plant Pathology' (Callow, J.A., ed.), pp. 137-162. John Wiley, Chichester.

Durbin, R.D. & Uchytil, T.F. (1971). The role of allicin in the resistance of garlic to *Penicillium* spp. Phytopathologica Mediterranea 10, 227-230.

Durbin, R.D. & Uchytil, T.F. (1977). A survey of plant insensitivity to tentoxin. Phytopathology 67, 602-603.

Fry, W.E. & Evans, P.H. (1977). Association of formamide hydro-lyase with fungal pathogenicity to cyanogenic plants. Phytopathology 67, 1001-1006.

Fry, W.E. & Munch, D.C. (1975). Hydrogen cyanide detoxification by *Gloeocercospora sorghi*. Physiological Plant Pathology 7, 23-33.

Fulton, N.D., Bollenbacker, K. & Templeton, G.E. (1965). A metabolite from *Alternaria tenuis* that inhibits chlorophyll production. Phytopathology 55, 49-51.

Garas, N.A., Doke, N. & Kuć, J. (1979). Suppression of the hypersensitive reaction in potato tubers by mycelial components from *Phytophthora infestans*. Physiological Plant Pathology 15, 117-126.

Gilchrist, D.G. (1983). Molecular modes of action. In 'Toxins and Plant Pathogenesis' (Daly, J.M. & Deverall, B.J., eds), pp. 81-136. Academic Press, Sydney.

Gilchrist, D.G. & Grogan, R.G. (1976). Production and nature of a host-specific toxin from *Alternaria alternata* f.sp. *lycopersici*. Phytopathology 66, 165-171.

Gold, R.E. & Mendgen, K. (1983). Activation of teliospore germination in *Uromyces appendiculatus* var. *appendiculatus*. II. Light and host volatiles. Phytopathologische Zeitschrift 108, 281-293.

Grambow, H.J. & Riedel, S. (1977). The effect of morphogenically active factors from host and non-host plants on the *in vitro* differentiation of infection structures of *Puccinia graminis* f.sp. *tritici*. Physiological Plant Pathology 11, 213-224.

Hammerschmidt, R. (1984). Rapid deposition of lignin in potato tuber tissue as a response to fungi non-pathogenic on potato. Physiological Plant Pathology 24, 33-42.

Hammerschmidt, R;., Bonnen, A.M. & Bergstrom, G.C. (1983). Association of lignification with non-host resistance of cucurbits. Phytopathology 73, 829.

Hargreaves, J.A., Mansfield, J.W. & Rossal, S. (1977). Changes in phytoalexin concentrations in tissues of the broad bean plant (*Vicia faba* L.) following inoculation with species of *Botrytis*. Physiological Plant Pathology 11, 227-242.

Harper, A.M., Strange, R.N. & Langcake, P. (1981). Characterization of the nutrients required by *Botrytis cinerea* to infect broad bean leaves. Physiological Plant Pathology 19, 153-167.

Heath, M.C. (1972). Ultrastructure of host and non-host reactions to cowpea rust. Phytopathology 62, 27-38.

Heath, M.C. (1974). Light and electron microscope studies of the interactions of host and non-host plants with cowpea rust - *Uromyces phaseoli* var. *vignae*. Physiological Plant Pathology 4, 403-414.

Heath, M.C. (1977). A comparative study of non-host interactions with rust fungi. Physiological Plant Pathology 10, 73-88.

Heath, M.C. (1979a). Effects of heat shock, actinomycin D, cycloheximide and blasticidin S on non-host interactions with rust fungi. Physiological Plant Pathology 15, 211-218.

Heath, M.C. (1979b). Partial characterization of the electron opaque deposits formed in the non-host plant, French bean, after cowpea rust infection. Physiological Plant Pathology 15, 141-148.

Heath, M.C. (1981). The suppression of the development of silicon-containing deposits in French bean leaves by exudates of the bean rust fungus and extracts from bean rust-infected tissue. Physiological Plant Pathology 18, 149-155.

Heath, M.C. & Higgins, V.J. (1973). In vitro and in vivo conversion of phaseollin and pisatin by an alfalfa pathogen Stemphylium botryosum. Physiological Plant Pathology 3, 107-120.

Heuvel, J. van den, & Glazener, J.A. (1975). Comparative abilities of fungi pathogenic and non-pathogenic to bean (Phaseolus vulgaris) to metabolize phaseollin. Netherlands Journal of Plant Pathology 81, 125-137.

Higgins, V.J. & Millar, R.C. (1969). Comparative abilities of Stemphylium botryosum and Helminthosporium turcicum to induce and degrade a phytoalexin from alfalfa. Phytopathology 59, 1493-1499.

Hutson, R.A. & Mansfield, J.W. (1980). A genetical approach to the analysis of mechanisms of pathogenicity in Botrytis/Vicia faba interactions. Physiological Plant Pathology 17, 309-317.

Johnson, L.E.B. (1977). Cited in Bushnell, W.R. (1979). The nature of basic compatibility: comparisons between pistil-pollen and host-parasite interaction. In 'Recognition and Specificity in Plant Host-Parasite Interactions' (Daly, J.M. & Uritani, I., eds), pp. 211-227. Japan Scientific Societies Press, Tokyo.

Jones, D.R., Graham, W.G. & Ward, E.W.B. (1975a). Ultrastructural changes in pepper cells in an incompatible interaction with Phytophthora infestans. Phytopathology 65, 1274-1285.

Jones, D.R., Unwin, C.H. & Ward, E.W.B. (1975b). The significance of capsidiol induction in pepper fruit during an incompatible interaction with Phytophphora infestans. Phytopathology 65, 1286-1288.

King, J.E. & Coley-Smith, J.R. (1969). Production of volatile alkyl sulphides by microbial degradation of synthetic alliin and alliin-like compounds in relation to germination of sclerotia of Sclerotium cepivorum Berk. Annals of Applied Biology 64, 303-314.

Klisiewicz, J.M. (1972). Effect of host plant materials and temperature on germination of teliospores of Puccinia carthami. Phytopathology 62, 436-438.

Klisiewicz, J.M. (1973). Effect of volatile substances from safflower on germination of teliospores of Puccinia carthami. Phytopathology 63, 795.

Kojima, M. & Uritani, I. (1976). Possible involvement of furanoterpeniod phytoalexins in establishing host-parasite specificity between sweet potato and various strains of Ceratocystis fimbriata. Physiological Plant Pathology 8, 97-111.

Kunoh, H. & Ishizaki, H. (1975). Silicon levels near penetration sites of fungi on wheat, barley, cucumber and morning glory leaves. Physiological Plant Pathology 5, 283-287.

Leach, R. (1955). Recent observations on the Botrytis infection of beans. Transactions of the British Mycological Society 38, 171.

Leath, K.T. & Rowell, J.B. (1970). Nutritional and inhibitory factors in the resistance of Zea mays to Puccinia graminis. Phytopathology 60, 1097-1100.

Macfoy, C.A. & Smith, I.M. (1979). Phytoalexin production and degradation in relation to resistance of clover leaves to Sclerotinia and Botrytis spp. Physiological Plant Pathology 14, 99-111.

Mansfield, J.W. (1982). The role of phytoalexins in disease resistance. In 'Phytoalexins' (Bailey, J.A. & Mansfield, J.W., eds), pp. 253-282. Blackie, Glasgow.

Mansfield, J.W. (1983). Antimicrobial compounds. In 'Biochemical Plant Pathology' (Callow, J.A., ed.), pp. 237-265. John Wiley, Chichester.

Mansfield, J.W. & Bailey, J.A. (1982). Phytoalexins: current problems and future prospects. In 'Phytoalexins' (Bailey, J.E. & Mansfield, J.W., eds), pp. 319-322. Blackie, Glasgow.

Mansfield, J.W., Hargreaves, J.A. & Boyle, F.C. (1974). Phytoalexin production by live cells in broad bean leaves infected with Botrytis cinerea. Nature 252, 316-317.

Mansfield, J.W. & Hutson, R.A. (1980). Microscopical studies of fungal development and host responses in broad bean and tulip leaves inoculated with five species of Botrytis. Physiological Plant Pathology 17, 131-144.

Mansfield, J.W. & Richardson, A. (1981). The ultrastructure of interactions between Botrytis species and broad bean leaves. Physiological Plant Pathology 19, 41-48.

Martin, J.T. (1964). Role of cuticle in the defense against plant disease. Annual Review of Phytopathology 2, 81-100.

Matta, A. (1971). Microbial penetration and immunization of uncongenial host plants. Annual

Review of Phytopathology 9, 387-410.

Mendgen, K. (1978). Attachment of bean rust cell wall material to host and non-host plant tissue. Archives of Microbiology 119, 113-117.

Millar, R.L. & Higgins, V.J. (1970). Association of cyanide with infection of birdsfoot trefoil by Stemphylium loti. Phytopathology 60, 104-110.

Miller, S.A. & Maxwell, D.P. (1984). Ultrastructure of susceptible, host resistant and non-host resistant interactions of alfalfa with Phytophthora megasperma. Canadian Journal of Botany 62, 117-128.

Mitchell, J.W. (1976). The effects of roots on the activity of soil-borne plant pathogens. In 'Encyclopedia of Plant Physiology Vol. 4 Physiological Plant Pathology' (Heitefuss, R. & Williams, P.H., eds), pp. 104-128. Springer-Verlag, Heidelberg.

Müller, K.O. (1958). Studies on phytoalexins. I. The formation and the immunological significance of phytoalexin produced by Phaseolus vulgaris in response to infections with Sclerotinia fructicola and Phytophthora infestans. Australian Journal of Biological Science 11, 275-300.

Müller, K.O. & Börger, H. (1940). Experimentelle Untersuchungen über die Phytophthora - Resistenz der Kartoffel-Zugleich ein Beitrag zum Problem der "erworbenen Resistenz" im Pflanzenreich. Arbeiten Biologische Reichsanstalt für Land- und Forst-wirtschaft (Berlin-Dahlem) 23, 189-231.

Nishimura, S. & Kohmoto, K. (1983). Roles of toxins in pathogenesis. In 'Toxins and Plant Pathogenesis' (Daly, J.M. & Deverall, B.J., eds), pp. 137-157. Academic Press.

O'Neill, T.M. & Mansfield, J.W. (1982). Mechanisms of resistance to Botrytis in narcissus bulbs. Physiological Plant Pathology 20, 243-56.

Palmerley, R.A. & Callow, J.A.; (1978). Common antigens in extracts of Phytophthora infestans and potatoes. Physiological Plant Pathology 12, 241-248.

Pearce, R.B. & Ride, J.P. (1980). Specificity of induction of the lignification response in wounded wheat leaves. Physiological Plant Pathology 16, 197-204.

Perrin, D.R. (1964). The structure of phaseollin. Tetrahedron Letters 1, 29-35.

Perrin, D.R. & Bottomley, W. (1962). Studies on phytoalexins. V. The structure of pisatin from Pisum sativum L. Journal of the American Chemical Society 84, 1919-1922.

Pratt, R.G. (1978). Germination of oospores of Sclerospora sorghi in the presence of growing roots of host and non-host plants. Phytopathology 68, 1606-1613.

Pueppke, S.G. & Van Etten, H.D. (1974). Pisatin accumulation and lesion development in peas infected with Aphanomyces euteiches, Fusarium solani f.sp. pisi, or Rhizoctonia solani. Phytopathology 64, 1433-1440.

Pueppke, S.G. & Van Etten, H.D. (1976). The relation between pisatin and the development of Aphanomyces euteiches in diseased Pisum sativum. Phytopathology 66, 1174-1185.

Ride, J.P. (1975). Lignification in wounded wheat leaves in response to fungi and its possible role in resistance. Physiological Plant Pathology 5, 125-134.

Ride, J.P. (1980). The effect of induced lignification on the resistance of wheat cell walls to fungal degradation. Physiological Plant Pathology 16, 187-196.

Ride, J.P. (1983). Cell walls and other structural barriers in defence. In 'Biochemical Plant Pathology' (Callow, J.A., ed.), pp. 215-236. John Wiley, Chichester.

Ride, J.P. & Pearce, R.B. (1979). Lignification and papilla formation at sites of attempted penetration of wheat leaves by non-pathogenic fungi. Physiological Plant Pathology 15, 79-92.

Rossall, S. & Mansfield, J.W. (1980). Investigation of the causes of poor germination of Botrytis spp. on broad bean leaves (Vicia faba L.). Physiological Plant Pathology 16, 369-382.

Royle, D.J. (1976). Structural features of resistance to plant diseases. In 'Biochemical Aspects of Plant-Parasite Relationships' (Friend, J. & Threefall, D.R., eds), pp. 161-193. Academic Press, London.

Scheffer, R.P. (1983). Toxins as chemical determinants of plant disease. In 'Toxins and Plant Pathogenesis' (Daly, J.M. & Deverall, B.J., eds), pp. 1-40. Academic Press, Sydney.

Schroth, M.N. & Hendrix, F.F. (Jr) (1962). Influence of non-susceptible plants on the survival of Fusarium solani f. phaseoli in soil. Phytopathology 52, 906-909.

Schroth, M.N. & Snyder, W.C. (1961). Effect of host exudates on chlamydospore germination of the bean root rot fungus Fusarium solani f. phaseoli. Phytopathology 51, 389-393.

Sherwood, R.T. & Vance, C.P. (1976). Histochemistry of papillae formed in reed canarygrass leaves in response to noninfecting pathogenic fungi. Phytopathology 66, 503-510.

Sherwood, R.T. & Vance, C.P. (1980). Resistance to fungal penetration in Gramineae. Phytopathology 70, 273-279.

Shiraishi, T., Oku, H., Ouchi, S. & Tsuji, Y. (1977). Local accumulation of pisatin in tissues of pea seedlings infected by powdery mildew fungi. Phytopathologische Zeitschrift 88, 131-135.

Smalley, E.B. & Hansen, H.N. (1962). Penicillium decay of garlic. Phytopathology 52, 666-678.

Smith, D.A., Harrer, J.M. & Cleveland, T.E. (1982). Relation between production of extracellular kievitone hydratase by isolates of Fusarium and their pathogenicity on Phaseolus vulgaris. Phytopathology 72, 1319-1323.

Staub, T., Dahmen, H. & Schwinn, F.J. (1974). Light- and scanning electron microscopy of cucumber and barley powdery mildew on host and non-host plants. Phytopathology 64, 364-372.

Steele, J.A., Uchytil, T.F., Durbin, R.D., Bhatnagar, P. & Rich, D.H. (1976). Chloroplast coupling factor 1: a species-specific receptor for tentoxin. Proceedings of The National Academy of Sciences, U.S.A. 73, 2245-2248.

Stewart, A. & Mansfield, J.W. (1984). Fungal development and plant response in detached onion, onion bulb scales and leaves inoculated with Botrytis allii, B. cinerea, B. fabae and B. squamosa. Plant Pathology 33, 401-409.

Stoll, A. & Seebeck, E. (1951). Chemical investigations on alliin, the specific principal of garlic. Advances in Enzymology 11, 377-400.

Strange, R.N., Majer, J.R. & Smith, H. (1974). The isolation and identification of choline and betaine as the two major components in anthers and wheat germ that stimulate Fusarium graminearum in vitro. Physiological Plant Pathology 4, 277-290.

Strange, R.N. & Smith, H. (1978). Effects of choline, betaine and wheat-germ extract on growth of cereal pathogens. Transactions of The British Mycological Society 70, 193-199.

Tani, T., Yamamoto, H., Kadota, G. & Naito, N. (1976). Development of rust fungi in oat leaves treated with blasticidin S, a protein synthesis inhibitor. Technical Bulletin of Faculty of Agriculture Kagawa University 27 (Ser. 59), 95-103.

Tani, T., Yamashita, Y. & Yamamoto, H. (1980). Initiation of induced non-host resistance of oat leaves to rust infection. Phytopathology 70, 39-42.

Teasdale, J., Daniels, D., Davis, W.C., Eddy, R. Jr. & Hadwiger, L.A. (1974). Physiological and cytological similarities between disease resistance and cellular incompatibility responses. Plant Physiology 54, 690-695.

Tegtmeier, K.J. & Van Etten, H.D. (1982). The role of pisatin tolerance and degradation in the virulence of Nectria haematococca on peas: a genetic analysis. Phytopathology 72, 608-612.

Turner, E.M.C. (1961). An enzymic basis for pathogenic specificity in Ophiobolus graminis. Journal of Experimental Botany 12, 169-175.

Vance, C.P. & Sherwood, R.T. (1976). Cycloheximide treatments implicate papilla formation in resistance of reed canarygrass to fungi. Phytopathology 66, 498-502.

Van der Molen, G.E., Beckman, C.H. & Rodehorst, E. (1977). Vascular gelation: a general response phenomenon following infection. Physiological Plant Pathology 11, 95-100.

Van Etten, H.D. (1973). Differential sensitivity of fungi to pisatin and to phaseollin. Phytopathology 63, 1477-1482.

Van Etten, H.D., Matthews, D.E. & Smith, D.A. (1982). Metabolism of phytoalexins. In 'Phytoalexins' (Bailey, J.A. & Mansfield, J.W., eds), pp. 181-217. Blackie, Glasgow.

Van Etten, H.D., Matthews, P.S., Tegtmeier, K.J., Dietert, M.F. & Stein, J.I. (1980). The association of pisatin tolerance and demethylation with virulence on pea in Nectria haematococca. Physiological Plant Pathology 16, 257-268.

Ward, E.W.B. (1976). Capsidiol production in pepper leaves in incompatible interactions with fungi. Phytopathology 66, 175-176.

Wynn, W.K. (1976). Appressorium formation over stomates by the bean rust fungus: response to a surface contact stimulus. Phytopathology, 136-146.

Wynn, W.K. & Staples, R.C. (1981). Tropisms of fungi in host recognition. In 'Plant Disease Control. Resistance and Susceptibility' (Staples, R.C. & Toenniesson, G.H., eds), pp. 45-69. John Wiley & Sons, New York.

Yoshikawa, M., Yamanchi, K. & Masago, H. (1978). Glyceollin: its role in restricting fungal growth in resistant soybean hypocotyls infected with Phytophthora megasperma var sojae. Physiological Plant Pathology 12, 73-82.

Young, P.A. (1926). Penetration phenomena and facultative parasitism in Alternaria, Diplodia, and other fungi. Botanical Gazette 81, 258-279.

CHAPTER 4. GENETICS OF HOST RESISTANCE TO VIRUSES AND OF VIRULENCE

R. S. S. FRASER

4.1. INTRODUCTION

Much early work on plant virus diseases was directed towards finding hosts which gave varied and characteristic responses to infection, as aids to identification of particular viruses. This involved the screening of a wide range of cultivars of crop species, as well as related wild species. These investigations provided examples of varied host responses, some of which were clearly resistance mechanisms, in that the effects of the virus were contained, or were much less severe in some hosts than in others. From this starting point, research aimed to incorporate useful resistance into agricultural crops, and to determine the genetic basis of the resistance.

Since the pioneering work of Holmes (1938) on the genetics of resistance to TMV in Nicotiana glutinosa, sources of resistance have been identified in numerous species, and in many cases have been genetically characterized and successfully exploited in crop protection. It soon became apparent, however, that all resistance genes were not necessarily effective against all isolates of a particular virus, and that some resistance genes were quickly overcome by the emergence of apparently new, virulent forms of the pathogen. In this Chapter we are concerned with the genetic bases of resistance and virulence. The biochemical and physiological mechanisms involved are discussed in Chapter 6.

4.2. GENETICS OF RESISTANCE: MANY CASES ARE UNDER SIMPLE MENDELIAN CONTROL

Classically, genetic analysis of host resistance has involved crossing resistant and susceptible parents, then determining the reaction of the F_1; F_2 and, if necessary, backcross generations to inoculation with virus. Segregation ratios for different host reactions have mostly given good evidence for simple Mendelian control. However, the genetic control has sometimes appeared to be considerably more intricate, or other influences have tended to obscure simple segregation ratios. These more complex situations

Table 4.1. Examples of resistance controlled by a single dominant gene

Resistance gene	Host	Virus	References
N ts	N. glutinosa	TMV [b]	Holmes, 1938
	N. tabacum		Takahashi, 1975
N' ts	N. sylvestris	TMV [a]	Weber, 1951
	N. tabacum		Valleau, 1952, Fraser, 1983
Tm-1 ts/tr	tomato	TMV [a]	Pelham, 1972; Fraser & Loughlin, 1980
Tm-2 ts	tomato	TMV [a]	Pelham, 1972; Stobbs & Macneill, 1980
$Tm\text{-}2^2$ ts	tomato	TMV [a]	Cirulli & Alexander, 1969
I ts	Phaseolus vulgaris	BCMV [a]	Ali, 1950; Drijfhout, 1978
*	Lactuca saligna	CMV [a]	Provvidenti et al., 1980
*	cowpea	TRSV	de Zeeuw & Ballard, 1959
*	cowpea	CMV [a]	Sinclair & Walker, 1955
* ts	spinach	CMV [a]	Pound & Cheo, 1952
* ts	Phaseolus vulgaris	WMV	Provvidenti, 1974
SW1a, SW1b tr?	tomato	TSWV [a]	Finlay, 1953
*	Capsicum pendulum	PVX	Nagaich et al., 1968

* indicates an unnamed gene, ts that resistance is temperature-sensitive, and tr that resistance is temperature-resistant.

[a] virulent isolates of the virus capable of overcoming the resistance have been reported.

[b] No virulent isolates have been found, but large numbers of isolates have been tested.

will be considered in detail later.

4.2.1. Resistance controlled by single dominant genes. Table 4.1 lists examples of resistance where control is by a single dominant gene, i.e. where the F_1 displays the same or a closely similar resistance reaction to the homozygous resistant parent. The list, and also that in Table 4.2, are obviously not exhaustive, but consist of examples chosen at random, to give a sample from which general principles may be inferred. Thus some comparison may be made of the relative preponderance of dominant, incompletely dominant and apparently recessive mechanisms.

The two genes in tobacco show how probably allelic forms (Valleau, 1952) may be derived from related wild species: the *N* gene from *N. glutinosa* (Holmes, 1938), and the *N*' from *N. sylvestris* (Kunkel, 1934; Weber, 1951). The two SW1 alleles for resistance to TSWV in tomato were derived from different parent lines, and confer resistance to different groups of the virus (Finlay, 1953). Resistance to TMV in tomato gives an example of how separate genes, at different loci, may independently confer resistance to the same virus. However, when both *Tm-1* and a *Tm-2* or *Tm-2*^{2} allele are present, the effects may be cumulative (Pelham, 1972). There do not yet appear to be any fully substantiated cases of resistance mechanisms which involve obligate co-operation of dominant genes at more than one locus.

With the exception of the *Tm-1* gene in tomato (see below), the other examples of dominant genes cited in Table 4.1 appear to prevent establishment of detectable infection, or localize the virus at the site of infection: necrosis is a frequent but not universal accompaniment of localization. Where they have been tested, dominant resistance genes have also frequently been found to be temperature-sensitive in function. Above 28 - 32°C, depending on the gene, resistance breaks down (e.g. Takahashi, 1975; Pilowsky *et al.*, 1981). The tendency of dominant genes to localize virus and be temperature-sensitive has implications for the mode of action which will be returned to in Chapter 6.

The presence of the *Tm-1* gene in Table 4.1 cannot be completely justified. This gene suppresses formation of mosaic symptoms by TMV, and is normally completely effective in this function in the heterozygous configuration (Pelham, 1972; Fraser & Loughlin; 1980). However, examination of TMV multiplication showed that the virus spread systemically in resistant plants, and that multiplication was inhibited in a gene-dosage dependent manner (Fraser & Loughlin, 1980). Thus while the gene is normally completely dominant for suppression of mosaic formation, it is incompletely dominant for inhibition of multiplication. Definition of a gene as dominant or recessive in simple Mendelian terms is often easy when only visible symptoms are examined; more quantitative analysis of effects on virus multiplication or of the underlying mechanism may necessitate a more complex description of phenotypic relationships.

The dominant resistance to PVX in *Capsicum pendulum* appears to be cytoplasmically inherited (Nagaich *et al.*, 1968). This interesting system

Table 4.2. Examples of resistance controlled by incompletely dominant or recessive genes

Resistance gene	Host	Virus	References
Incompletely dominant genes			
L^1,L^3	Capsicum chinense	TMV[a]	Boukuma, 1980
y^a	Capsicum annuum	PVY[a]	Shifriss & Marco, 1980
*	tobacco	TVMV	Legg et al., 1980
*(3 genes)	melon	CMV	Karchi et al., 1975
Bm	sugarbeet	BeMV[b]	Lewellen, 1973
*	tomato	TYLCV	Pilowsky & Cohen, 1974
*	barley	BYDV	Catherall et al., 1970
Apparently recessive genes			
sw2, sw3, sw4, tr?	tomato	TSWV[a]	Finlay, 1953
mo ts	pea	BYMV + WMV-2	Schroeder & Provvidenti, 1971
am-1	alfalfa	AMV[a]	Crill et al., 1971
by-3 ts	Phaseolus vulgaris	BYMV[b]	Provvidenti & Schroeder, 1973
bc-1,bc-2,bc-3 tr	Phaseolus vulgaris	BCMV[a]	Drijfhout, 1978

Symbols are as explained in the footnote to Table 4.1.

should be further studied, in view of the comparative ease with which the chloroplast DNA can be analyzed using genetic manipulation techniques.

4.2.2. Resistance controlled by recessive or incompletely dominant genes. Table 4.2 gives several examples: it is notable that in many cases, resistance was described by authors as incompletely dominant, i.e. the F_1 heterozygotes were found to have symptoms intermediate between susceptible and homozygous resistant parents. In some cases where authors described resistance as 'recessive', symptoms in the F_1 were, however, reported to be less severe or slower to develop than in the susceptible parent (e.g. Legg et al., 1980; Karchi et al., 1975). In at least one case where resistance was classed as recessive on a symptom basis (Drijfhout, 1978), examination of virus multiplication showed clear gene-dosage dependence (Day, 1984). In the few cases where authors reported completely recessive behaviour on the basis of symptoms, possible differences in virus content between homozygous and heterozygous hosts were not fully examined (e.g. Schroeder & Provvidenti, 1971; Crill et al., 1971). Thus completely recessive behaviour of resistance genes may actually be very rare.

Recessive or incompletely dominant resistance genes are fairly frequently described in reports as conferring 'tolerance', in that they often reduce symptom severity, although to levels which may still be detectable.

In contrast to the majority of dominant resistance genes, none of the examples cited in Table 4.2 appeared to operate by localizing the virus in necrotic lesions. Incomplete localization was shown by Boukwma (1980) to be involved in resistance to TMV in Capsicum chinense, conferred by a series of incompletely dominant genes. The strong, possibly allelic gene L in C. annuum did localize TMV in lesions and is considered dominant (Holmes, 1954). Thompson et al. (1962) reported that lines of Scotia bean homozygous for the recessive gene tm did form local lesions on infection with TMV. However, this cannot be considered as an exception to the general rule, as the alternative reaction with a dominant allele was complete immunity. Some of the recessive or incompletely dominant genes in Table 4.2 prevent a necrosis which is part of the susceptible reaction.

A further contrast between dominant and 'recessive' genes is that temperature sensitivity has been recorded comparatively rarely for recessive or incompletely dominant genes (e.g. Schroeder & Provvidenti, 1971). However, temperature-sensitivity does not appear to have been tested for many host-virus combinations.

Although resistance controlled by a single pair of recessive or incompletely dominant genes is the most common in this group, there are several examples where more complex systems exist. Three recessive but independent genes control resistance to TSWV in tomato, in addition to the independent pair of dominant allelic genes (Finlay, 1953). In other cases, there are additive effects from different combinations of resistance genes (e.g. Karchi et al., 1975). In barley, five host genotypes, involving major genes at or near the same locus, control a variety of 'tolerance' responses to BYDV: the reactions were described as varying from complete dominance to recessiveness, depending on genotype and environment (Catherall et al., 1970). In the complex recessive resistance to BCMV described in more detail in section 4.5.1, there is good evidence that co-operative action between resistance genes at separate loci is required (Drijfhout, 1978).

An interesting converse case is that of a single gene in pea which confers resistance to two potyviruses, BYMV and WMV-2 (Schroeder & Provvidenti, 1971). However, these viruses are distantly related serologically (Van Regenmortel _et al._, 1962) so the apparent dual effect of this resistance gene may be more nominal than truly bifunctional.

4.2.3. _Is there polygenic resistance to plant viruses?_ Theoretically, we may distinguish between two models of polygenic resistance. In the first, resistance depends on co-operative or cumulative action of many genes, all directly and primarily involved in the resistance mechanism. Examples might include genes in a multistep pathway leading to a single functional end product. Alternatively, resistance might depend on co-operative interaction of the products of several genes with some stage of viral pathogenesis.

The second type of polygenic resistance is where several genes may modify or control expression of a few genes directly involved in the resistance mechanism, without themselves being directly involved in resistance as their prime function. Evidence for such indirect 'modifiers' is perhaps more compelling than for true polygenic resistance.

Polygenic resistance has generally been claimed where segregation of phenotypes after hybridization failed to show simple Mendelian ratios. Unfortunately, complex segregation ratios may be obtained for reasons other than complex genetic control. In at least two cases, apparently complex genetic controls became much simpler on re-examination.

Resistance to CMV in cucumber derived from the Chinese Long cultivar, originally proposed to be controlled by 3 complementary dominant genes plus modifiers (Shifris _et al._, 1942), was later shown to be controlled by a single dominant gene (Wasuwat & Walker, 1961). Baggett & Frazier (1957) suggested that resistance to BYMV in _Phaseolus vulgaris_ was controlled by two or three major recessive genes plus modifiers. Later work by Provvidenti & Schroeder (1973) showed control by a single recessive gene. The key difference in both pairs of studies is that the later, simpler results were obtained in experiments conducted under controlled environmental conditions. The more complex genetic systems proposed in the earlier studies were attempts to explain complex genotype-environment interactions in purely genetic terms.

Holmes (1955) examined cases of extremely effective resistance to virus, in which single genes were regarded as providing the final component of a polygenic control involving perhaps 20 to 40 other genes. Unfortunately there is no genetic evidence for these other genes. A negative explanation - that marginally 'susceptible' plants failed to support large amounts of virus multiplication because of lack of components essential to a productive host/virus interaction - cannot be excluded.

4.2.4. Interaction between resistance genes and other host genes. There are several documented cases where expression of a resistance phenotype is modified by other factors in the host genotype. In barley, an incompletely dominant gene conditioning 'tolerance' to BYDV was shown to be effective in rapidly growing hosts, but when transferred to slow-growing host genotypes, or when growth was slowed by environmental controls, resistance was ineffective (Jones & Catherall, 1970). In maize, accessory (B) chromosomes or an abnormal 10th chromosome altered resistance to WSMV. The aberrant ratio (AR) phenomenon strongly decreased susceptibility to WSMV (McGirr & Endrizzi, 1978). A single recessive gene for resistance to SMV was more effective when transferred to some soybean cultivars than in others (Kwon & Oh, 1980). An intriguing report is that resistance to TMV which was effective (as 'tolerance') in normal diploid tomato plants was ineffective in tetraploids prepared from them (Ghidoni *et al.*, 1977). The possibility that host genetic background may influence resistance gene function emphasizes the desirability of using near-isogenic lines when comparing resistant and susceptible hosts.

4.3. GENETIC DETERMINANTS OF VIRULENCE

Classical genetic analysis by hybridization and study of segregation ratios is of course impossible with plant viruses, the majority of which have single-stranded RNA genomes. Certain techniques, however, have allowed some mapping of virulence determinants to particular areas of the genome.

4.3.1. Viruses with multipartite genomes. Individual genomic components can be separated, either at the particle or RNA level, from isolates differing in virulence characteristics. They can then be recombined to give pseudorecombinants, to assess the functions of the different RNA species.

Lettuce breeding lines resistant to the B strain of CMV have been developed using CMV resistance from *Lactuca saligna* (Provvidenti *et al.*, 1980). The CMV-LsS strain infects such resistant lettuce lines systemically. Analysis of pseudorecombinants between CMV-B and CMV-LsS showed that ability to infect resistant lettuce systemically depended on both RNAs 2 and 3 of CMV-LsS (Edwards *et al.*, 1983), suggesting that virulence in this case involves two genetic determinants. Further studies have shown the importance of RNA 2 alone of CMV-B in determining the ability of this strain to infect legumes systemically. Infection was localized with CMV-LsS or pseudorecombinants with CMV-LsS RNA 2. Rao & Francki (1982) carried out a comprehensive examination of the localization of determinants of symptom type on the three genomic RNAs of CMV, using 18 pseudorecombinants and 10 test hosts. Although the link with virulence is somewhat loose, they found that reactions could be controlled by one, two or even all three RNAs.

Habili & Francki (1974) compared CMV, and the related virus TAV. Host reactions varied from immunity to systemic mosaic. Studies of pseudorecombinants showed that TAV RNAs 1 and 2 determined symptom types, i.e. the ability to overcome host resistance.

Wyatt & Kuhn (1980) showed that RNA 1 of a virulent strain of CCMV was responsible for its ability to invade cowpea systemically; this host is normally resistant to CCMV. Harrison et al. (1974) found that ability of RRV to invade the Lloyd George cultivar of raspberry, which is resistant to some isolates of the virus, was localized on RNA 1.

Clearly the genetic localization of virulence in multicomponent viruses is highly variable, and probably involves diverse functions. It also overlaps with the problem of determination of host range, where the interaction between host and viral genome may be more complex than a simple one between resistance gene and virulence gene. Further examples of localization of determinants of virulence in multicomponent viruses are discussed in Chapter 6.5, in the context of the mechanisms of virulence gene expression.

An interesting finding by Findlay (1953) was that virulent isolates of TSWV could facilitiate multiplication of and symptom formation by avirulent isolates, in mixed infections of resistant plants. It is not clear whether this involved formation of pseudorecombinants in vivo, or functional complementation between the two strains.

4.3.2. Viruses with monopartite genomes. With these viruses, genetic analysis is less simple, as pseudorecombinants cannot be made, and alternative approaches are limited. Kado & Knight (1966) stripped coat protein progressively from one end of TMV particles of a strain causing systemic mosaic on N' gene tobacco. Exposed RNA is more susceptible to mutation with nitrous acid. Frequency of mutation to the local lesion phenotype was measured in particles with various degrees of stripping. The results suggested that the determinant for local lesion formation was located towards the 3' end of the

RNA, though probably not in the coat protein gene itself which is very close to the 3' end. (The original report referred to the 5' end, but ends were wrongly attributed at the time of the study: see Wilson et al., 1976).

Another approach to mapping functions on monopartite viral RNAs is to try to associate altered function with alterations in known gene products. Some evidence from this approach suggests that altered patterns of interaction of TMV with the N' gene may be associated with changes in the biochemical properties of TMV coat protein (Tsugita, 1962; Fraser, 1983). These and related results are discussed in more detail in Chapter 6.5, in relation to the possible mode of action of the altered gene product.

Modern techniques of genetic manipulation may eventually offer a means of synthesizing artificial recombinants via DNA copies of RNA viruses. However, available methods of base sequencing, applied to isolates differing in virulence, would not necessarily offer a rapid biochemical explanation of the basis of virulence. Firstly, the inherent instability, or high mutation rate associated with RNA genomes (Reanney, 1982) would mean that many DNA clones would have to be sequenced to ensure that the differences between avirulent and virulent isolates were not simply due to non-representative mutation in particular clones. Secondly, the problem of random noise from the high background level of variability would make it difficult to assess which differences were responsible for virulence, especially if natural isolates of different virulence characteristics were being compared. For this reason, it would be preferable to compare mutants produced by chemical means from known wild types, if these could be made for the particular host-virus system involved. The final problem is that it is difficult to relate demonstrated changes in base sequence, and hence in amino acid sequence of protein products, to recognizable changes in biological function.

4.4. DOMINANCE, RECESSIVENESS AND MODIFIERS: THEORETICAL CONSIDERATIONS

4.4.1. Theoretical models of gene operation. Two types of model of resistance gene operation may be proposed.

In the 'positive' model, the resistance gene produces a product which interferes with virus replication or pathogenesis. If the wild-type (+) allele is a null (absent) allele, as might be expected where the resistance gene is on a chromosome or fragment derived from another species, then the resistance should be dominant or gene-dosage dependent. If the + allele is not null, it may or may not interact with the resistance gene product, and all possible resistance phenotypes on the spectrum from dominant to recessive may occur. Virulence in this model is represented by altered interaction between the virus and the resistance gene product, such that the resistance mechanism is evaded.

In the 'negative' model, a resistance gene is a null or mutant allele of a wild type (+) gene normally conferring susceptibility to the virus, e.g. specifying some function or state required by the virus for replication or pathogenesis. A null or mutant allele of a host gene for an RNA-dependent RNA polymerase or subunit would be a possibility. This type of resistance gene could be dominant, gene-dosage dependent or fully recessive, with the exception that the null-allele type could not be dominant. In this model, virulence would represent an altered viral function enabling it to accommodate to the altered host function or lack of it.

Temperature sensitivity of resistance could act directly in the positive model and in some examples of the negative model; the null-allele variant of the negative model, however, would not be expected to be temperature sensitive.

In the present state of knowledge, it is not possible to decide which model is more likely for any particular resistance system. Unequivocal evaluation may depend on future isolation and functional analysis of the active products of resistance genes. Cytogenetic demonstration that the resistance gene is located on a chromosome or fragment not present in the susceptible host (e.g. Wang & Liang, 1977) is strong but not conclusive evidence for a positive model. The negative model is perhaps most worth examining in cases where resistance has apparently arisen spontaneously in one individual in a susceptible population (e.g. de Zeeuw & Ballard, 1959).

4.4.2. Dominance and recessiveness. Clearly, the concepts of dominance and recessiveness in the resistance-virulence context are far from simple. Kacser & Burns (1981) have developed an analysis of the molecular basis of dominance which, although concerned with a model of consecutive enzymes in a metabolic pathway, has implications for resistance-virulence interactions.

A basic feature of the model is that 'dominance' is not a property of genes, but a shorthand description of relationships between phenotypes. Quantitative differences between phenotypes are not linearly related to quantitative changes in individual elements or their activity at the genetic level. Each component of the system has a 'sensitivity coefficient' (Z), which is defined as the fractional change in activity of the whole system, over the fractional change in activity of the component. The sum of all Zs in the system is unity.

Model systems with different numbers of components, and different values of Z for particular components, demonstrate the power of control available over the phenotype. A particularly important outcome is that it is not necessary to postulate 'modifiers' to maximize the fit of the phenotype to the genotype. As components of resistance and virulence systems become better characterized in quantitative terms, this type of analysis should be increasingly useful in assessing system complexity and interactions.

4.5. RESISTANCE AND VIRULENCE: THE GENE-FOR-GENE RELATIONSHIP

As shown in Tables 4.1 and 4.2, virulent strains of virus which can overcome host resistance genes are known in several cases. In some of these, a gene-for-gene relationship between resistance genes and corresponding virulence genes has been worked out, analogous to Flor's (1956) classical analysis of complementary genetic systems in higher plants and fungi. The term 'gene' is applied rather broadly in the viral context.

4.5.1. *Examples of gene-for-gene relationships*. In tobacco, the *N* gene has been effective under normal conditions against all isolates of TMV so far tested. In contrast, the *N'* gene localizes some TMV isolates and causes formation of necrotic lesions, but others can spread systemically without causing necrosis (Weber, 1951; Valleau, 1952). Gierer & Mundry (1958) showed that mutagenesis with nitrous acid could convert a systemic strain to a necrotic strain, and that the alteration could be ascribed to a single base change in the viral RNA. Possible natural mutation in the reverse direction has also been suggested (Fraser, 1983).

Resistance to TMV in tomato shows a more complex gene-for-gene relationship, which is set out in Table 4.3 in the form of a 'checkerboard' (Person, 1959). Virulent virus strains are numbered according to the host resistance genes which they can overcome (Pelham, 1972). Thus strain 0 isolates have no virulence genes. Strain 1 isolates, capable of overcoming *Tm-1* resistance, were detected in commercial crops within about 1 year of introduction of cultivars containing only *Tm-1* resistance (Pelham *et al*., 1970). Strain 2 types are rarer, and strain 2^2 virulence is very infrequent (Hall, 1980). So far, the few reports of strain 2^2 virulence seem to be of isolates which are partly defective in other functions, i.e. combining virulence with low aggressiveness in any host (Hall & Bowes, 1980). There do

Table 4.3. Genetic interactions between TMV-resistant tomato plants and strains of the virus

Host genotypes	Virus genotypes							
	0	1	2	2^2	1.2	1.2^2	2.2^2	$1.2.2^2$
+	M	M	M	M	M	M	m	m
Tm-1	R	M	R	R	M	M	r	m
Tm-2 [a]	R	R	M	R	M	R	m	m
Tm-2^2 [a]	R	R	R	M	R	M	m	m
Tm-1/Tm-2	R	R	R	R	M	R	r	m
Tm-1/Tm-2^2	R	R	R	R	R	M	r	m
Tm-2/Tm-2^2	R	R	R	R	R	R	m	m
Tm-1/Tm-2/Tm-2^2	R	R	R	R	R	R	r	m

M indicates systemic mosaic, R indicates resistance. m and r represent predicted responses with combinations of virulence genes which have not yet been demonstrated in virus isolates.

[a] Plants with these genotypes may show local and variable systemic necrosis when inoculated with avirulent strains (Hall, 1980).

not yet seem to have been any unequivocal reports of isolates with both 2 and 2^2 virulence genes. A virus strain reported to be $1,2,2^2$ (Hollings & Pawley, 1977) was later questioned (Hall, 1980).

Both the tomato and tobacco systems of resistance to TMV illustrate clearly how resistance genes may vary considerably in 'durability', i.e. in the likelihood of their being overcome by existing strains or mutations to virulence. Most commercial TMV-resistant tomato cultivars now have the genetic composition *Tm-1/Tm-1:Tm-2/Tm-2*^{2} or *Tm-1/+:Tm-2/Tm-2*^{2} and this combination appears to be highly durable. Clearly the requirement to overcome three resistance genes is a very difficult if not impossible evolutionary step for the virus.

The recessive system of resistance to BCMV in *Phaseolus vulgaris* shows one of the most complex gene-for-gene relationships which has yet been elucidated (Drijfhout, 1978). Resistance depends on co-operative action of the gene *bc-u* with one or more further genes from a group collectively termed *bc-x*. Thus plants which are *bc-u/bc-u:bc-x/bc-x* are resistant to 'strain 0' types of BCMV, while those with genotypes *bc-u*$^{+}$*/bc-u*$^{+}$*:bc-x/bc-x*; *bc-u/bc-u:bc-x*$^{+}$*/bc-x*$^{+}$ or any heterozygotes are susceptible and form mosaic.

The *bc-x* genes occur at three unlinked loci (*bc-1*, *bc-2* and *bc-3*). Allelic series have been found at the *bc-1* and *bc-2* loci.

BCMV exists as a complex series of strains which contain different combinations of virulence genes, each of which permits the virus to overcome one of the *bc-x* genes or alleles. The *bc-x* genes are therefore virus strain-specific resistances; in contrast the *bc-u* gene is strain non-specific, and in combination with a suitable strain-specific gene, confers resistance to all strains. For example, plants with resistance genes *bc-u/bc-u:bc-2/bc-2* are resistant to virus strains with no virulence genes or with virulence genes 1, 1^{2} and 2^{2} but are susceptible to strains containing the virulence gene 2.

Virus strains may have various combinations of virulence gene, e.g. 1, 1^{2} and 2, but certain combinations have not yet been found, or may be impossible for molecular reasons. No gene to overcome *bc-3* is yet known. No strain contains both 2 and 2^{2}: it may be that these determinants are located in the same functional region on the virus RNA, *i.e.* are allelic, and are mutually exclusive. Strains with virulence gene 1^{2} also show gene 1 type virulence, although the converse is not true. If 1 and 1^{2} are allelic, it could be that the alteration in function which confers 1^{2} virulence also inevitably includes ability to overcome *bc-1*. Note, however, that although a particular pair of host resistance genes may be allelic, the corresponding virulence genes in the virus need not necessarily be allelic.

Clearly the relationship between resistance and virulence genes in the BCMV/*P. vulgaris* system is extremely complex. Drijfhout (1978) has presented a complete gene-for-gene 'checkerboard' for all known genes and combinations: with 16 possible virus groups (omitting any containing the unknown virulence gene 3) and 18 host differential genotypes the pattern is obviously considerably more complex than the simple tomato example shown in Table 4.3. However, much remains to be discovered and verified about the genetics of the relationship.

4.5.2. *Are there 'vertical' and 'horizontal' resistances to viruses*? The concepts of vertical and horizontal resistance have been applied predominantly to fungal pathogens, and are discussed in that context in Chapters 3, 5 and 8. It is relevant to ask here whether they can be of use in examination of the plant/virus interaction.

Robinson (1976), in his provocative book on plant-pathogen interactions, considered that 'vertical resistance against viruses does not seem to occur...it seems that no Person differential interactions are known.' (i.e. in the sense used by Person (1959), to mean the existence of pathogen races capable of differentiating, by the reactions produced, all variation in resistance of the host population, and vice versa). The evidence presented in this Chapter, especially in the preceding section, indicates that differential reactions are now known to be common for virus diseases; many would be recognized as quite classical examples of vertical resistance in the Vanderplank (1968) interpretation. Robinson's argument that 'virus capacity to mutate to virulent forms overcoming vertical resistance is so great as to make this kind of resistance of limited survival value against them' does not invalidate the definition. Moreover, the demonstrated durability of some major resistance genes is obviously sufficient to make the vertical resistance conferred by them of high survival value to the host.

Horizontal resistance is a rather more nebulous concept and more difficult to establish by observation and experiment. The group of effects known as disease avoidance, which is discussed in Chapter 6.2, probably forms the best example for virus diseases, in that it is known to be effective against all isolates of the virus. Indeed some mechanisms are effective against several distinct viruses. Disease avoidance, operating by resistance to the vector or by vector non-preference, is obviously a completely indirect form of interaction between the host and viral genomes. Resistance to infection for physical or structural reasons, as appears to be the case for the non-specific resistance to several viruses in T.I. 245 tobacco (Holmes, 1961; Thomas & Fulton, 1968) could well be based genetically on complex determinants of morphogenesis in the host, and would not appear likely to select specifically virulent isolates of the virus.

For resistance mechanisms operating after infection, within the plant, no cases spring to mind where all the criteria of horizontal resistance have been unequivocally demonstrated.

Horizontal resistance, if polygenically inherited, might be lost during intensive breeding of crop species for agronomic characters determined by major genes. There are cases of new F_1 hybrid cultivars showing extreme sensitivity to viruses (Tomlinson et al., 1976). By extension of this argument, horizontal resistance might best be sought and tested experimentally in wild host species. For example, CMV causes serious disease in cultivated crops, and resistance strategies are vertical, based on major genes. In the wild species chickweed (Stellaria media), CMV causes symptomless infection, and there is evidence that this may have evolved from selection pressures acting on both host and viral genomes (Tomlinson & Walker, 1973). However, it is clear that horizontal resistance mechanisms would require to be rigorously proved and studied before any exploitation in crop protection would be possible. Furthermore, breeding for resistance based on a large number of genes is a daunting task by present day techniques. The ability to insert specific resistance genes into plants by recombinant DNA techniques would undoubtedly enhance the breeder's capacity to construct complex resistance systems, but this remains a very long-term prospect.

REFERENCES

Ali, M.A. (1950). Genetics of resistance to the common bean mosaic virus (bean virus 1) in the bean (Phaseolus vulgaris L.). Phytopathology 40, 69-79.

Baggett, H.J.R. & Frazier, W.A. (1957). The inheritance of resistance to bean yellow mosaic virus in Phaseolus vulgaris. Journal of the American Society of Horticultural Science 70, 325-333.

Boukwma, I.W. (1980). Allelism of genes controlling resistance to TMV in Capsicum L. Euphytica 29, 433-439.

Catherall, P.L., Jones, A.T. & Hayes, J.D. (1970). Inheritance and effectiveness of genes in barley that condition tolerance to barley yellow dwarf virus. Annals of Applied Biology 65, 153-161.

Cirulli, M. & Alexander, L.J. (1969). Influence of temperature and strain of tobacco mosaic virus on resistance of a tomato breeding line derived from Lycopersicon peruvianum. Phytopathology 59, 1287-1297.

Crill, P., Hanson, E.W. & Hagedorn, D.J. (1971). Resistance and tolerance to alfalfa mosaic virus in alfalfa. Phytopathology 61, 369-371.

Day, K.L. (1984). Resistance to bean common mosaic virus in Phaseolus vulgaris L. Ph. D. Thesis, University of Birmingham, U.K.

de Zeeuw, D.J. & Ballard, J.C. (1959). Inheritance in cowpea of resistance to tobacco ringspot virus. Phytopathology 49, 332-334.

Drijfhout, E. (1978). Genetic interaction between Phaseolus vulgaris and bean common mosaic virus with implications for strain identification and breeding for resistance. Agricultural Research Report 872; Centre for Agricultural Publishing and Documentation, Wageningen, the Netherlands.

Edwards, M.C., Gonsalves, D. & Provvidenti, R. (1983). Genetic analysis of cucumber mosaic virus in relation to host resistance: location of determinants for pathogenicity to certain legumes and Lactuca saligna. Phytopathology 73, 269-273.

Finlay, K.W. (1953). Inheritance of spotted wilt resistance in the tomato. II. Five genes controlling spotted wilt resistance in four tomato types. Australian Journal of Biological Science 6, 153-163.

Flor, H.H. (1956). The complementary genetic systems in flax and flax rust. Advances in Genetics 8, 29-54.

Fraser, R.S.S. (1983). Varying effectiveness of the N' gene for resistance to tobacco mosaic virus in tobacco infected with virus strains differing in coat protein properties. Physiological Plant Pathology 22, 109-119.

Fraser, R.S.S. & Loughlin, S.A.R. (1980). Resistance to tobacco mosaic virus in tomato: effects of the Tm-1 gene on virus multiplication. Journal of General Virology 48, 87-96.

Ghidoni, A., Fior, D. & Togliani, F. (1977). Reaction of diploid and tetraploid tomato (Lycopersicon esculentum Mill.) to infection by tobacco mosaic virus. Genetica Agraria 31, 347-354.

Gierer, A. & Mundry, K.-W. (1958). Production of mutants of tobacco mosaic virus by chemical alteration of its ribonucleic acid in vitro. Nature 182, 1457-1458.

Habili, N. & Francki, R.I.B. (1974). Comparative studies on tomato aspermy and cucumber mosaic viruses. III. Further studies on relationship and construction of a virus from parts of the two viral genomes. Virology 61, 443-449.

Hall, T.J. (1980). Resistance at the Tm-2 locus in the tomato to tomato mosaic virus. Euphytica 29, 189-197.

Hall, T.J. & Bowes, S.A. (1980). Disease resistance. Annual Report of the Glasshouse Crops Research Institute for 1979, pp. 49-53.

Harrison, B.D., Murant, A.F., Mayo, M.A. & Roberts, I.M. (1974). Distribution of determinants for symptom production, host range and nematode transmissibility between the two RNA components of raspberry ringspot virus. Journal of General Virology 22, 233-247.

Hollings, M. & Pawley, R.R. (1977). Glasshouse vegetable crops: tomato. Annual Report of the Glasshouse Crops Research Institute for 1976, p. 124.

Holmes, F.O. (1938). Inheritance of resistance to tobacco mosaic disease in tobacco. Phytopathology 28, 553-561.

Holmes, F.O. (1954). Inheritance of resistance to viral diseases in plants. Advances in Virus Research 2, 1-31.

Holmes, F.O. (1955). Additive resistance to specific viral diseases in plants. Annals of Applied Biology 42, 129-139.

Holmes, F.O. (1961). Concomitant inheritance of resistance to several virus diseases in tobacco. Virology 13, 409-413.

Jones, A.T. & Catherall, P.L. (1970). The relationship between growth rate and the expression of tolerance to barley yellow dwarf virus in barley. Annals of Applied Biology 65, 137-145.

Kacser, H. & Burns, J.A. (1981). The molecular basis of dominance. Genetics 97, 639-666.

Kado, C.I. & Knight, C.A. (1966). Location of a local lesion gene in tobacco mosaic virus RNA. Proceedings of the National Academy of Sciences, U.S.A. 55, 1276-1283.

Karchi, Z., Cohen, S. & Govers, A. (1975). Inheritance of resistance to cucumber mosaic virus in melons. Phytopathology 65, 479-481.

Kunkel, L.O. (1934). Studies on acquired immunity with tobacco and aucuba mosaic viruses. Phytopathology 24, 437-466.

Kwon, S.H. & Oh, J.H. (1980). Resistance to a necrotic strain of soybean mosaic virus in soybean. Crop Science 20, 403-404.

Legg, P.D., Litton, C.C. & Collins, G.B. (1980). Inheritance of resistance to the tobacco vein mottling virus in tobacco. Canadian Journal of Genetical Cytology 22, 21-20.

Lewellen, R.T. (1973). Inheritance of beet mosaic resistance in sugarbeet. Phytopathology 63, 877-881.

McGirr, S.C. & Endrizzi, J.E. (1978). The effects of B, K10 and AR chromosomes on the resistance of maize to viral infection. Genetics 90, 331-338.

Nagaich, B.B., Upadhya, M.D., Prakash, O. & Singh, S.J. (1968). Cytoplasmically-determined expression of symptoms of potato virus X in crosses between species of Capsicum. Nature 220, 1341-1342.

Pelham, J. (1972). Strain-genotype interaction of tobacco mosaic virus in tomato. Annals of Applied Biology 71, 219-228.

Pelham, J., Fletcher, J.T. & Hawkins, J.H. (1970). The establishment of a new strain of tobacco mosaic virus resulting from the use of resistant varieties of tomato. Annals of Applied Biology 65, 293-297.

Person, C. (1959). Gene-for-gene relationships in host:parasite systems. Canadian Journal of

Botany 37, 1101-1130.
Pilowsky, M. & Cohen, S. (1974). Inheritance of resistance to tomato yellow leaf curl virus in tomatoes. Phytopathology 64, 632-635.
Pilowsky, M., Frankel, R. & Cohen, S. (1981). Studies of the variable reaction at high temperature of F_1 hybrid tomato plants resistant to tobacco mosaic virus. Phytopathology 71, 319-323.
Pound, G.S. & Cheo, P.C. (1952). Studies on resistance to cucumber virus 1 in spinach. Phytopathology 42, 301-306.
Provvidenti, R. (1974). Inheritance of resistance to watermelon mosaic virus-2 in Phaseolus vulgaris. Phytopathology 64, 1448-1450.
Provvidenti, R., Robinson, R.W. & Shail, J.W. (1980). A source of resistance to a strain of cucumber mosaic virus in Lactuca saligna L. HortScience 15, 528-529.
Provvidenti, R. & Schroeder, W.T. (1973). Resistance in Phaseolus vulgaris to the severe strain of bean yellow mosaic virus. Phytopathology 63, 196-197.
Rao, L.N. & Francki, R.I.B. (1982). Distribution of determinants for symptom production and host range on the three RNA components of cucumber mosaic virus. Journal of General Virology 61, 197-205.
Reanney, D.C. (1982) The evolution of RNA viruses. Annual Review of Microbiology 36, 47-73.
Robinson, R.A. (1976). Plant Pathosystems. Springer Verlag, Berlin, Heidelberg & New York.
Schroeder, W.T. & Provvidenti, R. (1971). A common gene for resistance to bean yellow mosaic virus and watermelon mosaic virus-2 in Pisum sativum. Phytopathology 61, 846-848.
Shifriss, C. & Marco, S. (1980). Partial dominance of resistance to potato virus Y in Capsicum. Plant Disease 64, 57-59.
Shifriss, O., Myers, C.H. & Chupp, C. (1942). Resistance to mosaic virus in the cucumber. Phytopathology 32, 773-784.
Sinclair, J.B. & Walker, J.C. (1955). Inheritance of resistance to cucumber mosaic virus in cowpea. Phytopathology 45, 563-564.
Stobbs, L.W. & MacNeill, B.H. (1980). Response to tobacco mosaic virus of a tomato cultivar homozygous for gene Tm-2. Canadian Journal of Plant Pathology 2, 5-11.
Takahashi, T. (1975). Studies on viral pathogenesis in plant hosts. VIII. Systemic virus invasion and localization of infection in "Samsun-NN" tobacco plants resulting from tobacco mosaic virus infection. Phytopathologische Zeitschrift 84, 75-87.
Thomas, P. & Fulton, R.W. (1968). Correlation of the ectodesmata number with nonspecific resistance to initial virus infection. Virology 34, 459-469.
Thompson, A.E., Lower, R.L. & Thornberry, H.H. (1962). Inheritance in beans of the necrotic reaction to tobacco mosaic virus. Journal of Heredity 53, 89-91.
Tomlinson, J.A. & Walker, V.M. (1973). Further studies on seed transmission in the ecology of some aphid-transmitted viruses. Annals of Applied Biology 73, 293-298.
Tomlinson, J.A., Ward, C.M. & Webb, M.J.W. (1976). Brassica virus diseases. Annual Report of the National Vegetable Research Station for 1975, pp. 107-108.
Tsugita, A. (1962). The proteins of mutants of TMV: composition and structure of chemically evoked mutants of TMV RNA. Journal of Molecular Biology 5, 284-292.
Valleau, W.D. (1952). The evolution of susceptibility to tobacco mosaic in Nicotiana and the origin of the tobacco mosaic virus. Phytopathology 42, 40-42.
Van Regenmortel, M.H.V., Brandes, J. & Bercks, R. (1962). Investigations on the properties of watermelon mosaic virus. Phytopathologische Zeitschrift 45, 205-216.
Vanderplank, J.E. (1968). Disease Resistance in Plants. Academic Press, New York & London.
Wang, R.C. & Liang, G.H. (1977). Cytogenetic location of genes for resistance to wheat streak mosaic in an Agropyron substitution line. Journal of Heredity 68, 375-378.
Wasuwat, S.L. & Walker, J.C. (1961). Inheritance of resistance in cucumber to cucumber mosaic virus. Phytopathology 51, 423-426.
Weber, P.V.V. (1951). Inheritance of a necrotic lesion reaction to a mild strain of tobacco mosaic virus. Phytopathology 41, 593-609.
Wilson, T.M.A., Perham, R.N., Finch, J.T. & Butler, P.J.G. (1976). Polarity of the RNA in tobacco mosaic virus particles and the direction of protein stripping in sodium dodecyl sulphate.

FEBS Letters 64, 285-289.
Wyatt, S.D. & Kuhn, C.W. (1980). Derivation of a new strain of cowpea chlorotic mottle virus from resistant cowpeas. Journal of General Virology 49, 289-296.

CHAPTER 5. THE GENETIC BASES OF RELATIONSHIPS BETWEEN MICROBIAL PARASITES AND THEIR HOSTS

I. R. CRUTE

5.1. INTRODUCTION: AN OVERVIEW OF THE GENETICS OF HOST-MICROBIAL PARASITE ASSOCIATIONS

5.1.1. Historical perspective. Loss of reproductive capacity due to infectious disease is one of many selective forces operating in natural plant populations (Harlan, 1976; Levin, 1975). One result of the imposed selection is the evolution of genotypes exhibiting an enhanced capability to reproduce despite the prevalence of a disease-causing agent. There is now clear evidence that disease resistance and tolerance are important features of natural plant populations (Segal *et al.*, 1980; Ben Kalio & Clarke, 1979).

Since the times when man first cultivated crops, it is likely that infectious diseases have caused losses. Some early documentation of man's awareness of plant diseases is provided by Horsfall & Cowling (1977). The outcome of selection imposed on primitive crops may not be far removed from the natural situation although at some stage in his agricultural development man must have started to impose conscious selection for desirable characters. As a result of natural and artificial selection, crop genotypes less prone to disease were maintained. These selective processes probably changed little over thousands of years of agricultural civilization, based as it was on an ignorance of the mechanisms controlling heredity. The legacy of these years are many of the genes which plant breeders exploit today.

5.1.2. Disease resistance and the rediscovery of Mendel's laws. With the rediscovery of Mendel's work, a scientific basis for plant breeding was laid. The discovery of the causes of infectious plant disease provided another impetus to breeding for disease resistance. This Chapter is concerned with the genetic control of the relationships between plants and parasitic fungi and bacteria.

The first reported study of the genetics of disease resistance was by Biffen (1905; 1907; 1912) who examined the reaction of wheat to yellow rust (referred to then as *Puccinia glumarum* but now known as *Puccinia striiformis*). Crosses were made between resistant (Rivet and America Club) and susceptible (Michigan Bronze and Red King) cultivars and the performance of the F_1, F_2 and F_3 generations studied in field trials. The data indicated that a single gene controlled reaction to yellow rust with susceptibility dominant to resistance for both sources of resistance studied. In retrospect, these findings are exceptional since resistance is now known to be more often inherited as a dominant trait.

A further point of interest to emerge from these earliest studies is that, between the extremes of resistance and susceptibility, an intermediate category of plants was recognized. Ratios close to 1:2:1 at F_2 imply incomplete dominance. Quantitative variation between susceptible families was also evident as was transgressive segregation. Minor gene modifications of major gene effects were therefore apparent, although it must be remembered that the experiments were performed in an uncontrolled environment.

Biffen (1912) was confident that resistance would not fail due to 'advance in the parasitic properties of the rusts' and believed that the evolution of rusts would take place 'too slowly to influence the work of plant breeders'. In this prediction he was soon to be proved wrong, but it is revealing that his opinion was based on the observation that cvs Rivet and Einkorn had remained resistant despite being in cultivation for many years. The concept of 'durable' resistance today is based upon a similar premise (Johnson & Law, 1975; Johnson, 1978; Johnson, 1979).

5.1.3. **Physiologic specialization and the 'boom and bust' cycle.** Early work with rusts and powdery mildews of grasses demonstrated that morphologically indistinguishable 'biologic forms' of these fungi varied in their capacity to be pathogenic on different host species (Salmon, 1904; Stakman, 1914; Stakman & Piemeisel, 1917). These biologic forms, later to be called *formae speciales*, were considered to be sub-specific taxa identified by the resistance or susceptibility of a set of 'differential hosts'. At this time, it was still held by some that 'breeding for rust resistance can proceed with the considerable assurance that the same rust will not adapt itself quickly to new varieties' (Stakman *et al.*, 1918). Pathogenic variants within a fungal species or *forma specialis* came to be referred to as 'physiologic races' and it soon became clear that their evolution had profound implications for resistance breeding. The taxonomic identification of physiologic races using a set of differential cultivars became standard practice in many pathological studies following early work with Puccinia graminis (Stakman & Levine, 1922).

In numerous crop-parasite associations, the identification, utilization and widespead deployment of a particularly effective source of resistance was frequently followed by its failure to exert control. This was as a consequence of the emergence of a previously unrecognized physiologic race of the pathogen capable of rendering the resistance ineffective. This pattern, which continues to the present, has been referred to as the 'boom and bust' cycle after the way in which the popularity of a new resistant cultivar increases only to decline rapidly as the resistance is no longer effective.

5.1.4. The gene-for-gene relationship. A genetic comprehension of the events associated with the 'boom and bust' cycle only followed the pioneering work with flax and flax rust (Melampsora lini) by Flor (1935; 1940; 1941; 1942; 1946; 1947; 1954; 1955; 1956; 1965; 1971). This led to the development of the gene-for-gene hypothesis which has become central to further studies on the genetics of host-parasite relationships.

Prior to this, several investigations had confirmed that disease resistance in a number of different host-parasite associations could be inherited as a Mendelian character and there was evidence that this was also true for specific virulence (Johnson et al., 1934). In retrospect, early work on the resistance of Phaseolus beans to Colletotrichum lindemuthianum (anthracnose) (Barrus, 1911; 1915; Burkholder, 1918; McRostie, 1919; 1921) almost pre-empted Flor's exposition of the gene-for-gene hypothesis. Two races of the fungus were differentiated by a series of bean cultivars and genetic studies demonstrated that dominant alleles at two different loci specifically controlled resistance to these two races. Since then, further genes for resistance to C. lindemuthianum have been identified and additional races categorized (Krüger et al., 1977). However, due to the inability to conduct genetic studies with the parasite, evidence for a gene-for-gene relationship remains circumstantial.

The importance of Flor's study was that he systematically and simultaneously examined the genetics of resistance and virulence in both partners of the host-parasite association. Several general conclusions were drawn from his studies. It was evident that there existed in the parasite a complementary locus controlling specific virulence for each allele in the host controlling specific rust reaction (Figure 5.1). In the host, resistance was usually dominant to susceptibility with genes controlling higher degrees of resistance being epistatic to genes of lesser effectiveness. It was also

Host genotype	Pathogen genotype: $A_1A_2A_3$	$a_1A_2A_3$	$A_1a_2A_3$	$A_1A_2a_3$	$a_1a_2A_3$	$a_1A_2a_3$	$A_1a_2a_3$	$a_1a_2a_3$
$r_1 r_2 r_3$	+	+	+	+	+	+	+	+
$R_1 r_2 r_3$	-	+	-	-	+	+	-	+
$r_1 R_2 r_3$	-	-	+	-	+	-	+	+
$r_1 r_2 R_3$	-	-	-	+	-	+	+	+
$R_1 R_2 r_3$	-	-	-	-	+	-	-	+
$R_1 r_2 R_3$	-	-	-	-	-	+	-	+
$r_1 R_2 R_3$	-	-	-	-	-	-	+	+
$R_1 R_2 R_3$	-	-	-	-	-	-	-	+

FIGURE 5.1. A hypothetical gene-for-gene relationship between host and parasite involving three matching gene pairs. Resistance and avirulence are dominant over susceptibility and virulence respectively.
+ = susceptible/virulent/compatible; - = resistant/avirulent/incompatible.

apparent that resistance alleles conditioned characteristic reaction types recognized by the extent of fungal colonization. In the fungus, complementary avirulence was usually dominant. Linkage between resistance genes and also multiple allelism were frequently encountered and further work has confirmed the existence in flax of five series of closely linked or allelic genes. At least 29 different specific resistance alleles have been recognized and of these, up to 13 are located in the 'L' linkage group; seven in 'M'; three in 'N'; five in 'P' and one in 'K' (Flor, 1956; Flor & Comstock, 1972; Wicks & Hammond, 1978; Zimmer & Comstock, 1973; Shepherd & Mayo, 1972; Mayo & Shepherd, 1980). Linkage between loci conditioning specific virulence also occurs (Lawrence _et al._, 1981b), but appears to be less common and there is no evidence for multiple allelism in the fungus.

The generalizations outlined above accurately summarize most of the data and have been confirmed in subsequent studies (Lawrence _et al._, 1981a).

However, some exceptions to these general trends emerged in Flor's studies and these are both interesting and important. The expression of at least one resistance gene was environmentally influenced, being less complete under some conditions; there was also evidence for incomplete dominance in the case of some resistance genes (Flor, 1941). In one cross between rust isolates, virulence on cv. Willison Brown was found to be dominant to avirulence and the data could not be accounted for by postulating the involvement of a single locus (Flor, 1946). In a different cross, however, avirulence for the same cultivar was dominant, the data being in agreement with the operation of a single locus (Flor, 1965). Lawrence *et al*. (1981a) have confirmed this pattern of inheritance for virulence on this and other cultivars and explain it in terms of a non-allelic interaction between the locus for specific virulence and a dominant 'inhibitor' allele at a second locus (see section 5.4.2). In subsequent studies with other host-parasite associations, further exceptions to the generalized relationships elucidated by Flor have emerged. These include resistance inherited as a recessive or partially dominant trait; modification of both the observed host reaction-type and dominance depending upon the genetic background in which an allele is located; different expressions of a resistance allele depending upon parasite isolate (possibly due to different alleles for avirulence in the parasite); the existence of duplicate genes for avirulence on a single R-gene; the matching of several resistance alleles by a single virulence allele; complementary resistance alleles, and additive effects in the expression of resistance alleles. These exceptions are the subject of more detailed treatment later in this Chapter.

In *Melampsora lini* there are also indications that virulence in the monokaryotic stage (basidiospores) may sometimes be under the control of different genes to those operating in the dikaryotic stage (aeciospores and uredospores) (Flor, 1959; Statler & Gold, 1980). Basidiospores from an isolate whose uredospores were avirulent on the flax cultivar Bombay proved to be virulent, but the aeciospores produced subsequently were avirulent (Flor, 1959). This result was confirmed by Statler & Gold (1980) who found cases of homozygous virulent isolates which yielded avirulent basidiospores and heterozygous isolates which did not yield virulent basidiospores as expected. No evidence for this phenomenon was found by Kolmer *et al*. (1984) in similar investigations with *Uromyces appendiculatus* (rust) on beans.

In the years following the exposition of the 'gene-for-gene' or 'complementary gene' hypothesis, it has been suggested that similar principles

Table 5.1. Examples of host-parasite associations where a gene-for-gene relationship has been suggested based on genetic analysis of both partners.

Host	Parasite	Representative references
Triticum spp. (wheat)	*Puccinia graminis* f. sp. *tritici* (stem rust)	Green (1966) Roelfs & McVey (1979)
	Puccinia recondita f. sp. *tritici* (leaf rust)	Statler (1979) Browder (1980) Samborski & Dyck (1976)
	Erysiphe graminis f. sp. *tritici* (powdery mildew)	McIntosh (1978) Powers & Sando (1960) Bennett (1984)
Hordeum spp. (barley)	*Ustilago hordei* (covered smut)	Sidhu & Person (1971; 1972) Ebba & Person (1975)
	Erysiphe graminis f. sp. *hordei* (powdery mildew)	Moseman (1963) Wolfe (1972) McIntosh (1978)
Zea mays (maize)	*Puccinia sorghi* (rust)	Flangas & Dickson (1961a) Saxena & Hooker (1968) Wilkinson & Hooker (1968)
Avena spp. (oats)	*Puccinia graminis* f. sp. *avenae* (stem rust)	Martens *et al*. (1970)
	Puccinia coronata f. sp. *avenae* (crown rust)	Simons & Michel (1964) Zimmer *et al*. (1965)
	Ustilago avenae (smut)	Holton & Halisky (1960)
Solanum spp. (potato)	*Phytophthora infestans* (late blight)	Person (1959) Romero & Erwin (1969)
Linum usitatissimum (flax)	*Melampsora lini* (rust)	Flor (1956) Lawrence *et al*. (1981a)
Lactuca sativa (lettuce)	*Bremia lactucae* (downy mildew)	Johnson *et al*. (1978) Norwood & Crute (1984)
Helianthus annuus (sunflower)	*Puccinia helianthi* (rust)	Putt & Sackston (1963) Miah & Sackston (1970)
Malus spp. (apple)	*Venturia inaequalis* (scab)	Boone (1971)
Phaseolus vulgaris (bean)	*Uromyces phaseoli* (rust)	Christ & Groth (1982a, 1982b)

Table 5.2. Examples of host-parasite associations where a gene-for-gene relationship may be suggested on circumstantial evidence.

Host	Parasite	Representative references
Triticum spp. (wheat)	Puccinia striiformis (yellow rust)	Zadoks (1961) Macer (1972)
	Tilletia spp. (bunt)	Holton et al. (1968)
Hordeum spp. (barley)	Ustilago nuda (loose smut)	Metcalfe (1966)
	Rhynchosporium secalis (scald)	Habgood & Hayes (1971)
	Puccinia hordei (leaf rust)	Roane & Starling (1970) Parlevliet (1976b)
	Pyrenophora teres (net blotch)	Khan & Boyd (1969a)
Oryza sativa (rice)	Pyricularia oryzae (blast)	Atkins et al. (1967) Robinson (1979)
Lycopersicon esculentum (tomato)	Fulvia fulvum (leaf mould)	Kooistra (1964) Day (1956)
	Fusarium oxysporum f. sp. lycopersici (wilt)	Gabe (1975)
Phaseolus vulgaris (bean)	Colletotrichum lindemuthianum (anthracnose)	Albersheim et al. (1969) Kruger et al. (1977)
Glycine max (soybean)	Pseudomonas syringae pv. glycinea (bacterial blight)	Cross et al. (1966)
	Phytophthora megasperma f. sp. glycinea (root rot)	Ellingboe (1983)
	Peronospora manshurica (downy mildew)	Crute (1981)
Spinacea oleracea (spinach)	Peronospora farinosa f. sp. spinacea (downy mildew)	Crute (1981)
Brassica spp. (turnip, swede and rape)	Plasmodiophora brassicae (clubroot)	Crute et al. (1980, 1983)

Table 5.2 continued.

Zea mays (maize)	Helminthosporium turcicum (Northern leaf-blight)	Perkins & Hooker (1981)
Cucumis melo (melon)	Fusarium oxysporum f. sp. melonis (wilt)	Risser et al. (1976)
Pisum sativum (pea)	Pseudomonas syringae pv. pisi (bacterial blight)	Taylor (1972) Taylor et al. (1984)
	Peronospora viciae (downy mildew)	Crute (1981)
Coffea arabica (coffee)	Hemileia vastatrix (rust)	Noronha-Wagner & Bettencourt (1967)
Gossypium spp. (cotton)	Xanthomonas malvacearum (bacterial blight)	Brinkerhoff (1970) Innes (1983)
Helianthus annuus (sunflower)	Plasmodiophora halstedii (downy mildew)	Crute (1981)

apply to many host-parasite associations, involving fungi and bacteria (as well as viruses, nematodes, insects and parasitic higher plants) (Person, 1959; Flor, 1971; Person & Ebba, 1975; Day, 1974; Sidhu, 1975). Tables 5.1 and 5.2 list examples for microbial parasites, and viral examples are discussed in Chapter 4. In many instances, however, the existence of a gene-for-gene relationship in the strict sense is unproved since critical genetical analyses of both partners have not been conducted. The occurrence of race specific resistance controlled oligenically in the host is only circumstantial evidence for a gene-for-gene relationship. Even in the most extensively investigated associations, relatively few parasite isolates have been investigated genetically. Consequently the contention that virulence /avirulence to match specifically a particular R-gene always results from alleles at the same locus is largely unproved.

The flax-flax rust association is genetically better defined than any other host-parasite combination. However, the bias is still with the host and taking all studies into account, genotypes have been established for fewer than twenty parent flax rust isolates.

The mathematical symmetry of the gene-for-gene relationship and predictions that follow from it were first explored by Person (1959) and more recently by Robinson (1976; 1979). The general principles of the model allow predictions to be made about the number of host resistance phenotypes and parasite virulence phenotypes (races) that can be observed (2^n, where n = the number of gene pairs). The mathematical logic enables hypothetical genotypes for host cultivars and parasite isolates to be derived by computer-based analysis of reaction-type data when different parasite isolates are combined with different host cultivars (Loegering *et al.*, 1971; Loegering & Burton, 1974; Browder & Eversmeyer, 1980; 1982; Knott & Johnson, 1981; 1983; Roelfs *et al.* 1982). Person & Mayo (1974) have, however, drawn attention to the importance of the conditional nature of R-genes and the epistatic relationship between functional R-genes with respect to those 'concealed' due either to the presence of matching virulence alleles or because they condition a less completely resistant phenotype. Because of this, it is never possible to be certain that undetected genes in either the host or the parasite are not contributing to the outcome of any host-parasite combination.

5.1.5. *Quantitatively inherited resistance*. While many early studies demonstrated that disease resistance could be determined by single genes of major effect, some evidence suggested that this might not always be the case. It became apparent that, in segregating populations, resistance sometimes behaved as a quantitative trait with a continuous distribution. The notion emerged that there were two different forms of resistance to parasites, distinguishable on the basis of their genetic control.

This idea is typified by studies on resistance to *Fusarium oxysporum* (yellows) in cabbage (Walker, 1930; Anderson, 1933) as outlined by Walker (1969). Resistance in some cabbage lines was clearly controlled by a single dominant gene which was effective up to temperatures of 26°C. In other resistant lines, inheritance was quantitative and the expression of resistance was markedly more influenced by environment. Walker (1969) referred to these as 'Type A' and 'Type B' resistance.

In some host-parasite associations, it was readily possible to observe a contrast between cultivars which were consistently less severely attacked and those which were either disease-free or severely attacked depending on the race of the parasite they encountered. This was noted particularly in potato cultivars in response to *Phytophthora infestans* (late blight). Disease control contributed by major genes derived from *Solanum demissum* was short-lived and this was contrasted with the more predictable 'field resistance' of some traditional European cultivars (Müller & Haigh, 1953; Deshmukh & Howard, 1956).

Genetic studies, although far from comprehensive, indicated that field resistance to P. infestans was inherited as a quantitative character. Stevenson et al. (1937) examined progeny from crosses between cultivars of Solanum tuberosum and concluded that resistance was 'a recessive character controlled probably by multiple genes'. In a more extensive investigation, Toxopeus (1959) also obtained evidence for recessive inheritance as well as transgressive segregation in crosses between resistant parents. He concluded that field resistance was 'governed by a series of genes'. Graham (1963) investigated blight resistance in two diploid Solanum species (S. bulbocastanum and S. verrucosum) and again the quantitative nature of the trait was confirmed.

A different type of study involved the comparative behaviour of P. infestans in combination with cultivars considered to differ in their field susceptibility to the fungus (Van der Zaag, 1959; Guzman, 1964). It became apparent that resistance could be described in terms of several components affecting different stages of fungal colonization such as frequency of infection, growth rate (latent period) and sporulation efficiency. Since these components were not always correlated, it was assumed that they must be under separate genetic control.

More recently, it has been suggested that detailed genetic and statistical analysis of the various components of resistance may assist in the location of resistance genes which operate at different stages of host colonization. By this means, it may be possible to identify additive genes and possibly enhance resistance levels by appropriate gene combinations (Zadoks & Schein, 1970; Jeger, 1980; Neervoort & Parlevliet, 1978).

Further discussion of the genetics of quantitatively inherited resistance and its components follows in section 5.3.

5.1.6. Genes determining toxin sensitivity. Some fungi and bacteria produce toxins whose action in some cases is responsible for the symptoms of disease and also enables the parasite to colonize the host (see Chapter 7.2.2). Hence, host insensitivity to these pathotoxins is an attribute which may confer both disease resistance (i.e. to symptom expression) and resistance to the parasite (i.e. to growth). Where only certain cultivars of a host species are highly sensitive to a particular toxin, it is referred to as host-selective or host-specific; the genes which result in insensitivity in other cultivars can be regarded as genes for resistance.

In oat cultivars, a single gene is responsible for response to HV toxin produced by *Helminthosporium victoriae*. Cultivars lacking the dominant *Vb* allele are resistant to the parasite and insensitive to the toxin (Scheffer, 1976). In sorghum cultivars, resistance to *Periconia circinata* and insensitivity to PC toxin which this fungus produces is also controlled by one gene (*Pc*) exhibiting incomplete dominance. Three categories of plant can be recognized: homozygous-sensitive/susceptible, heterozygous/intermediate and homozygous-insensitive/resistant (Schertz & Tai, 1969). A single locus in tomato also controls sensitivity to the toxin produced by *Alternaria alternata* f. sp. *lycopersici* and resistance to this fungus (Gilchrist & Grogan, 1976).

In contrast, the inheritance of insensitivity in cultivars of maize to HC toxin produced by *Helminthosporium carbonum* Race 1 is more complex (Nelson & Ullstrup, 1964; Kuo *et al.*, 1970) and apparently involves two interacting genes located on different chromosomes. At the *Hm* locus, which exerts the greatest effect, several alleles have been identified. The dominant *Hm* allele confers complete insensitivity and is also epistatic over the other locus. Two other alleles Hm^A and Hm^B confer different degrees of intermediate sensitivity while plants carrying two copies of the *hm* allele are most sensitive. The presence of the dominant *Hm*-2 allele at the second locus decreases the level of sensitivity in plants not carrying *Hm*. The level of resistance to the parasite and sensitivity to the toxin appear fully correlated.

The sensitivity of maize cultivars to HMT-toxin produced by *Helminthosporium maydis* Race T and PM-toxin produced by *Phyllosticta maydis* is strongly associated. Only those cultivars carrying Texas male-sterile (Tms) cytoplasm are highly sensitive to these toxins and particularly susceptible to the fungi which produce them. The gene or genes involved are apparently carried on the mitochondrial DNA although there is some evidence that in the case of HMT toxin, nuclear genes can modify the response of plants to both the toxin and the fungus (Comstock *et al.*, 1973; Yoder, 1973; Sheffer, 1976; Payne & Yoder, 1978; Leaver & Forde, 1980) (see Chapter 7.2.2).

It has also been established that the difference between the translation products of normal (N) and Texas male sterile (T) mitochondria resides in a single polypeptide. A 21,000 M_r peptide from N cytoplasm is replaced by one of 13,000 in T cytoplasm. The synthesis of the 13,000 M_r polypeptide can be specifically suppressed by the presence of certain nuclear alleles (Rf alleles which restore male fertility and apparently result in reduced sensitivity to T-toxin (Barratt & Flavell, 1975). Because of the possibility of linkage between nuclear loci, it has yet to be established with certainty that identical alleles suppress both the 13,000 M_r polypeptide and also T-toxin sensitivity. If this turns out to be the case, however, the product of a gene for specific susceptibility will have been identified for the first time.

Although there are some other cases where maternal influences do affect resistance to fungi or bacteria (Lunsford *et al.*, 1975), this is the clearest example of resistance to microbial parasites under the control of cytoplasmic genes.

5.2. TYPES AND STABILITY OF RESISTANCE

5.2.1. *Two types of resistance?* Accumulated evidence on both the genetic and epidemiological characteristics of disease resistance, particularly of potatoes to *Phytophthora infestans*, enabled Vanderplank (1963; 1968) to formulate the influential idea that resistance to parasites can be exclusively classified as either 'vertical' or 'horizontal'. This idea and some of the practical and theoretical implications pertaining to it have been further elucidated by Vanderplank (1975; 1978; 1982) and Robinson (1969; 1971; 1973; 1976; 1979).

The fact that genotypic, phenotypic, epidemiological and mechanistic features of disease resistance are often correlated has however led to some misconceptions and to a lack of precision in the way the terminology has been applied. The basis for distinguishing vertical (differential) and horizontal (non-differential) resistance is solely statistical. When variation in the extent of host colonization by a parasite can be explained in terms of main effects with no significant host x parasite interaction, the relationship is horizontal. Put another way, there is a constant ranking relationship in the degree of resistance between host genotypes regardless of parasite genotype and the same is true for the degree of parasite virulence with respect to host genotypes. Because of problems of the scale on which measurements may have been taken, transformation of the data may be necessary to establish that any statistical interactions are due to inversions in ranking.

In contrast, when variation can only be accounted for in terms of a statistically significant host genotype x parasite genotype interaction, the relationship is said to be vertical. The vertical relationship follows the principles of the gene-for-gene relationship, but resistance may be complete or partial and the genes involved in its control may have either major or minor effects. The designation of resistance as horizontal or vertical carries with it no inevitable mechanistic or genetic implications. At least two interacting gene pairs in the host and parasite are required to demonstrate conclusively that there is a vertical relationship.

Horizontal resistance may also be either complete or partial, under the control of one, several or many genes and under some circumstances may appear race specific. However, the contention is that there is no specific gene-for-gene relationship underlying the horizontal host-parasite association. It is envisaged that in any given host-parasite association, the observed extent of host colonization may be due simultaneously to both horizontal and vertical effects.

Because vertical resistance is often controlled by single major genes; tends to be completely expressed; is often associated with hypersensitivity, and is race specific, terms describing these separate but correlated phenomena have been used almost synonymously. Similarly, there is a tendency for partial resistance to be under more complex genetic control, to be race non-specific, 'rate-reducing' and apparently horizontal. Table 5.3 shows some antonyms related to characteristics of disease resistance which derive their origins from different types of study. The use of terms out of context by extrapolating to correlated phenomena has caused a good deal of confusion and unnecessarily complex debate. Scott et al. (1980) provide a useful discussion of these concepts which have also been explored in a different context by Crute (1985).

Beyond the theoretical concepts involved, several questions of importance have been raised which relate to whether the facts substantiate the existence of two fundamentally different types of resistance.

Parlevliet & Zadoks (1977) examined the characteristics of a polygenic model which comprised five loci in both the host and the parasite; each locus exerted an equal and additive effect. The interaction between host and parasite loci was considered in one situation to follow the specific complementary gene principles of the gene-for-gene relationship. In another situation, host and parasite genes did not interact specifically and their effects were additive regardless of identity. Analysis of the model demonstrated that the statistical interaction in the 'gene-for-gene' model would be too small to discern, given the usual error variance in genuine

Table 5.3. Terminologies derived from different types of investigation which have been used to describe different features of resistance to parasites

Criterion for classification[a]	Antonyms	
Statistical	Vertical Differential	Horizontal Non-differential
Genotypic	Mono/oligogenic Qualitatively inherited	Polygenic Quantitatively inherited
Epidemiological	(Race) specific	(Race) non-specific Rate-reducing General/Uniform
Developmental	Seedling	Adult plant
Phenotypic	Complete	Incomplete/Partial Field
Mechanistic	Hypersensitive Active Escape	Non-hypersensitive Passive True

[a]Terms derived on the basis of different criteria cannot be used synonymously.

experiments of this type. Hence, such a relationship would appear horizontal despite being the result of a gene-for-gene relationship. It was concluded that horizontal and vertical resistance were not distinct phenomena and merely reflected the results of a gene-for-gene relationship between polygenic and monogenic/oligogenic systems respectively.

Experimental evidence for this assertion has been presented by Parlevliet (1976a; 1977; 1978a; 1978b) from studies on partial resistance in barley to Puccinia hordei (leaf rust). The resistance studied was quantitatively inherited, and a hypothesis for the involvement of at least seven loci was advanced. However, significant host genotype x parasite genotype interactions were evident when latent period was examined as a measure of resistance. These observations confirmed the findings of Clifford & Clothier (1974), who had previously investigated the same host-parasite association and reported significant isolate x cultivar interactions in a situation where resistance was considered to be quantitatively inherited.

In similar studies, Jeffrey et al. (1962); Jinks & Grindle (1963) and Caten (1974) demonstrated that isolates of Phytophthora infestans were apparently differentially adapted to the cultivars from which they were obtained despite the fact that none of these cultivars carried R-genes and all were classed as susceptible to the isolates studied.

Rouse et al. (1980) obtained evidence that isolates of Erysiphe graminis f. sp. tritici (powdery mildew) were differentially adapted to cultivars of wheat lacking known major genes and differing in 'rate-reducing' resistance. Statistical interactions due to changes in ranking relationships were observed when two components of resistance (infection efficiency and sporulation capacity) were examined. It was concluded that there was a potential for this type of resistance to be eroded.

On the basis of slightly different arguments, Ellingboe (1981) also contends that all resistance follows the principles of the gene-for-gene theory and is therefore vertical. It is suggested that horizontal resistance is resistance 'that has not yet been shown to be vertical'. The results of two types of experiment lead to this conclusion. In cultivars of wheat which exhibited partial resistance to Erysiphe graminis (powdery mildew), the inheritance of resistance appeared quantitative under uncontrolled environmental conditions. However, this character was readily attributed to a single gene when tests were made in a strictly controlled environment. Furthermore, a single gene for resistance to powdery mildew in wheat (Pm 4) controlled complete resistance to some isolates of E. graminis, to others the reaction was completely susceptible and to others, two types of reduced susceptibility were recorded. The argument was advanced that not only does all resistance follow a gene-for-gene relationship but that it is always attributable to the action of a small number of genes. The quantitative expression of resistance under simple genetic control is discussed in section 5.3.

From yet another standpoint, Nelson (1978) has also argued that genes for resistance do not differ in quality as envisaged by the concepts of horizontal and vertical resistance. It is suggested that single resistance genes of major or minor effect may be rendered less than fully effective by virulent parasite variants but that some residual effect of the resistance gene can still be discerned ('ghost effect'). Hence it is argued that when several genes are combined, regardless of their individual effects, resistance is observed regardless of parasite virulence. In this respect no distinction is drawn between major genes which are demonstrably race-specific in isolation and genes of small effect. The validity of the conclusion reached from experiments which purport to demonstrate the existence of 'ghost-effects' (Nass *et al.*, 1981) has been challenged (Anderson, 1982) on the basis that there may be a range of alleles at a parasite locus conditioning virulence or that modifier genes may affect the virulence of some isolates. There is also ample evidence from a number of host-parasite associations that cultivars carrying many R-genes can be rendered highly susceptible by a parasite isolate with an appropriately matching genotype. However, there is historical evidence which demonstrates that those wheat cultivars which have, over a long period, exhibited resistance to *Puccinia graminis* f. sp. *tritici* (stem rust) in Canada do appear to have several identifyable major resistance genes and these genes apparently exert their effects in different ways (Green & Campbell, 1979). Similarly, the apparently stable resistance to *Albugo candida* (white blister) in *Brassica napus* (rape) cultivars grown in Canada can be attributed to genes at three loci; the presence of a dominant allele at any locus provides resistance (Fan *et al.*, 1983).

A further suggestion is that horizontal resistance is an artifact of 'parasitic epistasis' (section 5.4.2; Sidhu & Webster, 1983). The argument is that while resistance to a single parasite may be race specific and under simple genetic control, the reaction of the host can be modified by the presence of other organisms. Hence, in field situations, resistance may appear to be inherited quantitatively and have some of the attributes of horizontal resistance.

In contrast to the preceding discussion, an apparent artifactual differential relationship between host cultivars and parasite isolates may also result from host genotype x environment or parasite genotype x environment interactions (Kulkarni & Chopra, 1982). Hence it may be that a host parasite relationship is erroneously interpreted as vertical. Data presented by Scott & Hollins (1977) with *Cercosporella herpatrichoides* on wheat, Jenns *et al.* (1982) with *Bipolaris maydis* and *Colletotrichum graminicola* on maize and Norwood *et al.* (1984) with *Bremia lactucae* on lettuce illustrates this point. In all these studies, significant isolate x cultivar interactions were recorded in individual experiments, but these were inconsistent between experiments and disappeared when data for several experiments were examined together. The implication is that uncontrolled features of the environment were responsible for suggesting the existence of a vertical relationship. In reality, this is more likely to be due to the differential adaptation of particular host or parasite genotypes to environment.

On the basis of studies with Xanthomonas malvacearum (bacterial blight) on cotton, Arnold & Brown (1968) have argued that the complexity of interactions between environment, parasite virulence and host resistance means that distinguishing horizontal and vertical effects as fundamentally different phenomena is not possible. They contend that 'such descriptions (vertical and horizontal) represent particular cases of a general system, rather than that they are fundamentally different'.

As a result of studies on resistance in natural populations of Populus to Melampsora spp., Heather & Chandrashekar (1982) have also concluded that the results of associations between particular host and parasite genotypes are highly environmentally labile and considered that insufficient attention has been paid to this component of variation in other studies.

Whether all host-parasite variation for resistance and virulence conforms to the principles of the gene-for-gene relationship is still an open question. What is clear is that some resistance genes are more stable over time than others and that both qualitatively inherited and quantitatively inherited resistance and virulence are observed. The correlation between these two features may be strong but not absolute.

5.2.2. Stability of resistance. The practical relevance of the foregoing discussion relates to the stability of resistance when encountering a potentially variable parasite. Clearly, simple major resistance genes matched by parasite genes in a gene-for-gene relationship are by definition potentially unstable. There are, however, cases where resistance genes remain effective in commercial practice even when matching virulence is known to occur.

On the basis of different theoretical models, both Parlevliet & Zadoks (1977) and Fleming & Person (1982) concluded that associations where both resistance and virulence are polygenically determined should prove more stable. Person et al.(1983) and Minogue (1983) have, however, questioned the validity of both analyses. Regardless of theory, there is no evidence that resistance which is genuinely under additive polygenic control (as distinct from being quantitatively expressed; see section 5.3) can be readily rendered ineffective by marked parasite adaptation.

Vanderplank (1971) presented evidence to suggest that despite indications

Table 5.4. Pleiotropic effects of genes for scale colour and resistance to Colletotrichum circinans in onion (Walker, 1969).

Gene notation:

I - partially dominant gene inhibiting pigmentation

C - dominant gene responsible for red or yellow pigmentation

R - dominant allele for red pigmentation in the presence of C

r - recessive allele for yellow pigmentation in the presence of C

	Plant Genotype	Colour	Resistance
ii	RRCC, RrCC, RRCc, RrCc	Red	High
	rrCC, rrCc	Yellow	
Ii	RRCC, RrCC, RRCc, RrCc	Pink	Intermediate
	rrCC, rrCc	Cream	

All other genotypes white with low resistance

of some adaptation to potato cultivars lacking R-genes (see section 5.2) for resistance to Phytophthora infestans, there was little to suggest that horizontal resistance was in fact eroded. More recently, James & Fry (1983) have provided further experimental evidence to substantiate this claim.

It is clear that 'stable' or 'durable' resistance can also be under simple genetic control. Eenink (1976) and Johnson (1983) provide several examples of single genes for resistance which have not been rendered ineffective by parasite variants. One suggestion is that stability results from the absence of a gene-for-gene relationship, or in other words that changes in a number of parasite genes would be required to overcome the effects of a single host resistance gene.

An example where host resistance seems to be understood both genetically and mechanistically serves to illustrate this point. Skin colour in onions is controlled by three major genes (Rieman, 1931; Clarke et al., 1944) (Table 5.4). Cultivars with genotypes resulting in red or yellow skins are also resistant to Colletotrichum circinans (smudge) by virtue of the concentrations of catechol and protocatechuic acid they contain (see Chapter 7). Pink or cream skinned onions have an intermediate level of resistance and unpigmented cultivars are susceptible (Walker, 1969). These three major genes therefore control a form of resistance which is stable probably because the development of insensitivity to the fungitoxic compounds in the pigmented onions is beyond the genetic competence of the parasite.

5.3. DETAILED CONSIDERATION OF QUANTITATIVELY INHERITED RESISTANCE AND VIRULENCE

5.3.1. Resistance. Resistance which is inherited as a quantitative trait need not necessarily be controlled by large numbers of genes, although most commonly, it has been assumed that continuous variability and polygenic inheritance are synonymous. Vanderplank (1982) pointed out that in addition to a polygenic model, continuously variable resistance can result from relatively few genes if they have unequal effects or if genes of small effect are linked. In addition, where the expression of a single gene is markedly influenced by environment, resistance may appear to be a continuous variable as a result of large environmental variance. On the basis of studies with the environmentally sensitive Pm4 gene for resistance to Erysiphe graminis (powdery mildew) in wheat, Ellingboe (1981) concluded that much of the resistance thought to be polygenically inherited, may in fact be under single gene control (see section 5.2). Quantitative expression of resistance controlled by a single gene has been observed for several other fungal and bacterial host-parasite associations (Hogenboom, 1970; Alan et al., 1974; Fallik et al., 1983).

Vanderplank (1982) tentatively suggests that continuously variable resistance is a result of additive genetic variance as distinct from dominance and epistasis, thus implying that the number of genes involved is of lesser importance and that in practice they may be few. The fact that quantitatively inherited resistance can often respond readily to selection in breeding programmes also implies that the number of genes involved may not always be large. Simons (1972) has examined some of the genetic evidence for polygenically controlled resistance and a few further case studies serve to illustrate the contention that not all resistance is attributable to simple genetic systems.

In a genetic study of the resistance of wheat to Septoria nodorum, Scott et al. (1982) could find no evidence for the involvement of major genes, variation being continuously distributed. Inheritance was primarily additive and greater resistance was strongly associated with long straw and to a lesser extent with late ear emergence. The latter two characters were also continuous variables. It was concluded that this association was more likely to be due to pleiotropic effects than to linkage. Reasons for this are discussed elsewhere (Chapter 7.2.1).

Habgood (1974) studied the inheritance of resistance to Rhynchosporium secalis (leaf blotch) in a diallel cross between six barley cultivars without major genes and varying in their field susceptibility. The data could only be explained in terms of complex inheritance with evidence for the action of both dominant and recessive genes. Additive genetic effects were important and transgressive segregation indicated that the most resistant cultivars did not carry all the available alleles for resistance.

The genetics of the quantitatively assessed character of 'slow-rusting' has been studied in several cereal-rust associations and Wilcoxon (1981) has reviewed the available information. This type of resistance has primarily been studied in field trials in the absence of effective race-specific major genes. In most cases, the level of resistance in segregating populations has been shown to vary continuously and is often normally distributed. Estimates of heritable variance are frequently high with strong evidence for additive genetic effects, both dominance and epistasis apparently being of lesser importance. The frequent occurrence of transgressive segregation is also a feature of these studies. Occasional associations between 'slow-rusting' and major genes for rust resistance have been identified (Skovmand et al., 1978) but these effects were probably due to linkage. Estimates of the numbers of genes controlling slow rusting range from two to twelve depending upon both the crosses studied and the host-parasite combination. In any particular host species, genes for this form of resistance appear to operate only against a single species of rust.

The resistance of wheat cultivars to Puccinia striiformis (yellow or stripe rust) can be controlled by a series of race specific dominant genes (Macer, 1972) of major qualitative effect. Some cultivars, however, express resistance which is only recognized quantitatively. This resistance is of particular interest because it seems to be durable. Sharp and his collaborators (Lewellen *et al.*, 1967; Lewellen & Sharp, 1968; Sharp & Volin, 1970; Sharp *et al.*, 1976; Krupinsky & Sharp, 1979) have identified genes from several sources which appear to be race non-specific and act additively. In crosses between different cultivars, transgressive segregation was frequently observed. Using a different range of cultivars, Johnson and his collaborators also identified resistance in this host-parasite association which appears to be controlled by several genes which are mainly effective in adult plants (Lupton & Johnson, 1970; Johnson & Law, 1975; Pink *et al.*, 1983; Wallwork & Johnson, 1984). By use of chromosome substitution lines, it has been shown that some but not all of the genes involved are associated on the short arm of chromosome 5. Interestingly, genes on the long arm of this chromosome appear to be responsible for increases in susceptibility. In common with other studies, in crosses between several cultivars with different levels of resistance, trangressive segregation was commonly observed, indicating additive gene action.

As has been discussed previously (section 5.2) there is evidence that at least some components of quantitatively inherited resistance can be race specific. In addition, because many of the studies in which resistance has been suggested to be under polygenic control have been conducted in uncontrolled environmental conditions, this may have resulted in some overestimates of the numbers of genes involved.

5.3.2. Virulence. It is well established that qualitative virulence to match single race specific resistance genes is controlled by single complementary loci in the parasite (section 5.1.4). There is however much less information on the genetic basis for differences in quantitative virulence (referred to as aggressiveness by Vanderplank, 1968) which are sometimes observed between isolates of a parasite. The asssumption is that this variation will be controlled by several genes. Whether there is a complementary gene relationship between host and parasite as is envisaged by Parlevliet & Zadoks (1977) (see section 5.2) remains to be demonstrated.

In the models expounded by Fleming & Person (1982), polygenic inheritance of resistance and virulence is assumed and the outcome of the association between the two partners is considered to result from either an additive or a multiplicative combination of resistance or virulence attributes. Specific complementation is not envisaged. In both models quantitative variation for either resistance or virulence results in a relatively smaller change in the actual outcome of the host-parasite association. This phenomenon has been termed 'phenotypic damping' and it is thought that it could contribute to the stability of polygenically inherited systems.

Some evidence that quantitative variation for virulence may be polygenically inherited comes from work with *Ustilago hordei* (smut) on barley (Emara, 1972; Emara & Sidhu, 1974; Person *et al*., 1983). The degree of virulence on a single barley cultivar carrying a race specific resistance factor was determined in F_2 dikaryons from a single parental F_1 dikaryon. Each F_2 dikaryon was either homozygous virulent, homozygous avirulent or heterozygous for virulence to the major R-factor. However, within these three classes, there was a further range of variability which was attributable to polygenes. Dikaryons conformed to a constant ranking relationship with respect to quantitative virulence in accordance with the expectation from the mean performance of each contributing sporidial (haploid) line.

In a continuation of these studies Caten *et al*. (1984) employed biometrical analyses to examine segregation from three parental dikaryons which differed in their quantitative virulence and also the degree of heterozygosity for genes affecting this character. Analyses confirmed that the character was apparently under polygenic control with both additive and non-additive genetic effects being important. Non-additive variation was due to ambidirectional dominance rather than epistasis. The suggestion was made that the ambidirectional dominance could reflect a history of stabilizing selection for types of intermediate virulence. Should this be so, then there could be a directional response towards greater quantitative virulence in response to the use of hosts with higher levels of resistance.

Polygenic control of quantitative virulence is also evident from studies with *Ceratocystis ulmi* (Dutch Elm disease) (Brasier & Gibbs, 1976; Brasier, 1977). Crosses between 'aggressive' and 'non-aggressive' isolates resulted in progeny expressing a range of 'aggressiveness'. However, no progeny could be classified as comparable to the 'aggressive' parent implying non-additive genetic effects. It was not possible from the data to distinguish between dominance and epistasis.

Blanch *et al*. (1981) investigated the inheritance of differences in quantitative virulence on wheat between isolates of *Gaeumannomyces graminis* var. *tritici* (take-all). Polygenic control was evident from the data and genetic variation was predominantly additive, although in two crosses there was also evidence of non-additive effects. Based on the assumptions of equal additive effect, no linkage and no interaction, an estimate of 5-9 loci controlling differences in virulence was arrived at.

5.4. DETAILED CONSIDERATION OF GENE-FOR-GENE RELATIONSHIPS

5.4.1. Dominance relationships. In gene-for-gene controlled host-parasite associations, resistance and avirulence are usually dominant traits (see section 5.1.4) but this is not invariably the situation. Further discussion of dominance relationships is particularly important because at least one mechanistic hypothesis to explain gene-for-gene associations has the dominance of alleles for resistance and avirulence as an initial premise (Ellingboe, 1976) (see Chapter 7.6.2). In fact, recessive alleles for resistance are not particularly uncommon in gene-for-gene associations and incomplete dominance and gene-dosage effects are frequent in the case of both resistance and avirulence loci. Furthermore, both environment (particularly temperature) and non-specific genetic background effects as well as specific non-allelic interactions can influence the dominance of alleles for both resistance and virulence with examples of reversals in dominance being documented (see section 5.4.2). Some selected examples serve to illustrate these points.

Specific resistance to *Puccinia recondita* f. sp. *tritici* (leaf rust) in wheat can be under the control of dominant alleles, partially dominant alleles or recessive alleles (Rajaram *et al.*, 1971; Browder, 1980; Bartos *et al.*, 1969). The observed resistance may be effective from the seedling stage onwards or may only be effective in adult plants. Furthermore, the dominance of several alleles and their phenotypic expression can be affected by growth stage (Anderson, 1963); environment, particularly temperature (see section 5.4.3) (Anderson, 1961; Johnson & Heyne, 1964; Johnson & Schafer, 1965; Dyck & Johnson, 1983); host genetic background (see section 5.4.2) (Anderson, 1963; Dyck *et al.*, 1966; Dyck & Samborski, 1968), and the genetic constitution of the fungus isolate used to identify the resistance (Samborski, 1963; Dyck & Samborski, 1974).

Published data on the genetics of resistance in wheat to *Puccinia graminis* f. sp. *tritici* have been collated by Roelfs & McVey (1979). Of the numerous *Sr* alleles identified, some appear completely dominant, a large number are incompletely dominant and some have been reported as recessive particularly under certain temperature conditions, at some stages of development, and in some genetic backgrounds. For example the temperature sensitive allele *Sr6* may appear recessive in some crosses, dominant in others and is usually seen as incompletely dominant in adult plants. Alleles *Sr7*, *Sr8*, *Sr9* and *Sr10* all behave as partially dominant (Knott & Anderson, 1956).

A clear example of the way in which both environment and genetic background can affect the expression and dominance relationships of a single gene is provided by work on resistance to *Pyrenophora teres* (net blotch) in barley (Khan, 1969; Khan & Boyd, 1969a; 1969b; 1969c). A putative gene-for-gene relationship exists in this host-parasite association. The expression of a resistance gene located in several Manchurian genotypes (e.g. CI 2330, CI 4797 and CI 4407-1) was markedly influenced by environment and genetic background and could appear to segregate as a completely recessive or completely dominant character. A similar result was obtained with a second gene of more complete expression, identified in two Ethiopian genotypes (CI 5791 and CI 9819). In crosses to one susceptible cultivar the gene was completely dominant while when a different susceptible parent was used it appeared to be partially dominant.

Incomplete dominance in barley for resistance to *Puccinia hordei* (leaf rust) was clear for alleles at some loci when comparisons were made between resistant parents and F_1 progeny (Parlevliet, 1976b). A similar situation has been observed in oats with respect to both *Puccinia graminis* f. sp. *avenae* (stem rust) and *Puccinia coronata* f. sp. *avenae* (crown rust) (Martens *et al.*, 1980; Wong *et al.*, 1983; Harder *et al.*, 1984).

Resistance in barley to *Erysiphe graminis* f. sp. *hordei* is also frequently observed to be incompletely dominant. Starling *et al.* (1963) found evidence that the F_1 of several crosses expressed a different reaction type to the resistant parent. Alleles *Ml-a2* and *Ml-a3* have also been shown to be incompletely dominant, their expression being dependent upon environment but with heterozygotes being more sensitive in this respect than homozygotes (Schaller & Briggs, 1955; Moseman & Schaller, 1960). In contrast, *Ml-a1* consistently behaved as a completely dominant allele.

Incomplete dominance of avirulence is implicit in the studies of Flangas & Dickson (1961a; 1961b) with *Puccinia sorghi* (maize rust). A range of reaction phenotypes was observed on host lines carrying different resistance genes following inoculation with genotypically different fungal isolates. This variation was thought to be a consequence of different combinations of host and parasite loci in either the homozygous or heterozygous condition. For example, the most incompatible reaction type resulted from the association between RR and AA while RR and aa resulted in the most compatible reaction type. A mesothetic or intermediate reaction type was thought to result from the Rr and Aa genotype combination.

Samborski (1963) and Dyck & Samborski (1974) provided clear evidence that avirulence in *Puccinia recondita* f. sp. *tritici* to certain *Lr* alleles in wheat (*Lr*2 and *Lr*9) was incompletely dominant. The reaction type in host-parasite combinations involving isolates heterozygous at the virulence locus was less incompatible than those involving homozygous avirulent isolates.

Data on the inheritance of virulence in *Puccinia coronata* f. sp. *avenae* (crown rust) on oats also suggests incomplete dominance at some loci in addition to the occurrence of epistatic modification of dominance (see section 5.4.2) (Zimmer *et al*., 1965).

In *Puccinia graminis* f. sp. *tritici*, avirulence to *Sr* alleles in wheat has usually been observed to be dominant to virulence. Cases of dominant virulence are recorded which may be due to non-allelic interactions (see section 5.4.2) (Johnson, 1954; Luig & Watson, 1961; Green, 1964; 1966). There is also good evidence that isolates heterozygous at the virulence locus behave differently on some *Sr* alleles compared to homozgyous avirulent isolates (Loegering & Powers, 1962; Williams *et al*., 1966; Loegering & Harmon, 1969).

In the association between *Ustilago hordei* (covered smut) and barley, original analyses of the genetics of resistance and virulence confirmed a classical gene-for-gene relationship (Sidhu & Person, 1971; 1972). However, some important deviations from the model were subsequently noted (Ebba & Person, 1975). Virulence for two cultivars (Keystone and Himalaya) was found to be controlled by duplicate loci. It was also noted that avirulence for cv. Excelsior (UhR_2) was incompletely dominant and that heterozygous isolates expressed avirulence/virulence to varying degrees depending upon environment and the presence of modifier genes in different genetic backgrounds. The modifiers appeared specific for this virulence locus and did not generally influence the expression of avirulence at other loci (see section 5.4.2). It is also of interest that identical loci for virulence/avirulence for certain barley cultivars have been located in crosses between the closely-related smut species *Ustilago hordei* (covered smut) and *Ustilago nigra* (false loose smut) (Fullerton & Nielsen, 1974).

Day *et al*. (1983) discussed the question of dominance in host-parasite genetic relationships. They concluded that the fact that resistance genes are often dominant 'has little bearing on the mechanisms by which they bring about their effects' and that dominant avirulence 'is unlikely to have any biochemical or functional significance in terms of mechanisms'. This conclusion is based on the following arguments. A resistance gene would only be located if it was relatively frequent and would only reach this frequency if it was capable of being selected as a heterozygote. Modifier genes promoting dominance would also tend to be selected and this would apply for both natural and artificial selection. In a diploid or dikaryotic parasite, it is envisaged that intense selection for virulence among recessive mutants would result in the selection of homozygous genotypes with no opportunity for the evolution of dominance.

From a different standpoint, Kacser & Burns (1981) present a general model to explain dominance at a molecular level. Their explanation of dominance relationships makes it unnecessary to invoke the selection of 'modifiers' to explain the maximized fit of genotype to phenotype. On the basis of enzyme kinetics, it is argued that heterozygosity will usually be observed in terms of the activity of individual enzymes (the level of gene product) but will rarely be observed in the final phenotype resulting from a biosynthetic pathway of different enzymic reactions. This is because the flux through the pathway is not totally regulated at any indivdual step, but is controlled as a property of the pathway as a whole. The extension of this argument to resistance and virulence is that the product of a locus at which the heterozygote is phenotypically expressed either has an overall controlling effect on a synthetic pathway, or the pathway is very short, or the gene product is not involved in an enzymic function.

It is clear from the evidence presented above that incomplete dominance at loci controlling resistance and virulence in gene-for-gene relationships is common. Because of the tendency to 'lump' reaction categories together in genetic studies, in order to obtain interpretable ratios, it may be even more common than is recognized.

On the basis of genetic studies, Ellingboe (1976; 1981) has suggested that incompatibility in gene-for-gene systems results directly from a structural protein dimer formed between the products of the specifically interacting alleles for avirulence and resistance (see Chapter 7.6.2). One expectation of this model might be that more of the hypothetical dimer molecule would result from a combination involving two homozygous partners compared to two heterozgyous partners. In other words, it might be expected that resistance and avirulence would invariably be seen as incompletely dominant characters (Figure 5.2).

	Parasite genotype		
	AA	Aa	aa
Host genotype	Colonization		
RR	1	2	4
Rr	2	3	4
rr	4	4	4

FIGURE 5.2. A hypothetical 'quadratic-check' arrangement for a host-parasite association between two diploid partners in which both avirulence and resistance are partially dominant. Colonization is assessed in arbitary units on a scale of 1-4.

In summary therefore, Ellingboe's model together with Kacser & Burn's interpretation of dominance suggest that resistance alleles and avirulence alleles are probably not involved in synthesis (at least in their role as alleles conditioning host-parasite specificity). Furthermore, incomplete dominance would always be expected with different phenotypes resulting from each of the combinations AARR, AaRR, AaRr and AARr. Detailed studies are likely to be required to demonstrate these differences and to eliminate variation due to environmental and background genetic effects. However, this interpretation is in good agreement with much of the available data discussed above and there is no requirement that avirulence and resistance alleles should be expected to be completely dominant, only that there should be heterozygous expression. Non-allelic interactions with other loci (see section 5.4.2) can also be accommodated on the basis that other gene products may interfere with the envisaged protein-protein recognition. In these regards, the hypothesis is not in conflict with the views on dominance expressed by Day _et al_. (1983).

5.4.2. Non-allelic interaction.

Epistatic ratios. Resistance which is demonstrably race specific is usually under simple genetic control, as has been amply illustrated earlier in this Chapter. However, segregation ratios implying epistatic relationships between loci controlling resistance have often been reported. Commonly, a 9:7 ratio suggesting the involvement of complementary genes has been recorded (Sidhu & Webster, 1977). For example, Habgood & Hayes (1971) provide good evidence for the existence of two complementary recessive genes (rh_5 and rh_6) which contribute resistance to Rhynchosporium secalis (scald) in barley and further examples are cited by Sidhu & Webster (1977). While some of these reports may reflect genuine complementation, there are also artifactual reasons why the segregation of single independent resistance genes may result in apparent epistatic ratios.

Loegering & Seers (1963) studied resistance to Puccinia graminis f. sp. tritici in the F_2 of the cross between wheat cultivars Chinese Spring x Timstein and obtained a 9:7 ratio when a 3:1 ratio for the segregation of gene Sr11 was expected. This disturbed ratio was shown to be due to a linked allele for 'pollen-killing' (ki) in Chinese Spring which reduced the transmission of Sr11.

When a single gene for resistance is incompletely dominant; is incomplete in its expression, and tests on segregating F_2 populations are conducted in an uncontrolled environment, it is possible that some of the heterozygotes may be classed as resistant and others as susceptible. Hence, an expected 3:1 or 1:2:1 ratio may in fact be recorded as not significantly different from 9:7. Both Globerson et al. (1974) and Johnson et al. (1977) reported that a factor (R7) for resistance to Bremia lactucae (downy mildew) in lettuce was inherited as a pair of independent dominant complementary genes. However, further studies suggested that only a single dominant gene was involved. The earlier reports probably resulted from the incomplete expression of this gene in the heterozygous condition, particularly in some genetic backgrounds and when exposed to high inoculum concentrations (Crute & Norwood, 1978; Norwood & Crute, 1984).

A 9:7 ratio for resistant:susceptible can also result at F_2 in a cross involving the segregation of two independent dominant genes if the plant population is inoculated with a mixture of two isolates each one being avirulent on only one of the resistance genes in question. Only host genotypes carrying dominant alleles at both resistance loci would prove resistant to the inoculum. It is possible that at an early stage in an investigation with a new host-parasite association, the individual components of a mixed inoculum may not be recognizable.

A slightly different cause of apparent epistasis may result from induced resistance or induced susceptibility phenomena when segregating progeny are inoculated with mixtures of pathotypes of a single pathogen species or a mixture of entirely different pathogens. One component of the inoculum alters the host response to the other resulting in disturbed segregation ratios which are absent when unmixed inoculum is used. This phenomenon, called parasitic epistasis, has been researched and discussed fully by Sidhu (1979; 1984) and Sidhu & Webster (1979), who consider it to be common under natural circumstances (see section 5.1.3).

Additivity. In gene-for-gene controlled resistance, a commonly held assumption is that the relationship between individual resistance genes present in the same plant is solely epistatic. In other words, when two or more genes are functional, the phenotype observed is that associated with the gene of greatest effect. On available evidence, however, this assumption is not invariably valid and examples of different gene combinations resulting in altered phenotypes are well documented.

Loegering & Geis (1957) studied phenotypes associated with the action of three genes for resistance to *Puccinia graminis* f. sp. *tritici* in wheat. Genes *Sr*6, *Sr*8 and *Sr*9 from cv. Red Egyptian were each present individually in three chromosome substitution lines from Red Egyptian into the susceptible cv. Chinese Spring. It was demonstrated that the parent line carrying all three *Sr* genes was considerably more resistant to one isolate of the fungus at high incubation temperatures than were any of the substitution lines. This suggested additive gene action involving at least two genes, possibly *Sr*6 and *Sr*8 which individually conditioned a less completely resistant phenotype under these conditions.

In a study of resistance to the same fungus involving a different range of *Sr* genes, Knott (1957) also obtained data which suggested the additive modification of the expression of one *Sr* gene (*Sr*7) by the presence of others (particularly *Sr*9). More recently, in the course of an investigation of the effect of temperature and genetic background on the expression of *Sr* genes in wheat, Luig & Rajaram (1972) also obtained evidence for additive gene action.

Similar observations have been made in studies of resistance to *Puccinia recondita* (leaf rust) in wheat. Schafer *et al*. (1963) demonstrated enhanced resistance to two isolates of the fungus in progeny from crosses involving the combination of different individual *Lr* genes when compared to either parent. In a study with the same fungus, Dyck (1977) also found that different combinations of *Lr* genes resulted in different phenotypes. Resistance of a wheat introduction (PI 268454b) was shown to be under the control of two genes. These genes individually gave reaction types indicative of greater fungal colonization than the reaction type resulting when they were combined in the same genotype. Recently this has been shown to occur more generally

(Samborski & Dyck, 1982) and several pairs of *Lr* genes, in combination, contribute a higher level of resistance than either does alone.

Modification of resistance alleles. Mention has already been made of the effect that genetic background may have on the expression and apparent dominance of single resistance alleles (section 5.4.1) in various host-parasite associations. Often this seems to be a non-specific effect, in that less-susceptible backgrounds have the same effect on more than one specific resistance allele. It is nevertheless an important type of non-allelic interaction.

There are also examples of more specific non-allelic modifications of the expression of particular resistance alleles which can sometimes be attributed to single loci or at least localized on a particular chromosome. For example, Kerber & Dyck (1977) and Kerber & Green (1980) localized an 'inhibitor' of resistance to *Puccinia graminis* f. sp. *tritici* on chromosome 7B of the hexaploid wheat cultivar Canthatch. The gene or genes involved in the suppression of one or more resistance genes were probably derived from the species *Aegilops squarrosa*. Hare & McIntosh (1979) presented data which suggests that allele *Sr*2 in wheat for resistance to the same parasite is rendered less effective in the presence of a modifier allele on a different chromosome.

Haggag & Dyck (1973) found evidence for the non-allelic suppression of resistance controlled by alleles at the *Lr*3 locus in wheat for resistance to *Puccinia recondita* f. sp. *tritici*. Resistance was unexpectedly not expressed in some crosses. A different type of non-allelic interaction has been reported in the same host-parasite association (Samborski & Dyck, 1982). Alleles *Lr*13 and *Lr*16 control resistance which is expressed respectively at the adult plant stage only and at all growth stages including seedlings. Interestingly, when these alleles are combined in a single genotype, *Lr*13 is expressed at the seedling stage even in combinations with isolates carrying virulence to match *Lr*16.

Law & Johnson (1967) provided further evidence for non-allelic interactions in the control of resistance to *P. recondita* in wheat. Using chromosome substitution lines, it was shown that the expression of a major gene located on the long arm of chromosome 7B (*Lr*14) could be modified by a gene at a second locus on the short arm of 7B in addition to genes present on at least one other chromosome.

'Inhibition' of avirulence. Non-allelic 'inhibition' of avirulence has

already been mentioned in discussion of the gene-for-gene relationship between *Melampsora lini* (rust) and flax (section 5.1.4) (Lawrence *et al.*, 1981a). The occurrence of dominant loci capable of restoring parasite virulence to isolates which would otherwise be genotypically avirulent has particular importance for understanding the molecular mechanisms underlying gene-for-gene relationships.

It may in future be possible to use these 'inhibitor' loci as tools to probe molecular events in gene-for-gene systems. The study with *Melampsora lini* is the most comprehensive conducted so far. The evidence is that one 'inhibitor' locus (or several closely linked loci) influences the expression of avirulence on several, but not all, of the resistance alleles at the *L* locus in flax. This 'inhibitor' locus is also linked to the locus responsible for modifying the expression of avirulence on resistance allele M^{1}. In some cases, the avirulence loci involved are linked to other avirulence loci which are apparently not subject to modification. Furthermore, the 'inhibitor' locus does not modify the expression of avirulence in the combination L^{x} - *A* x yet it does affect the combination L^{6} - *A* 6. This is despite the fact that L^{x} was derived from L^{6} by rare recombination and that the *A* alleles are also closely linked. All these observations emphasize forcibly the highly specific nature of the molecular events which occur between the products of these alleles.

There is increasing evidence that non-allelic 'inhibition' of avirulence may in fact be a common occurrence indicative of an additional tier of genetic control above the frequently recognized gene-for-gene association.

Samborski & Dyck (1968) found that virulence in *Puccinia recondita* f. sp. *tritici* (leaf rust) for *Lr*3a in wheat could be inherited as either a recessive or dominant character depending upon the isolate studied. The implication was that virulence could result from the homozygous recessive condition of the virulence locus or as a result of a non-allelic interaction involving a dominant allele at a second locus. Haggag *et al.* (1973) reported similar findings and also found evidence that virulence as well as avirulence could be modified by non-allelic interaction. This observation was confirmed by Dyck & Samborski (1974) with respect to virulence for alleles at the *Lr*2 locus. In a continuation of these studies, Samborski & Dyck (1976) also found that virulence on *Lr*2b could be inherited as a dominant character. They comment that when virulence is dominant the reaction is 'variable and difficult to classify'. A feature of the studies with *Melampsora lini* (Lawrence *et al.*,

1981a) was also that virulence involving 'inhibitor' loci was phenotypically different to that involving homozygous recessive alleles at the virulence locus.

In addition to dominant avirulence alleles in Uromyces appendiculatus (rust) to match specific resistance alleles in bean, Christ & Groth (1982a; 1982b) also identified a dominant allele for virulence. It was suggested that avirulence for a bean line carrying the resistance allele Up1 could be controlled by duplicate loci. The hypothesis was that avirulence resulted either when the first locus was homozygous recessive or when a dominant allele was present at the second locus regardless of the genotype at the first. In fact, the data presented can also be intepreted in terms of an identical non-allelic interaction to that described above to account for dominant virulence in Melampsora lini on flax. The two parental isolates virulent for Up1 are probably both heterozygous at the 'inhibitor' locus but they differ at the virulence locus, one being homozygous dominant and the other homozygous recessive.

For three out of the eleven loci (A1, A4 and A5/8) for avirulence studied in Bremia lactucae (downy mildew) to match resistance alleles in lettuce, there is evidence for the occurrence in some isolates of dominant 'inhibitors' of avirulence (Michelmore et al., 1984; Norwood & Crute, 1984).

5.4.3. Temperature sensitivity. As has already been mentioned, both the expression and dominance relationships of alleles for resistance can be conditional on temperature (section 5.4.1). Both Ellingboe (1976) and Vanderplank (1978; 1982) consider temperature sensitivity to be indicative of protein-protein recognition in host-parasite gene-for-gene relationships although their molecular interpretations of the phenomenon (which will not be repeated here) differ markedly (see Chapter 7.6.2).

Vanderplank (1982) provides numerous examples of alleles for resistance, particularly to rust diseases of cereals, which become less effective the higher the temperature. Two of the best researched examples are Sr6 for resistance in wheat to Puccinia graminis f. sp. tritici (Watson & Luig, 1968; Mayama et al., 1975; Skipp & Samborski, 1974), and Lr20 for resistance to P. recondita f. sp. tritici (Jones & Deverall, 1977). There are however, also examples of resistance alleles in the same host-parasite associations which are more effective at higher temperatures than lower temperatures (Rajaram et al., 1971) over the same temperature range (see Chapter 7.6.2).

Dyck & Johnson (1983) have recently made a thorough investigation of this phenomenon in wheat genotypes carrying different Lr alleles for resistance to Puccinia recondita (leaf rust). All alleles were located in the same genetic background. Only one allele (Lr19) gave an identical reaction type at all temperatures studied (10°-25°C). Some alleles (e.g. Lr18 and Lr14a) became markedly less effective at higher temperatures while others (e.g. Lr16, Lr17 and Lr23) became more effective at higher temperatures. These effects sometimes also depended upon the parasite isolate used.

The marked effect that temperature has on some specific gene-for-gene combinations implies either a possible influence on the tertiary structure of the protein product which affects recognition characteristics or alternatively the existence of different cardinal points for some enzymatic function. More experimentation seems appropriate to test the hypotheses which are currently based on apparent over-generalization.

5.4.4. Linkage and allelism. Host loci conditioning gene-for-gene resistance are frequently linked and commonly exist as multiple alleles or pseudo-alleles, for example in resistance to Melampsora lini in flax (section 5.1.4). Work by Shepherd & Mayo (1972) and Mayo & Shepherd (1980) suggested that in some cases, reported allelism in fact represented closely linked but functionally independent loci (e.g. 'alleles' at the 'M' locus). In another case (with alleles L^{2} and L^{10}), the evidence from recombination studies suggested that these were genuine alleles of the same gene. Whilst susceptible recombinants were recovered from the cross $L^{2} \times L^{10}$, no recombinants indicating the presence of both resistance alleles were found. This implies the inability of the gene product to express both resistance functions simultaneously. It is however possible that the '$L^{2} + L^{10}$' alleles might express a totally different race specificity.

The full significance of linkage and allelism is uncertain but some further examples serve to illustrate the frequency of its occurrence. In barley, genes controlling resistance to Erysiphe graminis f. sp. hordei (powdery mildew) are located on two chromosomes. On chromosome 5, there are several distinct loci which determine response to the parasite. At the Ml-a locus at least 11 alleles have been recognized on the basis of their specificity with respect to different fungal isolates (Starling et al., 1963; Moseman & Schaller, 1960; Moseman & Jørgensen, 1971; 1973). It is suggested that there is probably a cluster of linked loci near Ml-a, each with two or more alleles. The Ml-k locus also occurs in this region. Attempts to obtain susceptible recombinants between Ml-a, Ml-a3, Ml-a8 and Ml-a9 proved unsuccessful (Jørgensen & Moseman, 1972). Despite this, it is more likely that these observations reflect functional as distinct from structural allelism, given the differences in specificity.

Two loci controlling resistance to Erysiphe graminis f. sp. hordei also occur on the long arm of chromosome 4 (Ml-g and ml-o). Completely recessive resistance alleles at the ml-o locus were identified in a series of 10 host lines following mutagenesis, although subsequently a single recessive allele at the same locus was also identified in an existing host genotype (Favret, 1960; 1971; Jørgensen, 1971a; 1971b; 1977). Alleles at the ml-o locus may not in fact control gene-for-gene resistance, since no fungus isolate with matching virulence has been found. It has been suggested that this may be a 'structural gene' as distinct from a 'regulator gene' which is envisaged for other dominant Ml alleles (Favret, 1971). There have also been recent suggestions that genes controlling specificity in gene-for-gene systems have a regulatory function (Tepper & Anderson, 1984).

Each ml-o allele results from a separate mutational event and although the phenotypes they condition are essentially the same (Jørgensen & Mortensen, 1977), there is evidence from recombination studies that the mutations are at different sites within the ml-o locus. This indicates functional rather than structural allelism (Jørgensen, 1976; 1977).

In wheat, loci controlling resistance to E. graminis f. sp. tritici are distributed on at least nine chromosomes although alleles at two loci (Pm3 and Pm4) with distinct specificities have been identified (Bennett, 1984).

Several of the many resistance loci identified in wheat to Puccinia graminis f. sp. tritici (stem rust) are also linked and allelism has been reported (Luig & McIntosh, 1968; Knott, 1971).

In maize, there are at least three loci (Rp_1, Rp_5 and Rp_6) on the terminal region of the short arm of chromosome 10 which control resistance to Puccinia sorghi (rust) (Russell & Hooker, 1959; Hooker & Russell, 1962; Wilkinson & Hooker, 1968; Saxena & Hooker, 1968). At the Rp locus, fourteen alleles have been recognized on the basis of specific reaction to different fungus isolates. In a classical study of gene fine-structure, Saxena & Hooker (1968) looked for susceptible plants in crosses between homozygous recessive plants and those heterozygous for two dominant resistance alleles at the Rp_1 locus; progeny were inoculated with a fungal isolate avirulent on both dominant alleles. When susceptible progeny occurred at a frequency greater than the rate of mutation from Rp_1 to rp_1, it was taken to be indicative of recombination. Eight dominant *Rp*-1 alleles were investigated and since recombination between several 'alleles' was detected it was clear that the locus was either a complex of genes or there were several altered sites within a single functional gene. It was also noted that a resistant recombinant between two dominant alleles resulted in the resistance characteristics of a third putative allele. On this evidence, it could be suggested that Rp_1^d could result from two changes in different parts of the Rp_1 locus each separately identified as a different allele.

There are also examples of linkage between genes controlling resistance to different parasites. The *Ml-a* locus for resistance in barley to *Erysiphe graminis* f. sp. *hordei* is linked to a locus controlling resistance to *Puccinia hordei* (*Pa*4) (Moseman & Reid, 1961; Roane & Starling, 1970).

In wheat, resistance genes *Lr*20, *Sr*15 and *Pm*1 (for leaf rust, stem rust and powdery mildew respectively) are invariably associated. Recombination has never been noted (The & McIntosh, 1975). By use of mutagenesis, the molecular independence of *Pm*1 was demonstrated but the functions of *Lr*20 and *Sr*15 were not separated (McIntosh, 1977). The implication is that one host gene product is capable of being recognized separately by products of avirulence alleles in the two rust fungi; a genuine pleiotropic effect.

The numerous examples of tight linkage and allelism of resistance genes is consistent with the notion that they may have resulted from repeated duplication followed by separate evolution. Consequently, they express unique specificity or recognition characteristics yet may prove to code for isoenzymes with the same primary function.

In contrast, linkage between virulence loci, although recorded, does not seem to be as common. This may be genuine or an artifact resulting from the restricted amount of genetic data available for parasites. It can certainly be argued that the parasite benefits by having no limitation on its capacity to recombine specific virulence determinants. Allelism or tight linkage would impose such a limitation.

Several examples of linkage between loci controlling virulence are provided by Moseman and collaborators from work with *Erysiphe graminis* f. sp. *hordei* (barley powdery mildew) (Moseman, 1963; Moseman *et al*., 1965; Moseman & Jørgensen, 1971). In this fungus (which is haploid) linkage is to be expected since it is thought to have only two chromosomes (Kimber & Wolfe, 1966) although given the virulence loci that have been shown to segregate independently, the chromosome map length must be considerable.

In *Bremia lactucae* (lettuce downy mildew), following studies on the segregation of virulence to match 11 resistance factors, only two virulence loci (*A*2 and *A*11) proved to be linked (Norwood & Crute, 1984).

Studies on the inheritance of specific virulence in *Puccinia recondita* f. sp. *tritici* on wheat have suggested that a few loci are linked but the majority appear to segregate independently (Samborski & Dyck, 1968; 1976; Statler, 1977; 1979).

Linkage between the loci controlling virulence in *Melampsora lini* to resistance alleles M^1 and M^4 and also four alleles at the *P* locus in flax was reported by Lawrence *et al*. (1981a; 1981b). These studies also confirmed the close linkage of loci controlling virulence for several alleles at the *L* locus for resistance in flax.

In the host, alleles for resistance are primarily distinguished by phenotypic differences for reaction to different parasite isolates i.e. by their race specificity. Interestingly, there appear to be no substantiated examples of allelism in parasites where alleles at a locus condition specific virulence for different specific resistance genes/alleles in the host. However, it has been proposed that at an individual locus for virulence, many alleles may be possible which condition varying degrees of virulence or avirulence towards a particular single resistance allele. This is one reason to account for the different expression of a single resistance allele dependent upon parasite isolate.

Martin & Ellingboe (1976) have convincingly shown that isolates of *Erysiphe graminis* f. sp. *tritici* differ quantitatively in their degree of virulence for *Pm4* (see section 5.4.6). In some cases, virulence was not fully restored to the equivalent of that observed in the absence of the host resistance allele. Others have also noted what has come to be referred to as a 'ghost' effect of a 'defeated' resistance gene (Riley, 1973; Nass *et* al., 1981; Nelson, *et al*., 1982). However, the validity of the conclusions has been challenged (Anderson, 1982).

By analogy, there seems no reason to rule out the possibility of different alleles at an R-gene locus which control identical specificity with respect to the parasite but which condition quantitatively different effects. Alleles L^7 and L^x in flax for resistance to *Melampsora lini* (Lawrence *et al*., 1981a) could be of this type.

The existence of different alleles at a virulence locus which affect the quantitative aspects of the host-parasite relationship as distinct from simply regulating specificity has not been proved by genetic analyses. Phenotypically, it is not possible to differentiate between differences due to the presence of different alleles and those due to either specific or non-specific modifier genes present in the background of different parasite isolates (section 5.4.2).

5.4.5. Shared antigenic determinants. An intriguing observation of uncertain significance is the occurrence of antigenic determinants shared by hosts and microbial parasites: antibodies formed in response to antigens from the host and the parasite have similar specificity.

It is believed that animals may be immunologically unresponsive to certain parasites (particularly helminths) because they share common antigens with them (DeVay *et al.*, 1972). This has been called antigenic mimicry (DeVay & Adler, 1976; Vanderplank, 1978). Where the phenomenon has been studied in plants, there is generally a greater antigenic similarity between compatible host and parasite partners than with incompatible combinations.

Work by Doubly *et al.* (1960) appeared to indicate that *Melampsora lini* (rust) and its host, flax, shared common antigens. The occurrence of common antigenic determinants, determined by high titres for rust antiserum against flax antigen correlated with compatibility between four host lines and four fungus isolates. As DeVay *et al.* (1972) pointed out, however, the possibility of an artifact due to contamination of the fungus uredospores by host antigenic material could not be ruled out.

In an effort to avoid the artifact referred to above, DeVay and his collaborators (Schnathorst & DeVay, 1963; DeVay *et al.*, 1967; Wimalajeewa & DeVay, 1971) concentrated their efforts on bacterial and fungal parasites which could be cultured axenically. Greater antigenic similarity was observed between cotton cultivars and isolates of *Xanthomonas malvacearum* (bacterial blight) in compatible than in incompatible combinations. Although no gene-for-gene relationship could be investigated, *Ustilago maydis* (smut) was shown to share antigens in common with its host, maize, but not with barley, a non-host. Palmerly & Callow (1978) also demonstrated that *Phytophthora infestans* shared antigens in common with its hosts potato and tomato as well as the solanaceous non-host, tobacco. However, there was no antigenic relationship betwen *P. infestans* and non-solanaceous species or between several other fungi and potato.

Vanderplank (1978; 1982) has speculated that the occurrence of common antigens may be a consequence of a gene-for-gene relationship. The hypothesis is that the protein products of host genes for resistance and pathogen genes for virulence copolymerize to form a dimer. It is suggested that this is the basis for gene-for-gene recognition and this model is discussed in more detail in Chapter 7. In order for dimerization to occur, the surfaces of the proteins of host and parasite origin must be complementary and therefore antigenically dissimilar. However, it is envisaged that in addition, these host and parasite proteins could also have antigenically similar surfaces which would facilitate the formation of larger protein polymers.

5.4.6. Phenotypic expression in gene-for-gene relationships. A detailed treatment of the mechanisms associated with gene-for-gene controlled resistance is provided in Chapter 7. However, it is appropriate to discuss here some studies on host-parasite genetics as they relate to phenotype expression at early stages of the host-parasite association. These investigations not only provide insights into the way in which host and parasite genes may exert their effects but also assist in a detailed understanding of the genetic control.

Over a period of twenty years, Ellingboe and co-workers have conducted unequalled, in-depth investigations of the relationship between genetic control and phenotypic expression of compatibility and incompatibility in the powdery mildew (Erysiphe graminis) diseases of barley (Masri & Ellingboe, 1966; McCoy & Ellingboe, 1966) and wheat (Slesinski & Ellingboe, 1969; 1970; 1971; Hsu & Ellingboe, 1972; Yang et al., 1972; Wise & Ellingboe, 1983; Stuckey & Ellingboe, 1974; 1975; Martin & Ellingboe, 1976; Haywood & Ellingboe, 1979).

A firm basis for this work was provided by a thorough understanding of the optimal conditions of inoculation and environment necessary to allow predictable, synchronous and well defined stages of fungal development on the host. Throughout these studies the experimental design and interpretation of data is made by reference to the 'quadratic check' arrangement for host and parasite genotype combinations (Ellingboe, 1972; 1976; 1979; 1981) (see also Chapter 7.F and Figure 7.3).

A series of isolines of barley carrying different Ml genes were used in studies with E. graminis f. sp. hordei. It was demonstrated that each gene could have several pleiotropic effects on parasite development. These included exclusion of the parasite, delay in haustorial development, distortion or destruction of haustoria and suppression of sporulation. Different Ml genes could exert different effects at different stages of parasite ontogeny. For example, Ml-k and Ml-p did not influence the rate of haustorial development while Ml-a and Ml-g did. The presence of Ml-k also resulted in hypersensitive cell collapse which was not associated with the action of other genes. The major criterion for determining if a compatible host-parasite relationship had been established was the production of elongating (functional) secondary hyphae (ESH). All genes markedly reduced the proportion of parasite units which gave rise to ESH but there were differences between genes in the extent to which this occurred. Interestingly, in segregating populations, there were also differences between plant individuals in the proportion of parasite units which initiated secondary hyphae. However, this character was clearly shown to be independent of the presence of any Ml gene and it was inferred that other genes by which the isolines differed must be involved.

When the proportion of parasite units producing ESH was used as the criterion, it was also shown that the range for this value was greater in homozygous individuals produced from heterozygous parents than it was in those produced from homozygous parents. This occurred with both Ml Ml and ml ml plants. Hence an intriguing 'carry-over effect' from the heterozygote was identified.

In addition to effects on the production of ESH, the action of Ml genes markedly reduced the rate of transfer of ^{35}S-sulphate from host to parasite by comparison with a compatible combination. Of particular significance was the observation that, using lines isogenic for Ml-g and ml-g no difference between the transfer kinetics of ^{35}S could be detected for the three compatible genotype-genotype combinations (Ml-g/p-g, ml-g/P-g and Ml-g/P-g).

Approximately, 30 alleles or pseudo-alleles are known to occur at the Ml-a locus in barley. It was suggested that if these are genuine alleles, they should affect the same stage of parasite development. Six isolines carrying different Ml-a 'alleles' were investigated using a fungus isolate avirulent for them all. All the 'alleles' did affect the same ontogenetic stage of fungal development, but on the basis of differences in proportion of parasite units giving rise to ESH it was suggested that the six 'alleles' may reside in three cistrons. From differences in the rate of ESH development between two isolines both thought to carry Ml-a7 (Mu and LG) it was proposed that these were in fact alleles with different specificities and Ml-a7(LG) was tentatively designated Ml-a15.

In a complementary study, Jørgensen & Mortensen (1977) investigated the primary infection by E. graminis f. sp. hordei of a range of barley genotypes resistant by virtue of alleles at the ml-o locus (see section 5.4.4). Each of the lines used resulted from a separate mutational event and other evidence suggested that at least some of the mutants were distinct since recombination could occur (Jørgensen, 1976). All the lines with ml-o showed similar effects on primary infection. Some colony development occasionally occurred which was significantly variable between lines but could be accounted for by differences in genetic background.

Similar studies were made using lines of wheat isogenic for different Pm genes for resistance to E. graminis f. sp. tritici. Again, the proportion of parasite units resulting in ESH was shown to be indicative of the time of inhibition of the infection process.

Different Pm genes apparently exerted their influence at different stages of fungal development and possibly in different ways. For example, Pm1 and Pm4 acted early and a very small proportion of parasite units produced ESH. Pm4 could also be distinguished from Pm1 by an earlier cessation of growth of those ESH which did form. In contrast, alleles at the Pm3 locus, Pm3a and Pm3b, had a much less profound effect on the formation of ESH and those that did form elongated for longer before growth ceased. Interestingly, a further allele, Pm3c, did not inhibit formation of ESH at the seedling stage and was only functional in older leaves. A different situation again occurred when Pm2 was active: this gene had no effect on ESH formation but, much later, host cells were observed to be necrotic and chlorotic and some sparse sporulation occurred. From these comparative studies it was suggested that genes which controlled resistance operating immediately host-parasite contact was established could involve constitutive host characters. Genes which acted at later stages may rely upon the induction of new host metabolism.

Differences between the action of genes Pm1, Pm2, Pm3a and Pm4 were also clearly distinguished when observations on primary haustorial development and host cell response were made. For example, by 30 h after inoculation, more haustoria were distorted and collapsed when Pm1 was operative compared to Pm2.

In all investigations where ESH was examined, it was not possible to distinguish between the three host genotype-parasite genotype combinations which conditioned compatibility (i.e. Pm/p, pm/p and pm/P). In general, the rate of transfer of ^{35}S from host to parasite confirmed the observed differences between different Pm genes. Interestingly, however, with the Pm1 locus, transfer of ^{35}S was reduced in the compatible Pm/p1 combination compared to the compatible combinations pm1/p1 and pm1/P1. This suggested that the change from P1 to p1 in the particular isolate used did not restore full compatibility.

The above phenomemon was investigated further using three different isolates of *E. graminis* f. sp. *tritici* virulent for Pm4 in comparison with one avirulent isolate. It was clearly demonstrated that there were differences between the 'virulent' isolates when the Pm/p4 genotype-genotype combination was examined. For one isolate, the proportion of parasite units forming ESH was the same as combinations involving pm4 but infection occurred more slowly. With the other two isolates, the proportion of parasite units resulting in ESH was considerably reduced in the Pm4/p4 combination compared to those involving pm4. Nevertheless, successful penetrations still gave fully compatible reaction types.

These data illustrated that there may be a range of potentially different 'virulent' mutations at the P1 locus which differ in the degree of restored compatibility when studied in detail. Alternatively, the data may reflect other genetic differences between the virulent isolates which specifically modify the outcome of the Pm/p4 host-parasite combination.

5.4.7. The search for gene products. Probably the most challenging and yet tantalizingly elusive problem in plant pathology is the quest for the products produced by the complementary genes controlling specific resistance and virulence in gene-for-gene associations. Broadly, three separate approaches have been taken.

The first (and almost classical) approach involves attempting to find molecules of parasite origin which precisely reproduce the differential host responses to the particular parasite isolates from which those molecules were derived. Progress in this direction is discussed fully in Chapter 7.4. The logical basis is that by identifying a molecular component of the parasite which regulates the host response, it may prove possible to determine how this happens and thence identify the host molecules involved.

The second and more recent approach is to use recombinant DNA technology to clone either an R-gene or a v-gene thus making the identification of the gene product more feasible. The idea is to locate the appropriate piece of DNA by demonstrating the expected transformation in either the host or the parasite. Currently, convenient vectors for suitably defined host-parasite systems are only available for bacterial parasites and dicotyledonous hosts. Progress and likely future developments using this approach are discussed fully in Chapter

11.

A third, and equally recent approach has involved the use of high resolution two-dimensional electrophoretic mapping of polypeptides. The idea is to search for host or parasite polypeptides which correlate absolutely with the presence of particular resistance or virulence alleles in a range of genetically defined cultivars or isolates.

In the only reported study using the host plant, Gabriel & Ellingboe (1982a) extracted proteins from 13 isolines of wheat cv. Chancellor, which has no known genes for resistance to Erysiphe graminis f. sp. tritici (powdery mildew). Twelve further lines examined were each homozygous for one of 12 different resistance alleles. These lines had been produced by backcrossing eight times to cv. Chancellor. No differences among approximately 300 polypeptides were found which could be attributed to the different Pm genes although interestingly, two forms of cv. Chancellor were distinguishable on the basis of the mobility of a single polypeptide.

Other studies have concentrated effort on investigations of different fungal isolates. Gabriel & Ellingboe (1982b) investigated the proteins extracted from two isolates of Colletotrichum lindemuthianum (bean anthracnose). One isolate was a chemically induced mutant of the other and was incapable of growth at temperatures greater than 28°C in vitro while not being temperature sensitive in vivo. This was the only apparent modification to the phenotype resulting from mutagenesis. However, a change in the electrophoretic mobility of at least 30 polypeptides (approximately 10% of the total examined) was observed. This indicated that mutagenesis can affect many polypeptides despite the fact that only one phenotypic change was evident and provides a lesson for similar investigations.

Gabriel & Ellingboe (1982b) also studied the polypeptide profiles of two isolates of Erysiphe graminis f. sp. tritici which differed from one another at at least three virulence loci together with isolates of the eight possible virulence phenotypes resulting from the cross between the two isolates. In addition, three induced mutants of one of the parent isolates which expressed changed virulence at two virulence loci were investigated. In the latter case, no differences in electrophoretic mobility were observed between the mutants and the original isolate for approximately 600 polypeptide spots. The two parent isolates could consistently be differentiated from one another by the positions of five polypeptides and the eight sexual progeny isolates also had polypeptide spots in positions characteristic of one or other parent, but not both. However, the presence or absence of these spots did not correlate with the presence or absence of the three genes for virulence by which the parent isolates differed.

Also using E. graminis f. sp. tritici, Torp & Anderson (1982) studied the polypeptides from 7 fungal isolates differing in virulence. The polypeptide profile of a mutant virulent for the resistance gene Ml-g was indistinguishable from the isolate it was derived from. However, one polypeptide changed position consistently in isolates either virulent or avirulent for both genes Ml-a10 and Ml-k. It is thought that these avirulence loci are linked and therefore this observation may prove significant.

Howes et al. (1982) and Kim et al. (1982) studied detergent-soluble polypeptides extracted from isolates of Puccinia graminis f. sp. tritici. Of the 290 or so polypeptides visualized, 18 varied between the four isolates studied. However, there was no clear association with the virulence characteristics of the fungus. One of the 18 polypeptides altered in concentration as dormant spores germinated and differentiated. A number of polypeptides not differing between isolates were also shown either to increase or decrease depending on the stage of sporeling development. In a comparable study with P. graminis f. sp. avenae, the polypeptides from nine isolates were analyzed (Kim et al., 1984). There were greater differences in the polypeptide maps between the two formae speciales of P. graminis than there were between isolates of the same form (a total of 34 detected polypeptides). In P. graminis f. sp. avenae there were statistically significant associations between the presence of certain polypeptides and virulence or avirulence in the nine isolates for particular resistance genes. However, there was no absolute correlation and further data would be required to substantiate that any of the polypeptides identified was either the product of a virulence locus or one linked to it.

Despite the fact that no gene products have been positively identified using these techniques, Gabriel & Ellingboe (1982b) conclude that the procedure is not unrealistic. It may be possible to visualize many more polypeptides by expanding the pH gradient and altering extraction procedures. Furthermore, in many host-parasite systems, more host and parasite loci are potentially available for investigation than have hitherto been examined.

5.4.8. Some further considerations. One of the themes that runs through research on gene-for-gene assocations is the notion that there are some host genotypes that lack any major genes for resistance (usually referred to as universally susceptible cultivars). Furthermore there is a tendency to consider that the control of resistance and virulence is the primary function of alleles identified with this capacity.

The flax cultivar Bison was employed as a 'universally susceptible' cultivar throughout studies with *Melampsora lini* (rust) in the USA (Flor & Comstock, 1972). However, work in Australia revealed isolates of the fungus to which Bison was resistant, confirming that this cultivar did in fact carry at least one gene for resistance ($\underline{L}^9$) (Kerr, 1960). Clearly, however, avirulence towards this gene is almost invariably absent from *M. lini* isolates.

Crute & Lebeda (1981; 1983) have demonstrated the existence of several specific resistance factors to *Bremia lactucae* (downy mildew) in lettuce cultivars previously considered to lack any resistance. Isolates capable of revealing this resistance are clearly rare implying the almost universal occurrence of matching virulence. Lebeda (1984) has also shown that isolates of *B. lactucae* collected from *Lactuca serriola* (a possible wild progenitor of the cultivated lettuce, *L. sativa*), revealed in *L. sativa* the existence of additional specific resistance factors. In the same way, genes for specific resistance in *L. serriola* are readily identified by isolates of *B. lactucae* from *L. sativa* (Crute & Davis, 1977; Norwood *et al*., 1981).

Vanderplank (1975; 1978) has argued that it is likely that genes capable of expressing themselves as genes for resistance or virulence may be 'even more numerous than we are aware'. A potential resistance gene is only identified by the lack of matching virulence and if the particular resistance gene in question is commonly distributed throughout host genotypes, mutation to avirulence would frequently be lethal.

Vanderplank (1978) suggests that loci for resistance and virulence primarily control functions other than the determination of host-parasite specificity. Common R-genes and v-genes may code for proteins whose primary function is particularly important or which provide a fitness advantage.

It may be helpful to consider loci for specific resistance and virulence as producing protein products with a primary metabolic function but which secondarily have importance in host-parasite recognition. It is, however, by their secondary functions that they are more readily identified. Perhaps more realistically, reference should in future be made to genes for specific recognition rather than 'genes for resistance or virulence'. Vanderplank (1984) has recently speculated that peroxidase isoenzymes constitute the primary enzymatic function of products of alleles identified as specific resistance (recognition) factors. This is suggested on the basis that these enzymes are synthesized particularly actively when cells are damaged as in the case of parasitism and constitute a utilizable source of nutrient to a potential parasite.

5.5. GENETICS OF RESISTANCE AND COMPLEMENTARY VIRULENCE ABOVE THE CULTIVAR LEVEL

5.5.1 Hybridization studies. It has until now been difficult to obtain any information on the genetic control of the resistance exhibited by a plant species to parasites for which it is not normally a host (i.e. a non-host). Similarly, the same applies to the genetics of avirulence of a parasite towards a non-host species. The difficulties result from the natural barriers to hybridization which exist between higher plant species and also some parasites. Furthermore, some obligate parasites have a highly specific host range making crossing technically impossible even if there were not barriers to hybridization.

Different *formae speciales* of *Puccinia graminis* (stem rust) tend to be recognized by their specialization to different genera of gramineae, although genotypes of a particular host genus can sometimes be located with susceptibility to a *forma specialis* from a different genus. For example, *P. graminis* f. sp. *avenae*, f. sp. *tritici* and f. sp. *secalis* are respectively principally parasitic on oats, wheat and rye. However, some unusual genotypes of wheat are susceptible to *P. graminis* f. sp. *secalis* (Sanghi & Luig, 1971) and some isolates of *P. graminis* have more extensive virulence (Lopez *et al.*, 1974). Interestingly, virulence towards barley genotypes in *P. graminis* is less specialized and susceptibility to all these *formae speciales* is known. However, different major genes in barley have been identified which separately control resistance to *P. graminis* f. sp. *tritici*, *P. graminis* f. sp. *secalis* and *P. graminis* f. sp. *avenae* (Powers & Hines, 1933; Martens *et al.*, 1983; Steffenson *et al.*, 1984)

Hybrids between different *forma speciales* of *P. graminis* can sometimes be made since the sexual cycle is completed on a common host, barberry. Studies of the virulence of these hybrids together with the segregation of resistance in crosses between host genotypes have provided some information about the genetics of this type of specificity.

When crosses between wheat genotypes resistant and susceptible to *P.*

graminis f. sp. *secalis* were made, several major genes were identified which were previously unrecognized in studies employing *P. graminis* f. sp. *tritici*. These genes for resistance to *P. graminis* f. sp. *secalis* also appeared to enhance the action of genes for resistance to *P. graminis* f. sp. *tritici* (Sanghi & Luig, 1971; Luig & Rajaram, 1972). Hybrid isolates recovered from crosses between *P. graminis* f. sp. *avenae* and *P. graminis* f. sp. *agrostidis* were used to inoculate a range of oat cultivars and they also identified previously unrecognized resistance. Oat cultivars universally susceptible to *P. graminis* f. sp. *avenae* proved resistant to the hybrids.

Loss of virulence towards host genotypes of the genera parasitized by either parent is characteristic of F_1 hybrids between *formae speciales* of *P. graminis* (Johnson *et al.*, 1932; Stakman *et al.*, 1930; Cotter & Roberts, 1963). This indicates the dominance of avirulence but not the numbers of loci involved.

The most comprehensive study of the genetics of virulence in crosses between *formae speciales* of *P. graminis* was conducted by Green (1971). Controlled matings were made between isolates of *P. graminis* f. sp. *tritici* and *P. graminis* f. sp. *secalis* apparently homozygous for avirulence to rye and wheat respectively. As expected the F_1 exhibited low virulence for both parental hosts i.e. avirulence was dominant. However, rather more unexpected were the reactions of the F_2 isolates which were also generally lacking in virulence for either parental hosts. This was explained by suggesting that many genetic factors govern virulence on wheat and rye and that the chances of recovering highly virulent isolates following crosses were therefore low given the numbers of progeny tested. The suggestion was made that primitive forms of *P. graminis* may exhibit low virulence to a wide range of hosts and that evolution has resulted in the accumulation of genes conditioning high virulence and a narrow host range.

Erysiphe graminis (powdery mildew) also occurs as different *formae speciales* adapted to particular species or genera of gramineae. Studies on wild host species in Israel imply, however, that the distinctions between different *formae speciales* may not be as clear cut as they appear in an agricultural context and *E. graminis* f. sp. *hordei*, *E. graminis* f. sp. *tritici* and *E. graminis* f. sp. *agropyri* do share congenial hosts (Eshed & Wahl, 1970). Hiura (1962) demonstrated that crosses between some *formae speciales* were possible and subsequent studies have revealed interesting aspects of the genetic control of this level of specificity in the fungus

(Hiura, 1978).

Two isolates of E. graminis were hybridized. One was virulent on Triticum spp. but avirulent on Agropyron spp., and the other was virulent on Agropyron spp. but avirulent on Triticum spp. The virulence of 190 F_1 progeny isolates was subsequently analysed on 65 wheat genotypes and 25 Agropyron genotypes which were all susceptible to the respective parental parasite isolates. Despite this, different segregations occurred for virulence:avirulence among the F_1 isolates depending on the host genotype considered. Furthermore, the degree of virulence of the hybrid isolates towards many of the host lines was reduced compared to the parental isolates. This implied the segregation of genes of minor effect which modified the expression of major genes for virulence/avirulence. Progeny were classified as either virulent or avirulent in order to try and determine the numbers of major genes controlling virulence by which the two parents may have differed. In fact, taking wheat genotypes and Agropyron genotypes together, the segregation of progeny identified 3, or more than 4 genes for avirulence towards more than half of the 90 host lines (indicated by haploid segregation ratios of 7:1 and 15:1). The implication was that the two parent cultures differed by a 'considerable number' of avirulence genes. The analysis also revealed that many of the wheat cultivars probably shared genes for resistance in common as did lines of Agropyron. Furthermore, there was good evidence that some loci conditioning avirulence towards wheat cultivars were linked as were loci conditioning avirulence towards Agropyron. However, genes conditioning avirulence on the two host genera were apparently inherited independently.

Because E. graminis only has two chromosomes (Kimber & Wolfe, 1966) linkage between loci controlling virulence is to be expected. However, it is argued that in crosses between isolates of the same forma specialis linkage may not be noted. In order to be pathogenic on genotypes of a host species which share resistance genes in common, genes for virulence would also be common between parasite isolates. Any linkage would be impossible to detect except in crosses between different formae speciales.

5.5.2. Mutation studies. To investigate the genetic control of avirulence/virulence towards non-hosts/hosts above the cultivar level, Ellingboe & Gabriel (1977) and Gabriel et al. (1979) have taken a totally different approach. The argument is that the gene-for-gene relationship is superimposed upon a previously evolved 'basic compatibility' (see Chapter 7). It is suggested that genes conditioning this basic compatibility should be identifiable by the induction and detection of conditional mutants.

To identify genes in a parasite with a positive function in parasitism (e.g. in the control of differentiation of infection structures), high-temperature sensitive mutations were sought in two fungal parasites capable of growth in vitro: Phyllosticta maydis (maize leaf spot) and Colletotrichum lindemuthianum (bean anthracnose). Constitutive mutations in such genes are likely to be lethal (in obligate parasites) or severely debilitating. The hypothesis was that three classes of mutant should be found. Type 1 mutants should be temperature sensitive in vitro and in vivo. These were considered to be mutations of genes responsible for general metabolism such as macromolecular synthesis. Type II mutants should be temperature-sensitive in vitro but not in vivo. These would be mutations in genes responsible for synthesis of metabolites capable of being provided by the host. Type III mutants should be temperature sensitive in vivo but not in vitro. It is these mutants which are thought to affect genes specifically involved in parasitism.

Mutants of all three types were found in C. lindemuthianum following treatment with nitrosoguanidine but further studies on the two Type III mutants obtained have not been reported.

In studies with P. maydis, the results were not as predicted, in that two types of mutant were recovered which implied a positive function for avirulence equivalent to that suggested in a gene-for-gene relationship. These mutants failed to grow in vivo at the 'normal' temperature but were able to grow in vivo at the higher temperature used. In vitro, one type (IIb) was capable of growth at both temperatures whereas the other (Ib) failed to grow well at the higher temperature. It was suggested that the Ib type of mutation may be affecting genes with a primary function in basic metabolism and a secondary function in controlling avirulence. A further type of mutant with a generally enhanced constitutive virulence but without any temperature sensitivity was also recovered.

Using an obligate parasite (Erysiphe graminis f. sp. tritici), mutations to increased specific virulence on host resistance genes Pm1a and Pm4a were readily recovered (Gabriel et al., 1982). This agrees with previous studies with a range of other fungi (Flor, 1956; 1958; 1960; Watson, 1957; Schwinghammer, 1959; Zimmer et al., 1963) and implies the loss of a positive function for avirulence (see above). However, in addition, a number of temperature sensitive mutants only capable of growth at 'normal' temperature were also recovered. The specific virulences of these mutants were unchanged and it is considered that the genes involved controlled basic parasitic capability and development.

5.6. CONCLUSION

An attempt has been made in this Chapter to summarize the current knowledge concerning the genetic control of plant resistance to fungal and bacterial parasites, in addition to the genetics of virulence. The subject has its share of healthy controversy and debate. Throughout, efforts have been made to highlight those areas where a concensus of opinion has not yet been reached.

On the face of it, genetic mechanisms of disease resistance may appear to be more easily understood than the biochemical and physiological mechanisms that result from them. In reality, the complexity of the host-parasite relationship continues to be revealed in genetic studies. Unravelling the intimate genetic association between the two organisms remains one of the most challenging tasks facing geneticists. Undoubtedly, newly available gene transfer technology will further aid these investigations.

An understanding of genetic control is fundamental to establishing the mechanistic basis for resistance and virulence at both molecular and population levels (the subjects of Chapters 7 and 8). Furthermore, genetic studies underpin the vitally important exercise of plant breeding for disease resistance.

Because of their central role, of all the subjects in plant pathology from which benefit will continue to be derived, genetic studies must be rated amongst the most important.

REFERENCES

Albersheim, P., Jones, T.M. & English, P.D. (1969). Biochemistry of the cell wall in relation to infective processes. Annual Review of Phytopathology 7, 171-194.

Alan, H., Katan, J. & Keder, N. (1974). Factors affecting penetrance of resistance to Fusarium oxysporum f.sp. lycopersici on tomatoes. Phytopathology 64, 455-461.

Anderson, M.E. (1933). Fusarium resistance in Wisconsin Hollander cabbage. Journal of Agricultural Research 46, 639-661.

Anderson, M.G. (1982). Interpreting residual effects of "defeated" resistance genes. Phytopathology 72, 1383-1384.

Anderson, R.G. (1961). The inheritance of leaf rust resistance in seven varieties of common wheat. Canadian Journal of Plant Science 41, 342-359.

Anderson, R.G. (1963). Studies on the inheritance of resistance to leaf rust of wheat. Hereditas 2 (Supplement), 144-155.

Arnold, M.H. & Brown, S.J. (1968). Variation in the host-parasite relationship of a crop disease. Journal of Agricultural Science, Cambridge 71, 19-36.

Atkins, J.G., Robot, A.L., Adair, C.R., Goto, K., Kozaka, T., Yanagida, R., Yamada, M. & Matsumoto, S. (1967). An international set of rice varieties for differentiating races of Pyricularia oryzae. Phytopathology 57, 279-301.

Barratt, D.H.P. & Flavell, R.B. (1975). Alterations in mitochondria associated with cytoplasmic and nuclear genes concerned with male sterility in maize. Theoretical and Applied Genetics 45, 315-321.

Barrus, M.F. (1911). Variation of varieties of bean in their susceptibility to anthracnose. Phytopathology 1, 190-195.

Barrus, M.F. (1915). An anthracnose-resistant red kidney bean. Phytopathology 5, 303-311.

Bartos, P., Dyck, P.L. & Samborski,. D.J. (1969). Adult plant leaf rust resistance in Thatcher and Marquis wheat: a genetic analysis of the host-parasite interaction. Canadian Journal of Botany 47, 267-269.

Ben Kalio, V.D. & Clarke, D.D. (1979). Studies on tolerance in wild plants; effects of Erysiphe fischeri on the growth and development of Senecio vulgaris. Physiological Plant Pathology 14, 203-208.

Bennett, F.G.A. (1984). Resistance to powdery mildew in wheat: a review of its use in agriculture and breeding programmes. Plant Pathology 33, 279-300.

Biffen, R.H. (1905). Mendel's laws of inheritance and wheat breeding. Journal of Agricultural Science 1, 4-48.

Biffen, R.H. (1907). Studies in the inheritance of disease resistance. Journal of Agricultural Science 2, 104-128.

Biffen, R.H. (1912). Studies in the inheritance of disease resistance. II. Journal of Agricultural Science 4, 421-429.

Binkerhoff, L.A. (1970). Variation in Xanthomonas malvacearum and its relation to control. Annual Review of Phytopathology 8, 85-110.

Blanch, P.A., Asher, M.J.C. & Burnett, J.H. (1981). Inheritance of pathogenicity and cultural characters in Gaeumannomyces graminis var. tritici. Transactions of the British Mycological Society 77, 391-399.

Boone, D.M. (1971). Genetics of Venturia inaequalis. Annual Review of Phytopathology 9, 297-318.

Brasier, C.M. (1977). Inheritance of pathogenicity and cultural characters in Ceratocystis ulmi; hybridization of protoperithecial and non-aggressive strains. Transactions of the British Mycological Society 68, 45-52.

Brasier, C.M. & Gibbs, J.M. (1976). Inheritance of pathogenicity and cultural characters in Ceratocystis ulmi: hybridization of the aggressive and non-aggressive strains. Annals of Applied Biology 83, 31-37.

Browder, L.E. (1980). A compendium of information about named genes for low reaction to Puccinia recondita in wheat. Crop Science 20, 775-779.

Browder, L.E. & Eversmeyer, M.G. (1980). Sorting of Puccinia recondita:Triticum infection-type data sets towards the gene-for-gene model. Phytopathology 70, 666-670.

Browder, L.E. & Eversmeyer, M.G. (1982). Sorting of infection-type data sets towards the gene-for-gene model: a reply. Phytopathology 72, 458-460.

Burkholder, W.H. (1918). The production of an anthracnose-resistant white marrow bean. Phytopathology 8, 351-359.

Caten, C.E. (1974). Intra-racial variation in Phytophthora infestans and adaptation to field resistance for potato blight. Annals of Applied Biology 77, 259-270.

Caten, C.E., Person, C., Groth, J.V. & Dhahi, S.J. (1984). The genetics of pathogenic aggressiveness in three dikaryons of Ustilago hordei. Canadian Journal of Botany 62, 1209-1219.

Christ, B.J. & Groth, J.V. (1982a). Inheritance of virulence to three bean cultivars in three isolates of the bean rust pathogen. Phytopathology 72, 767-770.

Christ, B.J. & Groth, J.V. (1982b). Inheritance of resistance in three cultivars of beans to the bean rust pathogen and the interaction of virulence and resistance genes. Phytopathology 72, 771-773.

Clarke, A.E., Jones, H.A. & Little, T.M. (1944). Inheritance of bulb colour in the onion. Genetics 29, 569-575.

Clifford, B.C. & Clothier, R.B. (1974). Physiologic specialization of Puccinia hordei on barley. Transactions of the British Mycological Society 63, 421-430.

Comstock, J.C., Martinson, C.A. & Gengenbach, B.G. (1973). Host specificity of a toxin from Phyllosticta maydis for Texas cytoplasmically male-sterile maize. Phytopathology 63, 1357-1361.

Cotter, R.U. & Roberts, B.J. (1963). A synthetic hybrid of two varieties of Puccinia graminis. Phytopathology 53, 344-346.

Cross, J.E., Kennedy, B.W., Lambert, J.W. & Cooper, R.L. (1966). Pathogenic races of the bacterial blight pathogen on soybeans, Pseudomonas glycinea. Plant Disease Reporter 50, 557-560.

Crute, I.R. (1981). The host specificity of Peronosporaceous fungi and the genetics of the relationship between host and parasite. In 'The Downy Mildews' (Spencer, D.M., ed.), pp. 237-253. Academic Press, London & New York.

Crute, I.R. (1985). The relationship between Plasmodiophora brassicae and its hosts - the application of concepts relating to variation in interorganismal associations. Advances in Plant Pathology, in press.

Crute, I.R. & Davis, A.A. (1977). Specificity of Bremia lactucae from Lactuca sativa. Transactions of the British Mycological Society 69, 405-410.

Crute, I.R., Gray, A.R., Crisp, P. & Buczacki, S.T. (1980). Variation in Plasmodiophora brassicae and resistance to clubroot disease in brassicas and allied crops - a critical review. Plant Breeding Abstracts 50, 91-104.

Crute, I.R. & Lebeda, A. (1981). Evidence for a new specific resistance factor in some lettuce (Lactuca sativa L.) cultivars previously considered to be universally susceptible to Bremia lactucae Regel. Theoretical and Applied Genetics 60, 185-189.

Crute, I.R. & Lebeda, A. (1983). Two new specific resistance factors to Bremia lactucae identified in cultivars of lettuce. Tests of Agrochemicals and Cultivars No. 4 (Annals of Applied Biology 102, Supplement), 128-129.

Crute, I.R. & Norwood, J. M. (1978). Incomplete specific resistance to Bremia lactucae in lettuce. Annals of Applied Biology 89, 467-474.

Crute, I.R., Phelps, K., Barnes, A., Buczacki, S.T. & Crisp, P. (1983). The relationship between genotypes of three Brassica species and collections of Plasmodiophora brassicae. Plant Pathology 32, 405-420.

Day, P.R. (1956). Race names of Cladosporium fulvum. Tomato Genetics Co-operative Report 6, 13-14.

Day, P.R. (1974). 'Genetics of Host-Parasite Interaction' Freeman, San Francisco. 238 pp.

Day, P.R., Barrett, J.A. & Wolfe, M.S. (1983). The evolution of host-parasite interaction. In 'Genetic Engineering of Plants, an Agricultural Perspective' (Kosuge, T., Meredith, C.P. & Hollaender, A., eds), pp. 419-430. Plenum Press, New York.

Deshmukh, M.J. & Howard, H.W. (1956). Field resistance to potato blight (Phytophthora infestans). Nature 177, 794-795.

DeVay, J.E. & Adler, H.E. (1976). Antigens common to hosts and parasites. Annual Review of Microbiology 30, 147-168.

DeVay, J.E., Charudattan, R. & Wimalajeewa, D.L.S. (1972). Common antigenic determinants as a possible regulator of host-pathogen compatibility. American Naturalist 106, 185-194.

DeVay, J.E. Schnathorst, W.C. & Foda, M.S. (1967). Common antigens and host parasite interactions. In 'The Dynamic Role of Molecular Constituents in Plant-Parasite Interactions' (Minocha, C.J. & Uritani, I., eds), pp. 318-328. Bruce, Minneapolis.

Doubly, J.A., Flor, H.H. & Clagett, C.O. (1960). Relation of antigens of Melampsora lini and Linum usitatissimum to resistance and susceptibility. Science 131, 229.

Dyck, P.L. (1977). Genetics of leaf rust reaction in three introductions of common wheat. Canadian Journal of Genetics and Cytology 19, 711-716.

Dyck, P.L. & Johnson, R. (1983). Temperature sensitivity of genes for resistance in wheat to Puccinia recondita. Canadian Journal of Plant Pathology 5, 229-234.

Dyck, P.L. & Samborski, D.J. (1968). Genetics of resistance to leaf rust in the common wheat varieties Webster, Loos, Brevit, Corina, Malakov and Centenario. Canadian Journal of Genetics and Cytology 10, 7-17.

Dyck, P.L. & Samborski, D.J. (1974). Inheritance of virulence in *Puccinia recondita* on alleles at the *Lr2* locus for resistance in wheat. Canadian Journal of Genetics and Cytology 16, 323-332.

Dyck, P.L., Samborski, D.J. & Anderson, R.G. (1966). Inheritance of adult-plant leaf rust resistance derived from the common wheat varieties Exchange and Frontana. Canadian Journal of Genetics and Cytology 8, 665-671.

Ebba, T. & Person, C. (1975). Genetic control of virulence in *Ustilago hordei*. IV. Duplicate genes for virulence and environmental modification of a gene-for-gene relationship. Canadian Journal of Genetics and Cytology 17, 631-636.

Eenink, A.H. (1976). Genetics of host-parasite relationships and uniform and differential resistance. Netherlands Journal of Plant Pathology 82, 133-145.

Ellingboe, A.H. (1972). Genetics and physiology of primary infection by *Erysiphe graminis*. Phytopathology 62, 401-406.

Ellingboe, A.H. (1976). Genetics of host-parasite interactions. In 'Encyclopedia of Plant Physiology, Vol. 4. Physiological Plant Pathology' (Heitefuss, R. & Williams, P.H., eds), pp. 761-778. Springer Verlag, Berlin.

Ellingboe, A.H. (1979). Evidence from the quadratic check. Proceedings of Symposium IX, International Congress of Plant Protection Vol. 1, 194-198.

Ellingboe, A.H. (1981). Changing concepts in host-pathogen genetics. Annual Review of Phytopathology 19, 125-143.

Ellingboe, A.H. (1983). Genetical aspects of interaction between plant hosts and their soilborne pathogens. Phytopathology 73, 941-944.

Ellingboe, A.H. & Gabriel, D.W. (1977). Induced conditional mutations for studying host/pathogen interactions. In 'Induced Mutations Against Plant Disease' (International Atomic Energy Agency, ed.), pp. 35-45. IAEA, Vienna.

Emara, Y. (1972). Genetic control of aggressiveness in *Ustilago hordei*. I. Natural variability among physiological races. Canadian Journal of Genetics and Cytology 14, 919-924.

Emara, Y.A. & Sidhu, G. (1974). Polygenic inheritance of aggressiveness in *Ustilago hordei*. Heredity 32, 219-224.

Eshed, M. & Wahl, I. (1970). Host ranges and interrelations of *Erysiphe graminis hordei*, *E. graminis tritici* and *E. graminis avenae*. Phytopathology 60, 628-634.

Fallik, E., Basham, Y., Okon, Y. & Keder, M. (1983). Genetics of resistance to bacterial speck of tomato caused by *Pseudomonas syringae* pv. *tomato*. Annals of Applied Biology 104, 321-325.

Fan, Z., Rimmer, S.R. & Steffansson, B.R. (1983). Inheritance of resistance to *Albugo candida* in rape (*Brassica napus* L.). Canadian Journal of Genetics and Cytology 25, 420-424.

Favret, E. (1960). Spontaneous and induced mutations of barley for the reaction to mildew. Hereditas 46, 20-28.

Favret, E. (1971). The host-pathogen system and its genetic relationship. In 'Barley Genetics II; Proceedings of the 2nd International Barley Genetics Symposium' (Milan, R.A., ed.), pp. 457-471. Pullman, Washington.

Flangas, A.L. & Dickson, J.G. (1961a). Complementary genetic control of differential compatibility in rusts. Quarterly Review of Biology 36, 254-272.

Flangas, A.L. & Dickson, J.G. (1961b). The genetic control of pathogenicity, serotypes and variability in *Puccinia sorghi*. American Journal of Botany 48, 275-285.

Fleming, R.A. & Person, C.O. (1982). Consequences of polygenic determination of resistance and aggressiveness in nonspecific host:parasite relationships. Canadian Journal of Plant Pathology 4, 89-96.

Flor, H.H. (1935). Physiologic specialization of *Melampsora lini* on *Linum usitatissimum*. Journal of Agricultural Research 51, 819-837.

Flor, H.H. (1940). New physiologic races of flax rust. Journal of Agricultural Research 60, 573-591.

Flor, H.H. (1947). Inheritance of reaction to rust in flax. Journal of Agricultural Research 74, 241-262.

Flor, H.H. (1941). Inheritance of rust reaction in a cross between the flax varieties Buda and J.W.S. Journal of Agricultural Research 63, 369-388.

Flor, H.H. (1942). Inheritance of pathogenicity in Melampsora lini. Phytopathology 32, 653-669.

Flor, H.H. (1946). Genetics of pathogenicity in Melampsora lini. Journal of Agricultural Research 7, 335-357.

Flor, H.H. (1954). Identification of races of flax rust by lines with single rust-conditioning genes. Technical Bulletin No. 1087. U.S. Department of Agriculture. 25 pp.

Flor, H.H. (1955). Host-parasite interaction in flax rust - its genetics and other implications. Phytopathology 45, 680-685.

Flor, H.H. (1956). The complementary genetic systems in flax and flax rust. Advances in Genetics 8, 29-54.

Flor, H.H. (1958). Mutation to wider virulence in Melampsora lini. Phytopathology 48, 297-301.

Flor, H.H. (1959). Differential host range of the monocaryon and the dicaryons of an eu-autoecious rust. Phytopathology 49, 794-795.

Flor, H.H. (1960). Asexual variants of Melampsora lini. Phytopathology 50, 223-226.

Flor, H.H. (1965). Tests for allelism of rust-resistance genes in flax. Crop Science 5, 415-418.

Flor, H.H. (1971). Current status of the gene-for-gene concept. Annual Review of Phytopathology 9, 275-296.

Flor, H.H. & Comstock. V.E. (1972). Identification of rust-conditioning genes in flax cultivars. Crop Science 12, 800-804.

Fullerton, R.A. & Nielsen, J. (1974). Identical genes for virulence in the smuts Ustilago hordei and U. nigra and inheritance of virulence in the barley cultivars Keystone and Conquest. Canadian Journal of Plant Science 54, 253-257.

Gabe, H.L. (1975). Standardization of nomenclature for pathogenic races of Fusarium oxysporum f. sp. lycopersici. Transactions of the British Mycological Society 64, 156-159.

Gabriel, D.W. & Ellingboe, A.H. (1982a). High resolution two-dimensional electrophoresis of proteins from congenic wheat lines. Physiological Plant Pathology 20, 349-357.

Gabriel, D.W. & Ellingboe, A.H. (1982b). Polypeptide mapping by two dimensional electrophoresis and pathogenic variation in field isolates and induced mutants of Erysiphe graminis f. sp. tritici. Phytopathology 72, 1496-1499.

Gabriel, D.W., Ellingboe, A.H. & Rossman, E.C. (1979). Mutations affecting virulence in Phyllosticta maydis. Canadian Journal of Botany 57, 2639-2643.

Gabriel, D.W., Lisker, N. & Ellingboe, A.H. (1982). The induction and analysis of two classes of mutations affecting pathogenicity in obligate parasites. Phytopathology 72, 1026-1028.

Gilchrist, D.G. & Grogan, R.G. (1976). Production and nature of a host-specific toxin from Alternaria alternata f. sp. lycopersici. Phytopathology 66, 165-171.

Globerson, D., Netzer, D. & Tjillingii, F. (1974). Mode of inheritance of resistance in lettuce (Lactuca sativa L.) to three Israeli and four Dutch races of downy mildew (Bremia lactucae Reg.). Euphytica 23, 54-60.

Graham, K.M. (1963). Inheritance of resistance to Phytophthora infestans in two diploid Mexican Solanum species. Euphytica 12, 35-40.

Green, G.J. (1964). A color mutation, its inheritance, and the inheritance of pathogenicity in Puccinia graminis Pers. Canadian Journal of Botany 42, 1653-1664.

Green, G.J. (1966). Selfing studies with races 10 and 11 of wheat stem rust. Canadian Journal of Botany 44, 1255-1260.

Green, G.J. (1971). Hybridization between Puccinia graminis tritici and Puccinia graminis secalis and its evolutionary implications. Canadian Journal of Botany 49, 2089-2095.

Green, G.J. & Campbell, A.B. (1979). Wheat cultivars resistant to Puccinia graminis tritici in Western Canada: their development, performance and economic value. Canadian Journal of Plant Pathology 1, 3-11.

Guzman, J.N. (1964). Nature of partial resistance of certain clones of three Solanum species to Phytophthora infestans, Phytopathology 54, 1398-1404.

Habgood, R.M. (1974). The inheritance of resistance to Rhynchosporium secalis in some European spring barley cultivars. Annals of Applied Biology 77, 191-200.

Habgood, R.M. & Hayes, J.D. (1971). The inheritance of resistance to Rhynchosporium secalis in barley. Heredity 27, 25-37.

Haggag, M.E.A. & Dyck, P.L. (1973). The inheritance of leaf rust resistance in four common wheat varieties possessing genes at or near the Lr3 locus. Canadian Journal of Genetics and Cytology 15, 127-134.

Haggag, M.E.A., Samborski, D.J. & Dyck, P.L. (1973). Genetics of pathogenicity in three races of leaf rust on four wheat varieties. Canadian Journal of Genetics and Cytology 15, 73-82.

Harder, D.E., McKenzie, R.I.H. & Martens, J.W. (1984). Inheritance of adult plant resistance to crown rust in an accession of Avena sterilis. Phytopathology 74, 352-353.

Hare, R.A. & McIntosh, R.A. (1979). Genetic and cytogenetic studies of durable adult-plant resistances in 'Hope' and related cultivars to wheat rusts. Zeitschrift für Pflanzenzüchtung 83, 350-367.

Harlan, J.R. (1976). Diseases as a factor in plant evolution. Annual Review of Phytopathology 14, 31-51.

Haywood, M.J. & Ellingboe, A.H. (1979). Genetic control of primary haustorial development of Erysiphe graminis on wheat. Phytopathology 69, 48-53.

Heather, W.A. & Chandrashekar, M. (1982). Evolutionary, epidemiological and ecological implications of forms of resistance in Populus spp. to Melampsora leaf rusts. Australian Forest Research 12, 231-244.

Hiura, U. (1962). Hybridization between varieties of Erysiphe graminis. Phytopathology 52, 664-666.

Hiura, U. (1978). Genetic basis of forma speciales in Erysiphe graminis D.C. In 'The Powdery Mildews' (Spencer, D.M., ed.), pp. 101-128. Academic Press, London & New York.

Hogenboom, N. G. (1970). Inheritance of resistance to corky root in tomato (Lycopersicon esculentum Mill.). Euphytica 19, 413-425.

Holton, C.S. & Halisky, P.M. (1960). Dominance of avirulence and monogenic control of virulence in race hybrids of Ustilago avenae. Phytopathology 50, 766-770.

Holton, C.S., Hoffmann, J.A. & Duran, R. (1968). Variation in the smut fungi. Annual Review of Phytopathology 6, 213-242.

Hooker, A.L. & Russell, W.A. (1962). Inheritance of resistance to Puccinia sorghi in six corn inbred lines. Phytopathology 52, 122-128.

Horsfall, J.G. & Cowling, E.B. (1977). The sociology of plant pathology. In 'Plant Pathology, an Advanced Treatise' (Horsfall, J.G. & Cowling, E.B., eds), pp. 12-33. Academic Press, New York.

Howes, N.K., Kim, W.K. & Rohringer, R. (1982). Detergent-soluble polypeptides extracted from uredospores of four physiologic races of Puccinia graminis f. sp. tritici. Physiological Plant Pathology 21, 361-366.

Hsu, S. & Ellingboe, A.H. (1972). Elongation of secondary hyphae and transfer of ^{35}S from barley to Eryshiphe graminis f. sp. hordei during primary infection. Phytopathology 62, 876-882.

Innes, N.L. (1983). Bacterial blight of cotton. Biological Reviews 58, 157-176.

James, R.V. & Fry, W.E. (1983). Potential for Phytophthora infestans populations to adapt to potato cultivars with rate-reducing resistance. Phytopathology 73, 984-988.

Jeffrey, J.I.B., Jinks, J.L. & Grindle, M. (1962). Intraracial variation in Phytophthora infestans and field resistance to potato blight. Genetica 32, 323-338.

Jeger, M.J. (1980). Multivariate models of the components of partial resistance. Protection Ecology 2, 265-269.

Jenns, A.E., Leonard, K.J. & Moll, R.H. (1982). Variation in the expression of specificity in two maize diseases. Euphytica 31, 269-279.

Jinks, J.L. & Grindle, M. (1963). Changes induced by training in Phytophthora infestans. Heredity 18, 245-264.

Johnson, A.G., Crute, I.R. & Gordon, P.L. (1977). The genetics of race specific resistance in lettuce (Lactuca sativa) to downy mildew (Bremia lactucae). Annals of Applied Biology 86, 87-103.

Johnson, A.G. Laxton, S.A., Crute, I.R., Gordon, P.L. & Norwood, J.M. (1978). Further work on the genetics of race specific resistance in lettuce (Lactuca sativa) to downy mildew (Bremia lactucae). Annals of Applied Biology 89, 257-264.

Johnson, L.B. & Schafer, J.F. (1965). Identification of wheat leaf rust resistance combinations by differential temperature effects. Plant Disease Reporter 49, 222-224.

Johnson, R. (1978). Practical breeding for durable resistance to rust diseases of self pollinating cereals. Euphytica 27, 529-540.

Johnson, R. (1979). The concept of durable resistance. Phytopathology 69, 198-199.

Johnson, R. (1983). Genetic background of durable resistance. In 'Durable Resistance in Crops' (Lamberti, F., Waller, J.M. & Van der Graaf, M.A., eds), pp. 5-26. Plenum Press, London & New York.

Johnson, R. & Law, C.N. (1975) Genetic control of durable resistance to yellow rust (*Puccinia striiformis*) in the wheat cultivar Hybride de Bersee. Annals of Applied Biology 81, 385-391.

Johnson, T. (1954). Selfing studies with physiological races of wheat stem rust, *Puccinia graminis* var. *tritici*. Canadian Journal of Botany 32, 506-522.

Johnson, T., Newton, M. & Brown, A.M. (1932). Hybridization of *Puccinia graminis tritici* with *Puccinia graminis secalis* and *Puccinia graminis agrostidis*. Scientific Agriculture 13, 141-153.

Johnson, T., Newton, M. & Brown, A.M. (1934). Further studies of the inheritance of spore colour and pathogenicity in crosses between physiologic forms of *Puccinia graminis tritici*. Scientific Agriculture 14, 360-373.

Johnston, C.O. & Heyne, E.G. (1964). Whichita wheat back cross lines for differential hosts in identifying physiologic races of *Puccinia recondita*. Phytopathology 54, 385-388.

Jones, D.R. & Deverall, B.J. (1977). The effect of the *Lr20* resistance gene in wheat on the development of leaf rust, *Puccinia recondita*. Physiological Plant Pathology 10, 275-284.

Jørgensen, J.H. (1971a). An allelic series of mutant genes for powdery mildew in barley. In 'Barley Genetics II. Proceedings of the 2nd International Barley Genetics Symposium' (Milan, R.A., ed.), pp. 475-477. Pullman, Washington.

Jørgensen, J.H. (1971b). Comparison of induced mutant genes with spontaneous genes in barley conditioning resistance to powdery mildew. In 'Mutation Breeding for Disease Resistance' (International Atomic Energy Agency, ed.), pp. 117-124. IAEA, Vienna.

Jørgensen, J.H. (1976). Studies on recombination between alleles in the *ml-o* locus of barley and on pleiotropic effects of the alleles. In 'Induced Mutations for Disease Resistance in Crop Plants' (International Atomic Energy Agency, ed.), pp. 129-140. IAEA, Vienna.

Jørgensen, J.H. (1977). Location of the *ml-o* locus on barley chromosome 4. In 'Induced Mutations Against Plant Disease' (International Atomic Energy Agency, ed.), pp. 533-549. IAEA, Vienna.

Jørgensen, J.H. & Mortensen, K. (1977). Primary infection by *Erysiphe graminis* f. sp. *hordei* of barley mutants with resistance genes in the *ml-o* locus. Phytopathology 67, 678-685.

Jørgensen, J.H. & Moseman, J.G. (1972). Recombination at the *ml-o* locus in barley conditioning resistance to *Erysiphe graminis* f. sp. *hordei*. Canadian Journal of Genetics and Cytology 14, 43-48.

Kacser, H. & Burns, J.A. (1981). The molecular basis of dominance. Genetics 97, 639-666.

Kerber, E.R. & Dyck, P.L. (1977). Inhibition of stem rust resistance by chromosome $7D^L$ of Comthatch hexaploid wheat. Canadian Journal of Genetics and Cytology 19, 575-576.

Kerber, E.R. & Green, G.J. (1980). Suppression of stem rust resistance in the hexaploid wheat cv. Comthatch by chromosome $7D^L$. Canadian Journal of Botany 58, 1347-1350.

Kerr, H.B. (1960). The inheritance of resistance of *Linum usitissimum* L. to the Australian *Melampsora lini* (pers.) Lev. race complex. Proceedings of the Linnean Society of New South Wales 85, 273-321.

Khan, T.N. (1969). Inheritance of resistance to net blotch in barley. I. Factors affecting the penetrance and expressivity of gene(s) conditioning host resistance. Canadian Journal of Genetics and Cytology 11, 587-591.

Khan, T.N. & Boyd, W.J.R. (1969a). Physiologic specialization in *Drechslera teres*. Australian Journal of Biological Science 22, 1229-1235.

Khan, T.N. & Boyd, W.J.R. (1969b). Environmentally induced variability in the host reaction of barley to net blotch. Australian Journal of Biological Science 22, 1237-1244.

Khan, T.N. & Boyd, W.J.R. (1969c). Inheritance of resistance to neck blotch in barley II. Genes conditioning resistance against race WA-2. Canadian Journal of Genetics and Cytology 11,

592-597.

Kim, W.K., Howes, N.K. & Rohringer, R. (1982). Detergent soluble polypeptides in germinated and differentiated uredosporelings of wheat stem rust. Canadian Journal of Plant Pathology 4, 328-333.

Kim, W.K., Martens, J.W. & Howes, N.K. (1984). Electrophoretic analysis of detergent-soluble polypeptides of nine races of Puccinia graminis f. sp. avenae and their relation to P. graminis f. sp. tritici. Canadian Journal of Plant Pathology 6, 111-118.

Kimber, G. & Wolfe, M.S. (1966). The chromosome number of Erysiphe graminis. Nature 212, 318-319.

Knott, D.R. (1957). The inheritance of rust resistance. III. The inheritance of stem rust resistance in nine Kenya varieties of common wheat. Canadian Journal of Plant Science 37, 366-384.

Knott, D.R. (1971). Genes for stem rust resistance in wheat varieties Hope and H-44. Canadian Journal of Genetics and Cytology 13, 186-188.

Knott, D.R. & Anderson, R.G. (1956). The inheritance of rust resistance. I. The inheritance of stem rust resistance in ten varieties of common wheat. Canadian Journal of Agricultural Science 36, 174-195.

Knott, D.R. & Johnson, R. (1981). The sorting and analysis of infection types from Triticum aestivum/Puccinia recondita interactions. Phytopathology 71, 1010-1012.

Knott, D.R. & Johnson, R. (1983). Some additional comments on sorting infection-type data sets. Phytopathology 73, 514-515.

Kolmer, J.A., Christ, B.J. & Groth, J.V. (1984). Comparative virulence of monokaryotic and dikaryotic stages of five isolates of Uromyces appendiculatus. Phytopathology 74, 111-113.

Kooistra, E. (1964). Recent experience of breeding leaf mould resistant tomatoes. Euphytica 13, 103-109.

Krüger, J., Hoffman, C.M. & Hubbeling, N. (1977). The kappa race of Colletotrichum lindemuthianum and sources of resistance to anthracnose in Phaseolus beans. Euphytica 26, 23-25.

Krupinski, J.M. & Sharp, E.L. (1979). Reselection for improved resistance of wheat to stripe rust. Phytopathology 69, 400-404.

Kulkarni, R.N. & Chopra, V.L. (1982). Environment as the cause of differential interaction between host cultivars and pathogen races. Phytopathology 72, 1384-1386.

Kuo, M., Yoder, O.C. & Scheffer, R.P. (1970). Comparative specificity of the toxins of Helminthosporium victoriae. Phytopathology 60, 365-368.

Law, C.N. & Johnson, R. (1967). A genetic study of leaf rust resistance in wheat. Canadian Journal of Genetics and Cytology 9, 805-822.

Lawrence, G.J., Mayo, G.M.E. & Shepherd, K.W. (1981a). Interactions between genes controlling pathogencitiy in the flax rust fungus. Phytopathology 71, 12-19.

Lawrence, G.J., Shepherd, K.W. & Mayo, G.M.E. (1981b). Fine structure of genes controlling pathogenicity in flax rust, Melampsora lini. Heredity 46, 297-313.

Leaver, C.J. & Forde, B.G. (1980). Mitochondrial genome expression in higher plants. In 'Genome Organization and Expression in Plants' (Leaver, C.J., ed.), pp. 407-425. Plenum Press, New York & London.

Lebeda, A. (1984). Resistance of differential cultivars of Lactuca sativa to Bremia lactucae isolates from Lactuca serriola. Transactions of the British Mycological Society 83, 491-494.

Levin, D.A. (1975). Pest pressure and recombination systems in plants. American Naturalist 109, 437-451.

Lewellen, R.T. & Sharp, E.L. (1968). Inheritance of minor reaction gene combinations in wheat to Puccinia striiformis at two temperature profiles. Canadian Journal of Botany 46, 21-26.

Lewellen, R.T., Sharp, E.L. & Hehn, E.R. (1967). Major and minor genes in wheat for resistance to Puccinia striiformis and their responses to temperature changes. Canadian Journal of Botany 45, 2165-2172.

Loegering, W.Q. & Burton, C.H. (1974). Computer-generated hypothetical genotypes for reaction and pathogenicity of wheat cultivars and cultures of Puccinia graminis tritici. Phytopathology 64, 1380-1384.

Loegering, W.Q. & Geis, J.R. (1957). Independence in the action of three genes conditioning stem rust resistance in Red Egyptian wheat. Phytopathology 47, 740-741.

Loegering, W.Q. & Harmon, D.L. (1969). Wheat lines near-isogenic for reaction to Puccinia graminis tritici. Phytopathology 59, 456-459.

Loegering, W.Q., McIntosh, R.A. & Burton, C.H. (1971). Computer analysis of disease data to derive hypothetical genotypes for reaction of host varieties to pathogens. Canadian Journal of Genetics and Cytology 13, 742-748.

Loegering, W.Q. & Powers, H.R. (1962). Inheritance of pathogenicity in a cross of physiological races 111 and 36 of Puccinia graminis f. sp. tritici. Phytopathology 52, 547-554.

Loegering, W.Q. & Sears, E.R. (1963). Distorted inheritance of stem-rust resistance of Timstein wheat caused by a pollen-killing gene. Canadian Journal of Cytology and Genetics 5, 65-72.

Lopez, A., Rajaram, S. & de Bauer, L.I. (1974). Susceptibility of Triticale, rye and wheat to stem rust of these three hosts. Phytopathology 64, 266-267.

Luig, M.H. & Watson, I.A. (1961). A study of inheritance of pathogenicity in Puccinia graminis var. tritici. Proceedings of the Linnean Society of New South Wales 86, 217-229.

Luig, M.H. & McIntosh, R.A. (1968). Location and linkage of genes in wheat chromosome 2D. Canadian Journal of Genetics and Cytology 10, 99-105.

Luig, N.H. & Rajaram, S. (1972). The effect of temperature and genetic background on host gene expression and interaction to Puccinia graminis tritici. Phytopathology 62, 1171-1174.

Lunsford, J.N., Futrell, M.C. & Scott, G.E. (1975). Maternal influence on response of corn to Fusarium moniliforme. Phytopathology 65, 223-225.

Lupton, F.G.H. & Johnson, R. (1970). Breeding for mature plant resistance to yellow rust in wheat. Annals of Applied Biology 66, 137-143.

McCoy, M.S. & Ellingboe, A.H. (1966). Major genes for resistance and the formation of secondary hyphae by Erysiphe graminis f. sp. hordei. Phytopathology 56, 683-686.

Macer, R.C.F. (1972). The resistance of cereals to yellow rust and its exploitation by plant breeding. Proceedings of the Royal Society, London, Series B. 181, 281-301.

McIntosh, R.A. (1977). The nature of induced mutations affecting disease reaction in wheat. In 'Induced Mutations Against Plant Disease' (International Atomic Energy Agency, ed.), pp. 551-565. IAEA, Vienna.

McIntosh, R.A. (1978). Breeding for resistance to powdery mildew in the temperate cereals. In 'The Powdery Mildews' (Spencer, D.M., ed.), pp. 237-257. Academic Press, London & New York.

McRostie, G.P. (1919). Inheritance of anthracnose resistance as indicated by a cross between a resistant and a susceptible bean. Phytopathology 9, 139-148.

McRostie, G.P. (1921). Inheritance of disease resistance in the common bean. Journal of the American Society of Agronomy 13, 15-32.

Martens, J.W., Green, G.J. & Buchannon, K.W. (1983). Inheritance of resistance to Puccinia graminis f. sp. avenae in a Hordeum vulgare selection. Canadian Journal of Plant Pathology 5, 266-268.

Martens, J.W., McKenzie, R.I.H. & Green, G.J. (1970). Gene-for-gene relationships in the Avena:Puccinia graminis host-parasite system in Canada. Canadian Journal of Botany 48, 969-975.

Martens, J.W., McKenzie, R.I.H. & Horder, D.E. (1980). Resistance to Puccinia graminis avenae and P. coronata avenae in the wild and cultivated Avena populations of Iran, Iraq and Turkey. Canadian Journal of Genetics and Cytology 22, 641-649.

Martin, T.J. & Ellingboe, A.H. (1976). Differences between compatible parasite/host genotypes involving the Pm4 locus in wheat and the corresponding genes in Erysiphe graminis f. sp. tritici. Phytopathology 66, 1435-1438.

Masri, S.S. & Ellingboe, A.H. (1966). Primary infection of wheat and barley by Erysiphe graminis. Phytopathology 56, 389-395.

Mayama, S., Daly, J.M., Rehfeld, D.W. & Daly, C.R. (1975). Hypersensitive response of near-isogenic wheat carrying the temperature-sensitive Sr6 allele for resistance to stem rust. Physiological Plant Pathology 7, 35-47.

Mayo, G.M.E. & Shepherd, K.W. (1980). Studies of genes controlling specific host-parasite interactions in flax and its rust. Heredity 44, 211-227.

Metcalfe, D.R. (1966). Inheritance of loose smut resistance. III. Relationships between the

"Russian" and "Jet" genes for resistance and genes in 10 barley varieties of diverse origin. Canadian Journal of Plant Science 46, 487-495.

Miah, M.A.J. & Sackston, W.E. (1970). Genetics of pathogenicity in sunflower rust. Phytoprotection 51, 17-35.

Michelmore, R.W., Norwood, J.M., Ingram, D.S., Crute, I.R. & Nicholson, P. (1984). The inheritance of virulence in Bremia lactucae to match resistance factors 3,4,5,6,8,9,10 and 11 in lettuce (Lactuca sativa). Plant Pathology 33, 301-315.

Minogue, K.P. (1983). Do constant ranking and durability follow from polygenic inheritance? Canadian Journal of Plant Pathology 5, 54-55.

Moseman, J.G. (1963). Relationship of genes conditioning pathogenicity of Erysiphe graminis f. sp. hordei on barley. Phytopathology 53, 1326-1330.

Moseman, J.G. & Jørgensen, J.H. (1971). Identification of genes at the Ml-a locus in barley for resistance to Erysiphe graminis f. sp. hordei. Crop Science 11, 547-550.

Moseman, J.G. & Jørgensen, J.H. (1973). Differentiation of resistance genes at the Ml-a locus in six pairs of isogenic barley lines. Euphytica 22, 189-196.

Moseman, J.G. & Reid, D.A. (1961). Linkage relationship of genes conditioning resistance to leaf rust and powdery mildew in Frangor barley. Crop Science 1, 425-427.

Moseman, J.G. & Schaller, C.W. (1960). Genetics of the allelic series at the Ml-a locus in barley and cultures of Erysiphe graminis f. sp. hordei that differentiate these alleles. Phytopathology 50, 736-741.

Moseman, J.G., Wiberg, A. & Greely, L.W. (1965). Genes conditioning pathogenicity in Erysiphe graminis f. sp. hordei on 23 barley varieties. Phytopathologishe Zeitschrift 52, 209-221.

Müller, K.O. & Haigh, J.C. (1953). Nature of 'field resistance' of the potato to Phytophthora infestans de Bary. Nature 171, 781-783.

Nass, H.A., Pedersen, W.L., MacKenzie, D.R. & Nelson, R.R. (1981). The residual effects of some "defeated" powdery mildew resistance genes in isolines of winter wheat. Phytopathology 71, 1315-1318.

Nelson, O.E. & Ullstrup, A.J. (1964). Resistance to leaf spot in maize. Journal of Heredity 55, 195-199.

Nelson, R.R. (1978). Genetics of horizontal resistance in plant diseases. Annual Review of Phytopathology 16, 359-378.

Nelson, R.R., Pedersen, W.L. & MacKenzie, D.R. (1982). The effect of pyramiding "defeated" wheat powdery mildew resistance genes on components of "slow mildewing". Phytopathology 72, 932.

Neervoort, W.J. & Parlevliet, J.E. (1978). Partial resistance of barley to leaf rust, Puccinia hordei. V. Analysis of the components of partial resistance in eight barley cultivars. Euphytica 27, 33-39.

Noronha-Wagner, M. & Bettencourt, A.J. (1967). Genetic study of the resistance of Coffea spp. to leaf rust. I. Identification and behaviour of four factors conditioning disease reaction in Coffea arabica to twelve physiologic races of Hemileia vastrix. Canadian Journal of Botany 45, 2021-2031.

Norwood, J.M. & Crute, I.R. (1984). The genetic control and expression of specificity in Bremia lactucae (lettuce downy mildew). Plant Pathology 33, 385-399.

Norwood, J.M., Crute, I.R. & Lebeda, A. (1981). The location and characteristics of novel sources of resistance to Bremia lactucae Regel (downy mildew) in wild Lactuca L. species. Euphytica 30, 659-668.

Norwood, J.M., Crute, I.R. & Phelps, K. (1984). Variation in seedling susceptibility to Bremia lactucae (lettuce downy mildew) in the absence of effective major resistance genes. Annals of Applied Biology 105, 147-157.

Palmerley, R.A. & Callow, J.A. (1978). Common antigens in extracts of Phytophthora infestans and potatoes. Physiological Plant Pathology 12, 241-248.

Parlevliet, J.E. (1976a). Evaluation of the concept of horizontal resistance in the barley/Puccinia hordei host - pathogen relationship. Phytopathology 66, 494-497.

Parlevliet, J.E. (1976b). The genetics of seedling resistance to leaf rust, Puccinia hordei Otthl. in some spring barley cultivars. Euphytica 25, 249-254.

Parlevliet, J.E. (1977). Evidence of differential interaction in the polygenic Hordeum vulgare -

Puccinia hordei relation during epidemic development. Phytopathology 67, 776-778.
Parlevliet, J.E. (1978a). Further evidence of polygenic inheritance of partial resistance in barley to leaf rust, Puccinia hordei. Euphytica 27, 369-379.
Parlevliet, J.E. (1978b). Race-specific effects of polygenic resistance of barley to leaf rust, Puccinia hordei. Netherlands Journal of Plant Pathology 84, 121-126.
Parlevliet, J.E. & Zadoks, J.C. (1977). The integrated concept of disease resistance; a new view including horizontal and vertical resistance in plants. Euphytica 26, 5-21.
Payne, G.A. & Yoder, O.C. (1978). Effect of the nuclear genome of corn on the sensitivity to Helminthosporium maydis race T toxin and on susceptibility to H. maydis race T. Phytopathology 68, 331-337.
Perkins, J.M. & Hooker, A.L. (1981). Reactions of eighty-four sources of chlorotic lesion resistance in corn to three biotyopes of Helminthosporium turcicum. Plant Disease 65, 502-504.
Person, C. (1959). Gene-for-gene relationships in host:parasite systems. Canadian Journal of Botany 37, 1101-1130.
Person, C. & Ebba, T. (1975). Genetics of fungal pathogens. Genetics 79, 397-408.
Person, C., Fleming, R, Cargeeg, L. & Christ, B. (1983). Present knowledge and theories concerning durable resistance. In 'Durable Resistance in Crops' (Lamberti, F., Waller, J.M. & Van der Graff, N.A., eds), pp. 27-44. Plenum Press, London & New York.
Person, C. & Mayo, G.M.E. (1974). Genetic limitations on models of specific interactions between a host and its parasite. Canadian Journal of Botany 52, 1339-1347.
Pink, D.A.C., Bennett, F.G.A., Caten, C.E. & Law, C.M. (1983). Correlated effects of homeologous group 5 chromosomes upon infection of wheat by yellow rust and powdery mildew. Zeitschrift für Pflanzenzüchtung 91, 278-294.
Powers, L. & Hines, L. (1933). Inheritance of reaction to stem rust and barbing of awns in barley crosses. Journal of Agricultural Research 46, 1121-1129.
Powers, N.R. & Sando, W.J. (1960). Genetic control of the host-parasite relationship in wheat powdery mildew. Phytopathology 50, 454-457.
Putt, E.D. & Sackston, W.E. (1963). Studies on sunflower rust IV. Two genes, R_1 and R_2 for resistance in the host. Canadian Journal of Plant Science 43, 490-496.
Rajaram, S., Luig, N.H. & Watson, I.A. (1971). The inheritance of leaf rust resistance in four varieties of common wheat. Euphytica 20, 574-585.
Rieman, G.H. (1931). Genetic factors for pigmentation in the onion and their relation to disease resistance. Journal of Agricultural Research 42, 251-278.
Riley, R. (1973). Genetic changes in hosts and the significance of diseases. Annals of Applied Biology 75, 128-132.
Risser, G., Banihashemi, A. & Davis, D.W. (1976). A proposed nomenclature of Fusarium oxysporum f. sp. melonis races and resistance genes in Cucumis melo. Phytopathology 66, 1105-1106.
Roane, C.W. & Starling, T.M. (1970). Inheritance of reaction to Puccinia hordei in Barley. III. Genes in the cultivars Cebeda, Capa and Franger. Phytopathology 60, 788-790.
Robinson, R.A. (1969). Disease resistance terminology. Review of Applied Mycology 48, 593-606.
Robinson, R.A. (1971). Vertical resistance. Review of Plant Pathology 50, 233-239.
Robinson, R.A. (1973). Horizontal resistance. Review of Plant Pathology 52, 483-501.
Robinson, R.A. (1976). 'Plant Pathosystems'. Springer Verlag, Berlin. 184 pp.
Robinson, R.A. (1979). Permanent and impermanent resistance to crop parasites; a re-examination of the pathosystem concept with special reference to rice blast. Zeitschrift für Pflanzenzüchtung 83, 1-39.
Roelfs, A.P., Baker, F.D. & McVey, D.V. (1982). An interactive computer-based system for comparing cultures of Puccinia graminis and postulating Sr genotypes in wheat. Phytopathology 72, 596-600.
Roelfs, A.B. & McVey, D.V. (1979). Low infection types produced by Puccinia graminis f. sp. tritici and wheat lines with designated genes for resistance. Phytopathology 69, 722-730.
Romero, S. & Erwin, D.C. (1969). Variation in pathogenicity among single-oospore cultures of Phytophthora infestans. Phytopathology 59, 1310-1317.
Rouse, D.I., Nelson, R.R., MacKenzie, D.R. & Armitage, C.R. (1980). Components of rate-reducing

resistance in seedlings of four wheat cultivars and parasitic fitness in six isolates of Erysiphe graminis f. sp. tritici. Phytopathology 70, 1097-1100.

Russell, W.A. & Hooker, A.L. (1959). Inheritance of resistance in corn to rust, Puccinia sorghi Schw., and genetic relationships amoung different sources of resistance. Agronomy Journal 51, 21-24.

Salmon, E.S. (1904). On Erysiphe graminis D.C. and its adaptive parasitism within the genus Bromus. Annals of Mycology 2, 255-267.

Samborski, D.J. (1963). A mutation in Puccinia recondita Rob. ex Desm. f. sp. tritici to virulence on Transfer, Chinese Spring x Aegilops umbellulata. Zhuk. Canadian Journal of Botany 41, 475-479.

Samborski, D.J. & Dyck, P.L. (1968). Inheritance of virulence in wheat leaf rust on the standard differential wheat varieties. Canadian Journal of Genetics and Cytology 10, 24-32.

Samborski, D.J. & Dyck, P.L. (1976). Inheritance of virulence in Puccinia recondita on six backcross lines of wheat with single genes for resistance to leaf rust. Canadian Journal of Botany 54, 1666-1671.

Samborski, D.J. & Dyck, P.L. (1982). Enhancement of resistance to Puccinia recondita by interactions of resistance genes in wheat. Canadian Journal of Plant Pathology 4, 152-156.

Sanghi, A.K. & Luig, M.H. (1971). Resistance in wheat to formae speciales tritici and secalis of Puccinia graminis. Canadian Journal of Genetics and Cytology 13, 119-127.

Saxena, K.M.S. & Hooker, A.L. (1968). On the structure of a gene for disease resistance in maize. Proceedings of the National Academy of Sciences, U.S.A. 61, 1300-1305.

Schafer, J.F., Caldwell, R.M., Patterson, F.L. & Compton, L.E. (1963). Wheat leaf rust resistance combinations. Phytopathology 53, 569-573.

Schaller, C.W. & Briggs, F.N. (1955). Inheritance of resistance to mildew, Erysiphe graminis hordei, in the barley variety, Black Russian. Genetics 40, 421-428.

Scheffer, R.P. (1976). Host specific toxins in relation to pathogenesis and disease resistance. In 'Encyclopedia of Plant Physiology, Vol. 4. Physiological Plant Pathology' (Heitefuss, R. & Williams, P.H., eds), pp. 247-269. Springer Verlag, Berlin.

Schertz, K.F. & Tai, Y.B. (1969). Inheritance of reaction of Sorghum bicolor to toxin produced by Periconia circinata. Crop Science 9, 621-624.

Schnathorst, W.C. & DeVay, J.E. (1963). Common antigens in Xanthomonas malvacearum and Gossypium hirsutum and their possible relationship to host specificity and disease resistance. Phytopathology 53, 1142.

Schwinghamer, E.A. (1959). The relation between radiation dose and the frequency of mutations for pathogenicity in Melampsora lini. Phytopathology 49, 280-289.

Scott, P.R., Benedikz, P.W. & Cox, C.J. (1982). A genetic study of the relationship between height, time to ear emergence and resistance to Septoria nodorum in wheat. Plant Pathology 31, 45-60.

Scott, P.R. & Hollins, T.W. (1977). Interactions between cultivars of wheat and isolates of Cercosporella herpatrichoides. Transactions of the British Mycological Society 69, 397-403.

Scott, P.R., Johnson, R., Wolfe, M.S., Lowe, H.J.B. & Bennett, F.G.A. (1980). Host specificity in aerial parasites in relation to their control. In 'Applied Biology V' (Coaker, T.H., ed.), pp. 350-393. Academic Press, London & New York.

Segal, A., Manisterski, J., Fischbeck, G. & Wahl, I. (1980). How plant populations defend themselves in natural ecosystems. In 'Plant Disease, an Advanced Treatise' (Horsfall, J.G. & Cowling, E.B., eds), pp. 76-98. Academic Press, New York.

Sharp, E.L., Sally, B.K. & Taylor, G.A. (1976). Incorporation of additive genes for stripe rust resistance in winter wheat. Phytopathology 66, 794-797.

Sharp, E.L. & Volin, R.B. (1970). Additive genes in wheat conditioning resistance to stripe rust. Phytopathology 60, 1146-1147.

Shepherd, K.W. & Mayo, G.M.E. (1972). Genes conferring specific plant disease resistance. Science 175, 375-380.

Sidhu, G.S. (1975). Gene-for-gene relationships in plant parasitic systems. Science Progress, Oxford 62, 467-485.

Sidhu, G.S. (1979). Genetic resistance and disease complexes. Proceedings of Symposium IX,

International Congress of Plant Protection, Vol. 1. 182-186.

Sidhu, G.S. (1984). Parasitic epistasis. Phytopathology 74, 382-384.

Sidhu, G.S. & Person, C. (1971). Genetic control of virulence in Ustilago hordei. II. Segregations for higher levels of virulence. Canadian Journal of Genetics and Cytology 13, 173-178.

Sidhu, G.S. & Person, C. (1972). Genetic control of virulence in Ustilago hordei. III. Identification of genes for host resistance and demonstration of gene-for-gene relations. Canadian Journal of Genetics and Cytology 14, 209-213.

Sidhu, G.S. & Webster, J.M. (1977). Genetics of single and complex host-parasite interactions. In 'Induced Mutations Against Plant Diseases' (International Atomic Energy Agency, ed.), pp. 59-79. IAEA, Vienna.

Sidhu, G.S. & Webster, J.M. (1979). Genetics of tomato resistance to the Fusarium - Verticillium complex. Physiological Plant Pathology 15, 93-98.

Sidhu, G.S. & Webster, J.M. (1983). Horizontal resistance in tomato against the Meloidogyne - Fusarium complex: an artefact of parasitic epistasis. Crop Protection 2, 205-210.

Simons, M.D. (1972). Polygenic resistance to plant disease and its use in breeding resistant cultivars. Journal of Environmental Quality 1, 232-240.

Simons, M.D. & Michel, L.J. (1964). International register of pathogenic races of Puccinia coronata var. avenae. Plant Disease Reporter 48, 763-766.

Skipp, R.A. & Samborski, D.J. (1974). The effect of the Sr6 gene for host resistance on histological events during the development of stem rust in near-isogenic wheat lines. Canadian Journal of Botany 52, 1107-1115.

Skovmand, B., Roelfs, A.P. & Wilcoxon, R.D. (1978). The relationship between slow-rusting and some genes specific for stem rust resistance in wheat. Phytopathology 68, 491-499.

Slesinski, R.S. & Ellingboe, A.H. (1969). The genetic control of primary infection of wheat by Erysiphe graminis f. sp. tritici. Phytopathology 59, 1833-1837.

Slesinski, R.S. & Ellingboe, A.H. (1970). Gene-for-gene interactions during primary infection of wheat by Erysiphe graminis f. sp. tritici. Phytopathology 60, 1068-1070.

Slesinski, R.S. & Ellingboe, A.H. (1971). Transfer of ^{35}S from wheat to the powdery mildew fungus with compatible and incompatible parasite/host genotypes. Canadian Journal of Botany 49, 303-310.

Stakman, E.C. (1914). A study in cereal rusts: physiological races. Minnesota Agricultural Experimental Station Bulletin No. 138. 56 pp.

Stakman, E.C. & Levine, M.N. (1922). The determination of biologic forms of Puccinia graminis on Triticum spp. Minnesota Experimental Station Technical Bulletin No. 8, 10 pp.

Stakman, E.C., Levine, M.N. & Cotter, R.U. (1930). Origin of physiological forms of Puccinia graminis through hybridisation and mutation. Scientific Agriculture 10, 707-720.

Stakman, E.C. & Piemeisel, F.J. (1917). Biologic forms of Puccinia graminis on cereals and grasses. Journal of Agricultural Research 5, 429-495.

Stakman, E.C., Piemeisel, F.J. & Levine, M.N. (1918). Plasticity of biologic forms of Puccinia graminis. Journal of Agricultural Research 15, 221-150.

Starling, T.M., Pineda, C.R., Kuo-chun Chen & Moseman, J.G. (1963). Loci of genes conditioning resistance in several barley varieties to race 9 of Erysiphe graminis f. sp. hordei. Crop Science 3, 151-154.

Statler, G.D. (1977). Inheritance of virulence of culture 73-47 of Puccinia recondita. Phytopathology 67, 906-908.

Statler, G.D. (1979). Inheritance of pathogenicity of culture 70-1, Race 1 of Puccinia recondita tritici. Phytopathology 69, 661-663.

Statler, G.P. & Gold, R.E. (1980). Comparative virulence of basidiospores and uredospores of three races of Melampsora lini. Phytopathology 70, 528-530.

Steffenson, B.J., Wilcoxon, R.D. & Roelfs, A.P. (1984). Inheritance of resistance to Puccinia graminis f. sp. secalis in barley. Plant Disease 68, 762-763.

Stevenson, F.J., Schultz, E.S., Clark, C.F., Cash, L. & Bonde, R. (1937). Breeding for resistance to late blight in the potato. Phytopathology 27, 1059-1070.

Stuckey, R.E. & Ellingboe, A.H. (1974). Elongation of secondary hyphae of Erysiphe graminis f.

sp. tritici in wheat with compatible and incompatible parasite/host genotypes. Phytopathology 64, 530-533.

Stuckey, R.E. & Ellingboe, A.H. (1975). Effect of environmental conditions on ^{35}S uptake by Triticum aestivum and transfer to Erysiphe graminis f. sp. tritici during primary infection. Physiological Plant Pathology 5, 19-26.

Taylor, J.D. (1972). Races of Pseudomonas pisi and sources of resistance in field and garden peas. New Zealand Journal of Agricultural Research 15, 441-447.

Taylor, J.D., Vivian, A. & Malik, A.N. (1984). Race structure in Pseudomonas syringae pv. pisi. Proceedings of the 2nd Pseudomonas Working Group, Sounion, Greece, 90-91.

Tepper, C.S. & Anderson, A.J. (1984). The genetic basis of plant-pathogen interaction. Phytopathology 74, 1143-1145.

The, T.T. & McIntosh, R.A. (1975). Cytogenetical studies in wheat. VIII. Telocentric mapping and linkage studies involving Sr22 in chromosome 7AL. Australian Journal of Biological Science 28, 531-538.

Torp, J. & Anderson, B. (1982). Two dimensional electrophoresis of proteins from cultures of Erysiphe graminis f. sp. hordei. Physiological Plant Pathology 20, 349-357.

Toxopeus, H.J. (1959). Notes on the inheritance of field resistance of the foliage of Solanum tuberosum to Phytophthora infestans. Euphytica 8, 117-124.

Vanderplank, J.E. (1963). 'Plant Diseases: Epidemics and Control.' Academic Press, New York & London. 349 pp.

Vanderplank, J.E. (1968). 'Disease Resistance in Plants.' Academic Press, New York & London. 206 pp.

Vanderplank, J.E. (1971). Stability of resistance to Phytophthora infestans in cultivars without R genes. Potato Research 14, 263-270.

Vanderplank, J.E. (1975). 'Principles of Plant Infection.' Academic Press, New York & London. 216 pp.

Vanderplank, J.E. (1978). 'Genetic and Molecular Basis of Plant Pathogenesis.' Springer Verlag, Berlin. 167 pp.

Vanderplank, J.E. (1982). 'Host-Pathogen Interactions in Plant Disease.' Academic Press, New York & London. 207 pp.

Vanderplank, J.E. (1984). 'Disease Resistance in Plants' (2nd Edition). Academic Press, New York & London, 194 pp.

Van der Zaag, D.E. (1959). Some observations on breeding for resistance to Phytophthora infestans. European Potato Journal 2, 268-286.

Walker, J.C. (1930). Inheritance of Fusarium resistance in cabbage. Journal of Agricultural Research 40, 721-745.

Walker, J.C. (1969). 'Plant Pathology' (3rd Edition). McGraw Hill, New York, 819 pp.

Wallwork, H. & Johnson, R. (1984). Transgressive segregation for resistance to yellow rust in wheat. Euphytica 33, 123-132.

Watson, I.A. (1957). Mutation for increased pathogenicity in Puccinia graminis var. tritici. Phytopathology 47, 507-509.

Watson, I.A. & Luig, N.H. (1968). Progressive increase of virulence in Puccinia graminis f. sp. tritici. Phytopathology 58, 70-73.

Wicks, Z.W. & Hammond, J.J. (1978). Screening of flax species for new sources of resistance to Melampsora lini (Ehrenb.) Lev. Crop Science 18, 7-10.

Wilcoxon, R.D. (1981). Genetics of slow rusting in cereals. Phytopathology 71, 989-993.

Wilkinson, D.R. & Hooker, A.L. (1968). Genetics of reaction to Puccinia sorghi in ten corn inbred lines from Africa and Europe. Phytopathology 58, 605-608.

Williams, M.D., Gough, F.J. & Ronder, M.R. (1966). Interaction of pathogenicity genes in Puccinia graminis f. sp. tritici and reaction genes in Triticum aestivum ssp. vulgare 'Marquis' and 'Reliance'. Crop Science 6, 245-248.

Wimalajeewa, D.L.S. & DeVay, J.E. (1971). The occurrence and characterization of a common antigen relationship between Ustilago maydis and Zea mays. Physiological Plant Pathology 1, 523-535.

Wise, R.P., & Ellingboe, A.H. (1983). Infection kinetics of Erysiphe graminis f. sp. hordei on barley with different alleles at the Ml-a locus. Phytopathology 73, 1220-1222.

Wolfe, M.S. (1972). The genetics of barley mildew. Review of Plant Pathology 51, 507-522.

Wong, L.S.L., McKenzie, R.I.H., Harder, D.E. & Martens, J.W. (1983). The inheritance of resistance to Puccinia coronata and floret characteristics in Avena sterilis. Canadian Journal of Genetics and Cytology 25, 329-335.

Yang, S.L., Moseman, J.G. & Ellingboe, A.H. (1972). The formation of elongating secondary hyphae of Erysiphe graminis and the segregatioon of Ml genes in barley. Phytopathology 62, 1219-1223.

Yoder, O.C. (1973). A selective toxin produced by Phyllosticta maydis. Phytopathology 63, 1361-1366.

Zadoks, J.C. (1961). Yellow rust in wheat. Studies in epidemiology and physiologic specialization. Tijdschrift over Pflantezeikten 67, 69-256.

Zadoks J.C. & Schein, R.D. (1979). 'Epidemiology and Plant Disease Management.' Oxford University Press, New York. 427 pp.

Zimmer, D.E. & Comstock, V.E. (1973). New genes for rust resistance in flax. Phytopathology 63, 777-780.

Zimmer, D.E., Schafer, J.F. & Patterson, F.L. (1963). Mutations for virulence in Puccinia coronata. Phytopathology 53, 171-176.

Zimmer, D.E., Schafer, J.F. & Patterson, F.L. (1965). Nature of fertilisation and inheritance of virulence in Puccinia coronata. Phytopathology 55, 1320-1325.

CHAPTER 6. MECHANISMS INVOLVED IN GENETICALLY CONTROLLED RESISTANCE AND VIRULENCE: VIRUS DISEASES

R. S. S. FRASER

6.1. INTRODUCTION

Plants with heritable resistance to a virus normally affecting the species may display a wide variety of visible resistance phenotypes. Resistant plants can be entirely symptomless; can localize the virus in necrotic or chlorotic lesions, or can allow systemic spread of the virus but with reduced symptom expression. Virulent isolates of the virus, which overcome the specific resistance, can produce symptoms similar to those produced by common isolates on susceptible plants, or may be visibly different. In some cases, acquisition of virulence may be accompanied by other alterations in virus behaviour. The objectives of this Chapter are to discuss our present understanding of the physical, biochemical and physiological mechanisms involved in the action of resistance genes, and to consider the biochemical changes in viral functions which allow the virus to overcome a specific host resistance function.

A resistance mechanism may inhibit virus multiplication, and indeed does so in the vast majority of cases. The term multiplication is interpreted here in the broadest sense, to mean increase of virus in time and space. Thus inhibition of multiplication may operate at the cellular, plant or crop level. Resistance expressed as inhibition of development of pathogenic effects, such as visible disease symptoms and loss of yield, may be a secondary consequence of reduction of virus multiplication and spread, or may be an independent effect of the resistance gene.

A prime objective in studies of resistance gene function is to identify the target of the active product of the gene, i.e. the stage of virus multiplication or pathogenesis which is directly inhibited. Conversely, a prime objective in studies of virulence is to identify the biochemical change in the virus which alters the interaction with the host resistance gene product at the target site. The simpler plant viruses can now be completely characterized at certain biochemical levels: their genomes can be sequenced; translational strategies and products identified, and biochemically characterized. This now offers the attractive possibility of explanation of a biological property - virulence - in terms of alterations in the functional properties of the participating molecules.

The targets of host resistance may be classified in two broad groups. There are mechanisms which reduce the spread of virus through the crop, by restricting the chance of any individual plant becoming infected, or by slowing the rate of development of infection in the crop. These mechanisms may have purely physical, structural or chemical bases, and include the effect known as 'disease avoidance'. The second group of mechanisms involves antiviral function within the plant, and may represent direct biochemical inhibition of some phase of the virus replicative cycle, or of symptom development. In the following sections we consider examples of these types of mechanism, and in appropriate cases look at possible mechanisms of virulence.

The great structural diversity of plant viruses; the visible diversity of resistance mechanisms, and the diversity of genetic control of resistance and virulence discussed in Chapter 4, make it reasonable to expect that the mechanisms of resistance gene operation might be correspondingly diverse.

6.2. DISEASE AVOIDANCE

6.2.1. Mechanisms based on resistance to the vector. Disease avoidance mechanisms frequently involve vector-transmitted viruses, and are based on a variety of effects which act directly on the vector, not on the virus. Thus anything which makes the plant resistant to a vector may confer a degree of indirect resistance to viruses transmitted by that vector. Conversely, a change in the virus or any extraneous effect on it which alters its pattern of transmission may alter its pathogenicity or interaction with host species. For example, TBRV exists in forms which can, and cannot, be transmitted by nematodes. Experiments with pseudorecombinants between the two forms showed that nematode-transmissibility depended on the presence of RNA-2 of the transmissible strain (Hanada & Harrison, 1977). Some viruses are only transmitted by aphid vectors when acquired by the insect from a plant which is also infected by a helper virus (Murant, 1978; Rochow, 1982). It is thought that the RNA of the non-transmitted virus is heterologously encapsidated in the coat protein of the transmitted virus. These effects are in some ways analogous to the presence or absence of a virulence gene which directly affects that interaction with a host resistance gene: the end effects on the plant-virus interaction may be similar.

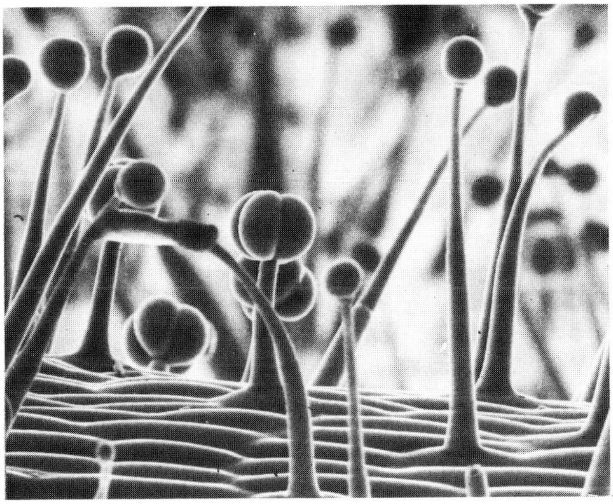

FIGURE 6.1. Scanning electron micrograph of a Solanum berthaultii leaf surface, showing two types of glandular hair. Type A has four lobes, and the taller Type B terminates in a single sticky droplet. Photograph by courtesy of Dr R A Gibson, Rothamsted Experimental Station.

Bar shows 100 μ

Plant resistance to the vector may be expressed without contact being made. Host features such as colour and hairiness may affect whether aphids and other flying vectors settle (reviewed in Gibson & Plumb, 1977). Aphids appear to prefer to settle on yellow rather than on green or red targets (Zimmerman-Gries, 1979; Hodgson, 1981) and it is possible that some degree of difference in resistance of cultivars to virus infection could reside in their colour differences.

The leaves of the wild potato Solanum berthaultii bear two types of hair. Type A hairs are glandular and release a sticky substance which traps aphids. Type B hairs are longer and entangle the insect (Tingey & Laubengayer, 1981) (Figure 6.1). Initially, it was thought that these physical effects were responsible for the resistance of S. berthaultii to a wide range of insect pests. However, an intriguing recent development is the report by Gibson & Pickett (1983) showing that an exudate from the hairs - probably the B type - releases the aphid alarm pheromone (E)-β-farnesene. This compound is normally produced in the cornicle droplets of aphids, for example when attacked by a predator. Other aphids sense the pheromone by receptors located on the antennae, and display various alterations in behaviour pattern, including movement away from a point source of the pheromone (Nault & Montgomery, 1977). The amounts of alarm pheromone released by leaves of S. berthaultii, and the concentration in the air around plants, were sufficient to disperse settled aphid colonies, and to cause avoidance of plants by passing insects (Gibson & Pickett, 1983). In laboratory experiments with the pheromone and its derivatives, acquisition and transmission of BYV and PVY by aphids were decreased (Dawson et al., 1982).

Further mechanisms of resistance may operate once the vector has contacted the plant. For example, Tarn & Adams (1982) postulated that host selection or rejection by aphids could be controlled by chemoreceptors on the tip of the labium, which respond to chemicals dissolved out of the cuticle by the insect's saliva. Thus on potato lines which were unacceptable hosts for Myzus persicae, the insect salivated but never completed penetration of the phloem by its stylet, and did not ingest (Adams & Wade, 1976). This type of mechanism of resistance to the aphid would be likely to prevent transmission of some viruses, especially those limited to the phloem.

6.2.2. Resistance to infection. There are several examples of resistance mechanisms which prevent or reduce the likelihood of the initial infection. The tobacco line T.I. 245 shows resistance to infection by several mechanically-transmitted viruses; these resistances were shown to be inherited concomitantly (Holmes, 1961). In a comparison of T.I. 245 and a susceptible line, Thomas & Fulton (1968a) reported a correlation between varietal differences in numbers of ectodesmata in the outer walls of the epidermal cells, and resistance to infection. Furthermore, experimental treatments of T.I. 245 which increased susceptibility to infection also increased numbers of ectodesmata. Thomas & Fulton therefore suggested that ectodesmata functioned as infection sites, and that their low frequency in T.I. 245 was responsible for its resistance to infection. Further experiments (Thomas & Fulton, 1968b) also suggested that cell-to-cell spread of viruses was slower in T.I. 245. The basis of this effect was not investigated, although it was shown that acquired systemic resistance (see Chapter 9) could not be induced in T.I. 245.

The resistance to infection of T.I. 245, and the indirect resistance to viruses through resistance to their vectors (section 6.2.1) are in some ways the best examples of horizontal resistance to viruses in plants. They meet many of Vanderplank's (1968) criteria of horizontal resistance. In particular, they are effective against all isolates of the pathogen, indeed against several pathogens, and they are unlikely to be overcome by virulent isolates of the viruses. The genetic bases of these effects are unknown at present, but might well involve complex determinants of morphogenesis.

Two further examples of mechanisms which appear to operate by preventing establishment of infection concern vector-transmitted viruses. Interestingly, in both systems the resistant plants were also less attractive to the vectors than susceptible plants. Thomas and Martin (1971a) showed that leafhoppers spent less time on tomato plants resistant to CTV than on susceptible plants, but further experiments suggested that resistance was principally because virus delivered by the vector failed to establish infection as often as in susceptible plants (Thomas & Martin, 1971b; 1972).

Lecoq et al. (1979) and Pitrat & Lecoq (1980) found that there was a relationship between non-preference of CMV-resistant melon (CS-melon) by Aphis gossypii and resistance to transmission. This resistance is controlled by a single dominant gene; is effective against all CMV isolates, and indeed also to transmission of WMV1 and WMV2 by A. gossypii (Lecoq et al., 1980). A separate system of resistance to CMV in melon, based on three recessive genes, is virus-strain specific (Karchi et al., 1975) and does not appear to act at transmission or establishment of infection. CMV is stylet-borne. A. gossipyii will probe CMV-infected CS-melon plants containing the single dominant resistance gene, and acquire the virus, which it can then transmit to susceptible cultivars, but not to healthy resistant cultivars. Other species of aphid can transmit CMV to the resistant cultivar. The basis of this intriguing resistance to transmission by A. gossypii, and the significance if any of the association with non-preference, remain to be discovered.

A special case of resistance to infection concerns the single recessive gene which confers resistance to seed transmission of BSMV in barley (Carroll et al., 1979).

6.2.3. Age effects: mature plant resistance. Plants of different ages or at different developmental stages may vary considerably in their susceptibility to infection or to the rate of development of disease. In particular, older plants have been described as supporting lower levels of virus multiplication, or developing less severe symptoms.

In the tobacco cv. Xanthi-nc, which forms necrotic local lesions after inoculation with TMV, leaves on old flowering plants each formed an average of 10 lesions, when inoculated with a concentration of TMV sufficient to produce several hundred lesions per leaf on younger plants (Fraser, 1981). This may have been partly caused by the fact that older leaves are tougher and more difficult to wound during inoculation than young leaves.

Analogous results have been reported for systemically-spreading viruses. Venekamp & Beemster (1980a; 1980b) and Venekamp et al. (1980) showed that mature potato plants were more resistant to PVX and PVY, and associated these changes with reduction in leaf ribosome content during maturation. Fraser

(1972) showed that the accumulation of TMV RNA in systemically-infected tobacco leaves decreased with increasing age of the leaf at the time of inoculation. The ribosomal RNA content of the tobacco leaf began to decline when leaf expansion was complete. The amount of TMV RNA accumulated was equal to the amount of ribosomal RNA degraded during the period of virus multiplication: this suggests that the lower multiplication of TMV when older leaves were infected was simply a result of reduced precursor supply.

These examples of mature plant resistance to systemically-spreading viruses are perhaps better regarded as mechanisms involving altered ability of the host to support virus multiplication, rather than as expression of a direct antiviral activity.

6.3. RESISTANCE OPERATING WITHIN THE PLANT: VIRUS LOCALIZING MECHANISMS

Many examples of resistance to viruses operate by preventing spread of the virus from the site of initial infection. Genetically, these are controlled by dominant, or less frequently by dosage-dependent genes. Necrotic localization is rarely, if ever, controlled by recessive genes (Chapter 4). Visible lesions form the basis for the classical bioassay method for many plant viruses: the infectivities of different virus preparations can be compared by measuring the numbers of lesions produced when they are inoculated to a localizing host.

Undoubtedly, the necrotic lesion response to TMV in *Nicotiana* species has been the most studied of all localizing mechanisms. Holmes (1938) identified a gene (*N*), in *N. glutinosa*. It has since been transferred to *N. tabacum* cvs such as Samsun-*NN* and Xanthi-nc. An advantage of these cvs for studies of the mechanisms of localization is that nearly isogenic lines (Samsun-*nn* and Xanthi) are available, which carry the *nn* genes, allowing TMV to spread systemically.

Under normal conditions, the *N* gene is effective against all TMV isolates so far tested. Some other localizing genes, in tobacco or in other species, only localize certain isolates of the relevant virus. For example, the *N'* gene from *N. sylvestris* (Valleau, 1943), which is thought to be allelomorphic with *N*, allows certain isolates of TMV to spread systemically without necrosis.

Some viruses, such as TNV and certain ringspot viruses, form predominantly necrotic or localized infections on all, or almost all their known natural and experimental hosts (e.g. Tomlinson *et al.*, 1983). Whether such localizing mechanisms should be regarded as 'resistance', in the absence of the completely 'susceptible' reaction, is a matter of semantics. What is important is that they may involve localization mechanisms similar to those of true resistance.

6.3.1. The relationship between virus localization and necrosis. Necrosis, or hypersensitivity, frequently results when resistant plants are infected by fungi or bacteria. There is considerable argument about whether the pathogen is actually localized as a consequence of cell death, or whether the necrosis is a secondary effect. The evidence is reviewed in Chapter 7. With localizing resistance to virus infection, the position is considerably clearer: the weight of the evidence suggests that necrosis is not centrally involved in restriction of virus spread.

Firstly, in many cases where the virus does not spread far from the site of infection, localization is achieved without any necrosis. The extreme example is restriction of TMV multiplication in cowpea to those cells which are directly inoculated (Sulzinski & Zaitlin, 1982). Virus may also spread sufficiently to form a detectable chlorotic or starch lesion (Kim, 1970; Cohen & Loebenstein, 1975; Bailiss *et al.*, 1977), but be unable to spread systemically.

Secondly, virus is frequently detectable in tissue surrounding the necrotic region in hypersensitive responses, either by electron microscopy (Hayashi & Matsui, 1965; da Graca & Martin, 1976) or by radiochemical or immunological methods (Konate *et al.*, 1983). The last authors showed that the concentration of TMV in non-necrotic cells surrounding lesions on tobacco leaves was much lower than in cells of a systemically-susceptible cultivar, and that TMV concentration declined with distance from the lesion. They concluded that with distance from the lesion centre, cells became more and more resistant to penetration by or multiplication of TMV.

Thirdly, it is possible to reduce the amount of necrosis without apparently causing a concomitant reduction in virus spread or multiplication, for example by treatment with cytokinins (Balazs *et al.*, 1976); with the antibiotic citrinin (antimycin) (Takusari & Takahashi, 1979), or as a result of a previous infection elsewhere on the plant (Fraser, 1979). Local lesions formed on TMV-infected callus cultures from Xanthi-nc tobacco, but these had little or no inhibitory effect on TMV synthesis (Beachy & Murakishi, 1973).

Fourthly, there are cases where necrosis only results when virus spread or multiplication are sufficient; in situations where virus multiplication is strongly inhibited little or no necrosis is seen. Thus in tomato plants homozygous for the *Tm-2* gene for TMV resistance, virus multiplication was inhibited by 99% and no necrosis occurred. In plants heterozygous for Tm-2, TMV multiplication was not so strongly inhibited, and necrotic areas developed on inoculated leaves (Cirulli & Alexander, 1969; Fraser & Airey, 1980). In *Phaseolus* beans containing the *I* gene for BCMV resistance, many strains of virus are strongly localized with little or no necrotic reaction. However, strains of the virus which can overcome the *I* gene spread systemically, and elicit systemic necrosis (Drijfhout, 1978).

These results tend to suggest that localization and necrosis are processes which are at least partly separable, and that necrosis does not play a central role in preventing virus spread from the edge of the infected area. This raises the question of what the significance of necrosis is. Its association with so many localizing resistance mechanisms would tend to support some useful role. One possibility is that necrosis might prevent the localized virus serving as inoculum for further infections. Virus crystals are detectable in necrotic cells by electron microscopy (Milne, 1966; da Graca & Martin, 1975) so cell death does not involve complete destruction of virus. However, the oxidation of phenolic compounds to quinones (Tanguy & Martin, 1972), and production of other toxic products during necrogenesis, might destroy the infectivity of virus particles in necrotic cells.

6.3.2. *Kinetics and conditions of localization*. Commonly, several hundred cells become infected before localization and necrosis become fully effective. Generally, localization appears to be an induced response, in that the mechanism is not present and active in healthy plants. This is supported by evidence showing that the early processes of TMV multiplication in Samsun *NN* and *nn* tobacco are quantitatively similar (Taniguchi, 1963; Otsuki *et al*., 1972; Takahashi, 1973). Furthermore, many viruses which elicit a localizing response, or which fail to cause detectable infection in leaves, can multiply well in isolated protoplasts (Otsuki *et al*., 1972; Beier *et al*., 1977; 1979; Loebenstein *et al*., 1982), although there are exceptions (e.g. Beier *et al*., 1977). Infection of protoplasts is synchronous, and only one cycle of virus multiplication is thought to occur. This probably gives less time for induction of effective resistance than in leaves, where successive cycles of replication occur as a result of progeny virus particles invading further cells. Finally, some experiments showing impairment of localization by inhibitors of transcription and translation (reviewed in section 6.3.4) also suggest that localization is an induced response. An exception to this conclusion is the severe localization of TMV in cowpea leaves (Sulzinski &

Zaitlin, 1982), where the prevention of virus spread from directly infected cells is effective so quickly as to suggest that the resistance to spread is probably not induced, but present in healthy leaves.

Generally, localization becomes more effective with time: necrosis appears a few days after inoculation and lesions grow in size for some days afterwards. The kinetics of lesion development depend on plant age, environmental conditions and virus strain (Rappaport & Wildman, 1957; Takahashi, 1974; Weststeijn, 1981). Mostly, lesions reach a final size and the virus is completely localized, but in some cases, the virus can continue to spread (Carroll, 1966), albeit at a much slower rate than in a systemically-infected host. In the case of resistance to TMV in tobacco, lesion growth ceased ten days after infection in Xanthi-nc plants containing the N gene, but continued indefinitely in White Burley plants with the N' gene (Fraser, 1983).

A further general conclusion is that virus localizing mechanisms tend to be temperature sensitive (Cirulli & Alexander, 1969; Van Loon, 1975; Drijfhout, 1978; Weststeijn, 1981; Fraser, 1983), with systemic spread of virus occurring without necrosis at temperatures above about 30°C. The failure of localization at high temperatures might be because virus multiplication is accelerated, allowing the virus to spread faster than the resistance mechanism could contain it. However, necrosis and resistance are normally restored when plants are transferred to a lower temperature (Weststeijn, 1981; Van Loon & Antoniw, 1982), so it is more likely that the resistance mechanism itself is disabled at the high temperature. However, there is also evidence that enhanced virus multiplication can switch off the necrotic reaction. Weintraub et al. (1961) showed that when clones of Dianthus barbatus normally hypersensitive to carnation mosaic virus were grafted on to infected susceptible clones, a systemic infection without necrosis resulted. This implies that a high level of inoculum may overcome resistance, and that under these circumstances the necrotic response is not triggered.

The general lack of inhibition of virus multiplication in protoplasts isolated from resistant plants, and their failure to become necrotic, led to suggestions that inhibition of multiplication and necrosis in the leaf might involve responses by cells which are not themselves infected. Supplementary evidence came from experiments showing that when leaves inoculated with TMV (Gulyas & Farkas, 1978) or TNV (Coutts, 1978) were floated on hypertonic solutions, lesion formation and virus multiplication were inhibited. This treatment plasmolysed cells and thus reduced cell-to-cell contact, which was therefore judged necessary for lesion formation. Time-course experiments (Kalpagam et al., 1977) suggested that cell-to-cell contact was only required in the first 24 hours after inoculation.

An alternative explanation of the protoplast experiments is that isolated protoplasts may not give inhibitory or necrotic responses, because sufficient concentrations of the compounds involved do not build up, due to the conditions of culture. Hooley & McCarthy (1980) showed that infected protoplasts of hypersensitive tobacco leaves did secrete toxic material which was detrimental to the survival of further healthy protoplasts. Haberbosch et al. (1981) showed that individual TMV lesions on Xanthi-nc tobacco leaves increased in size discontinuously, with near synchronous collapse of groups of neighbouring cells. Beier et al. (1979) found that CPMV multiplied well in protoplasts of a local lesion cowpea cv. when inoculation was in vitro. However, virus concentration declined with time in protoplasts isolated from inoculated leaves, suggesting that some antiviral effect had been initiated while cell contact was present.

Further evidence suggesting that localization and necrogenesis might involve a cooperative response between different cells in the leaf comes from experiments on the role of the epidermis. Generally, an epidermal cell becomes infected, and virus spreads to mesophyll cells a few hours later (Fry & Matthews, 1963; Takahashi, 1973), although there is evidence that mesophyll cells can sometimes be directly inoculated (Sulzinski & Zaitlin, 1982). Several workers have shown that early removal of the epidermis can prevent lesion formation on both the stripped epidermis and the remaining leaf, even though virus had spread to the mesophyll (Shimomura, 1977; Kasamo & Shimomura, 1977; Coutts, 1980). Later removal of the epidermis had no effect on lesion formation. In two of the reports, there appeared to be an interaction between removal of the epidermis, and effects of kinetin on lesion number (Kasamo & Shimomura, 1977; Coutts, 1980), but no consistent effects emerged.

6.3.3. Ultrastructural changes. Changes in ultrastructure during localization have been reported for a number of host and virus combinations, and many show common features. One difficulty is to decide which changes might be involved in localization, and which are merely secondary consequences of infection and necrosis.

Depending on the host and virus used, virus particles may be difficult to detect in sections from lesions. Crystalline packets of TMV were found in

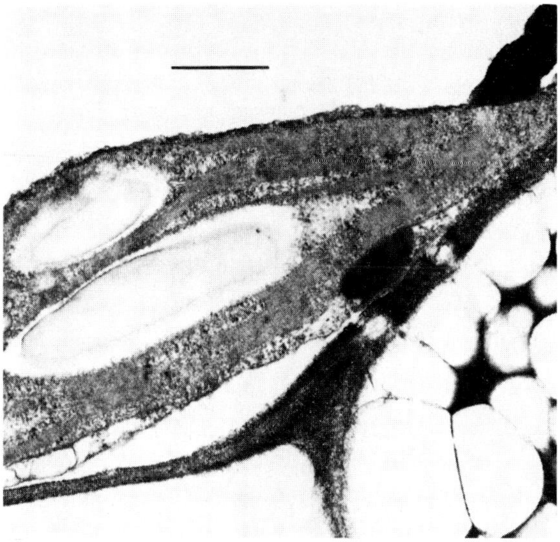

FIGURE 6.2. Electron micrograph of a thin section of tobacco leaf, from an area close to a necrotic lesion caused by TMV. The photograph shows a collapsed cell, and accumulation of starch grains in the chloroplast of the adjacent cell. The bar represents 500 nm. Electron micrograph by C M Clay, National Vegetable Research Station.

completely collapsed cells of N. glutinosa, while dispersed particles only were present in non-collapsed cells from the lesion edge (Milne, 1966; da Graca & Martin, 1976). It was suggested that crystallization occurred as a consequence of necrotization, perhaps because of the accompanying dehydration. Whether TMV was seen in sections from necrotic tissue depended critically on the fixation used (Israel & Ross, 1967). The outer edge of TMV lesions on Nicotiana leaves appears to consist of a mixture of uninfected, infected, and infected semi-collapsed cells. Milne (1966) and da Graca & Martin (1975) found virus in only about 5% of cells examined. Hayashi & Matsui (1965) and Ross & Israel (1970) reported TMV particles in margin cells at various stages of collapse. The evidence that TMV spreads to cells beyond the necrotic region is supported by experiments showing that when leaves with fully-expanded lesions are transferred to high temperatures where localization would not normally occur, the virus is able to spread systemically (Bawdon, 1964; Shimomura, 1971).

Virus particles have also been detected in non-necrotic cells at the lesion boundaries for other hosts and viruses, such as TMV-infected Pinto bean (Stobbs et al., 1977) and TBSV-infected G. globosa (Appiano et al., 1977).

The earliest changes in host components are detectable within a few hours of inoculation. Weintraub & Ragetli (1964) found that starch grains were increased in chloroplasts of TMV-infected N. glutinosa by 8 h, followed by rupture of the chloroplast membranes. Hayashi & Matsui (1965) noted disappearance of stroma in chloroplasts as the first observable change. In Samsun NN tobacco, da Graca & Martin (1975) found that cells with normal tonoplast and plasmalemma had swollen chloroplasts with distorted lamellae and grana of reduced size. The chloroplasts were also discharging ribosomes into the cytoplasm. Other cells had lost tonoplast and plasma membrane. Israel & Ross (1967) reported that in cells surrounding lesions on the same host, the chloroplasts became more numerous, as well as amoeboid in shape. Figure 6.2 shows some typical changes in TMV-infected hypersensitive tobacco.

Pennazio et al. (1979) found rupture of the plasma membrane of TBSV-infected G. globosa at 30 h after inoculation, some 8 h before necrosis became visible. However, damage to the plasma membrane probably started much earlier. Increased electrolyte leakage was detectable by 5 h after inoculation, and rose steadily thereafter.

A notable, and frequently reported feature of cells in the early stage of necrosis, or of those on the lesion edge, is structural evidence of higher metabolic activity. This includes increased number of mitochondria, smaller vacuoles and more cytoplasm, endoplasmic reticulum and ribosomes. (Weintraub & Ragetli, 1964; Hayashi & Matsui, 1965; Appiano et al., 1977). The evidence for increased metabolic activity extended in tobacco for up to 10 cell diameters from the lesion edge (Israel & Ross, 1967).

In the later stages, severe loss of structural integrity accompanies necrosis and cell death. By 7 days after inoculation, Israel & Ross (1967) found that some cells had collapsed, and had no vacuoles or identifiable components, other than dispersed ribosomes and starch grains. The nucleus in collapsing cells appeared to be stable until a comparatively late stage (Weintraub & Ragetli, 1964) and could show enlarged nucleoli (da Graca & Martin, 1965). However, in another study the nuclear ground substance was reported to disappear comparatively early (Hayashi & Matsui, 1965).

The majority of these changes probably represent degenerative effects accompanying necrosis and cell death, rather than having any specific influence on virus localization. However, those changes indicating increased metabolic activity, such as increase in number of mitochondria, might be involved in the activation and synthesis of a defence mechanism. Substances arising from degenerative changes during cell collapse might also serve as triggers for the activation of a localization mechanism in adjoining cells.

Many authors have suggested that localization might occur because of changes in the cell walls or plasmodesmata in the non-necrotic tissue surrounding the lesion, which may prevent cell-to-cell spread of the virus. There is certainly abundant evidence of such changes. Occlusion of plasmodesmata by callose deposition has been reported for many host and virus combinations (Hiruki & Tu, 1972; Shimomura & Dijkstra, 1975; Wu, 1973; Allison & Shalla, 1964, Stobbs et al., 1977). Treatments which prevent callose formation, such as ultraviolet light and darkness, can lead to larger lesions (Wu & Dimitman, 1970). Spencer & Kimmins (1971) reported that the plasmalemma became withdrawn and intercellular connections through plasmodesmata were severed, in cells of TMV-infected Pinto bean. These changes were observed up to three cell diameters from the lesion edge.

Other wall-associated changes include deposition of lignin and suberin on the outer cell walls of TBSV-infected G. globosa (Appiano et al., 1977), and in the intercellular spaces (Pennazio et al., 1978). Paramural bodies - membranous structures and vesicles sited between the plasmalemma and the cell wall, and probably involved in wall modification and deposition of wall material - have been reported frequently in virus-infected plants (Spencer & Kimmins, 1971; Hiruki & Tu, 1972). Faulkner & Kimmins (1978) found deposition of wall material and paramural bodies to be most active 3 - 4 days after infection of Pinto beans with TMV. Lesion growth ceased at this time. SBMV, which induces continuously-growing lesions on this host, caused only small, localized paramural bodies and little wall modification.

Wall changes may also be induced at considerable distance from the lesion. Favali et al. (1978) examined vascular bundles of TMV-infected N. glutinosa and found thickened walls and paramural bodies in the xylem parenchyma; callose-like deposits blocking the sieve plates and the plasmodesmata linking sieve tubes with companion cells, and evidence for lignification of the tracheid lumens. The surrounding mesophyll cells were indistinguishable from those of healthy leaves. They suggested that the changes in the vascular bundle cells might be capable of preventing long-distance spread of virus.

The evidence for involvement of wall changes in localization is quite comprehensive, and many authors have argued the case. However, there is also considerable evidence to the contrary. Russo et al. (1981) did find various thickenings and plugging of plasmosdesmata in cells surrounding BNYVV lesions, but concluded that their location and frequency were such as to exclude them as major factors in localization. Appiano et al. (1977) found infectious TBSV particles in cells of G. globosa leaves which had virus-induced wall thickenings and callose deposits. Pennazio et al. (1978) considered that lignin and suberin deposition in this system occurred too late to explain localization. They also found no evidence for callose or paramural bodies. Pennazio et al. (1981) also concluded that deposition of callose in PVX infections occurred too late to contribute to localization.

In Samsun NN tobacco, TSWV and TBRV both promoted strong deposition of callose around their local lesions, but both also spread systemically (Shimomura & Dijkstra, 1975).

Cucumber cotyledons infected with TMV form non-necrotic starch lesions, in which the virus is localized. Cohen & Loebenstein (1975) found that the cell walls of infected cotyledons were similar to those of healthy plants, with no callose. Plasmodesmata-like structures connected infected cells with neighbouring healthy cells. Kim (1970) also found no apparent wall changes when pokeweed mosaic virus was localized in chlorotic, non-necrotic lesions on *Chenopodium amaranticolor* leaves.

This evidence tends to support the view that in at least some host-virus combinations, gross wall changes are not the principal factor in virus localization. The evidence does not exclude the possibility that other, more subtle wall changes might be involved. Furthermore, the wall changes which do occur could still represent a considerable barrier to virus movement, although they might be indirect consequences of necrosis, rather than the direct mechanism of localization. A full appreciation of the part played by wall changes in localization will depend on a better understanding of how viruses move from cell to cell.

6.3.4. Metabolic changes. The considerable changes in ultrastructure during localization and necrogenesis are matched by equally dramatic changes in many areas of metabolism.

Respiration. Formation of local lesions on leaves of N-gene tobacco plants after inoculation with TMV was accompanied by an increase in the rate of respiration. The increase was detectable before appearance of necrosis (Sunderland & Merrett, 1965), and principally involved activation of glucose-6-phosphate dehydrogenase (Solymosy & Farkas, 1962) and the hexose monophosphate shunt pathway, with corresponding accumulation of pentoses (Solymosy & Farkas, 1963). Leaves bearing lesions had greater concentrations of ATP and ADP than healthy leaves (Sunderland & Merrett, 1964). This evidence of an activation of the energy supply system is consistent with the increase in numbers of mitochondria noted in ultrastructural studies (Weintraub et al., 1972). The areas around lesions were also shown to be active sinks for assimilates by studies of CO_2 fixation (Thrower, 1965), although results from this technique would be distorted by fixation of radioactive assimilate in comparatively permanent form during synthesis of cell wall material around the lesion. All of these studies give evidence for a general activation of metabolism associated with lesion formation, without implying that the changes are in any way associated with necrosis and localization of the virus.

Phenolic compounds and oxidative enzymes. During lesion formation, tobacco leaves accumulate a number of phenolic compounds and derivatives (Tanguy & Martin, 1972). Representative examples are shown in Figure 6.3. Some, such as chlorogenic acid, isochlorogenic acid and scopoletin have been implicated in necrogenesis (Cruikshank & Perrin, 1964). However, Tanguy and Martin (1972) found that the increases occurred only after lesions appeared, and therefore concluded that the phenolics were unlikely to have been responsible for the necrosis, but rather that their production was a secondary effect. Fritig *et al*. (1972) measured changes in concentrations of chlorogenic acid and scopoletin, and also attempted to measure their rates of synthesis, by incorporation of the precursor ^{14}C phenylalanine. Their results suggested that changes in the rate of catabolism of chlorogenic acid and scopoletin, presumed to be by polyphenoloxidase and peroxidase respectively, were important in controlling the concentrations. Furthermore, they suggested that changes both in concentrations and rates of synthesis of the two phenolics were established by an early stage, indeed perhaps before lesions appeared. Thus although this study did not provide positive evidence for a role of phenolics in control of necrosis, the possibility could not be excluded.

The activities of peroxidase and polyphenoloxidase are greatly increased during necrogenesis (Farkas *et al*., 1960; Solymosy *et al*., 1967). This led early workers to suggest that cell death might result from oxidation of phenols to toxic quinones, and that high levels of the enzymes might be involved in reducing the rate of lesion enlargement (Parish *et al*., 1965; Van Loon & Geelen, 1971). Quinones can also inactivate some viruses directly. However, time course studies suggested that the increases in enzyme activities occurred too late to be of major importance in localization and control of the onset of necrosis (Weststeijn, 1976; Fritig *et al*., 1972). Van Loon (1976) concluded that differences in peroxidase activity probably resulted from indirect effects of the physiological state of the plant, rather than being directly responsible for controlling the rate of lesion growth.

Phenylpropanoid metabolism. Phenylalanine is the precursor for a pathway leading to the phenylpropanoids, which are the immediate precursors of lignin (Figure 6.4). In view of ultrastructural evidence suggesting that lesion formation could involve lignification (Appiano et al., 1977; Faulkner & Kimmins, 1975; Vance et al., 1980) and the possibility that increased lignification might represent a barrier to virus movement, much effort has been devoted to studies of changes in activity of the phenylpropanoid pathway during necrotization.

Fritig et al. (1973) showed that higher levels of activity of the first enzyme in the pathway, phenylalanine ammonia lyase (PAL) were detectable before necrotic symptoms appeared. There was a sharp increase in PAL activity at the time of lesion appearance; the additional enzyme was located in a narrow ring around the lesion. Later experiments using density labelling showed that the increased activity was a result of increased synthesis and not activation of existing enzymes (Duchesne et al., 1977).

Subsequent enzymes in the pathway, cinnamic acid-4-hydrolase (CAH) and caffeic acid-O-methyltransferase (OMT), became activated in sequence (Legrande et al., 1976). Legrande et al. (1978) found three distinct forms of OMT which differed in substrate specificity. Two forms of the enzyme showed much greater increases in activity after infection than the third. Density labelling experiments again showed that synthesis of new protein was entirely responsible for the increase in enzyme activity (Collendavelloo et al., 1983).

These findings led Legrande et al. (1978) to suggest that thickening of cell walls could be involved in resistance to virus spread at the edge of the lesion. In support of this idea, α-aminooxyacetate (AOA), a competitive inhibitor of PAL activity, was shown by Massala et al. (1980) to increase the size of TMV local lesions by two to four fold, when applied 6 or more h before lesion appearance. The inhibitor did not suppress the normal increases in activities of the enzymes of the phenylpropanoid pathway after infection, but was shown to inhibit flux through the pathway *in vivo*. Interestingly, TMV was shown to spread to the edges of the larger lesions of AOA-treated leaves, but the total amount of virus per lesion was not increased. Thus the relative amount of virus per unit necrotized area was actually reduced by AOA treatment.

The authors proposed that accumulation of phenylpropanoids and their derivatives interfered with spread of virus from cell to cell, rather than with virus multiplication as such. Their model involves synthesis of eliciting substances in infected cells, and movement of these to neighbouring cells

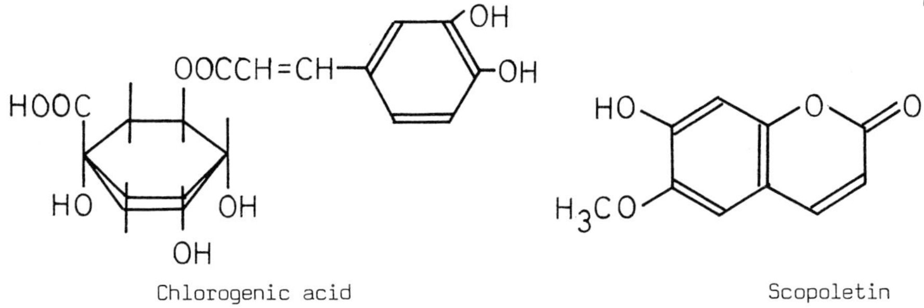

FIGURE 6.3. Phenolic compounds associated with lesion formation.

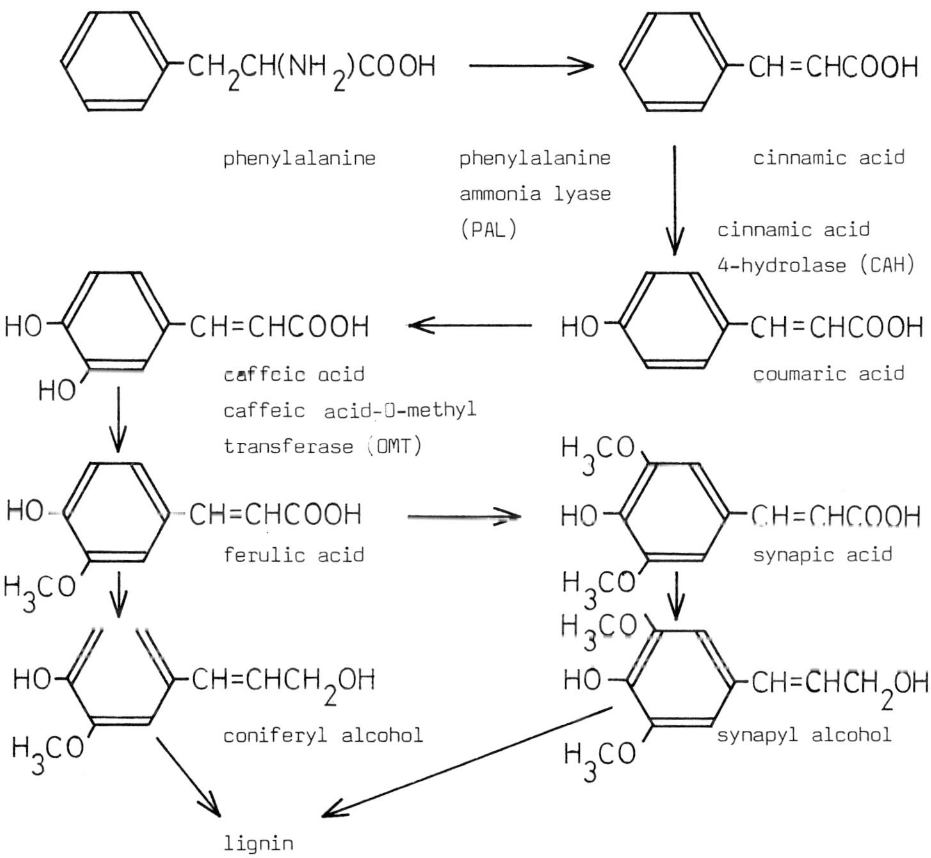

FIGURE 6.4. The phenylpropanoid pathway of lignin biosynthesis

which then become activated to some resistant state. Inhibition of phenylpropanoid synthesis by AOA thus allowed virus to spread to these neighbouring cells, but the authors suggest that the invaded cells quickly became necrotic, thus limiting virus multiplication within them. Interpretation of these experiments is clearly complex, and some possible side-effects of AOA or its action remain to be excluded, but the results do suggest that the phenylpropanoid pathway may play an important, though by no means the only part in regulation of the necrotic response.

Kopp et al. (1979) have shown that when protoplasts were prepared from leaves in which infection had been allowed to develop for some time, no increase in PAL activity was seen, in contrast to the increase found in whole leaves. Although these experiments also present difficulties of interpretation, they are consistent with the idea of diffusable stimulants which elicit PAL synthesis in cells distant from infected ones.

The phenylpropanoid pathway may also serve as a source of precursors for synthesis of aromatic amides. Martin-Tanguy et al. (1976) found that TMV infection of Xanthi-nc tobacco led to production of ferulyl putrescine and coumaryl putrescine in cells around the necrosis. Interestingly, some of these aromatic amides were shown to inhibit TMV multiplication strongly.

Plant growth regulators. Localization, especially if accompanied by necrosis, represents a wound or stress effect, and is thus likely to cause changes in stress-related growth regulator (or hormone) metabolism. Hormones might also be involved as chemical messengers between infected cells and neighbouring cells which show altered patterns of metabolism, whether as part of the localization process or as secondary effects. There has thus been considerable research on changes in hormones during localization, and many experiments have examined the effects of exogenous hormones. Wider aspects of plant growth regulator metabolism in virus-infected plants were reviewed by Fraser & Whenham (1982).

Generally, increased ethylene production appears to be associated with infection leading to necrotic or chlorotic reactions, but not with non-necrotic systemic infections. Ethylene production was stimulated 3 to 13 hours before the onset of necrosis in tobacco infected with TMV or TNV (Gaborjanyi et al., 1971; Pritchard & Ross, 1975; De Laat et al., 1981), although the largest increases occurred after the onset of necrosis. De Laat & Van Loon (1982) showed that the peak in ethylene production was preceded by a large increase in the concentration of the ethylene precursor

1-aminocyclopropane-1-carboxylic acid (ACC); the capacity to convert this to ethylene increased only after lesions became visible. These results raise the question of whether ethylene is involved in promoting the necrotic response, but the evidence is contradictory. In TMV-infected tobacco, lesion size was not altered by treatment with succinic acid 2,2'-dimethylhydrazine, an inhibitor of ethylene production (Pritchard & Ross, 1975). Ethylene production was also inhibited by aminoethoxyvinylglycine, but lesions were still formed (De Laat _et al._, 1981). Treatment of TMV-inoculated leaves with the ethylene-releasing compound 2-chloroethylphosphonic acid (Ethrel, ethephon) reduced the size of lesions formed (Van Loon, 1977). These results argue against a role of ethylene in causing necrosis.

In contrast, pricking healthy leaves with needles moistened with ethephon did cause necrotic spots similar to virus-induced lesions, and also caused similar changes in protein and peroxidase isoenzyme patterns (Van Loon, 1977). In _Tetragonia expansa_, BYMV normally causes chlorotic lesions, but these became necrotic when plants were exposed to ethylene (Bailiss _et al._, 1977). Exposure of plants to CO_2, an inhibitor of ethylene action, prevented both necrosis and chlorosis. These experiments provide evidence for an involvement of ethylene in control of the necrotic response, but do not prove that ethylene and viruses induce necrosis by the same mechanism: much of the stimulation of ethylene production during virus-induced necrosis is likely to be a result of non-specific wounding effects.

Abscisic acid (ABA) concentration was increased by both systemic and local lesion infection of tobacco by TMV, although the necrotic response caused the greater increase, of up to 30-fold (Whenham & Fraser, 1981). The increase in ABA occurred at the onset of necrosis: it is likely that water loss during necrotization, and disruptive effects of infection on the chloroplasts, caused release of ABA from the chloroplasts where it is normally sequestered, and derepressed further synthesis. Spraying tobacco leaves with sufficient ABA to cause a physiologically reasonable increase in ABA concentration caused a reduction in the size of lesions formed after a subsequent TMV infection (Fraser, 1982). It is therefore unlikely that ABA is involved in promotion of necrosis, although it could be involved in stimulating the localization mechanism. Other experiments suggested that ABA treatment increased lesion size, but these involved extremely high ABA concentrations which were shown to be phytotoxic (Balazs _et al._, 1973).

Van Loon & Berbee (1978) showed that appearance of TMV lesions on tobacco leaves was accompanied by an increase in the concentration of the auxin indoleacetic acid. This contrasts with reduced auxin concentration caused by systemic TMV infection (Pavillard, 1952), and suggests that the increase was a consequence of necrosis, rather than a direct response to TMV multiplication. Van Loon (1979) showed that various natural and synthetic auxins could reduce lesion size; indoleacetic acid was also effective in limiting TMV multiplication in young, systemically-infected leaves. He therefore suggested that this auxin might be involved directly and indirectly in stimulating localization, but recognised that additional factors were undoubtedly involved, as added auxins did not stimulate localization in all circumstances. It has been suggested that auxin-stimulated ethylene production might be responsible for inhibition of lesion growth (Van Loon, 1979).

Simons et al. (1972) had earlier shown for this host and virus that the synthetic auxin 2,4-dichlorophenoxyacetic acid (2,4-D) caused reduced lesion size at low concentrations, but when applied at high, near-phytotoxic levels caused an increase in lesion size of up to 15-fold. Loebenstein et al. (1980) found that 2,4-D enhanced TMV multiplication in protoplasts from N-gene plants, but inhibited it in those from n-gene plants. They suggested that the effects of 2,4-D in the incubation medium might explain the discrepancy between the low multiplication of TMV in leaves of N-gene plants, and the uninhibited multiplication found in protoplasts (Otsuki et al., 1972). The results further suggested an interaction between 2,4-D and the mechanism of action of the N-gene. However, the experiments by Loebenstein et al. (1980) were done with sub-optimal levels of infecting virus, and when higher multiplicities of infection were used with N-gene protoplasts, no effects of 2,4-D were found. Furthermore, the effects of 2,4-D on TMV multiplication in protoplasts appeared to depend critically on the way that the plants had been grown prior to protoplast preparation (Loebenstein & Gera, 1981). It thus remains possible that the effects of 2,4-D on TMV multiplication in N-gene cells can be explained at a more trivial level than that of a specific interaction between auxin and the N-gene mechanism.

Studies of changes in cytokinins during localization have generally used bioassays, which may be less than totally reliable for identification and quantification. Both systemic and local lesion infection of tobacco with TMV caused increases in cytokinins thought to include zeatin and zeatin riboside (Sziraki et al., 1980; Sziraki & Gaborjanyi, 1974; Sziraki & Balazs, 1975). There have been numerous reports of alterations in lesion development as a result of application of exogenous cytokinins; unfortunately the effects are highly contradictory and no clear role for cytokinins in the localization mechanism emerges. Generally, cytokinin treatment caused a reduction in lesion growth (Balazs <u>et al</u>., 1976; Selman, 1964) but this was not necessarily associated with any reduction in virus multiplication (Balazs <u>et al</u>., 1976; Milo & Srivastava, 1969). Time of application of the cytokinin relative to inoculation appeared to be important in several studies: either stimulation, or inhibition, of virus multiplication and lesion growth could be obtained with the same virus, depending on whether the cytokinin was applied early or late (Aldwinckle & Selman, 1967; Tavantzis <u>et al</u>., 1979).

Despite the complexities, the effects of exogenous cytokinins on virus multiplication and necrotization can be dramatic, and clearly deserve further study. The availability of new physical methods for the accurate identification and assay of cytokinins in small samples of leaf (Whenham, 1983) should permit more rigorous characterization of changes in endogenous cytokinins during localization and necrotization.

Changes in endogenous gibberellins, and effects of exogenous gibberellins on lesion development do not appear to have been extensively studied.

<u>Cell membranes and permeability changes</u>. Increased electrolyte leakage, indicative of increased permeability of the cell membrane, was found in tobacco infected with TMV (Weststeijn, 1978) and cowpea infected with TRV (Pennazio & Sapetti, 1982). In both cases, increased leakage was detectable in advance of necrosis. Ethylene synthesis is thought to occur at the plasmalemma (Lieberman, 1979); it is interesting that increases in permeability and in ethylene production occurred at roughly the same time.

At 30°C, TMV spreads systemically and does not cause necrosis on tobacco plants containing the <u>N</u> gene. Ruzicska <u>et al</u>. (1983) found that when TMV-infected Xanthi-nc tobacco plants were transferred from 33°C to 25°C to initiate the necrotic response, lipoxygenase activity rose, and a rapid increase in the saturation of phospholipid-bound fatty acids preceeded the increase in electrolyte leakage.

Pennazio & Sapetti (1982) considered that altered permeability was a non-specific marker of necrogenesis, but not directly correlated with virus localization. Further experimental evidence supporting an involvement of membrane changes in necrogenesis comes from experiments with <u>Nicotiana glutinosa</u> leaf discs. Treatment with membrane-damaging agents caused lesion formation, but this could be prevented by treatment with membrane stablizers such as polyamines or calcium ions (Martin & Martin-Tanguy, 1981; Ohashi & Shimomura, 1982).

A much earlier change in membranes after infection was the stimulation of membrane-bound, Mg^{2+}-activated ATPase of Samsun-NN tobacco between 0.5 and 3 h after inoculation with TMV (Kasamo & Shimomura, 1978). Although this change may have been a precursor to further membrane changes involved in necrogenesis, it could not itself have been the critical event triggering necrosis, as a similar response was observed when the systemic cv. Samsun nn was inoculated.

Changes in transcription and translation: inhibitor experiments. Several investigators have tested the effects of inhibitors of transcription or translation on necrogenesis and localization, to determine whether induction of these responses depends on activated expression of host genes. Kuhn (1971) showed that the base analogue 2-thiouracil increased lesion size and infectivity of CCMV lesions on soybeans. Actinomycin-D, an inhibitor of transcription, caused an increase in size and infectivity of TMV lesions on tobacco (Loebenstein et al., 1969). Treatment with ultraviolet light may inhibit transcription, and may specifically inhibit host protein synthesis (Sakai & Takebe, 1974). Ultraviolet light treatment partly suppressed localization of TMV in tobacco (Loebenstein et al., 1970) and in Pinto bean (Wu & Dimitman, 1970).

Sela et al. (1969) found that treatment of Nicotiana glutinosa with chloramphenicol, an inhibitor of protein synthesis, increased lesion size and TMV multiplication. Chloramphenicol affects primarily the chloroplast ribosomes. However, its action on plants is complex, and includes inhibition of processes other than protein synthesis (Ellis, 1981). The experiments of Sela et al. (1969) do not therefore provide strong evidence for involvement of translation at any location in localization of TMV.

A general problem with inhibitor experiments is that it is difficult to be sure that the effect observed is a specific one, and not merely some consequence of general phytotoxicity, or an indirect secondary effect. Interpretation of results may also be confused if the inhibitor interferes directly with virus multiplication. Thus 2-thiouracil actually inhibited infectivity in TMV lesions, in contrast to its effects on CCMV spread (Kuhn, 1971). Higher doses of ultraviolet light reduced infectivity in TMV lesions, probably by direct inhibition of virus multiplication (Loebenstein et al., 1979). Actinomycin-D could induce localized necrosis on leaves systemically-infected with TMV (Ohashi & Shimomura, 1972).

The results from inhibitor experiments are thus contradictory. While they provide some evidence that transcription and translation may be involved in activation of localization, there is a need for examination of the problem at the level of expression of specific genes, rather than at the overall level which is intrinsic to inhibitor experiments. More specific investigations have involved two types of approach: examination of changes in activity or synthesis of particular enzymes or proteins during localization, and attempts to identify new compounds with demonstrable antiviral activity. The latter are considered separately in section 6.3.5.

Protein and enzyme changes. Van Loon (1982) reviewed protein and enzyme changes during lesion formation. Changes in activity, generally increases, were found for many enzymes, especially those with oxidative or degradative functions. Activity of a purine-specific ribonuclease increased after TMV-infection of Xanthi-nc tobacco, but enzyme activity was also increased by mechanical injury (Wyen et al., 1972). There did not appear to be any correlation between virus multiplication and increase in nuclease level. Some other examples of changes in specific enzymes are discussed in detail elsewhere in this Chapter. None of the changes could be given a central role in localization. Most changes in activity appeared to mirror those occurring during senescence, and therefore probably reflected non-specific reactions involved in localized cell death.

The most dramatic change in protein, as opposed to enzyme metabolism, associated with lesion formation, is accumulation of several new proteins (Van Loon & Van Kammen, 1970; Gianinazzi et al., 1970). These proteins share certain properties: they tend to be soluble at low pH (Antoniw & Pierpoint, 1978), and to be highly resistant to degradation by proteolytic ezymes (Pierpoint et al., 1981; Van Loon, 1982). They have been referred to as 'b' proteins (Gianinazzi et al., 1970) but are now generally called 'pathogenesis-related' (PR) proteins (Antoniw et al., 1980). Up to ten PR proteins are associated with hypersensitive TMV infections of tobacco (Van Loon, 1982). Proteins with certain characteristics in common are found in necrotic responses to other viruses and in other hosts (reviewed by Redolfi, 1983).

PR proteins are also induced in uninoculated leaves of infected plants, and this has led several authors to suggest that they may be involved in the induced resistance to a subsequent infection in these leaves (Kassanis et al.,

1974; Gianinazzi, 1982). The possible role of PR proteins in induced resistance, and their biochemical properties, are considered in detail in Chapter 9.

Surprisingly, there have been few experiments testing whether PR proteins are involved in localization of the primary infection. Van Loon & Van Kammen (1970) reported that PR proteins became detectable in Samsun NN tobacco leaves 3 to 4 days after infection, at about the time of the onset of necrosis. It is likely that by this time, considerable localization of virus had already occurred (Taniguchi, 1963; Takahashi, 1973; Loebenstein et al., 1982). Van Loon (1983) has argued a central role for ethylene in the induction of PRs: it is likely that they are induced as a secondary response to the burst of ethylene synthesis occurring with or just before the onset of necrosis.

When Samsun NN plants were grown at 30°C, TMV spread systemically, and no PR proteins accumulated (Van Loon, 1975). Plants grown at 20°C localized the virus and accumulated PR proteins. When such plants were transferred to 30°C, the virus spread systemically, despite the continued presence of PR proteins.

Healthy Xanthi-nc tobacco plants accumulated PR proteins as they began to flower and as the lower leaves senesced (Fraser, 1981). Removal of the developing inflorescence, the lower leaves, or both, altered the amount of PR proteins accumulated in the remaining leaves, over a 20-fold range. The size of TMV lesions formed on subsequent inoculation of the different treatments bore no relationship to the PR protein concentration.

These lines of evidence therefore tend to suggest that PR proteins are not centrally involved in control of necrotization or localization of the primary infection.

6.3.5. Antiviral factors. An alternative to descriptive studies of metabolic changes during localization has been to seek to isolate defence molecules with demonstrable antiviral activities.

Antiviral activity has been measured in various ways. Perhaps the simplest involves mixing purified virus with preparations from infected plants or healthy plants (controls), and recording the reduction in infectivity on local lesion bioassay due to putative antiviral factors in the infected plant sap. Such experiments could, however, indicate an interference with the establishment of infection, such as by altering susceptibility of infection sites, rather than any direct inhibition of virus multiplication.

Demonstration of direct activity against virus multiplication requires measurements of virus concentration or rate of synthesis. The antiviral substance must be shown to have some specificity of action, and not merely to be acting as a non-specific metabolic poison. If possible, the demonstration of antiviral activity should include experiments in vitro on isolated parts of the replicative mechanism.

Further models have suggested that particular antiviral molecules might act as chemical messengers, inducing a direct antiviral activity in uninfected cells. In such cases, there is again a requirement to demonstrate some specificity of action, and also to establish the nature and mode of action of the induced activity.

Many investigators have reported situations where sap from infected plants inhibited viral infectivity more than sap from healthy plants. Inhibitory substances have been detected in leaves of several host species with localized infections caused by a variety of viruses. The inhibitory agents have also been found in uninoculated healthy leaves on infected plants (Sela & Applebaum, 1962; Loebenstein & Ross, 1963; Kimmins, 1969; Nagaich & Singh, 1970). Generally, the inhibitors showed little or no specificity for host species or virus, and in bioassays only inhibited virus infectivity to a maximum of 60 - 80%. Broadly similar inhibitory influences have been reported in systemically-infected plants (Chadha & MacNeill, 1969; Nienhaus & Janicka-Czarnecka, 1981). Uninfected plants have also been shown to contain substances with antiviral activity (Zaitlin & Siegel, 1963; Owens et al., 1973; Awasthi, 1981). Hooker & Kim (1962) found inhibitors of PVX infectivity in leaves of potato cultivars with hypersensitive or completely immune reactions to this virus. However, the inhibitors did not differ in quantity or quality from those in a cultivar allowing systemic spread.

The inhibitory agents in plants have generally not been fully characterized, but have been reported to include RNAs (Kimmins, 1969; Sela et al., 1966); proteins (Zaitlin & Siegel, 1963; Owens et al., 1973; Awasthi, 1981) or low molecular weight phenolics (Schuster & [illegible], 1982). As pointed out in a recent review (Sela, 1981), the same plant might contain several factors with antiviral activity. It is probable that the effects of many of these on virus infectivity will turn out to be trivial and non-specific.

The system which has been most researched and which features most in the literature is undoubtedly resistance to TMV in Nicotiana species containing the N gene. Sela et al. (1964) and Antignus et al. (1977) showed that an antiviral factor (AVF) from TMV-infected leaves of N. glutinosa inhibited TMV multiplication, measured by infectivity bioassay, in leaf discs of Samsun or

Samsun NN tobacco. Later tests showed that purified AVF at 70 ng protein / ml caused 60% inhibition of accumulation of TMV antigen, as measured by ELISA (Reichman et al., 1983). It has been claimed that antiviral activity could be detected at concentrations as low as 0.1 ng / ml, although full experimental data for such situations do not seem to have been published.

AVF has been partially purified by ion-exchange chromatography on DEAE cellulose (Sela et al., 1964) and by affinity chromatography on concanavalin-A-Sepharose, with elution by methyl-glycosides (Mozes et al., 1978). This, and carbohydrate analysis of purified AVF showing the presence of several sugars, suggested that AVF is a glycoprotein. However, most if not all of the sugars detected in AVF were also present in 'mock' AVF prepared from healthy leaves.

Partially-purified AVF fraction from infected leaves was shown to be phosphorylated (Antignus et al., 1975). The molecular weight of the de-phosphorylated protein was estimated to be 22,000 (Mozes et al., 1978). Column elution and gel profiles of AVF, measured either by the various chemical and radiochemical methods, or by biological activity, were generally very broad, or indicated the presence of several types of molecule. Sela (1981) suggested that this indicated heterogeneity and possibly differing levels of phosphorylation and carbohydrate. However, the presence of other proteins cannot be ruled out: the detailed biochemical composition of AVF still contains many areas of uncertainty. Mozes et al. (1978) and Sela (1981) have drawn attention to many similarities in purification methods and properties between AVF and interferon.

In a further series of papers, Sela and co-workers have proposed a complex scheme for the metabolism and mode of action of AVF (Figure 6.5). The scheme for metabolism was based on experiments in which AVF activity was synthesized in vitro using preparations from tobaccos with various genotypes. AVF was detected as activity interfering with TMV infectivity, and by its chemical properties. It was proposed that a gene P, present in both N and n-gene plants, codes for a precursor to AVF (pre-AVF). This is modified by a postulated enzyme N, the product of the N-gene (Sela & Harpaz, 1977; Sela et al., 1978). Phosphorylation appeared to play a part in activation of the enzyme N. AVF itself, although known to be phosphorylated in vivo, was not apparently posphorylated in vitro during activation from pre-AVF, although some other fraction, which could be separated from AVF, was phosphorylated.

It was further suggested that phosphorylation and activation of AVF was

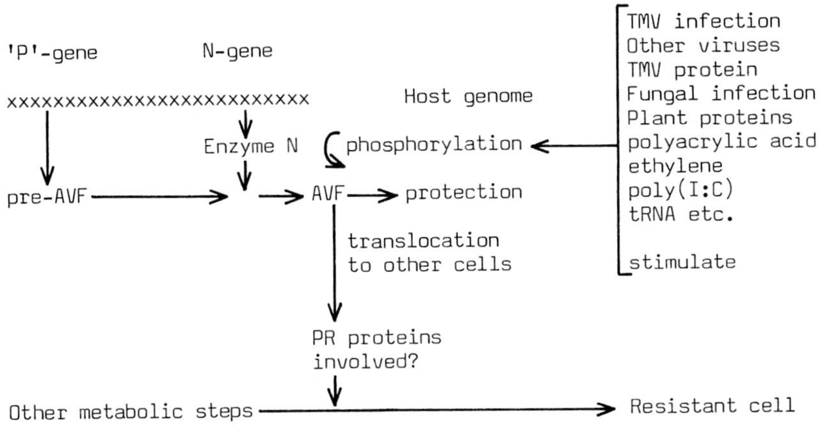

FIGURE 6.5. A putative mechanism for release and activation of antiviral factor (AVF) activity. From Sela, 1981. Reproduced from Advances in Virus Research by permission of Dr I Sela and Academic Press, New York.

promoted by TMV. This was presumed to involve the double-stranded replicative forms of TMV, but could also be stimulated by synthetic, exogenous double-stranded RNA (Gat-Edelbaum et al., 1983), and by all the biological or chemical agents causing acquired systemic resistance (Sela, 1981) (c.f. Chapter 9). The evidence from the in vitro experiments suggested a requirement for cyclic AMP (cAMP) and GMP (Sela et al., 1978; Gat-Edelbaum et al., 1983) for activation. The role of cyclic nucleotides in plants has been the subject of argument, but the evidence suggests that plants may contain all the components required for a cAMP regulatory system (reviewed in Brown & Newton, 1981). Demonstration of an effect of synthetic cyclic nucleotides in an in vitro system does not however provide conclusive proof of their involvement in vivo. The evidence presented by Sela et al. (1978) that cAMP concentration was elevated in TMV-infected plants was obtained by a rather indirect assay method.

Although the scheme in Figure 6.5 attempts to encompass the many complex pieces of experimental evidence on AVF, it has to be stressed that some of the steps are largely conjectural, and that none of the proposed mechanisms or constituents has yet been characterized in detail. The possibility of non-specific effects on infectivity, operating at trivial levels such as increased nuclease activity, are still to be excluded. It is a matter of some concern in this context that the biological activities of the AVF preparations activated *in vitro* rarely exceeded a 30% reduction of infectivity.

Sela's scheme also includes a number of possible ways in which AVF might exert its antiviral activity. Again, some of these are heavily based on the interferon model. A striking parallel between the interferon and AVF systems was drawn when Orchansky *et al.* (1982) demonstrated that mammalian interferon appeared to inhibit TMV multiplication in leaf discs. This result was surprising on two grounds: in mammals, interferons display a high level of species specificity; and in tobacco, antiviral activity seemed to require interferon concentrations much lower than those effective in animals. Reichman *et al.* (1983) suggested that interferon and AVF induced antiviral activity by similar mechanisms. Because of the very low concentrations of interferon and AVF (Mozes *et al.*, 1978) which conferred protection, and because protection may be conferred when leaves are treated with the chemicals before inoculation (Orchansky *et al.*, 1982; Reichman *et al.*, 1983) it has been postulated that both interferon and AVF might act in plants as hormones rather than as direct antiviral principles. It is perhaps remarkable that from the data presented by Reichman *et al.* (1983), human leukocyte γ3 interferon appeared to be over 10^4 times as active as tobacco leaf AVF, on a molecule for molecule basis, in suppression of TMV multiplication in tobacco leaves. This might suggest either that AVF is actually a comparatively inefficient inhibitor, or that AVF preparations are heavily contaminated with inactive compounds. Others have failed to detect activity of mammalian interferon preparations against plant viruses (e.g. Bawden & Pirie, 1959; Antoniw *et al.*, 1984). The possibility has also been raised (Pierpoint, 1983) that the reported inhibition of TMV multiplication by interferon (Orchansky *et al.*, 1982) might have been due to non-specific contaminating factors, and not to interferon itself.

If AVF does operate in the manner of interferon, how might the inhibition of virus multiplication actually be effected? In mammalian cells, interferon induces synthesis of an unusual oligonucleotide, ppp(A2'p5')$_n$A (n = 2 - 4) (abbreviated to 2,5-A). This activates a latent ribonuclease (Williams *et al.*, 1979) which inhibits virus multiplication. Synthetic 2,5-A was shown to inhibit TMV multiplication in leaf discs if applied soon after inoculation (Devash *et al.*, 1982). Reichman *et al.* (1983) presented evidence suggesting that plants contain a 2,5-A synthetase which would polymerize ATP into an oligonucleotide in the presence of a double stranded RNA inducer, which could be either synthetic (Gat-Edelbaum *et al.*, 1983) or the replicative form of virus RNA. However, Cayley *et al.* (1982) were unable to detect 2,5-A or associated proteins in higher plants.

Devash *et al.* (1981) reported that crude preparations of AVF contained another activity which they named discharging factor (DF). DF was separable from AVF, and was claimed to discharge the histidine from the tRNA-like 3' end of TMV RNA, with possible effects on infectivity. Inactive DF preparations could have their activity restored by incubation with a 'polymerized ATP' fraction, presumed to be analogous to 2,5-A. The authors suggested that 2,5-A might be involved in activation of DF *in vivo*. However, the question of whether DF action might involve a specific discharge of histidine, or merely a non-specific nucleolytic degradation of TMV RNA, was not rigorously examined. The latter seems probable, especially as crude AVF preparations were reported to contain more nuclease activity than the corresponding fractions from healthy plants. The status of DF and the role of the putative 2,5-A thus remain in some doubt.

Another possible mode of action of AVF suggested by Sela (1981) is through the 'pathogenesis-related' (PR) proteins, which are associated with acquired systemic resistance (c.f. Chapter 9). As yet there is no evidence linking PR proteins with inhibition of virus multiplication or spread, and no evidence linking AVF with PRs. The obvious experiments: investigation of the synthesis of PRs in healthy tissue treated with AVF, and comparison of the time courses of AVF and PR accumulation after infection, do not appear to have been reported.

On a more general note, the timing of all those events which have been claimed to be related to AVF and induction of localization, requires more detailed investigation. AVF has been detected as a radiophosphorus-labelled fraction by 48 h after inoculation, but little or none was present at 24 h (Antignus *et al.*, 1977). Inhibition of TMV multiplication was detectable from 24 h after inoculation (Otsuki *et* al., 1972). Although this does not preclude an involvement of AVF in restriction of virus multiplication, it may challenge whether it is a critically early event.

Faccioli & Capponi (1983) have described an antiviral factor, also a phosphorylated glycoprotein, in *Chenopodium amaranticolor* which had been locally infected with TNV. The method used for purification was similar to that used for the *Nicotiana* AVF.

A different type of antiviral agent associated with the *N* gene in *Nicotiana* was reported by Loebenstein & Gera (1981), and named 'inhibitor of virus replication' (IVR). This consists of two biologically-active principles, with molecular weights of about 26,000 and 57,000. They were released into the culture medium from TMV-infected protoplasts from N gene plants, but were not produced by healthy protoplasts, or by protoplasts from susceptible (n-gene) plants. IVR was partially purified by precipitation with zinc acetate. Purified IVR inhibited virus multiplication, as measured by infectivity or by ELISA, in N- or n-gene protoplasts, by about 60%. IVF was inhibitory if applied up to 18 h after inoculation. This excludes the possibility that it was acting against virus uptake or initiation of infection. However, the possibility that IVF is merely a non-specific metabolic inhibitor does not appear to have been tested. It would be interesting to compare its effects on aspects of host metabolism, especially RNA and protein synthesis, with its antiviral activity. In a later paper, Gera et al. (1983) showed that treatment of protoplasts from N-gene plants with actinomycin D or chloramphenicol increased TMV multiplication, and prevented accumulation of IVR.

6.3.6. Conclusion. Table 6.1 summarises some of the changes that occur during the localization of viruses, and relates them to the timings of inhibition of virus multiplication, and appearance of necrosis. It has to be stressed that the changes listed in the Table, and indeed all the events discussed in section 6.3, have been observed in different laboratories, under different experimental conditions, and using different combinations of host and virus. Thus the comparison of timings is at best an approximate guide to the order of things, and does not necessarily provide evidence for temporal relationships.

A second impediment to establishing temporal relationships is that the processes of inhibition of virus multiplication, and appearance of lesions on inoculated leaves are generally rather asynchronous. Thus inhibition of TMV multiplication in discs from N-gene tobacco leaves became apparent between 24 and 48 h after inoculation (Otsuki et al., 1972), and lesion number continued to increase from 48 to 120 h (Takahashi, 1974). Furthermore, in many investigations, the temporal relationships between metabolic changes with supposed antiviral significance, and lesion development or inhibition of virus multiplication, have not been studied.

The results discussed in section 6.3 have shown that there is a vast number of structural and metabolic changes associated with necrosis and virus localization. Indeed, it is difficult to pinpoint areas where metabolism and structure are not affected. Undoubtedly, this means that many of the changes observed must be secondary, and not directly involved in localization or

Table 6.1. A summary of the relative timing of some events during localization of virus and formation of necrotic lesions.

Early changes after inoculation:

1h Epidermal stripping of cowpea leaves no longer inhibits lesions (Coutts 1980)
3h Cold osmotic shock no longer inhibits lesions (Kasamo & Shimomura, 1982)
5h First signs of increased electrolyte leakage (Pennazio *et al.*, 1979)
8h Starch grains increased in chloroplasts (Weintraub & Ragetli, 1964)
9h Gramicidin-S treatment no longer inhibits lesions (Kasamo & Shimomura, 1982);
Increased infectivity in epidermis detectable (Takahashi, 1973)
15h Increased infectivity detectable in mesophyll (Takahashi, 1973)
24h Cell to cell contact no longer required for necrosis (Kalpagam *et al.*, 1977)

Timing of effects on virus multiplication and necrosis:

24 h - 48h (approx) Detectable inhibition of virus multiplication in comparison with systemic host (Otsuki *et al.*, 1972; Fraser, 1979)
30h - 72h (approx) First appearance of necrosis in various host / virus systems.
30h - 120 h (approx) Lesion numbers increase in various host / virus systems

Changes detectable shortly before necrosis appears:

Infected leaf extracts toxic to protoplast survival (Hooley & McCarthy, 1980)
AOA (PAL inhibitor) no longer increases lesion size (Massalo *et al.*, 1980)
Plasma membrane rupture and major increase in electrolyte leakage (Pennazio *et al.*, 1979; Weststeijn, 1978)
Increase in ethylene precursor ACC (De Laat & Van Loon, 1982)
First increase in ethylene production (De Laat *et al.*, 1981)
First signs of increases in PAL activity and concentrations of phenolics (Fritig *et al.* 1972; 1973).

Changes detectable at about the time when necrosis appears:

Increases in mitochondria, ribosomes and endoplasmic reticulum (Weintraub *et al.* 1972; Appiano *et al.*, 1977)
Increased OMT synthesis and activity detectable (Collendavelloo *et al.*, 1983)
ABA concentration up (Whenham & Fraser, 1981)
IAA concentration up (Van Loon & Berbee, 1978)
AVF detectable (Antignus *et al.*, 1977)

Changes detectable after necrosis has appeared:

PR proteins detectable (Van Loon & Van Kammen, 1970)
Major deposition of callose and wall material (Pennazio *et al.*, 1978; Faulkner & Kimmins, 1978)
Major ultrastructural evidence for cell collapse (Israel & Ross, 1967)
5 - 15 days Lesion growth may stop.

---

inhibition of virus multiplication. Some of the changes noted seem to occur too late to be of significance. However, although the major increase in factors such as PAL activity, ethylene synthesis or wall modification may occur with or after the onset of necrosis, smaller early changes in these factors could still be important in localization.

In no case does the experimental evidence support the idea that any one type of change plays a sole or dominant role in the localization of virus. On the other hand, several individual types of evidence persuade that the parameter examined could contribute to localization. Thus there is evidence that changes in phenolics, in the processes of wall modification, in membranes, and in plant growth regulators, could be involved, but perhaps in combination with other factors. Localization might involve several events, or a sequence of interlinked changes.

Evidence for molecules with demonstrable antiviral activity is at present weakened by the fact that many of the steps in their proposed mode of action are still conjectural, and by lack of proof that some of the effects are specifically antiviral. However, the suggestion that localization occurs because cells on the periphery have been induced to some antiviral state, which prevents infection or efficient establishment of multiplication, has attractions. This type of mechanism is also supported by studies of induced resistances, such as local acquired resistance and green islands (Chapter 9).

6.4. RESISTANCE MECHANISMS PERMITTING SOME SPREAD OF VIRUS

In some species, resistance to virus disease may be manifest as an inhibition of virus multiplication or of symptom development, which does not depend on a complete localization of the virus around the immediate site of infection, but which may permit spread within the inoculated leaf or systemically. The early literature, covering numerous examples of this type of mechanism, was reviewed by Holmes (1965). In many cases, the resistance is described as 'tolerance', in that visible symptoms of disease such as mosaic or stunting are prevented in the resistant plant, or are less severe than in the fully susceptible lines, while virus is nonetheless detectable in the resistant plants. In the narrower usage of the term 'tolerance', it refers only to amelioration of disease symptoms on the host, and does not imply any effect of the resistance genes on multiplication of the pathogen. In other words, this definition of tolerance refers to resistance to disease, not necessarily resistance to the pathogen. However, reduction of disease symptoms is not necessarily accompanied by inhibition of multiplication of the pathogen. Whether 'tolerance' should be restricted to description of disease effects on the host alone, and whether it should be classed as a mechanism of 'resistance' are purely semantic arguments. In this section, we are concerned with the mechanisms by which particular genes may prevent symptom development, and may inhibit virus multiplication. This requires a quantitative approach to measurement of symptoms, including growth effects, and virus multiplication. As an example of the pitfalls of too narrow a definition of 'tolerance', Kooistra (1968) reported lines of cucumber in which the mosaic symptoms of CGMMV were completely suppressed, although virus multiplication was the same as in susceptible lines. However, reductions of yield in the 'tolerant' lines as a result of infection were as severe as in the susceptible lines.

In practical crop protection, resistance mechanisms which permit some systemic multiplication of the virus are less attractive than those which give effective localization or no detectable multiplication. Firstly, systemic multiplication, although limited, may have effects on crop yield, and represents a reservoir of infective virus which may affect susceptible cultivars grown nearby. Secondly, the ability of the virus to multiply, albeit to limited extents, increases the chance of production of resistance-breaking strains.

However, resistance mechanisms which permit some virus spread may be the only types available. In some crops they have contributed significantly to virus restriction, for example in control of BCMV in Phaseolus beans. Systemically effective resistance has also proved useful in combination with localizing resistance, for example in control of TMV in tomato (Pelham, 1966; 1972; Hall, 1980). A particular beneficial effect of the combination of genes, apart from presenting a bigger barrier to evolution of resistance-breaking strains, is that the systemically-effective gene may reduce the amount of necrosis associated with operation of the localization gene. This may prevent secondary inhibitory effects of the necrotic response on plant growth (c.f. Fraser, 1979).

Comparatively, the systemically-effective types of resistance mechanism have been less studied than those clearly dependent on localization, which were discussed in section 6.3. Theoretically, systemically-effective resistance might operate by blocking any stage of symptom development; inhibition of host growth; or virus replication. The last could of course include inhibition of virus spread, although in the absence of visible accompaniments of complete localization such as lesions. In two of the examples discussed below, resistance to virus multiplication operates in isolated protoplasts, and is therefore most unlikely to involve solely an inhibition of virus spread. In other cases, resistance does appear to inhibit long-distance transport of virus, but not cell-to-cell spread.

6.4.1. Resistance to CMV in cucumber. Studies of mechanisms of resistance to CMV in cucumber are complicated by the fact that different workers have used cultivars with resistances derived from two sources: either from Chinese Long (Wasuwat & Walker, 1961a; Barbara & Wood, 1972; 1974; Amemiya & Misawa, 1977) or from Kyoto 3-foot (Nachman *et al.*, 1971; Levy *et al.*, 1974; Barnett & Wood, 1978; Maule *et al.*, 1980). These two types of mechanism may be genetically distinct (Coutts & Wood, 1977; Maule *et al.*, 1980), but results so far do not show any major differences in mode of action.

Resistance suppressed symptom formation and reduced virus multiplication to about one-tenth of that in susceptible plants, but did not eliminate systemic spread (Wasuwat & Walker, 1961b; Wood & Barbara, 1971; Amemiya & Misawa, 1977). The early patterns of virus accumulation in resistant and susceptible plants were identical, with inhibition only appearing in resistant plants 36 h after inoculation (Amemiya & Misawa, 1977). This suggested that the resistance mechanism has to be activated during the early stages of virus multiplication. Further support for this idea, and evidence that activation requires transcription of DNA, came from experiments showing that development of resistance was prevented by actinomycin-D (Nachman *et al.*, 1971; Barbara & Wood, 1974; Amemiya & Misawa, 1977); ultraviolet light (Levy *et al.*, 1974) or α-amanitin (Amemiya & Misawa, 1977). All three agents were effective only if applied not more than 12 or 24 h after inoculation; none affected virus multiplication in susceptible hosts.

A short heat treatment of resistant plants 6 h after inoculation also inhibited development of resistance, suggesting involvement of a thermolabile protein (Amemiya & Misawa, 1977). However, no antiviral agent has yet been identified. Cucumbers do contain an inhibitor of CMV infectivity on cowpea, a local lesion host (Sill & Walker, 1952). However, resistance to CMV in cucumber was shown not to be due to differences in concentration of this inhibitor (Wasuwat & Walker, 1961b). Activities of peroxidase isoenzymes and polyphenoloxidase were altered by CMV infection of cucumber (Wood, 1971; Wood & Barbara, 1971), but the changes appeared to be correlated with differences in virus multiplication and symptom development in resistant and susceptible cultivars, rather than with the resistance mechanism as such.

CMV-inoculated protoplasts from resistant (Kyoto) plants produced less progeny virus, and showed a lower proportion of cells in which virus antigen could be detected, than protoplasts from susceptible plants (Maule *et al.*, 1980). The apparent lower proportion of infected protoplasts could, however, have been due to difficulty in detecting those with very low levels of virus antigen, and need not indicate resistance to infection. Less extensive experiments also indicated that protoplasts from China plants were resistant. The evidence therefore indicates that resistance to CMV operates within the limits of the individual cell, and that it does not involve resistance to cell-to-cell spread; co-operative effects between cells, or induction of some antiviral state in uninfected cells as a result of infection of neighbouring cells. In contrast to whole plants, resistance in protoplasts did not allow an early phase of unrestricted virus multiplication (Maule *et al.*, 1980). A further difference from whole-plant resistance is that treatment with actinomycin-D and ultraviolet light did not prevent inhibition of virus multiplication in resistant protoplasts (Boulton *et al.*, 1984). These authors suggested that resistance in whole plants might consist of two components: one requiring to be activated after infection, and associated with a cell-to-cell interaction, and the other present at the level of the individual cell, and operating by limitation of virus multiplication. Only the second mechanism is thought to operate in protoplasts.

Resistance in protoplasts was also effective when CMV RNA was used as inoculum, suggesting that attachment or uncoating of virus could not be the target of the resistance mechanism. Furthermore, resistance could be detected when production of virus RNA was assayed, suggesting that the mechanism does not operate against assembly of progeny virus particles (Maule *et al.*, 1980). By elimination, these results therefore suggest that some phase of virus transcription or translation might be the target of the resistance mechanism, but what this is remains to be established.

6.4.2. *Resistance to TMV in tomato conferred by the* Tm-1 *gene*. Studies of the mechanism of *Tm-1* action, and indeed of the other TMV resistance genes in tomato, *Tm-2* and *Tm-2*^{2}, have been greatly facilitated by breeding of nearly isogenic lines of cv. Craigella containing different combinations of the resistance genes (Darby *et al.*, 1978).

The *Tm-1* gene in either homozygous or heterozygous configuration prevented

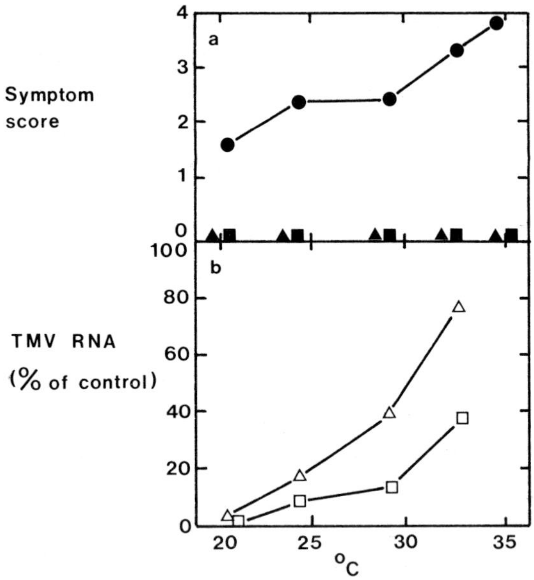

FIGURE 6.6. Effects of the Tm-1 gene on mosaic symptom formation and tobacco mosaic virus multiplication, in tomato plants grown at different temperatures. (a) Symptoms (relative scale) on susceptible (+/+)(●); heterozygous (Tm-1/+)(▲) and homozygous (Tm-1/Tm-1)(■) plants. (b) Virus multiplication (measured as TMV RNA accumulation, and expressed as a percentage of the level in susceptible plants grown at the same temperature) in Tm-1/+ (△) and Tm-1/Tm-1 (□) plants. From Fraser and Loughlin, 1982. Reproduced from Physiological Plant Pathology by permission of Academic Press, London.

formation of mosaic symptoms by strain-0 type TMV isolates (Pelham, 1972). However, the virus multiplied and spread systemically in resistant plants. At normal temperatures, multiplication was inhibited by 70% in plants heterozygous for Tm-1, and by 95% in plants with homozygous Tm-1 (Figure 6.6; Fraser & Loughlin, 1980). Thus suppression of symptom formation showed straightforward dominance, whereas inhibition of multiplication was gene-dosage dependent, presumably because two gene copies gave a higher concentration of the antiviral product.

In Tm-1 plants, there was a delay of 8 to 40 days between inoculation and the first detection of virus multiplication (Dawson, 1965; Fraser & Loughlin, 1980). This suggests that the antiviral principle was present at the time of inoculation, and did not require some type of activation during the early part of virus multiplication.

The effects of growth temperature on operation of the *Tm-1* gene were complex. Symptom formation was suppressed in resistant plants grown at constant temperatures between 20 and 35°C. In contrast, inhibition of virus multiplication was strongly temperature-dependent (Figure 6.6; Fraser & Loughlin, 1982). For example, in plants with heterozygous *Tm-1*, it was over 95% at 20°C, but only 20% at 33°C. This suggests that the product which inhibits virus multiplication is temperature sensitive. Experiments in which plants were shifted from a restrictive down to a permissive temperature for *Tm-1* activity suggested that antiviral function was quickly re-acquired (Fraser & Loughlin, 1982).

The lack of temperature and gene-dosage effects on symptom suppression are puzzling. They might imply that *Tm-1* actually consists of two separate mechanisms, either closely linked but separate genetic elements, or one gene with a branched pathway. Much remains to be discovered about how mosaic symptoms are caused, and what factors may influence severity in normal and resistant plants. However, the available data suggest that the effects of the *Tm-1* gene on symptoms cannot be explained solely as a secondary consequence of effects on virus multiplication.

TMV multiplication was inhibited in protoplasts from *Tm-1* plants (Motoyoshi & Oshima, 1977), indicating that resistance operates within the individual cell. In contrast, TMV multiplication was not inhibited in protoplasts from plants containing the *Tm-2*^{2} type of localizing resistance (Motoyoshi & Oshima, 1975).

Tm-1 resistance was expressed in protoplasts inoculated with TMV RNA, so the target of resistance must be at some stage after virus uptake or uncoating (Motoyoshi & Oshima, 1979). Evans (1982) and Evans *et al.* (1984) examined RNA-dependent RNA-polymerase activities in susceptible and *Tm-1*-resistant plants. Healthy plants of both genotypes contained high levels of enzyme activity. *Tm-1* resistance was not therefore simply due to lack of an enzyme system capable of copying RNA templates. Polymerase activity was increased by TMV strain 0 infection in susceptible plants, but not in resistant plants. When enzyme preparations from different types of plant were incubated with added TMV RNA as template and the reaction products were examined, only the enzyme from infected, susceptible plants produced the replicative form and replicative intermediate of TMV. Enzymes from healthy plants, or from infected *Tm-1* plants did not. These results could mean that *Tm-1* resistance operates by preventing synthesis or modification of enzyme activity to form a functional viral replicase, although other, more trivial explanations are also possible. Strains of TMV which overcame the *Tm-1* resistance did stimulate polymerase activity (Evans 1982).

Multiplication of TMV isolates with defective (temperature sensitive) coat proteins was especially strongly inhibited by *Tm-1* (Fraser & Loughlin, 1980; Fraser *et al.*, 1983), while isolates which overcame the inhibition of multiplication tended to form highly stable virus particles (Fraser *et al.*, 1984). These results are consistent with a model in which the *Tm-1* product may interfere with encapsidation of TMV RNA in coat protein to form progeny virus, although much of the mechanism still awaits experimental proof.

Selman & Yahampath (1973) compared various growth and physiological characteristics in non-isogenic susceptible and *Tm-1*-resistant cultivars. Bleeding sap from resistant plants had more gibberellin, more cytokinin and lower concentrations of amino acids than that from susceptible plants. It was not clear whether these differences were related to the resistance mechanism or were varietal.

Reciprocal grafting experiments showed that susceptible rootstocks stimulated virus multiplication in resistant scions. Selman & Yahampath (1973) suggested that this might be due to increased cytokinin and nutrient supply from the susceptible roots, but other interpretations are possible. Resistant rootstocks had no effect on virus multiplication in susceptible scions, so no inhibitor of virus multiplication was transported upward.

6.4.3. *Resistance to BCMV in Phaseolus vulgaris*. The recessive gene system of resistance to BCMV in *Phaseolus* beans requires cooperative action by the *bc-u* gene with one or more genes at the *bc-1*, *bc-2* or *bc-3* (collectively termed *bc-x*) loci. Virulent strains of the virus can overcome specific *bc-x* genes (Drijfhout, 1978).

Virus isolates lacking specific virulence genes generally did not spread systemically, or cause systemic symptoms in cvs homozygous for *bc-u* and a *bc-x* gene (Drijfhout, 1978). Day (1984) showed that plants homozygous for *bc-u*, but containing susceptible alleles of all the *bc-x* genes (i.e. *bc-x*$^{+}$) had no detectable resistance. However, plants with the genotype *bc-u*$^{+}$ *bc-u*$^{+}$/*bc-1* *bc-1* showed detectable systemic infectivity, but no systemic symptoms. These results suggest that *bc-u* is only effective in combination with a *bc-x* gene, but that *bc-1* alone might function against symptom formation.

Plants heterozygous for both *bc-u* and *bc-1*, and otherwise *bc-x*$^{+}$, showed no suppression of systemic mosaic. Thus the genes had fully recessive behaviour in this respect. Inhibition of systemic multiplication of virus was only 70 - 90% in heterozygotes, in contrast to the 100% inhibition obtained with homozygous resistance genes. Inhibition of multiplication was thus gene-dosage dependent.

Resistance by homozygous bc-u/bc-x was not effective in the inoculated leaves. This suggests either that a very long induction period was required, or that resistance did not inhibit multiplication within cells, or cell-to-cell movement of virus. However, no virus could be detected in the petiole of infected, resistant leaves, or the adjoining stem. BCMV is translocated in the phloem (Ekpo & Saettler, 1975). Day's (1984) results therefore suggest that resistance may have inhibited phloem loading of BCMV, thus preventing systemic spread.

6.4.4. Resistance to CCMV in cowpea. As with BCMV infection of resistant Phaseolus beans, CCMV multiplied in the inoculated leaves of resistant cowpea, but did not spread systemically (Wyatt & Kuhn, 1979). However, the rate of virus synthesis in inoculated resistant leaves was only 5 to 10% of that in susceptible leaves.

Wyatt & Kuhn (1979) found that the virus produced in resistant plants had, comparatively, very small amounts of one of the genomic RNAs, RNA-3. This deficiency disappeared when the virus was transferred to susceptible plants. RNA-3 of CCMV contains the coat protein gene, but this sequence is not thought to be translated in vivo (Lane, 1981). CCMV isolated from resistant cowpeas contained normal amounts of RNA-4, which is a monocistronic mRNA for coat protein. The resistance is thus unlikely to have operated by inhibition of coat protein synthesis. The other function specified by RNA-3 is a protein of weight 35000. The equivalent protein in BMV, another virus in the same group, may be associated with the viral replicase (Hariharasubramanian et al., 1973). The resistance of cowpea to CCMV may therefore be directed at replicase synthesis.

6.5. MECHANISMS OF VIRULENCE

There are two possible types of theoretical model for mechanisms of virulence. In the first, virulence involves a qualitative change in the virus, such that it fails to activate the host resistance mechanism, or fails to interact with an existing antiviral principle. In the second, virulence involves a quantitative change in virus multiplication, or in the amount or activity of one of its components. The consequence is that the virus multiplies and spreads faster than the host defence mechanism can be activated, or swamps an inhibitor of multiplication or spread which is present in limited quantities. In some forms of the second model, virulence is associated with increased aggressiveness.

Studies of mechanisms of virulence have generally tried to locate the property, either in a gene product with altered function, or in one of the RNAs of viruses with multicomponent genomes. The ability to create pseudorecombinants between virulent and avirulent strains in the latter case has provided a powerful approach, and has allowed more progress in understanding the genetics of virulence than has been possible for viruses with single-component genomes.

Kado & Knight (1966) showed that the ability of TMV to overcome the _N'_ gene of tobacco and spread systemically was located towards the 3' end of the viral RNA (Chapter 4). The gene for coat protein is close to the 3' end, then there is a gene coding for a protein of 30,000 molecular weight. The results from Kado & Knight's mutagenesis mapping technique would permit the virulence determinant to be located in this gene.

Nishiguchi _et al._ (1978) described a temperature-sensitive mutant of TMV which is defective in cell-to-cell movement in non-hypersensitive tobacco cvs at the restrictive temperature. It was shown that the coat protein of the mutant was identical to that of the parent wild type (Zaitlin & Leonard, 1982), but that it has a single amino acid substitution in its 30,000 protein (Ohno _et al._ 1983). Although the possibility that mutation elsewhere in the TMV RNA may have caused the loss of ability to move from cell to cell cannot be excluded, the results are consistent with an involvement of the 30,000 protein in control of virus movement. Both the mutant and the parent wild type formed local lesions on the _N'_ gene host _N. sylvestris_, therefore this particular mutation did not confer virulence. However, it would be interesting to examine the 30,000 proteins of strains able to overcome the _N'_ gene (Fraser, 1983) to see if they showed other alterations.

There is some evidence implicating the coat protein in control of the interaction with the _N'_ gene. Tsugita (1962) and Tsugita & Fraenkel-Conrat (1962) prepared mutants from wild type TMV by nitrous acid mutagenesis. The mutants were selected by their ability to cause necrotic lesions on _N. sylvestris_, on which the parent strain spread systemically. Almost all of the many mutants selected had differences in the amino acid compositions of their coat proteins. In contrast, only one out of 60 mutants which gave systemic symptoms differed from wild type in amino acid composition.

Fraser (1983) examined isolates of TMV which caused lesions of widely differing sizes on the N' gene tobacco cv. White Burley. Increased lesion size was strongly correlated with increased virus multiplication per lesion. There was no correlation between lesion size and intrinsic ability of each strain to multiply, as measured in systemic hosts. This would tend to argue against any link between increased virulence and increased aggressiveness in this case. There was, however, a strong correlation between lesion size and the temperature-sensitivity of the coat protein subunits. This does not imply that thermal denaturation as such of the coat protein played any part in the interaction with the resistance gene; the property was chosen as a readily measured index of alteration in biochemical properties. These results, together with those of Tsugita (1962) suggest that coat protein might influence the interaction with the localization mechanism.

Several authors have investigated TMV strains able to overcome the Tm-1 gene of tomato. Dawson et al. (1975) found that two isolates of TMV with strain 1 virulence differed from strain 0 in the amino acid composition of their coat proteins. In a later paper (Dawson et al., 1979), they suggested that there was no correlation between amino acid composition of an isolate and its ability to overcome Tm-1. However, this paper was mainly aimed at detecting unusual amino acid compositions, such as the presence of methionine, and the data as presented do not exclude the possibility that all strain 1 isolates examined could have differed from strain 0 in amino acid composition. McRitchie & Alexander (1963) found that strain 1 isolates could be differentiated serologically from strain 0.

Fraser et al. (1983) found that all 20 strain 1 isolates they examined differed from strain 0 in their coat protein properties. Most of the strain 1 isolates had been prepared from strain 0 by mutagenesis with nitrous acid under conditions designed to give a very low frequency of mutants. This forms a statistical argument against all the strain 1 isolates having some further mutation in a non-coat protein region which controlled virulence. However, the inherent instability of RNA genomes (Reanney, 1982) means that this possibility can not be entirely excluded. Fraser et al. (1984) reported that acquisition of strain 1 virulence was strongly associated with reduction in ability to multiply in susceptible hosts; i.e. virulence was linked to reduced aggressiveness. Virulence and reduced multiplication were both associated with high particle stability, and it was suggested that this factor might control both the interaction with the Tm-1 product, and the intrinsic ability of an isolate to multiply. In a completely different system, Hanada & Harrison (1977) found that isolates of RRV which could overcome the resistance of Lloyd George raspberry had low intrinsic ability to multiply in other hosts, and suggested that this explained the infrequency of resistance-breaking strains.

Valverde & Fulton (1982) found that four isolates of SBMV able to overcome the localizing resistance in *Phaseolus vulgaris* could be serologically distinguished from the avirulent type strain.

The apparent strong evidence for involvement of coat protein in control of virulence cited above should not be taken to indicate that other regions of the viral genome are not involved in virulence. With viruses with monopartite genomes, there is simply such great difficulty in studying the products and function of non-coat protein regions of the genome, that evidence for involvement in virulence is missing by default. Much of the evidence from pseudorecombinant experiments indicates that virulence in multicomponent viruses can be controlled by genes other than those coding for coat proteins. Examples where the analysis has been limited to mapping the gene for virulence are discussed in Chapter 4. Here we are concerned with studies of the biochemical expression of virulence genes.

Roosien *et al.* (1983) studied a mutant of AMV which can invade cowpea plants systemically. The wild type parent causes a hypersensitive reaction. The mutant was also replicated more efficiently in protoplasts than wild type, and comparison of viruses synthesised in protoplasts showed that the mutant contained a higher proportion of middle component (containing RNA-2) than the wild type. If mutant virus was supplemented with middle component of wild type virus, then wild type infection patterns were established in both plants and protoplasts. The results suggest that wild type RNA-2 suppressed RNA multiplication, and was also responsible for activating the local lesion response. In this case, therefore, virulence may be a negative property of the mutant RNA-2.

De Jager & Van Kammen (1970) prepared a nitrous acid mutant of CPMV, designated SbN3. The parent wild type, Sb, caused systemic infection of *P. vulgaris* at 22 - 24°C, whereas SbN3 caused chlorotic lesions and did not spread systemically, i.e. was apparently a mutation to avirulence. At the higher temperature of 30°C, both strains spread systemically. It is therefore unlikely that SbN3 was defective, unless it was cold-sensitive.

At 22 but not at 30°C, SbN3 formed comparatively large amounts of top component particles. Homologous and heterologous combinations of middle and bottom components between the two strains showed that middle component determined virulence. The authors suggested that inhibition of systemic invasion, and formation of excess top component, might have had mechanisms in common, perhaps controlled by the middle component.

De Jager & Wesseling (1981) studied two naturally-occurring variant strains of CPMV, which did not induce the normal local lesion reaction on cowpea cv. Early Red, but which spread systemically. Only one of the variants induced necrosis; the other caused mosaic, showing that the abilities to evade localization and to induce, or not to induce, necrosis were separate. In contrast to the reaction on P. vulgaris described above, the disease syndrome on cowpea was governed by the bottom component.

An intriguing report by Talianski et al. (1982) showed that in tomato plants with the TMV-localizing gene Tm-2, avirulent isolates of TMV were able to spread systemically if the plants were also infected with PVX. This may suggest either that the PVX infection was disabling the TMV resistance mechanism in some way, or that the TMV was able to make use of some PVX-specified factor to replace a function of its own which has been inhibited by the resistance. Such complementation-type experiments would also be interesting using virulent and avirulent isolates of the same virus, and may give further insights into mechanisms of resistance and virulence.

The major conclusion from all these studies of mechanisms of resistance and virulence is that despite intensive effort, we are still far from understanding how they work in the overwhelming majority of cases.

REFERENCES

Adams, J.B. & Wade, C.V. (1976). Aphid behaviour and host-plant preference demonstrated by electronic patterns of probing and feeding. American Potato Journal 53, 261-267.

Aldwinckle, H.S. & Selman, I.W. (1967). Some effects of supplying benzyladenine to leaves and plants inoculated with viruses. Annals of Applied Biology 60, 49-58.

Allison, A.V. & Shalla, T.A. (1964). The ultrastructure of local lesions induced by potato virus X: a sequence of cytological events in the course of infection. Phytopathology 54, 784-793.

Amemiya, Y. & Misawa, T. (1977). Studies on the resistance of cucumber to cucumber mosaic virus. II. Induction of resistance by infection. Tokohu Journal of Agricultural Research 28, 18-25.

Antignus, Y., Sela, I. & Harpaz, I. (1975). A phosphorus-containing fraction associated with antiviral activity in Nicotiana spp. carrying the gene for localization of TMV infection. Physiological Plant Pathology 6, 159-168.

Antignus, Y., Sela, I. & Harpaz, I. (1977). Further studies on the biology of an antiviral factor (AVF) from virus infected plants and its association with the N-gene of Nicotiana species. Journal of General Virology 35, 107-116.

Antoniw, J.F. & Pierpoint, W.S. (1978). The purification and properties of one of the 'b' proteins from virus-infected tobacco plants. Journal of General Virology 39, 343-350.

Antoniw, J.F., Ritter, C.E., Pierpoint, W.S. & Van Loon, L.C. (1980). Comparison of three pathogenesis-related proteins from plants of two cultivars of tobacco infected with TMV. Journal of General Virology 47, 79-87.

Antoniw, J.F., White, R.F. & Carr, J.P. (1984). An examination of the effect of human o-interferon on the infection and multiplication of tobacco mosaic virus in tobacco. Phytopathologische Zeitschrift 109, 367-371.

Appiano, A., Pennazio, S., D'Agostino, G. & Redolfi, P. (1977). Fine structure of necrotic local lesions induced by tomato bushy stunt virus in Gomphrena globosa leaves. Physiological Plant Pathology 11, 327-332.

Awasthi, L.P. (1981). The purification and nature of an antiviral protein from Cuscuta reflexa plants. Archives of Virology 70, 215-223.

Bailiss, K.W., Balazs, E. & Kiraly, Z. (1977). The role of ethylene and abscisic acid in TMV-induced symptoms in tobacco. Acta Phytopathologica Academiae Scientiarum Hungaricae 12, 133-140.

Balazs, E., Barna, B. & Kiraly, Z. (1976). Effect of kinetin on lesion development and infection sites in Xanthi-nc tobacco infected with TMV: single cell local lesions. Acta Phytopathologica Academiae Scientiarum Hungaricae 11, 1-9.

Balazs, E., Gaborjanyi, R. & Kiraly, Z. (1973). Leaf senescence and increased virus susceptibility in tobacco: the effect of abscisic acid. Physiological Plant Pathology 3, 341-346.

Barbara, D.J. & Wood, K.R. (1972). Virus multiplication, peroxidase and polyphenoloxidase activity in leaves of two cucumber (Cucumis sativus L.) cultivars inoculated with cucumber mosaic virus. Physiological Plant Pathology 2, 167-172.

Barbara, D.J. & Wood, K.R. (1974). The influence of actinomycin D on cucumber mosaic virus (strain W) multiplication in cucumber cultivars. Physiological Plant Pathology 4, 45-50.

Barnett, A. & Wood, K.R. (1978). Influence of actinomycin D, ethephon, cycloheximide and chloramphenicol on the infection of a resistant and a susceptible cucumber cultivar with cucumber mosaic virus (Price's No. 6 strain). Physiological Plant Pathology 12, 257-277.

Bawden, F.C. (1964). 'Plant Viruses and Virus Diseases', 4th Edition. Ronald Press, New York.

Bawden, F.C. & Pirie, N.W. (1959). The infectivity and inactivation of nucleic acid preparations from tobacco mosaic virus. Journal of General Microbiology 21, 438-450.

Beachy, R.N. & Murakishi, H.H. (1973). Effect of cycloheximide on tobacco mosaic virus synthesis in callus from hypersensitive tobacco. Virology 55, 320-328.

Beier, H., Bruening, G., Russell, M.L. & Tucker, C.L. (1979). Replication of cowpea mosaic virus in protoplasts isolated from immune lines of cowpeas. Virology 95, 165-175.

Beier, H., Siler, D.J., Russell, M.L. & Bruening, G. (1977). Survey of susceptibility to cowpea mosaic virus among protoplasts and intact plants from Vigna sinensis lines. Phytopathology 67, 917-921.

Boulton, M.I., Maule, A.J. & Wood, K.R. (1984). Effect of actinomycin D and UV irradiation on the replication of cucumber mosaic virus in protoplasts isolated from resistant and susceptible cucumber cultivars. Physiological Plant Pathology, in press.

Brown, E.G. & Newton, R.P. (1981). Cyclic AMP and higher plants. Phytochemistry 20, 2453-2463.

Carroll, T.W. (1966). Lesion development and distribution of tobacco mosaic virus in Datura stramonium. Phytopathology 56, 1348-1353.

Carroll. T.W., Gossel, P.L. & Hockett, E.A. (1979). Inheritance of resistance to seed transmission of barley stripe mosaic virus in barley. Phytopathology 69, 431-433.

Cayley, P.J., White, R.F., Antoniw, J.F., Walesby, N.J. & Kerr, I.M. (1982). Distribution of the $ppp(A2'p)_nA$-binding protein and interferon-related enzymes in animals, plants and lower organisms. Biochemical and Biophysical Research Communications 108, 1243-1250.

Chadha, K.C. & MacNeill, B.H. (1969). An antiviral principle from tomatoes systemically infected with tobacco mosaic virus. Indian Journal of Botany 47, 513-518.

Cirulli, M. & Alexander, L.J. (1969). Influence of temperature and strain of tobacco mosaic virus on resistance of tomato breeding lines derived from Lycopersicon peruvianum. Phytopathology

59, 1278-1297.

Cohen, J. & Loebenstein, G. (1975). An electron microscope study of starch lesions in cucumber cotyledons infected with tobacco mosaic virus. Phytopathology 65, 32-39.

Collendavelloo, J., Legrand, M. & Fritig, B. (1983). Plant disease and the regulation of enzymes involved in lignification. Plant Physiology 73, 550-554.

Coutts, R.H.A. (1978). Suppression of virus-induced local lesions in plasmolysed leaf tissue. Plant Science Letters 12, 77-85.

Coutts, R.H.A. (1980). Virus induced local lesions on cowpea leaves: the role of the epidermis and the effects of kinetin. Phytopathologische Zeitschrift 97, 307-316.

Coutts, R.H.A. & Wood, K.R. (1977). Inoculation of leaf mesophyll protoplasts from a resistant and a susceptible cucumber cultivar with cucumber mosaic virus. FEMS Microbiology Letters 1, 121-124.

Cruikshank, I.A.M., & Perrin, D.R. (1964). Pathological functions of phenolic compounds in plants. In 'Biochemistry of Phenolic Compounds' (J.B. Harborne, ed.), pp. 511-544. Academic Press, New York.

Darby, L.A., Ritchie, D.B. & Taylor, I.B. (1978). Isogenic lines of the tomato 'Ailsa Craig'. Report of the Glasshouse Crops Research Institute for 1977, pp. 168-184.

Dawson, G.W., Gibson, R.W., Griffiths, D.C., Pickett, J.A., Rice, A.D. & Woodcock, C.M. (1982). Aphid alarm pheromone derivatives affecting settling and transmission of plant viruses. Journal of Chemical Ecology 8, 1377-1387.

Dawson, J.R.O. (1965). Contrasting effects of resistant and susceptible tomato plants on tomato mosaic virus multiplication. Annals of Applied Biology 56, 485-491.

Dawson, J.R.O., Rees, M.W. & Short, M.N. (1975). Protein composition of unusual tobacco mosaic virus strains. Annals of Applied Biology 79, 189-194.

Dawson, J.R.O., Rees, M.W. & Short, M.N. (1979). Lack of correlation between the coat protein composition of tobacco mosaic virus isolates and their ability to infect resistant tomato plants. Annals of Applied Biology 91, 353-358.

Day, K.L. (1984). Resistance to bean common mosaic virus in Phaseolus vulgaris L. Ph. D. Thesis, University of Birmingham, U.K.

De Jager, C.P. & Van Kammen, A. (1970). The relationship between the components of cowpea mosaic virus. III. Location of genetic information for two biological functions in the middle component of CPMV. Virology 41, 281-287.

De Jager, C.P. & Wesseling, J.B.M. (1981). Spontaneous mutations in cowpea mosaic virus overcoming resistance due to hypersensitivity in cowpea. Physiological Plant Pathology 19, 347-350.

De Laat, A.A.M. & Van Loon, L.C. (1982). Regulation of ethylene biosynthesis in virus-infected tobacco leaves. II. Time course of levels of intermediates and in vivo conversion rates. Plant Physiology 69, 240-245.

De Laat, A.A.M., Van Loon, L.C. & Vonk, C.R. (1981). Regulation of ethylene biosynthesis in virus-infected tobacco leaves. I. Determination of the role of methionine as the precursor of ethylene. Plant Physiology 68, 256-261.

Devash, Y., Hauschner, A., Sela, I. & Chakraburtty, K. (1981). The antiviral factor (AVF) from virus infected plants induces discharge of histidinyl TMV-RNA. Virology 111, 103-112.

Drijfhout, E. (1978). Genetic interaction between Phaseolus vulgaris and bean common mosaic virus with implications for strain identification and breeding for resistance. Agricultural Research Report 872, Centre for Agricultural Publishing and Documentation, Wageningen, The Netherlands.

Duchesne, M., Fritig, B. & Hirth, L. (1977). Phenylalanine ammonia-lyase in tobacco mosaic virus-infected hypersensitive tobacco. Density labelling evidence of de novo synthesis. Biochimica et Biophysica Acta 485, 465-481.

Ekpo, E.J.A. & Saettler, A.W. (1975). Multiplication and distribution of bean common mosaic virus in Phaseolus vulgaris. Plant Disease Reporter 59, 939-943.

Ellis, R. J. (1981). Inhibitors for studying chloroplast transcription and translation in vivo. In 'Methods in Chloroplast Molecular Biology (Edelman, M., Hallick, R.B. & Chua, N.-H., eds), pp. 559-564. Elsevier / North Holland, Amsterdam.

Evans, D.M.A. (1982). RNA-dependent RNA polymerases in tomato (Lycopersicon esculentum Mill.). Ph. D. Thesis, University College, Cardiff.

Evans, D.M.A., Bryant, J.A. & Fraser, R.S.S. (1984). Characterization of RNA-dependent RNA polymerases in healthy and tobacco mosaic virus-infected plants. Annals of Botany 54, 271-281.

Faccioli, G. & Capponi, R. (1983). An antiviral factor present in plants of Chenopodium amaranticolor locally infected by tobacco necrosis virus. 1. Extraction, partial purification, biological and chemical properties. Phytopathologische Zeitschrift 106, 289-301.

Farkas, G.L., Kiraly, Z. & Solymosy, F. (1960). Role of oxidative metabolism in the localization of plant viruses. Virology 12, 408-421.

Faulkner, G. & Kimmins, W.C. (1975). Staining reactions to the tissue bordering virus lesions induced by wounding, tobacco mosaic virus and tobacco necrosis virus in bean. Phytopathology 65, 1396-1400.

Faulkner, G. & Kimmins, W.C. (1978). Fine structure of tissue bordering lesions induced by wounding and virus infection. Canadian Journal of Botany 56, 2990-2999.

Favali, M.A., Conti, G.G. & Bassi, M. (1978). Modifications of the vascular bundle ultrastructure in the "resistance zone" around necrotic lesions induced by tobacco mosaic virus. Physiological Plant Pathology 13, 247-251.

Fraser, R.S.S. (1972). Effects of two strains of tobacco mosaic virus on growth and RNA content of tobacco leaves. Virology 47, 261-269.

Fraser, R.S.S. (1979). Systemic consequences of the local lesion reaction to tobacco mosaic virus in a tobacco variety lacking the N gene for hypersensitivity. Physiological Plant Pathology 14, 383-394.

Fraser, R.S.S. (1981). Evidence for the occurrence of the pathogenesis-related proteins in leaves of healthy tobacco plants during flowering. Physiological Plant Pathology 19, 69-76.

Fraser, R.S.S. (1982). Are 'pathogenesis-related' proteins involved in acquired systemic resistance of tobacco plants to tobacco mosaic virus? Journal of General Virology 58, 305-313.

Fraser, R.S.S. (1983). Varying effectiveness of the N' gene for resistance to tobacco mosaic virus in tobacco infected with virus strains differing in coat protein properties. Physiological Plant Pathology 22, 109-119.

Fraser, R.S.S. & Airey, M.J. (1980). Resistance to tobacco mosaic virus in tomato: resistance gene Tm-2. Report of the National Vegetable Research Station for 1979, p.21.

Fraser, R.S.S., Gerwitz, A., Loughlin, S.A.R. & Leary, J.A. (1983). Resistance to tobacco mosaic virus in tomato plants. Report of the National Vegetable Research Station for 1982, pp. 18-19.

Fraser, R.S.S. & Loughlin, S.A.R. (1980). Resistance to tobacco mosaic virus in tomato: effects of the Tm-1 gene on virus multiplication. Journal of General Virology 48, 87-96.

Fraser, R.S.S. & Loughlin, S.A.R. (1982). Effects of temperature on the Tm-1 gene for resistance to tobacco mosaic virus in tomato. Physiological Plant Pathology 20, 109-117.

Fraser, R.S.S., Thomas, C.J.R., Gerwitz, A. & Payne, J.A. (1984). Resistance to tobacco mosaic virus in tomato. Report of the National Vegetable Research Station for 1983, pp. 22-23.

Fraser, R.S.S. & Whenham, R.J. (1982). Plant growth regulators and virus infection: a critical review. Plant Growth Regulation 1, 37-59.

Fritig, B., Gosse, J., Legrand, M. & Hirth, L. (1973). Changes in phenylalanine ammonia lyase during the hypersensitive reaction of tobacco to TMV. Virology 55, 371-379.

Fritig, B., Legrand, M. & Hirth, L. (1972). Changes in the metabolism of phenolic compounds during the hypersensitive reaction of tobacco to TMV. Virology 47, 845-848.

Fry, P.R. & Matthews, R.E.F. (1963). Timing of some early events following inoculation with tobacco mosaic virus. Virology 19, 461-469.

Gaborjanyi, R., Balazs, E. & Kiraly, Z. (1971). Ethylene production, tissue senescence and local virus infection. Acta Phytopathologica Academiae Scientiarum Hungaricae 6, 51-55.

Gat-Edelbaum, O., Altman, A. & Sela, I. (1983). Polyinosinic:polycytidylic acid in association with cyclic nucleotides activates the antiviral factor AVF in plant tissues. Journal of

General Virology 64, 211-214.
Gera, A., Loebenstein, G. & Shabtai, S. (1983). Enhanced tobacco mosaic virus production and suppressed synthesis of a virus inhibitor in protoplasts exposed to antibiotics. Virology 127, 475-478.
Gianinazzi, S. (1982). Antiviral agents and inducers of virus resistance: analogies with interferon. In 'Active Defense Mechanisms in Plants' (Wood, R.K.S., ed.), pp. 275-298. Plenum Publishing Corporation, New York.
Gianinazzi, S., Martin, C & Valee, J.-C. (1970). Hypersensibilite aux virus, temperature et proteines solubles chez le *Nicotiana* Xanthi n.c. Apparition de nouvelles macromolecules lors de la repression de la synthese virale. Compte rendu hebdomadaire des seances de l'Academie des Sciences de Paris 270, 2383-2386.
Gibson, R.W. & Pickett, J.A. (1983). Wild potato repels aphids by release of aphid alarm pheromone. Nature 302, 608-609.
Gibson, R.W. & Plumb, R.T. (1977). Breeding for resistance to aphid infestation. In 'Aphids as Virus Vectors' (Harris, K.F. & Maramorosch, K., eds), pp. 473-500. Academic Press, New York and London.
Graca, J.V. da, & Martin, M.M. (1975). Ultrastructural changes in tobacco mosaic virus-induced local lesions in *Nicotiana tabacum* L. cv. "Samsun *NN*". Physiological Plant Pathology 7, 287-291.
Graca, J.V. da & Martin, M.M. (1976). An electron microscope study of hypersensitive tobacco infected with tobacco mosaic virus at 32° C. Physiological Plant Pathology 8, 215-219.
Gulyas, A. & Farkas, G.L. (1978). Is cell to cell contact necessary for the expression of the N-gene in *Nicotiana tabacum* cv Xanthi nc plants infected by TMV? Phytopathologische Zeitschrift 91, 182-187.
Haberbosch, H., Nitschko, H., Strasser, R. & Mundry, K.W. (1981). Single lesion growth kinetics and early necrotic events in TMV-inoculated *Nicotiana tabacum* Xanthi nc leaves. Abstracts of the 5th International Congress of Virology, Strasbourg, p. 263.
Hall, T.J. (1980). Resistance at the *Tm-2* locus in tomato to tomato mosaic virus. Euphytica 29, 189-197.
Hanada, K. & Harrison, B.D. (1977). Effects of virus genotype and temperature on seed transmission of nepoviruses. Annals of Applied Biology 85, 79-92.
Hariharasubramanian, V., Hadidi, A., Singer, B. & Fraenkel-Conrat, H. (1973). Possible identification of a protein in brome mosaic virus-infected barley as a component of viral RNA polymerase. Virology 54, 190-198.
Hayashi, T. & Matsui, C. (1965). Fine structure of lesion periphery produced by tobacco mosaic virus. Phytopathology 55, 387-392.
Hiruki, C. & Tu, J.C. (1972). Light and electron microscopy of potato virus M lesions and marginal tissue in red kidney bean. Phytopathology 62, 77-85.
Hodgson, C.J. (1981). Effects of infection with cabbage black ringspot strain of turnip mosaic virus on turnip as a host to *Myzus persicae* and *Brevicoryne brassicae*. Annals of Applied Biology 98, 1-14.
Holmes, F.O. (1938). Inheritance of resistance to tobacco mosaic disease in tobacco. Phytopathology 28, 553-561.
Holmes, F.O. (1961). Concomitant inheritance of resistance to several virus diseases in tobacco. Virology 13, 409-413.
Holmes, F.O. (1965). Genetics of pathogenicity in viruses and of resistance in host plants. Advances in Virus Research 11, 139-161.
Hooker, W.J. & Kim, W.S. (1962). Inhibitors of potato virus X in potato leaves with different types of virus resistance. Phytopathology 52, 688-693.
Hooley, R. & McCarthy, D. (1980). Extracts from virus infected hypersensitive tobacco leaves are detrimental to protoplast survival. Physiological Plant Pathology 16, 25-38.
Israel, H.W. & Ross, A.F. (1967). The fine structure of local lesions induced by tobacco mosaic virus in tobacco. Virology 33, 272-286.
Kado, C.I. & Knight, C.A. (1966). Location of a local lesion gene in tobacco mosaic virus RNA. Proceedings of the National Academy of Sciences of the U.S.A. 55, 1276-1283.

Kalpagam, C., Foglein, F.J., Nyitrai, A., Premecz, G. & Farkas, G.L. (1977). Expression of the N gene in plasmolysed leaf tissues and isolated protoplasts of Nicotiana tabacum cv Xanthi nc infected by TMV. In Current Topics in Plant Pathology (Kiraly, Z., ed.), pp. 395-398. Akademiai Kiado, Budapest.

Karchi, Z., Cohen, S. & Govers, A. (1975). Inheritance of resistance to cucumber mosaic virus in melons. Phytopathology 65, 479-481.

Kasamo, K. & Shimomura, T. (1977). The role of the epidermis in local lesion formation and the multiplication of tobacco mosaic virus and its relation to kinetin. Virology 76, 12-18.

Kasamo, K. & Shimomura, T. (1978). Response of membrane-bound Mg^{2+}-activated ATPase of tobacco leaves to tobacco mosaic virus. Plant Physiology 62, 731-734.

Kasamo, K. & Shimomura, T. (1982). Interaction between tobacco mosaic virus and the plasma membrane of tobacco leaf cells in early infection processes. Effect of Gramicidin S and cold osmotic shock. Plant Science Letters 27, 255-264.

Kassanis, B., Gianinazzi. S. & White, R.F. (1974). A possible explanation of the resistance of virus-infected tobacco plants to second infection. Journal of General Virology 23, 11-16.

Kim, K.S. (1970). Subcellular responses to localized infection of Chenopodium quinoa by pokeweed mosaic virus. Virology 41, 179-183,

Kimmins, W.C. (1969). Isolation of a virus inhibitor from plants with localized infections. Canadian Journal of Botany 47, 1879-1886.

Konate, G., Kopp, M. & Fritig, B. (1983). Studies on TMV multiplication in systemically and hypersensitively reacting tobacco varieties by means of radiochemical and immunoenzymatic methods. Agronomie 3, 95.

Kooistra, E. (1968). Significance of the non-appearance of visible disease symptoms in cucumber (Cucumis sativus L.) after infection with cucumber virus 2. Euphytica 17, 136-140.

Kopp, M., Geoffroy, P. & Fritig, B. (1979). Phenylalanine ammonia-lyase levels in protoplasts isolated from hypersensitive tobacco pre-infected with tobacco mosaic virus. Planta 146, 451-457.

Kuhn, C.W. (1971). Cowpea chlorotic mottle virus local lesion area and infectivity increased by 2-thiouracil. Virology 43, 101-109.

Lane, L.C. (1981). Bromoviruses. In 'Handbook of Plant Virus Infection: Comparative Diagnosis' (Kurstak, E., ed.), pp. 333-376. Elsevier / North Holland, Amsterdam.

Lecoq, H., Cohen, S., Pitrat, M. & Labonne, G. (1979). Resistance to cucumber mosaic virus transmission by aphids in Cucumis melo. Phytopathology 69, 1223-1225.

Lecoq, H., Labonne, G. & Pitrat, M. (1980). Specificity of resistance to virus transmission by aphids in Cucumis melo. Annales de Phytopathologie 12, 139-144.

Legrande, M., Fritig, B. & Hirth, L. (1976). Enzymes of the phenylpropanoid pathway and the necrotic reaction of hypersensitive tobacco to tobacco mosaic virus. Phytochemistry 15, 1353-1359.

Legrand, M., Fritig, B. & Hirth, L. (1978). o-Diphenol O-methyltransferases of healthy and tobacco-mosaic-virus-infected hypersensitive tobacco. Planta 144, 101-108.

Levy, A., Loebenstein, G., Smookler, M. & Drovi, T. (1974). Partial suppression by uv irradiation of the mechanism of resistance to cucumber mosaic virus in a resistant cucumber cultivar. Virology 60, 37-44.

Lieberman, M. (1979). Biosynthesis and action of ethylene. Annual Review of Plant Physiology 30, 533-591.

Loebenstein, G., Chazan, R. & Eisenberg, M. (1970). Partial suppression of the localizing mechanism to tobacco mosaic virus by uv irradiation. Virology 41, 373-376.

Loebenstein, G. & Gera, A. (1981). Inhibitor of virus replication released from tobacco mosaic virus-infected protoplasts of a local lesion-responding tobacco cultivar. Virology 114, 132-139.

Loebenstein, G., Gera, A., Barnett, A., Shabtai, S & Cohen, J. (1980). Effect of 2,4-dichlorophenoxyacetic acid on multiplication of TMV in protoplasts from local lesion and systemic responding hosts. Virology 100, 110-115.

Loebenstein, G. & Ross, A.F. (1963). An extractable agent, induced in uninfected tissues by localized virus infections, that interferes with infection by tobacco mosaic virus. Virology

20, 507-517.
Loebenstein, G., Sela, I & Van Praagh, T. (1969). Increase of tobacco mosaic virus local lesion size and virus multiplication in hypersensitive hosts in the presence of Actinomycin-D. Virology 37, 42-48.
Loebenstein, G, Spiegel, S & Gera, A. (1982). Localized resistance and barrier substances. In 'Active Defense Mechanisms in Plants' (Wood, R.K.S., ed.), pp. 211-230. Plenum Publishing Corporation, New York.
McRitchie, J.J. & Alexander, L.J. (1963). Host-specific Lycopersicon strains of tobacco mosaic virus. Phytopathology 53, 394-398.
Martin, C. & Martin-Tanguy, J. (1981). Polyamines conjuguees et limitation de l'expansion virale chez les vegetaux. Compte rendu hebdomadaire des seances de l'Academie des Sciences de Paris 293, 249-251.
Martin-Tanguy, J., Martin, C., Gallet, M. & Vernoy, R. (1976). Sur de puissants inhibiteurs naturels de multiplication du virus de la mosaique du tabac. Compte rendu hebdomadaire des seances de l'Academie des Sciences de Paris 282, 2231-2234.
Massala, R., Legrande, M. & Fritig, B. (1980). Effect of α-aminoacetate, a competitive inhibitor of phenylalanine ammonia-lyase, on the hypersensitive resistance of tobacco to tobacco mosaic virus. Physiological Plant Pathology 16, 213-226.
Maule, A.J., Boulton, M.I. & Wood, K.R. (1980). Resistance of cucumber protoplasts to cucumber mosaic virus: a comparative study. Journal of General Virology 51, 271-279.
Milne, R.G. (1966). Electron microscopy of tobacco mosaic virus in leaves of Nicotiana glutinosa. Virology 28, 527-532.
Milo, G.E. & Srivastava, B.I.S. (1969). Effects of cytokinin on tobacco mosaic virus production in local lesion and systemic hosts. Virology 38, 26-31.
Motoyoshi, F. & Oshima, N. (1975). Infection with tobacco mosaic virus of leaf mesophyll protoplasts from susceptible and resistant lines of tomato. Journal of General Virology 29, 81-91.
Motoyoshi, F. & Oshima, N. (1977). Expression of genetically controlled resistance to tobacco mosaic virus infection in isolated tomato leaf mesophyll protoplasts. Journal of General Virology 34, 499-506.
Motoyoshi, F. & Oshima, N. (1979). Standardization in inoculation procedure and effect of a resistance gene on infection of tomato protoplasts with tobacco mosaic virus RNA. Journal of General Virology 44, 801-806.
Mozes, R., Antignus, Y., Sela, I & Harpaz, I. (1978). The chemical nature of an antiviral factor (AVF) from virus-infected plants. Journal of General Virology 38, 241-249.
Murant, A.F. (1978). Recent studies on association of two plant virus complexes with aphid vectors. In 'Plant Disease Epidemiology' (Scott, P.R. & Bainbridge, A., eds), pp 243-249. Blackwell, Oxford.
Nachman, I., Loebenstein, G., van Praagh, T. & Zelcher, A. (1971). Increased multiplication of cucumber mosaic virus in a resistant cucumber cultivar caused by actinomycin D. Physiological Plant Pathology 1, 67-71.
Nagaich, B.B. & Singh, S. (1970). An antiviral principle induced by potato virus X inoculation in Capsicum pendulum Willd. Virology 40, 267-271.
Nault, L.R. & Montgomery, M.E. (1977). Aphid pheromones. In 'Aphids as virus vectors (Harris, K.F. & Maramorosch, K., eds), pp. 528-546. Academic Press, New York and London.
Nienhaus, F. & Janicka-Czarnecka, I. (1981). Investigations on an extractable induced antiviral principle (AVP) in systemically diseased Phaseolus vulgaris upon tobacco mosaic virus infection. Zeitschrift fuer Pflanzenkrankheiten und Pflanzenschutz 88, 577-583.
Nishiguchi, M., Motoyoshi, F. & Oshima, N. (1978). Behaviour of a temperature-sensitive strain of tobacco mosaic virus in tomato leaves and protoplasts. Journal of General Virology 39, 53-62.
Ohashi, Y. & Shimomura, T. (1972). Induction of localised necrotic lesions by actinomycin D on leaves systemically infected with tobacco mosaic virus. Virology 48, 601-603.
Ohashi, Y. & Shimomura, T. (1982). Modification of cell membranes of leaves systemically infected with tobacco mosaic virus. Physiological Plant Pathology 20, 125-128.
Ohno, T., Takamatsu, N., Meshi, T., Okada, Y., Nishiguchi, M. & Kiho, Y. (1983). Single amino

acid substitution in 30K protein of TMV defective in transport function. Virology 131, 255-258.
Orchansky, P., Rubinstein, M. & Sela, I. (1982). Human interferons protect plants from virus infection. Proceedings of the National Academy of Sciences of the U.S.A. 79, 2278-2280.
Otsuki, Y., Shimomura, T. & Takebe, I. (1972). Tobacco mosaic virus multiplication and expression of the N gene in necrotic responding tobacco varieties. Virology 50, 45-50.
Owens, R.A., Bruening, G. & Shephard, R.J. (1973). A possible mechanism for inhibition of plant viruses by a peptide from Phytolacca americana. Virology 56, 390-393.
Parish, C.L., Zaitlin, M. & Siegel, A. (1965), A study of necrotic lesion formation by tobacco mosaic virus. Virology 26, 413-418.
Pavillard, J. (1952). Researches sur la croissance des plantes virosees; virus et auxines. Compte rendu hebdomadaire des seances de l'Academie des Sciences de Paris 235, 87-88.
Pelham, J. (1966). Resistance in tomato to tobacco mosaic virus. Euphytica 15, 258-267.
Pelham, J. (1972). Strain-genotype interaction of tobacco mosaic virus in tomato. Annals of Applied Biology 71, 219-228.
Pennazio, S., D'Agostino, G., Appiano, A. & Redolfi, P. (1978). Ultrastructure and histochemistry of the resistant tissue surrounding lesions of tomato bushy stunt virus in Gomphrena globosa leaves. Physiological Plant Pathology 13, 165-171.
Pennazio, S., Appiano, A. & Redolfi, P. (1979). Changes occurring in Gomphrena globasa leaves in advance of the appearance of tomato bushy stunt virus necrotic local lesions. Physiological Plant Pathology 15, 177-182.
Pennazio, S., Redolfi, P. & Sapetti, C. (1981). Callose formation and permeability changes during the partly localized reaction of Gomphrena globosa to potato virus X. Phytopathologische Zeitschrift 100, 172-191.
Pennazio, S. & Sapetti, C. (1982). Electrolyte leakage in relation to viral and abiotic stresses inducing necrosis in cowpea leaves. Biologia Plantarum 24, 218-225.
Pierpoint, W.S. (1983). Is there a phytointerferon? Trends in Biochemical Sciences 8, 5-7.
Pierpoint, W.S., Robinson, N.P. & Leason, M.B. (1981). The pathogenesis-related proteins of tobacco: their induction by viruses in intact plants and their induction by chemicals in detached leaves. Physiological Plant Pathology 19, 85-97.
Pitrat, M. & Lecoq, H. (1980). Inheritance of resistance to cucumber mosaic virus transmission by Aphis gossypii in Cucumis melo. Phytopathology 70, 958-961.
Pritchard, D.W. & Ross, A.F. (1975). The relationship of ethylene to formation of tobacco mosaic virus lesions in hypersensitive responding tobacco leaves with and without induced resistance. Virology 64, 295-307.
Rappaport, I & Wildman, S.G. (1957). A kinetic study of local lesion growth on Nicotiana glutinosa resulting from tobacco mosaic virus infection. Virology 4, 265-274.
Reanney, D.C. (1982). The evolution of RNA viruses. Annual Review of Microbiology 36, 47-73.
Redolfi, P. (1983). Occurrence of pathogenesis-related (b) and similar proteins in different plant species. Netherlands Journal of Plant Pathology 89, 245-254.
Reichman, M., Devash, Y., Suhadolnik, R.J. & Sela, I. (1983). Human leukocyte interferon and the antiviral factor (AVF) from virus-infected plants stimulate plant tissues to produce nucleotides with antiviral activity. Virology 128, 240-244.
Rochow, W.F. (1982). Dependent transmission by aphids of barley yellow dwarf luteoviruses from mixed infections. Phytopathology 72, 302-305.
Roosien, J., Sarachu, A.N., Alblas, F. & Van Vloten-Doting, L. (1983). An alfalfa mosaic virus RNA-2 mutant which does not induce a hypersensitive reaction in cowpea plants, is multiplied to a high concentration in cowpea protoplasts. Plant Molecular Biology 2, 85-88.
Ross, A.F. & Israel, H.W. (1970). Use of heat treatments in the study of acquired resistance to tobacco mosaic virus in hypersensitive tobacco. Phytopathology 60, 755-770.
Russo, M., Martelli, G.P. & Franco, A. Di (1981). The fine structure of local lesions of beet necrotic yellow vein virus in Chenopodium amaranticolor. Physiological Plant Pathology 19, 237-242.
Ruzicska, P., Gombos, Z. & Farkas, G.L. (1983). Modification of the fatty acid composition of phospholipids during the hypersensitive reaction in tobacco. Virology 128, 60-64.

Sakai, F. & Takebe, I. (1974). Protein synthesis in tobacco mesophyll protoplasts induced by tobacco mosaic virus infection. Virology 62, 426-433.

Schuster, G. & Wetzler, C. (1982). On virus-induced inhibitors from locally TMV-infected plants of Nicotiana glutinosa L. Phytopathologische Zeitschrift 104, 46-59.

Sela, I. (1981). Plant-virus interactions related to resistance and localization of viral infections. Advances in Virus Research 26, 201-238.

Sela, I & Applebaum, S.W. (1962). Occurrence of antiviral factor in virus-infected plants. Virology 17, 543-548.

Sela, I. & Harpaz, I. (1977). Further studies on the biology of an antiviral factor (AVF) from virus infected plants and its association with the N-gene of Nicotiana species. Journal of General Virology 35, 107-116.

Sela, I., Harpaz, I. & Birk, Y. (1964). Separation of a highly active antiviral factor from virus-infected plants. Virology 22, 446-451.

Sela, I., Harpaz, I. & Birk, Y. (1966). Identification of the active component of an antiviral factor isolated from virus-infected plants. Virology 28, 71-78.

Sela, I., Hauschner, A. & Mozes, R. (1978). The mechanism of stimulation of the antiviral factor (AVF) in Nicotiana leaves. The involvement of phosphorylation and the role of the N-gene. Virology 89, 1-6.

Sela, S., Loebenstein, G. & Van Praagh, T. (1969). Increase of tobacco mosaic virus multiplication and lesion size in hypersensitive hosts in the presence of chloramphenicol. Virology 39, 260-264.

Selman, I.W. (1964). The effects of kinetin on infection of petunia and tomato leaves with tomato spotted wilt virus. Annals of Applied Biology 53, 67-76.

Selman, I.W. & Yahampath, A.C.I. (1973). Some physiological characteristics of two tomato cultivars, one tolerant and one susceptible to tobacco mosaic virus. Annals of Botany 37, 853-865.

Shimomura, T. (1971). Necrosis and localization of infection in local lesion hosts. Phytopathologische Zeitschrift 70, 185-196.

Shimomura, T. (1977). The role of the epidermis in local lesion formation on Nicotiana glutinosa leaves caused by tobacco mosaic virus. Annals of the Phytopathological Society of Japan 43, 159-166.

Shimomura, T. & Dijkstra, J. (1975). The occurrence of callose during the process of local lesion formation. Netherlands Journal of Plant Pathology 81, 107-121.

Sill, W.H. & Walker, J.C. (1952). A virus inhibitor in cucumber in relation to mosaic resistance. Phytopathology 42, 349 352.

Simons, T.J., Israel, H.W. & Ross, A.F. (1972). Effect of 2,4-dichlorophenoxyacetic acid on tobacco mosaic virus lesions and on the fine structure of the adjacent cells. Virology 48, 502-515.

Solymosy, F. & Farkas, G.L. (1962). Simultaneous activation of pentose-phosphate shunt enzymes in a virus-infected local lesion host plant. Nature 195, 835.

Solymosy, F. & Farkas, G.L. (1963). Metabolic characteristics at the enzymatic levels of tobacco tissues exhibiting localized acquired resistance to viral infection. Virology 21, 210 221.

Solymosy, F., Szirmai, J., Beczner, L. & Farkas, G.L. (1967). Changes in peroxidase isozyme patterns induced by virus infection. Virology 32, 117-121.

Spencer, D.F. & Kimmins, W.C. (1971). Ultrastructure of tobacco mosaic virus lesions and surrounding tissue in Phaseolus vulgaris var Pinto.

Stobbs, L.W., Manocha, M.S. & Dias, H.F. (1977). Histological changes associated with virus localization in TMV-infected Pinto bean leaves. Physiological Plant Pathology 11, 87-94.

Sulzinski, M.A. & Zaitlin, M. (1982). Tobacco mosaic virus replication in resistant and susceptible plants: in some resistant species virus is confined to a small number of initially infected cells. Virology 121, 12-19.

Sunderland, D.W. & Merrett, M.J. (1964). Adenosine diphosphate and adenosine triphosphate concentrations in leaves showing necrotic local virus lesions. Virology 23, 274-276.

Sunderland, D.W. & Merrett, M.J. (1965). The respiration of leaves showing necrotic local lesions following infection by tobacco mosaic virus. Annals of Applied Biology 56, 477-484.

Sziraki, I. & Balazs, E. (1975). The effect of infection by TMV on cytokinin level of tobacco plants, and cytokinins in TMV RNA. In 'Current Topics in Plant Pathology' (Kiraly, Z., ed.), pp. 345-352. Akademiai Kiado, Budapest.

Sziraki, I., Balazs, E. & Kiraly, Z. (1980). Role of different stresses in inducing systemic acquired resistance to TMV and increasing cytokinin levels in tobacco. Physiological Plant Pathology 16, 277-284.

Sziraki, I & Gaborjanyi, R. (1974). Effect of systemic TMV infection on cytokinin level of tobacco leaves and stems. Acta Phytopathologica Academiae Scientiarum Hungaricae 9, 195-199.

Takahashi, T. (1973). Studies on viral pathogenesis in plant hosts. IV. Comparison of early process of tobacco mosaic virus infection in the leaves of 'Samsun NN' and 'Samsun' tobacco plants. Phytopathologische Zeitshrift 77, 157-168.

Takahashi, T. (1974). Studies on viral pathogenesis in plant hosts: VI. The rate of primary lesion growth in the leaves of Samsun NN tobacco to tobacco mosaic virus. Phytopathologische Zeitschrift 79, 53-66.

Takusari, H. & Takahashi, T. (1979). Studies on viral pathogenesis in plant hosts. IX. Effect of citrinin on the formation of necrotic lesions and virus localization in the leaves of 'Samsun NN' tobacco plants after tobacco mosaic virus infection. Phytopathologische Zeitshrift 96, 324-329.

Taliansky, M.E., Malyshenko,, S.I., Pshennikova, E.S. & Atabekov, J.G. (1982). Plant virus-specific transport function. II. A factor controlling virus host range. Virology 122, 327-331.

Tanguy, J. & Martin, C. (1972). Phenolic compounds and the hypersensitivity reaction in Nicotiana tabacum infected with tobacco mosaic virus. Phytochemistry 11, 19-28.

Taniguchi, T. (1963). Similarity in the accumulation of tobacco mosaic virus in systemic and local necrotic infection. Virology 19, 237-238.

Tarn, T.R. & Adams, J.B. (1982). Aphid probing and feeding, electronic monitoring and plant breeding. In 'Pathogens, Vectors and Plant Diseases. Approaches to Control.' (Harris, K.F. & Maramorosch, K., eds), pp. 221-246. Academic Press, New York.

Tavantzis, S.M., Smith, S.H. & Whitham, F.H. (1979). The influence of kinetin on tobacco ringspot virus infectivity and the effect of virus infection on the cytokinin activity in intact leaves of Nicotiana glutinosa L. Physiological Plant Pathology 14, 227-233.

Thomas, P.E. & Fulton, R.W. (1968a). Correlation of the ectodesmata number with nonspecific resistance to initial virus infection. Virology 34, 459-469.

Thomas, P.E. & Fulton, R.W. (1968b). Resistance to spread of virus from cell to cell in T.I. 245 tobacco. Virology 35, 108-111.

Thomas, P.E. & Martin, M.W. (1971a). Vector preference, a factor of resistance to curly top virus in certain tomato cultivars. Phytopathology 61, 1257-1260.

Thomas, P.E. & Martin, M.W. (1971b). Apparent resistance to establishment of infection by curly top virus in tomato breeding lines. Phytopathology 61, 550-551.

Thomas, P.E. & Martin, M.W. (1972). Characterization of a factor of resistance in curly top virus-resistant tomatoes. Phytopathology 62, 954-958.

Thrower, L.B. (1965). A radioautographic study of the formation of local lesions by tobacco mosaic virus. Phytopathology 55, 558-562.

Tingey, W.M. & Laubengayer, J.E. (1981). Defense against the green peach aphid and potato leafhopper by glandular trichomes of Solanum berthaultii. Journal of Economic Entomology 74, 721-725.

Tomlinson, J.A., Faithfull, E.M., Webb, M.J.W., Fraser, R.S.S. & Seeley, N.D. (1983). Chenopodium necrosis: a distinctive strain of tobacco necrosis virus isolated from river water. Annals of Applied Biology 102, 135-147.

Tsugita, A. (1962). The proteins of mutants of TMV: composition and structure of chemically evoked mutants of TMV RNA. Journal of Molecular Biology 5, 284-292.

Tsugita, A. & Fraenkel-Conrat, H. (1960). The amino acid composition and C-terminal sequence of a chemically evoked mutant of TMV. Proceedings of the National Academy of Sciences of the U.S.A. 46, 636-642.

Valleau, W.D. (1943). The relative positions of the N and N' factors in Nicotiana tabacum

chromosomes. Phytopathology 33, 14.
Valverde, R.A. & Fulton, J.P. (1982). Characterization and variability of strains of southern bean mosaic virus. Phytopathology 72, 1265-1268.
Vance, C.P., Kirk, T.K. & Sherwood, R.Y. (1980). Lignification as a mechanism of disease resistance. Annual Review of Phytopathology 18, 259-298.
Vanderplank, J.E. (1968). 'Disease Resistance in Plants.' Academic Press, New York & London.
Van Loon, L.C. (1975). Polyacrylamide disk electrophoresis of the soluble leaf proteins from Nicotiana tabacum var. Samsun and Samsun NN. Physiological Plant Pathology 6, 289-300.
Van Loon, L.C. (1976). Systemic acquired resistance, peroxidase activity and lesion size in tobacco reacting hypersensitively to TMV. Physiological Plant Pathology 8, 231-242.
Van Loon, L.C. (1977). Induction by 2-chloroethylphosphonic acid of viral-like lesions, associated proteins and systemic resistance in tobacco. Virology 80, 417-420,
Van Loon, L.C. (1979). Effects of auxin on the localization of tobacco mosaic virus in hypersensitively-reacting tobacco. Physiological Plant Pathology 14, 213-226.
Van Loon, L.C. (1982). Regulation of changes in proteins and enzymes associated with active defence against virus infection. In 'Active Defense Mechanisms in Plants' (Wood, R.K.S., ed.), pp. 247-274. Plenum Press, New York & London.
Van Loon, L.C. (1983). The induction of pathogenesis-related proteins by pathogens and specific chemicals. Netherlands Journal of Plant Pathology 89, 265-274.
Van Loon, L.C. & Antoniw, J.F. (1982). Comparison of the effects of salicylic acid and ethephon with virus induced hypersensitivity and acquired resistance in tobacco. Netherlands Journal of Plant Pathology 88, 237-256.
Van Loon, L.C. & Berbee, A.T. (1978). Endogenous levels of indoleacetic acid in leaves of tobacco reacting hypersensitively to tobacco mosaic virus. Zeitschrift fuer Pflanzenphysiologie 89, 373-375.
Van Loon, L.C. & Geelen, J.L.M.C. (1971). The relation of polyphenoloxidase and peroxidase to symptom expression in tobacco var 'Samsun NN' after infection with tobacco mosaic virus. Acta Phytopathologica Academiae Scientiarum Hungaricae 6, 9-20.
Van Loon, L.C. & Van Kammen, A. (1970). Polyacrylamide disc electrophoresis of the soluble leaf proteins from N. tabacum var Samsun and Samsun NN. II. Changes in protein constitution after infection with TMV. Virology 40, 199-211.
Venekamp, J.H. & Beemster, A.B.R. (1980a). Mature plant resistance of potato against some virus diseases. I. Concurrence of development of mature plant resistance against potato virus X and decrease of ribosome and RNA content. Netherlands Journal of Plant Pathology 86, 1-10.
Venekamp, J.H. & Beemster, A.B.R. (1980b). Mature plant resistance of potato against some virus diseases. II. Mature plant resistance and the influence of temperature on the ribosomal and RNA content in leaves. Netherlands Journal of Plant Pathology 86, 11-16.
Venekamp, J.H., Schepers, A. & Bus, C.B. (1980). Mature plant resistance against some virus diseases. III. Mature plant resistance against potato virus Y^N, indicated by decrease in ribosome content in ageing potato plants under field conditions. Netherlands Journal of Plant Pathology 86, 301-309.
Wasuwat, S. L. & Walker, J.C. (1961a). Inheritance of resistance in cucumber to cucumber mosaic virus. Phytopathology 51, 423-428.
Wasuwat, S.L. & Walker, J.C. (1961b). Relative concentration of cucumber mosaic virus in a resistant and a susceptible cucumber variety. Phytopathology 51, 614-615.
Weintraub, M., Kemp, W.G. & Ragetli, H.W.J. (1961). Some observations on hypersensitivity to plant viruses. Phytopathology 51, 290-293.
Weintraub, M. & Ragetli, H.W.J. (1964). An electron microscope study of tobacco mosaic virus lesions in Nicotiana glutinosa L. Journal of Cell Biology 23, 499-509.
Weintraub, M., Ragetli, H.W.J. & Lo, E. (1972). Mitochondrial content and respiration in leaves with localized virus infections. Virology 50, 841-850.
Weststeijn, E.M. (1976). Peroxidase activity in leaves of Nicotiana tabacum var. Xanthi nc before and after infection with tobacco mosaic virus. Physiological Plant Pathology 8, 63-71.
Weststeijn, E.M. (1978). Permeability changes in the hypersensitive reaction of Nicotiana tabacum cv. Xanthi nc after infection with tobacco mosaic virus. Physiological Plant

Pathology 13, 253-258.

Weststeijn, E.A. (1981). Lesion growth and virus localization in leaves of Nicotiana tabacum cv. Xanthi nc after inoculation with tobacco mosaic virus and incubated alternately at 22°C and 32°C. Physiological Plant Pathology 18, 356-368.

Whenham, R.J. (1983). Evaluation of selective detectors for the rapid and sensitive gas chromatographic assay of cytokinins, and application to the analysis of cytokinins in plant extracts. Planta 157, 554-560.

Whenham, R.J. & Fraser, R.S.S. (1981). Effect of systemic and local-lesion-forming strains of tobacco mosaic virus on abscisic acid concentration in tobacco leaves: consequences for the control of leaf growth. Physiological Plant Pathology 18, 267-278.

Williams, B.R.G., Golgher, R.R. & Kerr, I.M. (1979). Activation of a nuclease by pppA2'p5'A2'p5'A in intact cells. FEBS Letters 105, 47-51.

Wood, K.R. (1971). Peroxidase isoenzymes in leaves of cucumber (Cucumis sativus L.) cultivars systemically infected with the W strain of cucumber mosaic virus. Physiological Plant Pathology 1, 133-139.

Wood, K.R. & Barbara, D.J. (1971). Virus multiplication and peroxidase activity in leaves of cucumber (Cucumis sativus L.) cultivars systemically infected with the W strain of cucumber mosaic virus. Physiological Plant Pathology 1, 73-81.

Wu, J.H. (1973). Wound healing as a factor in limiting the size of lesions in Nicotiana glutinosa leaves infected with the very mild strain of tobacco mosaic virus (TMV-VM). Virology 51, 474-484.

Wu. J.H. & Dimitman, J.E. (1970). Leaf structure and callose formation as determinants of tobacco mosaic virus movement in bean leaves as revealed by UV irradiation studies. Virology 40, 820-827.

Wyatt, S.D. & Kuhn, C.W. (1979). Replication and properties of cowpea chlorotic mottle virus in resistant cowpeas. Phytopathology 69, 125-129.

Wyen, N.V., Udvary, J., Erdei, S. & Farkas, G.L. (1972). The level of a relatively purine sensitive ribonuclease increases in virus-infected hypersensitive or mechanically injured tobacco leaves. Virology 48, 337-341.

Zaitlin, M. & Leonard, D.A. (1982). A temperature-sensitive strain of tobacco mosaic virus defective in cell-to-cell movement generates an altered viral-coded protein. Virology 117, 416-424.

Zaitlin, M. & Siegel, A. (1963). A virus inhibitor from tobacco. Phytopathology 53, 224-227.

Zimmerman-Gries, S. (1979). Reducing the spread of potato leaf roll virus, alfalfa mosaic virus and potato virus Y in seed potatoes by trapping aphids on sticky yellow polyethylene sheets. Potato Research 22, 123-131.

CHAPTER 7. MECHANISMS BY WHICH GENETICALLY CONTROLLED RESISTANCE AND VIRULENCE INFLUENCE HOST COLONIZATION BY FUNGAL AND BACTERIAL PARASITES

I. R. CRUTE, P. J. G. M. DE WIT and M. WADE

7.1. INTRODUCTION

Within plant species, individuals exhibit variation for susceptibility to their microbial parasites; these in turn vary in their parasitic capability. The mechanisms that cause this variation are the subject of this Chapter.

7.1.1. Evolution and co-evolution. In the course of evolution, plants have developed many mechanisms which enable their tissues to resist invasion by potentially damaging microbes. Different species may have features in common which contribute to resistance whilst others will be unique. Likewise, evolutionary adaptation has resulted in some pathogens having the capacity to breach the defences of some plants but not others. The development of a particular parasitic capability by a microbe may permit it to colonize one, several or many species of plant. The existence of a host-parasite relationship between micro-organism and plant is therefore the starting position for most of the subsequent discussion. It is, however, worth speculating on the evolution of processes that may have led to the establishment of what has come to be called 'basic compatibility' (Ellingboe, 1976; Heath, 1981a; 1981b; 1982).

The evolution of a parasite tends to follow the evolution of its host because the host plays an essential role in the life cycle of the parasite. Similarly, host evolution may be influenced by its parasites. As a consequence, co-evolution occurs: host and parasite evolution are dependent upon each other.

An imagined example of co-evolution is given by Parlevliet (1979). At the starting point, a number of host species are all equally parasitized by a number of different facultative biotrophs. Both host and parasite populations are assumed to be genetically heterogeneous. This variation means that some individuals of some host species are less easy to exploit than the average host plant in the population. Such plants will be at a selective advantage.

Thus the characteristic which causes this advantage, a general defence mechanism, becomes more frequent, particularly since it is effective against all parasite species. Hence, a particular host species becomes less easy to exploit by particular parasites. In the course of evolution, defence mechanisms start to differ between host species as they develop independently. Parasites better adapted to host species with certain defence mechanisms tend to prevail on these species thus leading to an increased specialization of parasites to particular hosts. Parasites therefore become more and more tightly associated with a small range of host species. The rate at which this co-evolution takes place may vary. Some defence mechanisms are more likely to be rendered ineffective by parasite variants than others. Co-evolution may be so intimate that 'races' of a parasite emerge which are exclusively adapted to certain genotypes of a host species. At this stage a gene-for-gene relationship may begin to operate (Flor, 1956).

7.1.2. Scope. The primary objective of this Chapter is to examine the mechanistic relationship between host and parasite characters that influence the extent to which the host is colonized given 'basic compatibility'. In this context, there will be no attempt to explain the basis for variation in the severity of disease symptoms that may sometimes result independently of colonization (e.g. as a result of toxin production). Neither will there be discussion of the mechanisms that may be responsible for variation in how host growth or yield may be affected as a result of becoming diseased (tolerance). In terms of a previous analysis (Crute, 1985), the concern is with the causes of variation for resistance to the parasite and for parasitic ability.

The treatment of this subject will only be concerned with mechanisms which contribute to an understanding of differences in resistance between hosts of different genotype. Alterations in host resistance with age, development, environment or tissue type will not be discussed unless they help to explain genotypically determined resistance. Mechanisms which are involved in the modification of host response to a parasite as a result of a prior biotic or abiotic treatment will also only be discussed in so far as they relate to genotypically determined resistance. They are considered as resistance mechanisms in their own right in Chapter 10.

7.1.3. Establishing 'cause and effect'. One of the reasons why there is more debate associated with the study of disease resistance than any other area of plant pathology is because it is difficult to provide incontrovertible evidence for a causal relationship between any putative mechanism and the phenomena of resistance or virulence. The difficulty comes in demonstrating that any host or parasite character which is associated with resistance or virulence is in fact of primary importance rather than simply being a correlated or secondary phenomenon. In discussing the various mechanisms which could contribute to the outcome of host-parasite associations, particular attention will be given to this problem.

Unless there is variation for a particular characteristic between genetically different individuals of either host or parasite, it will not usually be possible to reach a conclusion on the importance of this characteristic in influencing either host resistance or parasite virulence.

It is difficult to study mechanisms of resistance in isolation from mechanisms influencing virulence. Understanding the mechanisms which underlie variation for virulence may help to establish the importance of certain host characteristics in providing resistance; similarly, understanding what causes the difference in resistance between hosts of different genotype may identify those parasite characteristics which contribute towards virulence.

The identification of phenotypic variation between host or parasite individuals of different genotype is usually the starting point for an investigation of 'cause and effect' in host-parasite relationships. Hence, this Chapter is concerned with identifying the mechanisms that explain the observed phenotypic differences between individuals in relation to what is known about the genetic basis for such differences.

Knowledge of the genetic basis for observed phenotypic differences in resistance or virulence is of fundamental importance for a full mechanistic understanding. It is rarely possible to reach sound conclusions in ignorance of the genetic control of the host or parasite function under investigation. Thus, there is much to be gained from the use of near-isogenic lines of both host and parasite in such studies (Daly, 1972; Johnson, 1976). This Chapter attempts to appraise critically whether proposed resistance and virulence mechanisms are in fact causal and for this reason, where possible, emphasis will be placed on studies which have employed genetically well-defined host and parasite partners.

7.1.4. <u>Different forms of parasitic nutrition and growth</u>. Whilst plant parasitic microbes share the ability to obtain their nutritional requirements from living plants, the processes that have evolved to allow this are diverse. No discussion of host resistance and parasite virulence can proceed without due consideration of these nutritional processes.

Plant parasites can be classified as necrotrophs (obtaining their nutrients from dead cells, usually previously killed by parasite metabolites) or biotrophs (obtaining their nutrition from living cells) (Lewis, 1973). While this distinction is important, the tendency to categorize parasites in this way may be misleading. It is probably more realistic to consider parasites as obtaining their nutrients either by biotrophic or necrotrophic processes. One parasite, at different stages of its association with a host may exploit both biotrophic and necrotrophic processes. For example, at the perimeter of a spreading colony of Phytophthora infestans on a potato leaf, the host-parasite relationship is briefly biotrophic. At the centre of the same lesion, however, it is said that the fungus thrives by means of necrotrophy. In contrast, toxin-producing plant parasites like Helminthosporium spp. or Alternaria spp. are probably solely necrotrophic, while some rust fungi are probably exclusively biotrophic.

The mode of parasite nutrition at any stage of development in a host-parasite association is critically important in determining the form that a potential resistance mechanisms might take. A parasite that can live only by virtue of killing host cells will be resisted by a mechanism which stops this from happening. In contrast, rapid host-cell death in response to a parasite which depends upon the living processes of its host must be expected to hinder further colonization. It is clear therefore that mechanisms of resistance to different types of organism or an organism at different developmental stages cannot always be meaningfully equated. Indeed, the same physiological and biochemical processes provoked in the host may have opposite consequences depending on circumstances.

There are two theories on the way in which different forms of parasitism may have evolved. The more commonly held opinion is that free-living saprophytic microbes evolved to become opportunist colonizers of dead tissues associated with otherwise healthy plants. Subsequently, further adaptation allowed for necrotrophic parasitism such that organisms developed the capability to kill previously healthy tissues. It is then envisaged that biotrophy evolved in response to the establishment of host mechanisms which counteracted the wholesale destruction of tissues. Thus biotrophy is seen as a more advanced trait than necrotrophy (Lewis, 1973; 1974).

The opposite view is that microbes first entered the terrestrial environment closely and harmoniously associated with plants. Such an association implies biotrophy. Evolution, it is suggested, has subsequently been in the direction of loss of dependence on the host culminating in a free-living saprophytic existence via necrotrophic parasitism (Cooke & Whipps, 1980). This latter suggestion begs the question of how biotrophic relationships evolved in pre-terrestrial times.

There is some evidence for both scenarios. Different forms of parasitic strategy are found in both bacteria and fungi and within phylogenetically unrelated groups (e.g. Phycomycetes and Basidiomycetes). This suggests that plant parasites have undergone both convergent and divergent evolution. What is important, however, is that there exist highly successful and well adapted parasites which are exclusively biotrophic, or exclusively necrotrophic, or which can express a spectrum of reliance upon both traits.

In addition to the mode of nutrition, it is also important to consider the growth habit which the parasite adopts in its host. The location of the parasite within the host and the way it colonizes host tissue are likely to be important in determining the form that any mechanism of resistance might take. The types of growth habit that microbial plant parasites adopt can be placed into eight categories:

a) growth between the cuticle and epidermal cell (e.g. *Venturia*);
b) growth over the leaf surface with haustoria in epidermal cells (e.g. *Erysiphe*);
c) intercellular growth only (e.g. *Pseudomonas* and *Cladosporium*);
d) growth between the cell wall and plasmalemma (e.g. *Colletotrichum*);
e) intercellular growth with intracellular haustoria (e.g. *Puccinia* and *Peronospora*);
f) growth within the vascular tissue (e.g. *Verticillium*);
g) growth involving intimate protoplast contact (e.g. *Plasmodiophora*), and
h) growth on tissue macerated from a distance (e.g. *Botrytis* and *Erwinia*).

Clearly, mechanisms which influence host colonization by vascular parasites and powdery mildews are likely to differ. However, the same may also be true for parasite types whose growth habits are not so strongly contrasting. An effort needs to be made to avoid generalizations or extrapolations between systems which may be biologically quite distinct.

7.1.5. Different forms of host-parasite association and the classification of resistance mechanisms. The extent of host colonization by a potential microbial parasite is a function of the host's resistance and the microbe's parasitic capability. Host and parasite individuals are likely to differ genetically for such characters and hence the outcome of differing associations between partners will be variable. The form that this variation takes has been the subject of much theorising but broadly speaking, two types of host-parasite association are possible.

In the first, all the variation in the association between host and pathogen genotypes can be explained statistically, following analysis of variance, in terms of host or parasite main effects. Host variants maintain the same relative susceptibility to one another regardless of the parasite variant considered. Similarly, regardless of host genotype on which it is assessed, parasite variants maintain the same relative degree of parasitic capability to one another. This type of association was named 'horizontal' by Vanderplank (1968). Although not an absolute rule, host resistance and parasite functions which contribute to horizontal variation are most often not under simple genetic control. This therefore has important connotations when considering the mechanistic basis for such variation.

In contrast, it is frequently the case that variation in a host-parasite association cannot be attributed solely to main effects, and a significant genotype x genotype interaction is found on analysis of variance. If this interaction is not simply due to the scale on which data are recorded, specific combinations of host and parasite variants result in degrees of colonization which cannot be predicted from the overall mean performance of either partner. This type of association is what Vanderplank (1968) called 'vertical' and it is explained by the principles of the gene-for-gene relationship first reported by Flor (1956) and further elucidated by Person (1959). Whilst the vertical association is defined in statistical terms, practical proof of its occurrence and therefore the likelihood of a gene-for-gene relationship only comes from a demonstration of the so-called 'double reciprocal check' (discussed in more detail in section 7.6; see Figure 7.4) (Vanderplank, 1982). The fact that complementary single genes control the outcome of particular genotype x genotype combinations must be accommodated by any mechanistic explanation of such events.

Some workers have suggested that all variation in host-parasite associations is under gene-for-gene control (Parlevliet & Zadoks, 1977; Ellingboe, 1981). Whilst this could be true, in this Chapter a distinction will be drawn between host-parasite associations which clearly exhibit the 'double-reciprocal check' and those for which this has not been clearly demonstrated. In any host-parasite association, genes contributing to both vertical and horizontal variation are likely to be operating simultaneously, but independently (Niks & Kuiper, 1983). This must be remembered when cytological, physiological or biochemical data are interpreted to provide a mechanistic explanation of observed events.

The foregoing perspectives on host-parasite associations owe their origins primarily to epidemiological and genetical studies. The concepts are not based on and are indeed essentially independent of any mechanistic considerations. Naturally therefore, some have felt the need to classify resistance and analagous parasite functions in mechanistic terms.

One symptom recognised for decades and apparently correlated with host resistance to microbial parasites is the so-called hypersensitive response (HR). Resistance of a plant species to a particular organism is not, however, always associated with HR. This observation led to the view that resistance could be classed as hypersensitive or non-hypersensitive. The former was considered to be a feature of gene-for-gene controlled resistance whilst the latter was not. The role and importance of hypersensitivity in host resistance is discussed later (section 7.3.3). Suffice to say here that its significance is still unclear and that there is evidence for its occurrence under diverse circumstances. It would therefore appear that this classification has little to commend it.

In a different type of mechanistic classification, Parlevliet (1977) drew a distinction between 'true' resistance and 'escape' resistance. The latter was considered to be the result of any mechanism which operated 'before an intimate relationship between the host tissue and parasite can develop'. In contrast, the former was envisaged as being due to mechanisms operating 'after intimate contact between host tissue and parasite has been established'. Parlevliet envisaged that all 'true' resistance was under gene-for-gene control whilst this was not so for 'escape' resistance. At first sight this categorization has its merits, but it will not be adopted here for three reasons. Firstly, some resistance mechanisms appear to operate at a number of stages of parasite development both before and after host penetration (Masri & Ellingboe, 1966; Ellingboe, 1972; Heath, 1974). Secondly, there is some evidence that gene-for-gene resistance can sometimes be expressed prior to 'intimate' contact (Wehtje _et al._, 1979) and thirdly, many parasites cannot be considered to have an 'intimate' relationship with their host, in that they may cause extensive cellular disruption mediated some distance away from the organism itself.

Mechanisms of resistance to fungi and bacteria have frequently been classified as either 'passive' or 'active' and such a distinction is followed in this Chapter. Both types of mechanism can operate either before or after a parasite penetrates host tissue. They are distinguished by the fact that a passive mechanism is a constitutive part of the plant's make up and requires no further energy input once it is established; in contrast an active mechanism requires energy and is only brought into operation as a result of some challenge to a healthy plant.

Passive and active defence mechanisms will be discussed in turn together with analogous parasite functions which assist in demonstrating that such mechanisms contribute to differences in resistance between genotypes.

7.2. PASSIVE RESISTANCE MECHANISMS

Man distinguishes plants one from another on the basis of consistent constitutive differences in their morphology and chemistry. At the same time, he develops some concepts of 'relatedness' by grouping into taxa those individuals with features in common. Parasitic microbes distinguish between plants in that some are satisfactory hosts and others are not. It is not surprising therefore that attempts have been made to correlate differences in the chemical or morphological make-up of plants with their relative susceptibility or resistance to attack by microbial parasites.

In order to parasitize a particular species, an organism must be physiologically adapted to gain access to the plant tissues and to grow within the chemical environment it encounters there. Therefore, both external and internal morphological and chemical features of a plant must play an important role in determining what organisms can successfully colonize it. In this sense, passive defence is probably an important component of a plant's 'non-host' resistance (see Chapter 3).

It is less certain how important passive defence mechanisms are in determining the relative susceptibility of genetically different plants of the same species to known parasites of that species. What follows is a critical evaluation of the evidence for the involvement of passive mechanisms in resistance of this type.

7.2.1. Morphological features. Investigating the role that morphological features may play in the comparative resistance between host cultivars to microbial parasites is no longer a fashionable area of research. As a consequence, there has been little significant advance since the subject was excellently reviewed by Royle (1976). The assembled evidence created a convincing case for explanation of many examples of 'age-related' and 'tissue-related' resistances by morphological features. Similarly, there is some good evidence that differences in 'environment mediated' resistance, resulting for example from growth at different levels of mineral nutrient, can be due to morphological modifications. The evidence for morphological causes of differences in resistance between host cultivars is less convincing.

Microclimate and the deposition and retention of propagules. A prerequisite for infection, particularly of the aerial parts of plants, is that propagules of a potential parasite are deposited and retained on plant surfaces appropriate for their further development. Thus, any morphological features which reduce the efficiency of inoculum deposition or retention on otherwise susceptible tissue will contribute towards resistance. Similarly, any features of the plant which affect microclimate in the vicinity of these tissues will influence cultivar susceptibility.

Differences in the gross morphology of cultivars of a species have sometimes been invoked to explain differences in their susceptibility to fungi. There is a strong association between plant height and susceptibility to attack by Septoria nodorum in wheat cultivars (Scott, 1973). Those with short straw are generally more severely affected than taller types. Part of the explanation for this is that the susceptible ears and flag leaves of shorter types are nearer the source of inoculum at the base of the plant and are in a region of high humidity conducive to infection by the splash dispersed spores. In a thorough genetic analysis of this association between plant morphology and resistance, Scott et al. (1982) concluded that it resulted from numerous genes of small effect which had pleiotropic influences on height and resistance to fungal infection. This analysis provided good evidence for a 'cause and effect' relationship although there was also some variation for susceptibility which was independent of plant height.

The growth habit of barley cultivars has been implicated in their relative resistance to Erysiphe graminis (Russell, 1975). Spore deposition was less on erect types than on those with a more prostrate habit.

Six-rowed barley cultivars with an erect spike posture were found to be more frequently infected by Helminthosporium sorokinianum than those with a 'nodding' posture. It was suggested that this could be due to prolonged moisture retention on erect spikes which would favour spore germination and infection (Banttari et al., 1975).

Surface features such as cuticle wettability and the occurrence and density of hairs have frequently been suggested as characters likely to influence host susceptibility. However, only meagre evidence is available to support this contention with respect to differences between cultivars.

Wilson & Jarvis (1963) reported that leaves of a wax-free mutant of Brussels sprout were more severely attacked by Albugo candida (white blister) than those of normal glaucous types. In wheat, single dominant genes can result in the absence of wax on stems, leaf sheaths and glumes (Jensen & Driscol, 1962). In passing, it was noted that waxless types were more susceptible to unspecified 'leaf and stem diseases'. Troughton & Hall (1967) demonstrated variation in the wettability of leaves of different wheat cultivars, but apart from commenting that two of the most wettable cultivars were also field susceptible to leaf rust (Puccinia recondita), there is no information on the pathological significance. Proof that surface wax can affect cultivar susceptibility is lacking from observations such as these.

Jennings (1962) has provided the most convincing evidence that phenotypic differences in surface morphology can affect susceptibility to fungal infection as a consequence of differences in inoculum retention. Individual canes from numerous segregating families of raspberries derived from a range of parental material were assessed for susceptibility to Didymella applanata (spur-blight), Botrytis cinerea (grey mould) and Elsinoe veneta (cane spot). Differences in hairiness, spininess and waxiness, all controlled by major genes, were also recorded. Canes with spines were more prone to attack by all three fungi than spine-free canes. Hairy leaves were associated with increased severity of attack by E. veneta but hairless canes were more prone to attack by the other two fungi. Reduced waxiness favoured infection by B. cinerea and E. veneta but attack by D. applanata was more prevalent on waxy canes. Whilst these observations may reflect linkage between genes for resistance and surface morphology rather than genuine pleiotropic effects, some further experimental evidence of a 'cause and effect' relationship was provided. The wettability of different cane types was assessed by using an aqueous dye. The presence of hairs appeared to allow little water to adhere at the nodes, the sites of infection by D. applanata. This explains the reduced suceptibility of hairy canes to this fungus. The effect of hairiness could be overridden by the effect of waxiness and to a lesser extent spininess. Both these characters appeared to impede run-off, possibly allowing infection droplets to be retained.

The resistance of cleistogamous barley cultivars to Ustilago nuda is one of the best documented examples of a morphological feature interfering with inoculum deposition on susceptible tissues. In some two-rowed barley cultivars, pollination occurs before the inflorescence emerges from the surrounding leaf sheath. As a consequence, spores of the smut fungus, which normally enter the embryo via the pistil, are physically excluded from the tissues. Artificial inoculation by injection of spores under the surrounding sheath proved that cultivars possessing this morphological protection would otherwise be as susceptible as open-flowering types (Appel, 1915; Malik & Batts, 1960). A similar exclusion mechanism was proposed to explain the high resistance of self-fertile barley and wheat cultivars to Claviceps purpurea (ergot). This fungus was found to affect male-sterile lines more frequently: their florets remained open for longer and were thus more vulnerable to infection (Puranik & Mathre, 1971).

Waller (1970) has shown that differences in the resistance of sugarcane cultivars to Ustilago scitaminea is correlated with a number of morphological characteristics of the buds through which the fungus primarily infects the plant. In one cultivar, susceptibility was greater when it was grown at a higher altitude, under conditions which also resulted in the production of a bud form which allowed spores to be more readily deposited beneath the bud scales.

Invasion. There is good evidence that the failure of tropic responses, such as those necessary to stimulate appressorium formation, greatly contributes to both 'non-host' and 'tissue-related' resistance (Wynn & Staples, 1981; Heath, 1974). There are, however, only a few indications of genotypic differences within species for characters that might so influence parasite invasion.

The best evidence comes from the study of appressorium formation by Erysiphe graminis on eceriform mutants of barley (Yang & Ellingboe, 1972). Surface waxes on the leaves of mutants have an altered physical structure and chemical composition. Abnormal appressoria were more frequently observed on mutant than on normal plants and other evidence pointed to a physical cause. It was concluded that variation in 'field resistance' might be party due to differences in surface features.

Physical exclusion may be more important than tropic failures in cultivar resistance, and is likely to involve the cuticle (cutin plus cuticular waxes) or other surface layers, as well as natural pores such as stomata and lenticels. A further mechanism whereby surface features could contribute to resistance has been elucidated by Carver & Bushnell (1983). Adherence of primary germ tubes of Erysiphe graminis to the leaf surface is essential for water uptake by the fungus, particularly under arid conditions. Establishment of the fungus could be affected by any feature which prevents adherence and water uptake.

The cuticle is probably a mechanical barrier only to those fungi which gain access to plants by non-enzymic means; its potential as a mechanical barrier was reviewed by Dickinson (1960). (The role of the cuticle as a chemical barrier is discussed in section 7.2.2.).

Wolkow et al. (1983) gained evidence for mechanical effects using a different experimental approach. They investigated the effects of inhibitors of melanin biosynthesis on infection of beans by Colletotrichum lindemuthianum. Tricyclazole and two other related compounds prevented melanization of fungal cell walls. Unmelanized appressoria appeared to lack the rigidity needed to facilitate penetration. The same fungicides failed to give any protection against artificial puncture inoculations.

Undoubtedly, differences in the strength or thickness of surface layers of plants can frequently explain 'age-related', 'environment-related' and 'tissue-related' resistances (Akai, 1959; Royle, 1976; Wang & Pinkard, 1973). Evidence that physical properties of surface layers contribute to differences in resistance between cultivars is more difficult to find.

Characteristics of the cuticle which made penetration more difficult were considered to explaining the observed differences in susceptibility between strawberry cultivars to Sphaerotheca macularis (powdery mildew) (Peries, 1962). Whilst no conclusions could be drawn on whether this effect was due to physical or chemical features, in a similar study employing two Fragaria species, Jhooty & McKeen (1965) concluded that cuticular thickness contributed to resistance.

Nagdy & Boyd (1965) obtained a strong correlation between the surface tuber infection of thirty potato cultivars with Oospora pustulans (skin spot) and periderm thickness. The protective effect of the periderm appeared to be related to the number of cells, the thickness of the suberin layer and percentage of crude fibre in the periderm.

The silicon content of rice epidermal cell walls clearly contributes to their strength and also to their resistance to penetration by Pyricularia oryzae (blast) (Akai, 1959). The extent of silicon deposition varies with cell type and is influenced by leaf age and growing conditions, particularly the availability of silicon (Volk et al., 1958; Akai & Fukutomi, 1980). It is less certain whether cultivar differences in the silicon content of epidermal cell walls also influence their resistance.

Many microbial parasites rely upon natural pores or wounds for entry; for bacteria they are essential. Differences in the morphological characteristics of stomata, lenticels and other types of natural pore, together with their number, arrangement and periodic behaviour have consequently been implicated in defence mechanisms.

In a thorough review, Royle (1976) showed that claims that cultivar resistance can be explained by pore frequency or distribution are often conflicting. He further concluded that it is difficult to substantiate the suggestion that stomatal structure, as distinct from physiological function, contributes to resistance.

There is a single example where stomatal structure alone is thought to control resistance. This relates to differences between two species, Citrus nobilis (mandarin) and C. grandis (grapefruit), in susceptibility to Pseudomonas citri (canker). If the bacterium is inoculated through wounds or forced into the sub-stomatal cavity under pressure, both species are susceptible. However, the cuticular lips which surround the stomatal vestibule in C. nobilis are thought to exclude bacteria under natural circumstances (McLean, 1921; McLean & Lee, 1922).

An interaction between guard cell structure and physiological function was thought to explain resistance to penetration by Puccinia recondita of stomata on the sheaths and culms of wheat (Romig & Caldwell, 1964). Stomata on leaves were more readily penetrated. Since profuse uredia occurred on the sheaths and poduncles of one cultivar, it was suggested that differing resistance of cultivars to the penetration mechanism might explain the between-cultivar difference. Better evidence for the contribution of a stomatal exclusion mechanism to differences in cultivar susceptibility was obtained by Kochman & Brown (1975). They found that on leaves of two oat cultivars differing in susceptibility to Puccinia coronata, appressoria were formed with equal frequency over stomata of both, but penetration occurred less frequently on the more resistant cultivar. In earlier work, however, Brown & Shipton (1964) had illustrated that frequency of penetration of wheat cultivars by different isolates of Puccinia graminis was not a consistent feature of either cultivar or isolate and there was a clear cultivar x isolate interaction.

Wilson & Coffey (1980) studied three potato cultivars which differed in their 'field' resistance to Phytophthora infestans (blight). One resistant cultivar had a lower stomatal frequency and since penetration tended to occur into cells adjacent to stomata this was considered to contribute to its overall low susceptibility.

Internal development. Having gained access to host tissues, it is necessary for some types of parasitic microbe to spread within them. Any physical features which restrict spread contribute towards resistance. Pathogens that grow intercellularly will be restricted by different features from those which grow intracellularly. Whilst structural features may help in explaining why some plant tissues are extensively colonized and others are not, there is little evidence to substantiate earlier suggestions that differences of this type between cultivars are due to the physical strength of their tissues, be it cell walls or the intercellular matrix (Royle, 1976; Akai & Fukutomi, 1980).

Murray & Bruehl (1981, 1982) presented evidence that the greater resistance of some wheat cultivars to Pseudocercosporella herpatrichoides (eyespot) resulted from a thicker, more heavily lignified stem hypodermis. Whether this provided a physical or chemical barrier is not known.

Pathogens causing wilt diseases colonize the vascular tissue and attempts have been made to explain differences in cultivar resistance in terms of anatomical differences in the xylem vessels. Vessel length and diameter and the size of vessel groupings differ between types of elm. These differences are thought to be correlated with resistance to the spread of Ceratocystis ulmi spores (Elgersma, 1970; McNabb et al., 1970). The contention is that spores move less readily through shorter, narrower vessels and that movement in small groups of vessels would be more readily restricted by the formation of tyloses and other blockages. A similar mechanism was proposed by Cho et al. (1973) as part of the explanation for the differences in resistance of alfalfa cultivars to Corynebacterium insidiosum. In this case resistant plants were found to have fewer vascular bundles, shorter vessels and a thicker cortex, features considered to make vascular invasion more difficult. However, resistance could not be ascribed solely to these features as it was also shown that aqueous extracts from resistant plants were inhibitory to bacterial growth.

Ratoon stunting disease of sugarcane is thought to result from infection of the vascular tissue by a small coryneform bacterium. The vascular anatomy of 'immune' or 'tolerant' clones differed from that of 'susceptible' and 'field resistant' clones. The former types had a greater impedance to water flow which appeared to be related to fewer large vascular bundles traversing the nodes (Teakle et al., 1975).

Dissemination. A potentially important contribution to resistance could involve restriction of spread of the parasite either from plant to plant or between parts of the same plant (e.g. leaf to leaf spread). There may be physical features of some plants which make it more difficult for a parasite to be disseminated than would otherwise be the case.

For example, in dissemination of rust fungi the surface layers of the plant must rupture to allow the developing pustules to release their spores. It has been suggested that differences in the thickness and degree of lignification of the epidermis could account for the restricted development of uredosori in some cultivars of flax infected by Melampsora lini (Scharvelle, 1936) and of wheat infected by Puccinia graminis (Hart, 1931). Peduncles of wheat cultivars can vary in their anatomy for such characters as epidermal thickness and quantity of collenchyma, the tissue in which uredosori develop. Skovmand et al. (1977) have shown that these differences are genetically controlled characters with low environmental variability. The suggestion is that a relatively large quantity of sclerenchyma in comparison to collenchyma, and a thick epidermis should result in less extensive fungal development. This should give smaller pustules, less able to erupt through the surface. Wilcoxon (1958) also provided evidence that the sclerenchyma tissue of some cultivars contained more abundant intercellular spaces than others. The contention is that this causes the tissue to be weaker and accounts for greater tissue damage and more rapid appearance of larger uredia.

7.2.2. Chemical features.

Absence of a chemical stimulus. There are good examples of parasites whose propagules (particularly resting structures) are stimulated to germinate by the presence of their host (Schroth & Hildebrand, 1964; Coley-Smith & Holt, 1966; Coley-Smith et al., 1967; Morgan, 1983). Similarly, with parasites possessing a motile stage, plant exudates may attract them to their host. Chemical stimulation of attachment to a receptive host surface has also been reported (Kerr & Flentje, 1957; Flentje et al., 1963). In some cases the identity of the exuded chemical stimulant is known; more often it is uncharacterized. Frequently, chemical stimuli are non-specific and do not explain host specificity of particular parasites (DeSilva & Wood, 1964). However, when the stimulus is specific, this is a clear case of parasitic adaptation to an appropriately receptive host and absence of the necessary stimulus must therefore be considered to be a component of 'non-host' resistance. The question is whether variation exists within a host species in ability so to attract its parasites, and whether such variation could contribute to cultivar resistance.

Buxton (1957a; 1957b) presented evidence that uncharacterized components of the root exudates from pea cultivars differing in their specific resistance to Fusarium oxysporum f. sp. pisi (wilt) differentially stimulated the germination of different races of the fungus. Similar claims were made for banana cultivars resistant and susceptible to races of Fusarium oxysporum f. sp. cubense (Buxton, 1962). However, in independent studies using different pea cultivars, these effects could not be reproduced (Kommendahl, 1966; Whalley & Taylor, 1973). Furthermore, Lacey & Horner (1966) could find no evidence that resistant and susceptible species of mint differentially affected the rhizosphere populations of Verticillium dahliae (wilt).

Schroth & Cook (1964) found a correlation between the amount of exudate from pea seeds of different cultivars during germination, and their relative susceptibility to pre-emergence damping-off by both Rhizoctonia solani and Pythium spp. The variation appeared to result from characteristics of the seed-coat, which rarely cracked prior to radicle emergence in the most resistant cultivar. With this exception, and despite reports of qualitative and quantitative differences in exudates from different genotypes of several species, there seems little reason to believe that a lack of chemical stimulation commonly contributes to cultivar resistance.

Lack of nutrients. A plant species must provide an appropriate nutritional substrate if a micro-organism is to parasitize it. Non-availability of an essential vitamin or other micronutrient could restrict colonization. In early work, evidence was sought for the control of differences in susceptibility between cultivars by nutrients (Garber 1956), but little was found. This is not surprising: crude sterilized aqueous extracts from plants often provide an ideal culture medium for micro-organisms regardless of whether they are parasitic on that plant or not.

Several workers showed that fungal and bacterial parasites which had mutated to auxotrophy for specific vitamins or amino-acids had also, although not always, lost pathogenicity. Sometimes, pathogenicity was lost for only certain cultivars or species, the implication being that some were better suppliers of the essential nutrient than others. Pathogenicity could sometimes be restored by providing the essential nutrient exogenously (Garber et al.,

1956; Garber & Shaeffer, 1957; Kline et al., 1957; Garber & Heggestad, 1958; Tinline, 1963). These studies, however, provided no evidence that natural variation in virulence and resistance is commonly controlled in this way (Williams & Boone, 1963).

Only a few studies imply a role for specific nutrients in the control of cultivar resistance. Lukezic & DeVay (1964) showed that Rhodosticta quercina, which causes canker on plums, has a vitamin requirement for myo-inositol. A highly susceptible plum cultivar had much more of the vitamin in its bark than two resistant cultivars. Booth (1969, 1974) found that root exudates from a cotton cultivar susceptible to Verticillium albo-atrum contained a higher concentration of alanine while more choline exuded from roots of a resistant cultivar. In vitro, growth of the fungus was increased by the presence of alanine, which also stimulated pectolytic enzyme activity. These effects were counteracted by choline. Treating the resistant cultivar with alanine increased its susceptibility, suggesting that resistance could be due to alanine deficiency.

Recently, Temmen et al. (1980), have made the intriguing histological observation that resistance of Ribes cultivars to Spaerotheca mors-uvae (powdery mildew) appears to correlate with a 'lack of sufficient cytoplasm' in the epidermal cells. The contention is that in some cultivars and tissues, epidermal cells provide insufficient nutrients to the fungus via its haustoria for the completion of its life cycle.

Unsuitable pH. The growth of micro-organisms can be greatly influenced in vitro by pH, and different potential parasites have different pH optima for growth. As sap from different plants varies in pH, there have been attempts to correlate resistance with pH differences. The effects could be direct, in that the enzyme systems of a potential parasite might be adversely affected outside a certain pH range. Alternatively, pH could exert an indirect effect by influencng the toxicity of some host antimicrobial metabolite.

Wood (1967) considered that the direct effect of pH was probably an important determinant of whether particular plant species were suitable hosts, especially for bacterial parasites. There is also evidence that pH can contribute indirectly to 'non-host', 'tissue-related' and 'age-related' resistance (Smith et al., 1946; Schlösser, 1975).

The limited evidence that pH can play a role in cultivar resistance is far from convincing, being based primarily on comparisons between a few resistant and susceptible cultivars. Braun (1942), for example, suggested that the resistance of a water melon cultivar to Fusarium oxysporum f. niveum (wilt) was due to more acetic acid in the shoots than in a susceptible cultivar while Hafiz (1952) concluded that the resistance of chickpea cultivars to Mycosphaerella rabiei (blight) was due to the inhibition of spore germination by low pH at the leaf surface. This was considered to result from a higher concentration of malic acid secreted by a larger population of glandular hairs on resistant compared to susceptible cultivars.

Osmotic pressure. Schnathorst (1959) provided evidence that resistance in lettuce to Erysiphe cichoracearum (powdery mildew) can be due to the high osmotic value of leaf sap. Osmotic values depended upon host genotype, leaf age and position, and photosynthetic activity. The implication was that uptake of water by the fungus was inhibited. This hypothesis appeared to explain: the field resistance of certain cultivars; differences in susceptibility between leaves of different ages; increased susceptibility of leaves infected by lettuce mosaic virus; alterations in susceptibility brought about by mineral nutrition and light intensity, and the susceptible response of field-resistant cultivars when grown under some conditions.

Whilst the findings explained the resistance of some cultivars to a 'lettuce-attacking' pathotype of E. cichoracearum, they did not account for the resistance of all lettuce cultivars to a pathotype of the fungus occurring on Lactuca serriola. However, the observations assume new relevance in the light of findings by Carver & Bushnell (1983), that water uptake in the early stages of infection is vital to the establishment of infection by germlings of Erisyphe graminis. The suggestion is that resistance would be contributed by any host feature which interferes with the chances of germling survival, particularly under arid conditions.

Encouragement of antagonistic organisms. It is well established that organisms at or near plant surfaces can influence infection by potential parasites (Schroth & Hildebrand, 1964; Baker, 1968; Fokkema, 1976; Blakeman & Brodie, 1976; Blakeman & Fokkema, 1982). Chemical exudates from plants also influence the nature of the microflora in their vicinity (Godfrey, 1976). In some cases, chemical treatments can be used to modify this microflora to the detriment of a parasite (Burchill & Cook, 1971). Against this background, it has been suggested that plants may differ in resistance to a potential parasite as a consequence of differing associated microflora; those individuals specifically favouring a particularly strong antagonist would be the more resistant.

Harper (1950) recorded that cultivars of banana resistant to *Fusarium oxysporum* f. sp. *cubense* (Panama disease) had a quantitatively and qualitatively different rhizosphere microflora to susceptible cultivars. A bacterium antagonistic to the parasite was only isolated from the rhizosphere of a resistant cultivar. However, these observations could not be verified (Rombouts, 1953).

Differential effects of cultivars on their rhizosphere microflora have also been invoked to explain the resistance of peas to *Fusarium oxysporum* f. sp. *pisi* (Buxton, 1957a; 1957b) and tomatoes to *Verticillium albo-atrum* (Subba-Rao & Bailey, 1961). However, the importance of these effects have not been independently established. Considering more recent information on post-penetration mechanisms of resistance to wilt fungi, it would appear that the observed effects could reflect coincidental association.

The most comprehensive analysis of the possible interactions between host genotype, rhizosphere and parasite has been made by Atkinson and his co-workers (Neal *et al.*, 1970; 1973; Larson & Atkinson, 1970; Atkinson *et al.*, 1974; 1975) who studied the resistance of wheat to *Cochliobolus sativus* (common root rot). By utilising chromosome substitution lines, it was shown that chromosome 5B had a major effect in determining resistance or susceptibility. This chromosome also appeared to exert an influence on the rhizosphere microflora. Lines carrying the chromosome from the resistant donor had a smaller bacterial population associated with their roots than did susceptible lines. Also, a higher proportion of those bacteria associated with resistant lines exhibited antibiotic activity against the parasite. However, the addition of antagonistic bacteria to the susceptible lines failed to alter their reaction to *Cochliobolus*, and it was also shown that under some circumstances, susceptible lines did not always have a low proportion of antagonistic bacteria in their associated microflora. It was concluded that resistance was not mediated through bacterial antagonists. Nevertheless, this careful study emphasises the impact that a plant's genotype can have on its associated microflora. Whether this can in turn ever be responsible for variation in susceptibility to parasites remains to be conclusively demonstrated.

Swinburne (1981) has recently highlighted another intriguing means whereby interactions between micro-organisms in the vicinity of a plant may influence its response to parasites. Spore germination of some fungi was enhanced in the presence of iron chelating agents or when the spores were produced under iron-deficient conditions. Iron deficiency also appeared to influence virulence: a preliminary investiation revealed that resistant bean cultivars reacted as susceptible to a normally incompatible isolate of Colletotrichum lindemuthianum (anthracnose), when spores were applied in the presence of a chelating agent. It was suggested that the iron deficient fungus may in some way circumvent normal host resistance. Further investigation seems appropriate, considering that the fluorescent pigment characteristic of many pseudomonads can chelate iron (Kloepper et al., 1980; Misaghi et al., 1982). In these studies iron deficiency appeared to explain the growth inhibition of several microbial parasites.

Chemical constitution of the host cuticle and cell walls. This section evaluates the evidence that cuticle and cell wall chemistry may contribute to cultivar resistance by differences in their response to degradation by a potential parasite. Degradation may be necessary for entry of the parasite or to release nutrients. Resistance contributed by toxic materials in the surface layers of plants or cell walls is dealt with in sections 7.2.7, 7.3.5 and 7.3.7.

Evidence that the cuticle can form a chemical barrier to infection by fungi has been elegantly provided by Shaykh et al. (1977) and Maiti & Kolattukudy (1979). Fusarium solani f. sp. pisi secreted a cutinase during the penetration of pea cuticle. Penetration was prevented both by an antibody to the cutinase, and by inhibition of the enzyme by diisopropyl fluorophosphate, although germination and germ tube growth were unaffected. Despite these findings, there appears to be no evidence for differences in cuticle chemistry being responsible for variation in cultivar resistance.

Albersheim et al. (1969) stated that all microbial plant parasites have the ability to produce polysaccharide degrading enzymes; a view later supported by others (Albersheim & Anderson-Prouty, 1975; Bateman & Basham, 1976). Different races of Colletotrichum lindemuthianum were shown to secrete different amounts of α-galactosidase, depending on whether the cell walls on which they were grown had been isolated from hypocotyls of resistant or susceptible beans (English & Albersheim, 1969). It was therefore suggested that differences in cell wall chemistry could explain gene-for-gene controlled resistance through enzyme regulation (Albersheim et al., 1969). Mussell & Green (1970) also provided evidence that tissues of resistant cultivars of tomato and cotton may regulate the production of polygalacturonase production by the wilt fungi Verticillium albo-atrum and Fusarium oxysporum f. sp. lycopersici. However, subsequent research has revealed how similar the primary cell walls of plants are in their construction and this hypothesis is now considered untenable (Albersheim & Anderson-Prouty, 1975).

In a study of the hydrolytic enzymes produced by the same wilt fungi, Cooper & Wood (1975) and Cooper et al. (1978) found no differences in the amounts or types of enzymes produced when the fungi were grown on cell wall preparations from resistant or susceptible tomato cultivars. There appeared to be no differences in the ease with which degradation of resistant or susceptible tissues occurred. As Cooper et al. (1978) point out, different genes control resistance to the two different wilt fungi and it is implausible to suggest that cell wall polysaccharides could control this specificity when enzyme action and the induction of enzyme synthesis in the two fungi is so similar.

No strong evidence exists therefore to indicate that cultivar resistance can be mediated through differences in the chemistry of host cell walls.

Preformed antimicrobial substances. Healthy plants contain and exude a wide range of substances with antimicrobial activity (Swain, 1977; Blakeman & Atkinson, 1976). The involvement of these 'secondary metabolites' in defence has been the subject of considerable research.

For an organism to parasitize a host, it must presumably be insensitive to any toxic metabolites present; avoid contact with them, or be able to inactivate them. There is every reason to believe that antimicrobial metabolites have a major role in non-host resistance.

Ingham (1973) divided plant antibiotics into four categories. Metabolites with activity, and present prior to infection were termed 'prohibitins'; those formed after infection from pre-existing substrates simply as a consequence of hydrolysis or oxidation were termed 'post-inhibitins'. A similar distinction was drawn by Bell (1981). These two categories were considered to constitute passive defence. In contrast, active defence could be based either upon 'inhibitins' found in tissues prior to infection but increasing markedly in invaded tissues, or on phytoalexins whose formation requires either gene derepression or the activation of a latent enzyme system. The role of prohibitins and post-inhibitins in cultivar resistance is examined in this section.

Three classes of prohibitins might have a role in cultivar resistance: phenolics, saponins and terpenes. From research on the resistance of onion to Colletotrichum circinans (smudge) (Angell et al., 1930; Link et al., 1929;

Link & Walker, 1933; Walker & Stahmann, 1955), evidence was assembled to indicate that the phenolic compounds catechol and protocatechuic acid in the outer scales of red or yellow onion cultivars were responsible for resistance. Unpigmented cultivars, and unpigmented areas of yellow or red onions were susceptible, and lacked these phenolics. In a later study, Ramsey et al. (1946) suggested that resistance to Diplodia natalensis could be accounted for by the same compounds. However, subsequent investigations with other host-parasite combinations have invariably failed to show that pre-formed inhibitors clearly contribute to differences in cultivar resistance. The early work by Walker and collaborators remains a 'land-mark', but as Schlösser (1980) pointed out, a re-examination using modern methodology would be worthwhile.

Relative concentrations of chlorogenic acid or its quinone auto-oxidation product have also been implicated in resistance (Schaal & Johnson, 1955). The periderm of a potato cultivar resistant to Streptomyces scabies (scab) was found to have a higher concentration of chlorogenic acid than a susceptible cultivar (Johnson & Schaal, 1957). A similar correlation with concentrations in the vascular tissue of potato and tobacco cultivars was established for resistance to the wilt fungi Verticillium albo-atrum and V. dahliae respectively (McLean et al., 1961; Sheppard & Peterson, 1976). Differences in cultivar susceptibility to Cercospora beticola (leaf spot) in sugar-beet has also been attributed to differing concentrations of an uncharacterized phenolic substance extracted from leaves (Harrison et al., 1961). In these studies it has been suggested that screening for phenolic content may aid selection of resistant breeding material, but the lack of critical genetic data allows no conclusion to be reached on whether or not the association is genuinely causal.

In contrast to the above, the investigation of the role in resistance to Erwinia chrysanthemi of an antibacterial cyclic hydroxamate in some maize lines by Lacey et al. (1969) reached conclusions which are more typical of studies on pre-formed inhibitors. Strains of the bacterium parasitic on maize were generally less sensitive to this metabolite than types non-parasitic on maize, but there was no evidence that host lines carrying a gene for hydroxamate production were any less susceptible.

Avenacin A and B, triterpene saponins produced by oats, function as prohibitins, through a well-documented membranolytic action (Schönbeck & Schlösser, 1976). Turner (1961) showed that the reason why Ophiobolus *graminis* var. *avenae* (take-all) could attack oat roots, while *O. graminis* from wheat could not, was because the former enzymatically inactivated avenacin. Several root and foliar parasites of oat are also able to convert the active saponin to various non-active forms (Schönbeck & Schlösser, 1976). Despite these findings, there appears to be no evidence for avenacin production contributing to differences between oat genotypes in resistance to any parasite.

α-tomatin is an alkaloid saponin prohibitin produced by tomatoes. As with avenacin, some tomato parasites are considered capable of enzymatic inactivation. There is also evidence of insensitivity to tomatin due to a low sterol content (Defago *et al*., 1983). Alternatively, some parasites grow in the presence of tomatin because they lower the pH in their vicinity to a level at which the saponin is not toxic. In addition to its role in non-host resistance, tomatin appears to control both 'age-related' and 'tissue-related' resistance (Schlösser, 1975; Défago & Kern, 1983; Défago *et al*., 1983).

There is conflicting information on the involvement of tomatin in determining the relative parasite resistance of tomato cultivars. Mohanakumaran *et al*. (1969) suggested that breeding lines resistant to *Pseudomonas solanacearum* (wilt) produced higher levels of tomatin than a susceptible cultivar. Resistance to race 1 of *Fusarium oxysporum* f. sp. *lycopersici* was not considered by Drysdale & Langcake (1973) to be due to tomatin, which they found increased to equivalent levels in susceptible and resistant cultivars. However, Hammerschlag & Mace (1975), using different cultivars, showed that regardless of whether plants were inoculated or not, a resistant line produced a significantly greater amount of tomatin than a susceptible line. If tomatin is important in determining specific resistance to race 1, isolates of race 2 should be less sensitive, but neither study examined this possibility.

Dow & Callow (1978) suggested a way in which tomatin could be important in gene-for-gene controlled resistance to *Cladosporium fulvum* (leaf mould). It was envisaged that the increased leakage from host cells characteristic of incompatible race-cultivar combinations could result in the accumulation of toxic levels of tomatin in the intercellular spaces where the fungus was located. Considering that phytoalexins are also produced under these circumstances (see section 7.3.4), the validity of this hypothesis is uncertain.

High concentrations of the monoterpene, β-phelladrene, have been shown to be significantly associated with resistance in slash pine to Cronartium fusiforme (rust), although the correlations quoted were not strong (Rockwood, 1974).

Cruickshank et al. (1977) concluded that a fungitoxic diterpene associated with the epicuticular wax of leaves of some Nicotiana species played no role in determining resistance to Peronospora hyoscyami f. sp. tabacina (downy mildew).

In addition to compounds of known identity there are a few cases of unidentified chemicals with prohibitin properties which have been implicated in cultivar resistance. For example, Lampard & Carter (1973) found that the activity of a chloroform soluble antifungal compound which co-extracted with cuticular wax from berries of a range of coffee cultivars was highly correlated with resistance to Colletotrichum coffeanum (anthracnose). This confirmed earlier reports that the cuticle was important in determining cultivar resistance (Nutman & Roberts, 1960). The inhibitory effects of leaf exudates on the germination of spores of Colletotrichum graminicola (anthracnose) were shown by Sharma & Sinha (1971) to differ between sorghum cultivars with differing resistance but the compounds responsible were not identified.

Four classes of post-inhibitins may play a role in cultivar resistance: unsaturated lactones, cyanogenic glycosides, phenolic glycosides and sulphur-containing compounds.

Unsaturated lactones occur in plants as inactive glycosides. Active fungitoxic compounds are liberated by the action of β-glucosidases following tissue damage (Schönbeck & Schlösser, 1976). Tuliposides A and B are two such glycosides found in tulips, which have been implicated in the control of resistance to bulb-rotting fungi. There is good evidence that tuliposides are important in controlling the non-host resistance of tulips to Botrytis cinerea (Schönbeck & Schroeder, 1972). In comparison, the pathogenic B. tulipae is less sensitive to the lactone, causes less tuliposide to be released from tissue and this is primarily converted to the stimulatory hydroxycarboxylic acid rather than to the lactone. In addition, tuliposides play a role in 'age-related' resistance to Fusarium oxysporium f. tulipae (bulb rot) (Bergman & Beijersbergen, 1968) but there is no evidence that differences in susceptibility between cultivars to either of these diseases are attributable to tuliposides.

Cyanogenic glycosides in plants release hydrogen cyanide following tissue damage and contact with hydrolases. There have been claims that resistance to certain parasites is correlated with the HCN content of tissues (Timonin, 1941) but these have not been substantiated (Ludtke & Hahn, 1953; Trione, 1960). In contrast, there is clear evidence that cyanide tolerance may be important for pathogens of cyanogenic plants (Fry & Millar, 1971a; 1971b).

Mustard oils (esters of isothiocyanic acid) occur as glycosides in cruciferous plants and have antimicrobial properties (Schönbeck & Schlösser, 1976). Whilst it has been suggested that they have a role in non-host resistance (Davis, 1964), their involvement in cultivar resistance is uncertain. Greenhalgh & Mitchell (1976) found that a cabbage cultivar resistant to Peronospora parasitica (downy mildew) had higher concentrations of allyl isothiocyanate in its tissues than susceptible cultivars. Similarly, amongst wild populations of Brassica oleracea, high mustard oil concentration appeared to correlate with a greater proportion of seedlings resistant to the fungus. More work, particularly with hybrid populations segregating for resistance and mustard oil concentrations, is needed to verify these suggestions.

Phenolic glucosides present in plant tissues release phenols after the action of β-glucosidases. In apples and pears, phloridzin and arbutin have respectively been implicated in resistance to Venturia inaequalis (apple scab) and Erwinia amylovora (fire-blight).

In apples it has been shown that phloridzin is converted to the aglucone, phloretin by the action of β-glucosidase. Both compounds are converted to a fungitoxic but unstable o-quinone by the action of polyphenoloxidase. There appears to be no difference between cultivars in their potential to produce the fungitoxic product. However, the contention is that in incompatible cultivar-isolate combinations, membrane damage leads to cellular disorganisation (hypersensitivity) and hence to the formation of toxic metabolites which in turn stop fungal growth. The cause of the initial host cell response is however still obscure (Raa & Overeem, 1968; Noveroske et al., 1964; Raa & Kaars Sijpesteijn, 1968).

In pears, oxidation of arbutin gives rise to an antibacterial hydroquinone and it is suggested that this compound may confer resistance to Erwinia amylovora. Resistant cultivars were shown to contain larger amounts of arbutin and in all cultivars, tissues more susceptible to bacterial attack had lower concentrations. Since neither cultivar nor tissue resistance were completely correlated with β-glucosidase activity, it may be that other antibacterial metabolites associated with arbutin were also involved (Hildebrand & Schroth, 1964; Smale & Keil, 1966; Hildebrand et al. 1969).

The ability to degrade plant cell walls to varying degrees is probably an essential characteristic of all microbial parasites (Albersheim & Anderson-Prouty, 1975). It has been suggested therefore that inhibitors of the enzymes involved, particularly polygalacturonase, might form the basis of a host defence mechanism. Good evidence has been produced for host phenolics and proteins which are inhibitory, but there are no positive indications that differences in cultivar susceptibility can be mediated in this way (Patil & Dimond, 1967; Anderson & Albersheim, 1972; Jones et al., 1972; Fischer et al., 1973).

In summary, much effort has been devoted to the identification and characterization of pre-formed, anti-microbial metabolites in plants. Such compounds appear to have a role to play in the determination of non-host resistance but in only a few instances are there any indications that they are of primary importance in determining differences in susceptibility between cultivars of a host species.

Insensitivity to toxins. Phytotoxins are usually defined as non-enzymic, diffusable products of bacteria, fungi, or of a host-parasite interaction, which are harmful to plants at low concentrations. Those that have an important role in causing disease are sometimes called pathotoxins.

Pathotoxins may be selective or non-selective. Selective pathotoxins are markedly more toxic to plants susceptible to the parasite which produces them than to resistant plants. In contrast, non-selective toxins are also toxic to some plants which are resistant to the parasite which produces them. Misaghi (1982) and Wheeler (1975) provide examples of these different types of toxin; Rudolph (1976) lists toxin-producing parasites. Recent reviews include those by Wood et al., 1972; Durbin, 1981; Scheffer, 1976; Strobel, 1974a; 1976; 1977; Rudolph, 1976; Patil, 1974; and Yoder, 1980.

If the action of a toxin is essential to or influences successful parasitic colonization, then insensitivity of a plant to the toxin must result in resistance to the organism concerned. There is incontrovertible evidence that this is so in several host-parasite combinations involving host-selective toxins. This, together with information on the basis of the insensitivity of some cultivars to toxins is considered in this section. In addition, more limited evidence that insensitivity to non-selective pathotoxins may also contribute to cultivar resistance is assessed. The significance of toxins in mechanisms of non-host resistance is considered in Chapter 3.

There are thirteen fungal diseases where good evidence exists that the parasite produces one or more host-selective toxins. These are six forms of Alternaria alternata (syn. A. tenuis) specialized to different host species (Nishimura et al., 1982); four graminicolous species of Helminthosporium (H. victoriae on oats, H. carbonum and H. maydis on maize and H. sacchari on sugar cane), and the species Periconia circinata (on sorghum), Phyllosticta maydis (on maize) (Scheffer & Yoder, 1972) and Corynespora cassiicola (on tomato) (Onesirosan et al., 1975). Evidence that the toxins produced are essential for extensive host colonization and thereby that toxin insensitivity can be equated with resistance is more compelling for some host-parasite combinations than others.

Primary evidence for a causal relationship between susceptibility to the parasite and sensitivity to the toxin comes from the agronomic background and the way in which the disease in question became prominent. Scheffer (1976) reviewed several cases of 'man-made' diseases which became important because of the introduction of cultivars which subsequently proved to be particularly sensitive to a fungal toxin. The most familiar case concerns the disastrous maize leaf-blight epidemic in the USA in 1970. This resulted from the extensive use of F_1 hybrid cultivars carrying Texas male-sterile (Tms) cytoplasm. These cultivars were highly suceptible to disease caused by a 'new' race (T) of Helminthosporium maydis compared to the 'original' race (O). Cultivars carrying Tms cytoplasm proved selectively sensitive to T-toxin which was produced solely by race T of the fungus.

However, there does appear to be a distinction between susceptibility to H. maydis and sensitivity to the toxin. Race T isolates will grow on maize cultivars insensitive to T-toxin to much the same extent as race O isolates, and race O isolates also grow on cultivars sensitive to T-toxin. Sensitivity to T-toxin conditions high susceptibility to disease (i.e. leaf-blight) rather than to the parasite itself. This separation of effects on parasite colonization as distinct from symptom production is also apparent from genetic studies (Yoder & Gracen, 1975; Yoder, 1980).

Extreme selectivity in toxin sensitivity between cultivars has been observed experimentally for several host-parasite combinations. Usually, but not invariably, an absolute correlation between susceptibility to the fungus and toxin sensitivity has been recorded (Scheffer & Pringle, 1961; Luke & Wheeler, 1964; Scheffer & Ullstrup, 1965; Schertz & Tai, 1969; Steiner &

Byther, 1971; Steiner & Strobel, 1971; Scheffer & Yoder, 1972; Comstock et al., 1973; Yoder, 1973; Nishimura et al., 1979; 1982; Gilchrist & Grogan, 1976; Kono & Daley, 1979; Scheffer & Livingston, 1980). It is on this basis that toxins have been defined as host-selective. The degree of selectivity depends upon the toxin, the most striking example being victorin (HV toxin) from H. victoriae where resistant oat cultivars are more than 40,000 times less sensitive than susceptible cultivars. In contrast, the ratio for HC toxin from H. carbonum on maize cultivars is only 100. In addition to the toxic selectivity to intact plants or tissues, a similar phenomenon has been observed in some cases with naked protoplasts (Samadder & Scheffer, 1968; Earl et al., 1978; Pelcher et al., 1975), and in the case of T-toxin, with mitochondria (Miller & Koeppe, 1971; Comstock et al., 1973; Gregory et al., 1980; Matthews et al., 1979; Payne et al., 1980; Bednarski et al., 1977).

In some cases, a toxin has been shown to reproduce many, if not all, of the macroscopic, histological and biochemical symptoms associated with successful fungal colonization (Kuo & Scheffer, 1970; Steiner & Byther, 1971; Strobel & Steiner, 1972) although this in itself is not proof that the toxin is causally responsible for susceptiblity to the fungus as distinct from the disease.

More convincing evidence for a cause and effect relationship comes from comparative studies using virulent and avirulent isolates of the parasite. In such cases, toxin-producing isolates are invariably virulent on toxin-sensitive cultivars, while strains incapable of toxin production prove avirulent (Scheffer & Pringle, 1967; Odvody et al., 1977). Parasite isolates incapable of toxin production will, however, colonize toxin-sensitive cultivars when the toxin is applied exogenously (Yoder & Scheffer, 1969; Comstock & Scheffer, 1973).

Genetic data entirely support the contention that cultivar susceptibility to the fungus is due to toxin sensitivity (Yoder, 1980). For example, Scheffer et al. (1967) demonstrated that progeny from crosses between H. victorae (parasitic on oats and producer of HV toxin) and H. carbonum (parasitic on maize and producer of HC toxin) were either parasitic on one, both or neither host species. Toxin production was invariably associated with parasitic ability and the segregation ratios indicated that different single loci controlled toxin production in each species.

Host sensitivity to the toxin may be dominant, semidominant or recessive to insensitivity, depending on the host-parasite combination. In the case of nuclear inheritance, one or several loci can control the character (Nelson &

Ullstrup, 1964; Payne & Yoder, 1978; Schertz & Tai, 1969; Pringle & Scheffer, 1964; Kuo et al., 1970). The ability of the fungus to produce toxin may also be under simple genetic control (Nelson et al., 1963; Scheffer et al., 1967; Yoder & Gracen, 1975; Yoder, 1976; Gilchrist & Grogan, 1976).

It has been possible to select plants for toxin insensitivity by using tissue culture techniques, or within predominantly sensitive populations (Gracen et al., 1971; Gengenbach et al., 1977; Scheffer, 1976). Plants selected for toxin insensitivity have invariably proved to be resistant to the fungus.

The evidence leads to the conclusion that cultivar resistance is a direct consequence of insensitivity to the toxin. To understand what the basis for this insensitivity is, it is necessary to know how toxins exert their effect on susceptible cultivars. However, since the chemical structure of host-selective toxins is still in doubt, there are no cases where the precise molecular basis for toxin action has been established with certainty, although damage to host membranes is strongly suspected to be a primary effect. The biochemical mode of action of toxins has been comprehensively discussed by Daly (1981) and is beyond the scope of this Chapter.

It has long been suggested that resistant cultivars lack a specific toxin receptor which is present in susceptible cultivars (Pringle & Scheffer 1964). This is a plausible working hypothesis, but proof of the existence of such a receptor and its location have proved elusive. Work by Strobel (1973a; 1973b; 1974a; 1974b; 1975; 1976) purports to show that sugarcane clones sensitive to HS-toxin produced by Helminthosporium saccharri carry a specific binding protein associated with the plasmalemma. The particular protein was not found in toxin-insensitive clones. The protein was said to be responsible for the transport of α-galactosides but its interaction with HS-toxins was considered to disrupt ion transport. Various aspects of this work have subsequently been strongly challenged (Daly, 1981) and the conclusions remain to be verified.

Earlier circumstantial evidence for a passive mechanism of resistance based on the lack of a toxin receptor came from work with H. victoriae and HV toxin by Samaddar & Scheffer (1970). Dormant aleurone cells from resting seeds of sensitive and insensitive cultivars were treated briefly with toxin prior to exposure to gibberellic acid, which promotes α-amylase synthesis. Resistant treated, and untreated aleurones produced α-amylase, while sensitive treated aleuorones did not. This implied that the toxin exerted its effect in the absence of metabolic activity, and that resistance to the toxin was also therefore a passive process.

This model of toxin action implies that the induction of host susceptibility is the specific event, and is supported by genetic data indicating that susceptibility and virulence are under the control of dominant alleles in the host and parasite respectively. As Wheeler (1975) points out, such a situation is exactly opposite to that normally encountered with gene-for-gene controlled host-parasite systems, where it is resistance and avirulence that are under dominant control, with resistance considered to be the specific event (Ellingboe, 1981).

An interesting perspective which goes some way to explaining the contrast outlined above was initially envisaged by Litzenberger (1949) and has been further expounded by Kaars Sijpesteijn and co-workers (Raa & Kaars Sijpesteijn, 1968; Van Dijkman & Kaars Sijpesteijn, 1971; 1973; Van Dijkman et al., 1973; Kaars Sijpesteign & Van Dijkman, 1973; Kaars Sijpesteijn, 1976). As a result of work with gene-for-gene controlled associations between Venturia inaequalis (scab) and apple, and Cladosporium fulvum (leaf mould) and tomato, it was suggested that avirulent fungal isolates produced toxins (possibly proteins) to which only resistant cultivars were receptive and sensitive. As a consequence, host cell death (hypersensitivity) occurred in such cultivar-isolate combinations, which it was thought might cause the accumulation of fungitoxic substances (Dow & Callow, 1978). Specific elicitors of hypersensitivity (see sections 7.4 and 7.6.2) are therefore conceived to be host-selective toxins. The notion that toxin-induced cell death may result in host susceptibility in the case of one type of parasite (e.g. Helminthosporium victoriae) but lead to resistance in another (e.g. Cladosporium fulvum) was rejected by Wheeler (1975) on histological evidence, particularly in that resistance in oat cultivars to H. victoriae was also expressed as reduced colonization, restricted to hypersensitive lesions (Jennings & Ullstrup, 1959; Yoder & Scheffer, 1969).

Nevertheless, it is possible that the dynamics of the host-parasite relationship at the early stages of invasion could be all-important in determining the eventual outcome. For example, toxin production by necrotrophic Helminthosporium species on a sensitive cultivar may debilitate the natural processes of active defence (see section 7.3), thus resulting in unfettered colonization and more toxin production. In contrast, active defence processes may proceed normally following invasion by a non-toxin producing isolate or invasion of a toxin-insensitive cultivar. Colonization would thus be restricted and localized in hypersensitive lesions. It seems necessary for species such as Cladosporium fulvum to establish a brief biotrophic relationship during the early stages of host colonization (although this may also be true of Helminthosporium spp.). In such a case, host sensitivity to toxin may be responsible for reduced growth allowing active defence mechanisms to function and severely affect subsequent colonization. If, on the other hand, host cells initially remain healthy in the absence of toxin or as a result of insensitivity, fungal growth may proceed rapidly, such that subsequent extensive cellular disruption and necrosis impairs active defence.

These suggestions are consistent with observations that phytoalexin accumulation (as an example of active defence), is usually associated with early restriction of colonization despite the fact that host cell necrosis may eventually be greater in susceptible host-parasite combinations (see section 7.3.4).

There is one further interesting contrast between those systems where toxin sensitivity conditions susceptibility, and those where there are some suggestions that toxin sensitivity conditions resistance. In none of the former cases, is there evidence for the double-reciprocal check arrangement (Figure 7.4) which is commonplace with associations of the latter type. Theoretically, there is no reason why this should not occur in both situations since it only requires a fungus to produce a minimum of two selective toxins which discriminate different cultivars. Above the cultivar level, there is a double reciprocal arrangement between oats and maize sensitive respectively to HV and HC toxins from H. victoriae and H. carbonum.

Helminthosporium turcicum, which causes Northern corn leaf-blight of maize, clearly does exhibit a double-reciprocal arrangement in the association between races and cultivars (Perkins & Hooker, 1981). Although this fungus induces symptoms more akin to a wilt than related species, it would be of considerable interest to know whether the cultivar-race relationship is also mediated through host selective toxins. However, toxin production by H. turcicum does not seem to have been examined.

In the case of toxins that are host non-selective, any role that they may have in mediating cultivar susceptibility or resistance to the organism producing them is much less clear. There are cases where cultivars do exhibit different degrees of sensitivity to a non-selective toxin which is manifest in reduced symptom expression (Goodman et al., 1975) Gray & Chamberlain, 1975; Noyes & Hancock, 1981). This is particularly the case with wilt diseases but it is likely that this is often a reflection of resistance to the disease as opposed to resistance to the parasite itself. There is little evidence that strains of a parasite producing non-selective toxins grow more extensively than non-toxin producing strains, even though the disease symptoms they cause may be more severe. Similarly, there is little evidence that parasitic colonization is greater in a host cultivar particularly sensitive to a toxin of this type than in one which is less sensitive. This distinction between resistance to the parasite and resistance to the disease is well illustrated by the work of Carlson (1973), who used tissue culture techniques to select tobacco cells resistant to tabtoxin produced by *Pseudomonas tabaci* (wildfire). The plants subsequently regenerated from these cells were resistant to the effects of the toxin, but there was no effect on the extent of bacterial multiplication.

A different potential role for this sort of toxin was suggested by Keyworth (1963; 1964), who used grafting to study resistance in tomatoes to *Fusarium oxysporum* f. sp. *lycopersici*. Scions from monogenically-resistant cultivars were more sensitive to an uncharacterized toxin produced by the fungus in a susceptible rootstock than were susceptible scions. The implication is that the toxin was eliciting a hypersensitive response along similar lines to that discussed previously for selective toxins.

A further role for non-selective toxins has been suggested by Patil (1974), based on studies with *Pseudomonas phaseolicola* (halo-blight) on beans. The hypothesis is that the toxin may prevent the expression of active host defence mechanims, so that uninhibited multiplication can occur. In support of this, application of exogenous toxin to resistant cultivars resulted in greater multiplication than normal (Ganamanickam & Patil, 1977; Rudolph, 1972). If this suggestion is correct, the primary target of host resistance to the bacterium is toxin production. The situation is confused however by the observation that a nontoxigenic mutant of *P. phaseolicola* multiplied *in vivo* as readily as the wild-type although it did not induce chlorosis (Patil *et al.*, 1974). In addition, there are indications that host sensitivity to the toxin and resistance to the bacterium are under different genetic controls (Hill *et al.*, 1972).

Inactivation of toxins. Where pathotoxin activity is needed to facilitate parasite growth, different capacities for inactivation by host cultivars could result in differences in susceptibility to the parasite. Patil (1980) has reviewed the evidence: there is only a modest amount and all of it is circumstantial (Wheeler, 1969). As Patil points out, resistance or susceptibility to host-selective toxins is usually expressed rapidly, implying that inactivation is not the basis for cultivar specificity, although it could account for quantitative differences in susceptibility between host cultivars.

Lytic enzymes. It has been suggested that resistance, particularly to fungi, may be contributed by the ability of the host to lyse hyphae (Wargo, 1975; Nichols *et al.*, 1980). Work on this has tended to concentrate on two glucanhydrolases; chitinase and β-1,3-glucanase, which hydrolyze common components of fungal walls.

Nitzsch (1983) suggested that the capacity to synthesize chitinase could be transferred to higher plants lacking this capability by 'genetic manipulation' thereby enhancing their resistance to fungal attack. The reduced severity of infection of wheat by *Puccinia striiformis* (yellow rust) following inoculation in the presence of chitinase was presented as evidence that such a character could be of value. However, there is little direct evidence that resistance to fungi can be attributed to lytic enzymes.

Enzymes capable of lysing hyphae do occur naturally in healthy plant cells (Clarke & Stone, 1962; Abeles *et al.*, 1971; Wargo, 1975). In some cases, it would appear that they are released from a compartmented state after infection (Wilson, 1973). There is no evidence that metabolic inhibitors interfere with the production of lytic enzymes, and as they are found in healthy plants, they have been considered here to constitute a passive defence mechanism. However, it should be noted that enhanced chitinase synthesis can be induced in plants either by ethylene treatment or by pathogenesis (Boller *et al.*, 1983a; 1983b). This enhanced synthesis is thought to play a role in resistance to fungi and if this is so, the mechanism is 'active' under these circumstances.

The relationship between hyphal lysis and both chitinase and β-1,3-glucanase in cultivars of tomato monogenically resistant or susceptible to *Verticillium albo-atrum* (wilt) has been investigated in detail (Dixon & Pegg, 1969; Pegg & Vessey, 1973; Vessey & Pegg, 1973; Pegg, 1976; Pegg & Young, 1981; 1982; Young & Pegg, 1982). Indications that lysis may contribute to resistance was obtained from a resistant cultivar-isolate combination, where the number of vessels containing hyphae began to decline 14 days after inoculation. Possibly, this may have been the result of autolysis since in pure culture, the fungus produced chitinase which appeared to influence its growth. However, enzyme preparations from healthy plants were capable of hydrolyzing chitin and purified host β-1,3,-glucanase and chitinase from infected plants degraded the hyphal walls of *V. albo-atrum*.

It was concluded that lytic enzymes may be relevant components of a general defence but they do not provide an explanation of major gene cultivar resistance. This contention was further strengthened by a thorough investigation of both chitinase and β-1,3-glucanase activity in near-isogenic resistant and susceptible tomato lines, either inoculated or uninoculated. The resistant cultivar possessed no greater capacity for enzyme synthesis, and the data implied that enzyme levels were proportional to the biomass of fungus present in infected parts of the plant.

The best evidence that lytic enzymes may be involved in cultivar resistance has been provided by Netzer *et al.* (1979). They investigated levels of β-1,3-glucanase in two near-isogenic lines of muskmelon resistant or susceptibe to race 0 of *Fusarium oxysporum* f. sp. *melonis* (wilt). The enzyme was detected in healthy plants of both lines but activity markedly increased after inoculation. However, in the resistant line, the activity was twice that in the susceptible line. Treatment with the substrate laminarin also markedly increased enzyme activity in uninoculated plants and again the response of the resistant line was considerably greater. Furthermore, the substrate-induced activity was paralleled by reductions in the amount of fungus and disease severity in the susceptible line, compared to untreated controls.

These findings are however confused by an earlier observation (Rabenantoandro *et al.*, 1976). In a different muskmelon cultivar, a similar increase in β-1,3-glucanase which followed inoculation with *Colletotrichum lagenarium* (anthracnose) was clearly accounted for by an endoenzyme of fungal origin.

Apart from the direct role that lytic enzymes may have in defence, Young & Pegg (1982) pointed out that they may also function to release degradation products which function as phytotoxins or elicitors of other defence responses from fungal walls (section 7.4).

7.2.3. *Parasite variation and passive resistance mechanisms.* It is difficult to conceive of a means whereby a defence mechanism, which operates by reducing inoculum deposition or retention, could impose selection pressure on a parasite such that forms able to counteract the host's resistance would evolve. However, once there is actual contact between the parasite and the host or its metabolites, the possibility exists that variants may be selected from the parasite population against which a particular defence mechanism is less effective.

Having said this, there is surprisingly little evidence for parasitic variation in respect of passive defence mechanisms. Whether this is because such evidence has not been sought or because such variation does not exist is not clear. It is conceivable that adaptation to overcome structural features of a plant may be beyond a parasite's physiological competence, but it seems unlikely that there would be no variation for sensitivity to chemical features.

There is no theoretical reason why passive mechanisms of defence should not control a 'double-reciprocal' relationship between host and parasite variants. However, in practice there is no evidence for this type of association when passive processes explain host variation. Thus as it is understood at present, passive resistance would all appear to be 'horizontal' or 'non-differential' as defined by Vanderplank (1968).

7.2.4. <u>Some conclusions on passive resistance mechanisms</u>. Much of the evidence for the involvement of passive processes in cultivar resistance is based on correlation, sometimes from examination of as few as two cultivars. This is not a sound logical nor statistical basis for proof of a causal relationship. Occasionally, direct observation of the host-parasite association has given additional support to a particular hypothesis, but critical experimental tests have been rare.

The problems of relying too heavily upon correlation have been highlighted by Wood (1967). Firstly the characteristic which is thought to be causally related to resistance may in reality be of no consequence, but might actually be closely associated with a third characteristic which is. Secondly, the discovery of a cultivar whose characteristics do not fit the correlation does not necessarily invalidate the hypothesis since resistance may result from the involvement of more than one character. A difficulty in testing hypotheses which are based on passive mechanisms is that there is no evidence for 'double-reciprocal' arrangements resulting from specific relationships between host and parasite variants. A genetic approach similar to that adopted by Scott <u>et al</u>. (1982) and recommended by Yoder (1981) (see sections 7.2.1. and 7.2.2) is probably the best way of establishing 'cause and effect'. Hybridization of resistant and susceptible individuals, followed by one or two further generations of genetic reassortment has the effect of 'randomising' the host genetic background. This overcomes the problem of the many other differences between cultivars other than the one causing resistance. Even studies with near-isogenic lines do not fully counteract this difficulty.

If, in the segregating populations created by hybridization, resistance is still invariably associated with the characteristic postulated as its cause, the hypothesis whilst not proved must be considered sound. Provided population sizes are large enough and both resistance and the other character in question can be accurately determined, it should even be possible in the absence of recombinant types, to discount the involvement of genetic linkage. Clearly this approach is only applicable to sexually reproducing host species with a relatively short generation time, and to cases where resistance and the character postulated to cause it can be assessed on the same individual.

For an organism to be parasitic on a particular host species, it must have the capacity to breach, be insensitive to, or render ineffective the armoury of passive defences outlined in the foregoing sections. Since all microbial parasites are restricted in their host range, it is probable that passive mechanisms have a considerable role in 'non-host' resistance. The same is probably also true for 'age-related', 'tissue-related' and some forms of 'environment-related' resistance.

It must be concluded, however, that major differences in the susceptibility of cultivars to potential parasites can rarely be attributed to a passive defence mechanism alone. But to be able to investigate resistance mechanisms, differences between cultivars need to be relatively great. Passive processes may be more often responsible for the smaller but nevertheless significant differences in susceptibility to parasites which are invariably observed between cultivars when careful quantitative assessments are made. For obvious reasons, mechanistic studies on these quantititative differences have rarely been undertaken.

7.3. ACTIVE RESISTANCE MECHANISMS

Active resistance mechanisms are energy-requiring processes and may be divided into two temporally and functionally distinct phases: a determinative phase during which recognition is thought to take place, leading to a phase in which the resistance mechanism is expressed (Keen & Bruegger, 1977). Inhibition of the expression phase with inhibitors of nucleic acid and protein synthesis is possible, and can turn an incompatible (resistance) association into a compatible (susceptibility) one. This suggests that the expression phase has to be induced (Yoshikawa *et al*., 1978; Keen *et al*., 1981; Keen, 1982) although there are contrary suggestions that it is susceptibility which is the induced phenomenon (see section 7.4).

7.3.1. Exclusion and containment of parasites. A number of active host responses appear to cause physical containment or exclusion of the parasite. These may occur in different ways depending on the type of parasite. Evidence that these mechanisms contribute to cultivar resistance is discussed in this section.

Papilla formation. One of the earliest morphological responses of plants to directly penetrating fungi is the formation of papillae. These occur directly beneath the penetration peg of the infecting organism and are considered synonymous with lignitubers or callosities (Aist, 1976). They are thickenings on the inner surfaces of plant cell walls adjacent to the cell membrane. Most ultrastructural studies suggest that papillae are formed before physical penetration of the host cell wall occurs.

The chemical composition of papillae is heterogenous (Aist, 1976; Heath, 1980a). They commonly contain large proportions of callose (a β-1,3 glucan) which has recently been claimed to be an intermediate in cellulose synthesis (Meier *et al.*, 1981). Lignin has also been identified as a major component. Phenolics, suberin, gums, proteins and silicon are also reported (Aist, 1976).

Papillae have been reported to block fungal penetration of host cells. Sherwood & Vance (1980) showed that after inhibition of papilla formation with cycloheximide, twelve species in the Gramineae were penetrated by three isolates of *Erisyphe graminis* normally unable to do so. However, as Aist (1976) pointed out, the use of a protein synthesis inhibitor like cycloheximide may not only inhibit papilla formation but also any other active resistance mechanism. Hence the role of papillae in providing resistance is unproved.

Papilla formation occurs in susceptible as well as in resistant hosts but seems to be more pronounced in the latter, the response being quantitative rather than qualitative. Papilla formation has been associated with gene-for-gene controlled resistance in tomato to *Cladosporium fulvum* (Lazarovits & Higgins, 1976; De Wit, 1977), potato to *Phytophthora infestans* (Hohl & Stössel, 1976) and barley to *Erisyphe graminis* f. sp. *hordei* (Skou, 1982; Koga *et al.*, 1980; 1983).

Envelopment and immobilization of bacteria. The response of plants to heterologous bacteria (i.e. those which do not multiply substantially in the plant) has been the subject of intense research over the last ten years. Three types of heterologous combination can be recognised (although the distinctions have often been overlooked):

a) between plants and non-parasitic bacteria, either saprophytic species or mutants of parasitic species;

b) between non-hosts and bacteria pathogenic on a different host species either closely or distantly related to the non-host, and

c) between a resistant cultivar of a host species and a race of a bacterial parasite avirulent on it.

The first two types of association have been the most extensively studied. This Chapter is primarily concerned with the third type of association but the processes involved in restricting bacterial multiplication may be common to all three.

It has been suggested that the bacteria to which a host is susceptible multiply in intercellular fluids and do not come into close contact with host cell surfaces. In contrast, bacterial multiplication may be inhibited in an unconducive host because they become bound to the plant cell wall, and enveloped or encapsulated in a granular and fibrillar material apparently of host origin (Huang et al., 1975; Goodman et al., 1976; Sing & Schroth, 1977; Sequeira et al., 1977).

This general hypothesis has been challenged for a number of reasons. Hildebrand et al. (1980) proposed that the apparent active attachment and encapsulation of heterologous bacteria was an artifact which resulted after the substomatal cavities dried out following inoculation. In bean leaves, they found that water alone dissolved material from the cell wall surfaces and both heterologous and homologous bacteria appeared to become entrapped in this substance if drying occurred. No attachment was apparent if water soaking conditions were maintained. Daub & Hagedorn (1980) also concluded that encapsulation was non-specific and occurred in bean cultivars both resistant and susceptible to Pseudomonas syringae pv. syringae. They could find no evidence from growth studies that immobilization of bacteria was an important mechanism of resistance. Using a well defined, race-specific system, Fett & Jones (1982) found that some envelopment of Pseudomonas syringae pv. glycinea occurred in both compatible and incompatible cultivar-race combinations. Envelopment did not appear to be necessary for the expression of hypersensitivity and no specific and active host response was indicated.

In the compatible association between tobacco and Pseudomonas syringae pv. tabaci, Mazzuchi et al. (1982) noted that bacteria became encapsulated at the cell wall surface and later trapped in a fibrillar material within the intercellular spaces. Encapsulation was interpreted as a phase which enabled survival until active growth commenced.

Since modifications of the host cell wall are often observed in the vicinity of bacterial cells apparently attached to them, the notion that envelopment is always a passive process may be an oversimplification. What is important is whether attachment and envelopment alone play a role in restricting multiplication. Since many bacteria in a heterologous combination did not become associated with host cell walls, Smith & Mansfield (1981; 1982) concluded that it was unlikely that the process was important in resistance. Roebuck *et al*. (1977) in their study of resistance to *Pseudomonas syringae* pv. *phaseolicola* in beans, and Carson *et al*. (1978) examining the race-specific relationship between *Xanthomonas malvacearum* and cotton cultivars, both recorded attachment of bacteria to cell walls in resistant combinations, although the encapsulating material in these two studies appeared to have different properties. However, both sets of workers were unconvinced that immobilization was in any way responsible for the cultivar resistance observed. The hypersensitive collapse of host cells which followed attachment was considered to be of more significance. Clear evidence that heterologous bacteria apparently entrapped at host cell surfaces in tobacco leaves are metabolically active and can undergo division has been provided by Al-Issa & Sigee (1982).

In homologous combinations, it has been suggested that bacteria are able to accumulate large amounts of extracellular polysaccharide (EPS) which prevents contact with the host cell wall and maintains the water-soaked conditions necessary for muliplication. The absence of EPS in heterologous combinations might therefore result in contact between bacteria and host cells, leading to a host response (Smith & Mansfield, 1982). EPS from *Pseudomonas syringae* pv. *phaseolicola* has been shown to induce the water-soaking symptom characteristic of infection of susceptible hosts (El-Banoby & Rudolph, 1979; 1980; El Banoby *et al*, 1980; 1981). The water-soaking phenomena induced by purified EPSs from different races of the bacterium generally paralleled their cultivar specificity. Resistant cultivars degraded and inactivated EPS whereas susceptible cultivars did not. Evidence for the importance of EPS in determining susceptibility is however, conflicting, since heterologous strains of *Pseudomonas solanacearum* which produce copious amounts of EPS still become attached and enveloped at the surface of tobacco mesophyll cells (Sequeira, 1981).

On balance, there appears to be little cause to believe that immobilization and envelopment of bacteria alone can explain cultivar resistance. There is more reason to conclude that contact between bacteria and host-cells may be required for the expression of symptoms associated with resistance such as hypersensitivity (Stall & Cook, 1979). Hypersensitivity and the mechanisms which may be responsible for the agglutination of bacteria to cell walls are discussed in sections 7.3.3. and 7.5.1.

Vascular occlusion. Fungi and bacteria that infect the vascular tissue of plants frequently cause the vessels to become occluded by pectic gels, gums and tyloses. The last are balloon-like enlargements of vascular parenchyma cells, which project through pits in the vessel walls and into the lumen. Processes involved in tylose formation have been discussed by Beckman (1971). It has been suggested that vascular blockages constitute a defence mechanism which inhibits the systemic spread of the parasite. It has also been suggested that such responses cause water stress and eventual wilting (Struckmeyer et al, 1954; Robb & Busch, 1983). In wilt diseases, however, resistance to the parasite and resistance to the disease may not always be equated (Pegg & Dixon, 1969).

Vascular occlusion is clearly a non-specific response, in that it may be induced by different organisms and abiotic factors (Powers, 1954; Beckman & Halmos, 1962; Beckman, 1966; Van der Molen et al., 1977; Gardner et al., 1983). The formation of gums and tyloses may play a role in non-host resistance to vascular invasion (Davis, 1966).

How efficient vascular occlusions are in limiting parasite spread within a plant depends on how rapidly they form (Talboys, 1972). Rapid occlusion followed by sufficient hyperplasia of the xylem to compensate for the blocked vessels could result in reduced parasite spread and symptom severity. Delayed occlusion would not prevent parasite spread even though the number of blocked vessels could eventually be greater. Uncertainties about the importance of occlusions in resistance to vascular parasites are well illustrated by a few case studies.

Monogenic resistance to Verticillium albo-atrum in tomato was associated with formation of more tyloses than in a susceptible cultivar (Blackhurst & Wood, 1963). This could explain why invasion of leaf petioles in resistant plants was much less than in susceptible plants, even though stems were initially equally colonized. Sinha & Wood (1967) confirmed these findings using isogenic lines. Not only were more tyloses formed in resistant lines following infection, but occlusion was faster. However, indications that antifungal substances might also contribute to the more restricted parasite colonization were confirmed by Tjamos & Smith (1974; 1975), who concluded that tylose formation was of secondary importance to phytoalexin production.

Hutson & Smith (1980) showed convincingly that phytoalexin production preceded tylose formation in a monogenic resistant tomato line infected by V. albo-atrum, although Elgersma (1980) found that the phytoalexin, rishitin, accumulated to the same extent in both susceptible and resistant cultivars. Gel formation was variable and appeared unrelated to resistance. The involvement of phytoalexins agrees with the findings of Dixon & Pegg (1969). Before knowledge about the production of phytoalexins in this system, they concluded that tyloses did not form soon enough in resistant tomato cultivars to inhibit systemic spread of the fungus. In addition, tyloses formed extensively only after infection by some fungal strains, and did not correlate well with resistance.

A similar uncertainty exists in explaining monogenic resistance of tomatoes to Fusarium oxysporum f. sp. lycopersici. Beckman (1968) concluded that rapid vascular occlusion and restriction of the fungus, initially by gels, was a likely explanation of cultivar resistance. Subsequent studies using isogenic lines (Elgersma et al., 1972; Beckman et al., 1972) suggested that the prevention of systemic invasion, characteristic of monogenic resistance, could be explained solely by more rapid tylose formation. In contrast, however, Stromberg & Corden (1980) could find no evidence for containment of this fungus by tyloses when resistant and susceptible tomato lines were examined using scanning electron microscopy. The SEM revealed that hyphae and spores had collapsed in resistant plants, suggesting that fungitoxic substances might be of primary importance. Direct evidence for the involvement of phytoalexins in resistance prior to processes of occlusion was provided by Hutson & Smith (1980).

In the case of wilt fungi of tomatoes, therefore, it would appear that if vascular occlusion has a role in monogenic cultivar resistance, it is by confining the fungus, thus allowing phytoalexins more readily to exert their effects.

In sweet potato clones resistant to F. oxysporum f. sp. batatas, more tyloses formed after infection than in susceptible clones (Collins & Nielsen, 1976). However, because of the timing of the response, and the poor growth of hyphae in vessels of resistant lines, it was considered unlikely that occlusions alone could account for resistance.

Stuehling & Nelson (1981) concluded that neither gums nor tyloses played any part in the partial resistance of a cultivar of chrysanthemum to F. oxysporum f. sp. chrysanthemi since the fungus spread readily in the plant. Resistance was associated with less profuse mycelial growth.

In contrast to these last two reports, resistance of both elms and oaks to infection by Ceratocystis species has been associated with more rapid and extensive vascular occlusion, particularly by tyloses (Elgersma, 1973; Jacobi & MacDonald, 1980). Although direct observations were not carried out, Sinclair et al. (1975) provided good circumstantial evidence that infections were localized in resistant elms.

In comparison with susceptible cultivars, vascular occlusions by both gels and tyloses have been reported to occur more rapidly in cotton cultivars resistant to Fusarium oxysporum f. sp. vasinfectum or Verticillium dahliae (Bugbee, 1979; Mace, 1978; Harrison & Beckman, 1982). Differences in response to the two fungal species were found, but since phytoalexins only accumulated at infected sites after occlusion had occurred, the latter process was implicated in reducing systemic spread. However, the sequential integration of occlusion mechanisms and phytoalexin production was considered to be of critical importance, because the fungi were thought to have the enzymatic capability to degrade both gels and tyloses.

Pierce's disease of grapevines is thought to be caused by a bacterial infection of the vascular system. In vine species 'tolerant' of the disease, Mollenhauer & Hopkins (1979) recorded a greater incidence of occlusion by both gums and tyloses, which were infrequent in susceptible types. The bacteria were encapsulated in these occlusions and it was suggested that this limited systemic spread.

Beckman (1980) proposed a model for resistance to vascular parasites involving occlusion, based particularly on studies on banana cultivars infected by F. oxysporum f. sp. cubense (Beckman & Halmos, 1962; Beckman et al., 1962). First, gels are deposited above the vessel end-plate which has trapped spores travelling in the transpiration stream, and which the fungus must penetrate. This gelation cuts off the mass-flow within the vessel, thus restricting spore dissemination. When occlusion is complete, a second spore generation may have been formed, but gels may also trap them (Van der Molen et al., 1977). Subsequently, tyloses form, and the walls of adjacent parenchyma cells become modified. These events occur some distance from the penetrated end-plate, and tyloses may fuse to block the vessel lumen. Phenolic substances then polymerize with the gels and cell wall constituents of tyloses to form an enzyme-resistant barrier which physically contains the original infection site. Since gels and tyloses can probably be readily degraded by wilt parasites, the accumulation of antimicrobial compounds in the vicinity of infection sites may be critical for functional occlusions (Bell & Mace, 1980).

The suggestion is made that in susceptible cultivars, gels and tyloses form more slowly and may subsequently be degraded. In resistant cultivars, or in susceptible cultivars invaded by incompatible forms of Fusarium, occlusion is rapid and permanent. An interpretation of this scenario is that in a susceptible cultivar, incompatible lines of the fungus fail to trigger the normal non-specific defence mechanisms.

It seems clear that vascular occlusion can be an important non-specific component of resistance to vascular parasites, but that integration with other forms of defence may be needed for full effectiveness. There are, however, doubts that such a mechanism is of primary importance in resistance apparently controlled by specifically interacting major genes in the host and parasite. In such cases, it is likely that some form of specific recognition occurs, followed by a series of non-specific events of which vascular occlusion is just one. Whether occlusion actually contributes to resistance under these circumstances is unclear.

It is surprising that the considerable parasitic variability within vascular parasite species or formae speciales has not been exploited more to probe the importance of responses such as vascular occlusion. Parasitic variability occurs either as the horizontal (e.g. 'aggressive'/'non-aggressive' strains of Ceratocystis ulmi and 'fluctuating'/'progressive' strains of Verticillium albo-atrum), or the apparently vertical type (e.g. races of F. oxysporum f. sp. lycopersici; Gabe, 1975). There is, however, little information on how strongly proposed resistance mechanisms correlate with parasite colonization in different cultivar - isolate combinations. At the same time, 'parasitic epistasis' resulting from induced resistance appears to occur frequently in plants which are hosts of more than one vascular parasite (Sidhu & Webster, 1979, 1983). This is strong evidence for a general defence mechanism and since the phenomenon is genetically well defined, it could provide a particularly useful means of identifying events of primary importance in resistance.

Haustorial encasement. The partial or complete encasement (as defined by

Bracker & Littlefield, 1973) of fungal haustoria by apposition of cell-wall material has frequently been reported as a symptom associated with the colonization of resistant hosts. The phenomenon is most clearly observed when there is appreciable fungal development. Encasement occurs after a haustorium or incipient haustorium has penetrated the host cell wall; the encased haustorium is commonly necrotic or vacuolated.

It has been suggested that encasements, particularly if they are largely composed of callose, could impede the movement of small molecules between host and pathogen (Davison, 1968; Hardwick et al., 1971) thus starving the fungus. In addition, slower growth after encasement might make a fungus more susceptible to phytoalexins (Allen & Friend, 1983). One hypothesis implying a role in resistance is therefore that encasement causes the haustorium to become non-functional.

A contrary hypothesis is that encasements form after haustoria have already ceased to function normally (Heath & Heath, 1971; Pring & Richmond, 1976; Coffey, 1976; Hickey & Coffey, 1980) and may therefore merely be a consequence of resistance.

Haustorial encasements have primarily been observed in host associations with parasites belonging to the Urediniales (rusts) or Peronosporales (downy mildews, white blisters, blight). These unrelated groups will be discussed separately, since it is not certain that the phenomena involved in encasement have the same significance for both.

In the Peronosporales, there is good evidence that encasements can occur in compatible as well as in incompatible host-cultivar combinations. Berlin & Bowen (1964) found that about 1% of haustoria of Albugo candida (white blister) on radish were 'walled-off'. The same cell could contain both a healthy and a necrotic, encased haustorium. Davison (1968) observed that 'some' haustoria of Peronospora parasitica (downy mildew) on a susceptible wallflower cultivar were encased by a substance reacting histochemically like callose. Hickey & Coffey (1977) also noted that in a compatible association between Peronospora pisi (downy mildew) and pea, a small proportion of haustoria were encased. When plants infected by this fungus were examined 5 days after treatment with the fungicide metalaxyl, many haustoria had become vacuolated and encased in a 'callose-like' material (Hickey & Coffey, 1980). All these observations strongly suggest that encasement is a non-specific host response which occurs following haustorial malfunction, regardless of how this is induced.

This suggestion can also be applied to comparative studies of the association between *Phytophthora infestans* (blight) and resistant and susceptible potato cultivars, although in this case a more positive role in resistance is envisaged (Hohl & Stoessel, 1976; Hohl & Suter, 1976; Hohl *et al.*, 1980; Allen & Friend, 1983). In tubers of potatoes with resistance genes R1 or R3, it is clear that haustorial encasement of an avirulent isolate of the fungus is an early event associated with incompatibility, and accompanying cessation of growth. Significant numbers of haustoria become encased by 24 hours after inoculation, and heavily encased haustoria are invariably moribund. In a susceptible combination, encasement occurs to a lesser extent, and only several days after inoculation when the host-parasite relationship becomes more necrotrophic. Encasements appear to consist of both cellulose and callose.

In contrast, the distinction between leaves of resistant and susceptible cultivars is less clear, as encasements are observed in both types. However, no detailed quantitative data are available on incidence or rates of formation.

The contention is that encasement is a non-specific host response to invasion, but that a virulent isolate of the fungus prevents encasement by the action of extracellular glucanases. In a resistant cultivar, these glucanases may be inhibited by phytoalexins. Under these circumstances, encasement might contribute to haustorial malfunction rather than merely being a consequence of it.

In a susceptible bean cultivar infected by *Uromyces phaseoli* (= *U. appendiculatus*) (rust), Hardwick *et al.* (1971) observed that occasionally, haustoria were totally enclosed by a 'callose-like' encasement. They concluded that this encasement was the product of an active, though non-specific, host response since there were clear modifications to the adjacent host plasmalemma and an abundance of endoplasmic reticulum in the vicinity. Encased haustoria were also observed in advanced infections of a susceptible cowpea cultivar by *Uromyces phaseoli vigna*. However, in a cultivar exhibiting an extreme form of resistance to this rust fungus, haustorial encasement was observed in the first cell penetrated if rapid death of the host cell had not occurred (Heath, 1971; Heath & Heath, 1971). Hypersensitivity and haustorial encasement appeared to represent two alternative manifestations of incompatibility in the same cultivar, although other events associated with incompatibility, particularly the response of the host membrane surrounding the haustorium, were considered to be of greater significance in determining the outcome of the association. In this context, it is relevant that Mendgen (1977; 1981) has shown that there is less uptake of ^{3}H-lysine by haustoria of Uromyces phaseoli on beans in at least one incompatible cultivar-isolate combination than in a susceptible combination, well before symptoms of incompatibility can be observed microscopically. Labelling studies also showed no evidence of damage to fungal membranes at this time.

In an investigation of structural aspects of race-specific resistance in near-isogenic flax lines to Melampsora lini (rust), Coffey (1976) and Coffey & Allen (1983) concluded that haustorial encasement was not directly involved. The K gene in flax conditions an intermediate level of resistance with appreciable fungal growth in the host. By 9 days after inoculation, 22% of haustoria were encased in the incompatible combination, compared to none in the compatible. However, since encasement occurred after incompatibility was determined, it was interpreted to be of no consequence. In lines carrying other genes, and when lines with the K gene were examined at earlier stages after inoculation, there was no evidence that more encasement was associated with incompatibility than with compatibility. Less than 10% of haustoria usually became encased regardless of the isolate-cultivar combination.

It therefore appears that encasement is a normal, non-specific host response to haustorial fungi. Indeed, a similar reponse was noted in tobacco cells when a glass needle was inserted and left in position (Nims et al., 1967). However, as Bushnell (1972) has pointed out, a successful parasite must have some means of limiting this host response. When this parasite function is impaired, by age, by the host or by an external influence, then encasement may ensue. A causal role in resistance, however, remains to be demonstrated.

7.3.2. 'Pathogenesis-related' ('b') proteins. The role that 'pathogenesis-related' (PR) or 'b' proteins may play in induced resistance to viral, bacterial and fungal infections is discussed in Chapters 9 and 10. These proteins can be induced in plants by primary infections by bacteria and fungi. For example PR proteins are formed in tobacco following infection by Thielaviopsis basicola, which causes necrotic localized lesions on the leaves (Gianinazzi et al., 1980). PR proteins have also been shown to accumulate in tomatoes in response to infection by Cladosporium fulvum (Camacho-Henriquez & Sänger, 1982; de Wit & Bakker, 1980), and in cucumber following infection by various fungal and bacterial parasites (Gessler & Kuć, 1982). In none of these studies was there any evidence to suggest that PR protein synthesis was involved in the resistance mechanism.

Leach et al. (1983) found that PR-proteins accumulated in tobacco following infiltration with bacterial lipopolysaccharide or avirulent and virulent isolates of Pseudomonas solanacearum. There was no evidence that the occurrence of PR-proteins was related to the mechanism of resistance to bacterial multiplication.

7.3.3. The hypersensitive response (HR). Hypersensitivity is an ill-defined term: 'One man's HR is another man's lesion'. In this section, HR refers to host cell necrosis which is associated with less extensive parasite colonization or growth than in a more susceptible host. HR is a symptom of host-parasite incompatibility.

HR is a widespread reaction of plants to fungal and bacterial parasites, and was first connected with host resistance nearly eighty years ago (Ward, 1905; Marryat, 1907), although it was Stakman (1915) who coined the term. The common association between HR and race-specific resistance to biotrophic parasites, led to the assumption that HR caused resistance because biotrophs could not grow in dead cells. This interpretation did not explain resistance to necrotrophic parasites, where hypersensitivity is nevertheless common. As a consequence, the concept of an association of hypersensitivity with phytoalexin production has become commonly accepted (Müller, 1959). It is traditionally held that host cell death precedes cessation of parasite growth. The opposing view, that HR is solely a symptom and not responsible for resistance was first voiced by Brown et al. (1966) and the debate this caused will be discussed below.

It has also been implied that HR is solely an expression of race-specific resistance controlled by major genes, and that this type of resistance is invariably expressed as HR. This view now appears to be unfounded and evidence for the diverse occurrence of HR will be discussed.

HR and bacterial parasites. Generally, plant parasitic bacteria have the capability to induce HR while saprophytes do not. In addition, those bacterial parasites which induce HR tend to be host-specific while those with an extensive host range tend not to induce (Klement, 1982).

The events which lead to cell collapse have been well documented in several ultrastructural studies (Roebuck et al., 1978; Sigee & Epton, 1976; Sequeira et al., 1977). Attachment of the bacterium to the host cell wall appears to be necessary to initiate the process.

The time taken to induce HR varies between 1.5 and 5 h after inoculation with an avirulent pathogen (Keen et al., 1981). The length of this period is not influenced by inoculum concentration but the extent of macroscopic symptoms is. Wyman & Van Etten (1982) found that resistant beans developed confluent HR when they were injected with a suspension of 10^8 cells ml^{-1} of Xanthomonas campestris pv. phaseoli, but that after infiltration with a suspension of 10^6 cells ml^{-1}, no HR was visible macroscopically. However, microscopic studies by Turner & Novacky (1974) with Pseudomonas syringae pv. pisi in tobacco, Essenberg et al. (1979) with X. campestris pv. malvacearum in cotton and Holliday et al. (1981) with P. syringae pv. glycinea in soybeans showed that at low inoculum concentrations, necrosis did occur at the cellular level. Apparently the number of host cells which undergo necrosis is much smaller when low concentrations of bacterial cells are used.

HR and fungal parasites. It has sometimes been assumed that all gene-for-gene controlled resistance conditions HR and vice versa. There are indeed many examples of host-parasite relationships where specific gene action and HR are associated. These include Phytophthora infestans and potato (Kitazawa & Tomiyama, 1969; Shimony & Friend, 1975; Doke, 1982); Erysiphe graminis and wheat (Slesinski & Ellingboe, 1969); Bremia lactucae and lettuce (Maclean & Tommerup, 1979); Puccinia graminis and wheat (Brown et al., 1966; Harder et al., 1979a; 1979b; Samborski et al., 1977; Rohringer et al., 1979); Cladosporium fulvum and tomato (Lazarovits & Higgins, 1976; De Wit, 1977), and Melampsora lini and flax (Littlefield, 1973; Littlefield & Aronson, 1969; Coffey, 1976; Coffey & Allen, 1983).

However, there are also examples where genes controlling race-specific resistance operate without HR (Rohringer et al., 1979; Hohl & Suter, 1976; Masri & Ellingboe, 1966; Jørgensen & Mortensen, 1977; Johnson et al., 1979). It is also true that HR can be associated with resistance which is considered not to be under qualitative genetic control. Furthermore, even in compatible host-cultivar combinations a proportion of penetrations may result in HR (Mercer et al., 1974; Asher & Thomas, 1983; Wilson & Coffey, 1980; Coffey & Wilson, 1983).

Although generalizations are dangerous, there are nevertheless cases where qualitative major gene resistance and resistance which is expressed quantitatively do appear to operate through different mechanisms, HR being characteristic of the former. Niks & Kuiper (1983) and Niks (1983a; 1983b) found that partial resistance in barley to Puccinia hordei was characterized by a reduced rate of epidemic development in spite of a susceptible infection type. Resistance appeared to result from the early abortion of a significant proportion of primary infections as haustorial mother cells failed to develop haustoria. Abortion was not, however, associated with the collapse of host cells. The frequent occurrence of early abortion in several unrelated barley genotypes indicates that genes controlling this form of resistance are common and widely distributed. Major genes controlling gene-for-gene resistance in barley to *P. hordei* are expressed at the time haustorial mother cells are produced (Pa_3) or at the time haustorial mother cells give rise to haustoria (Pa_7). HR is associated with the action of both of these genes. When Pa_3 or Pa_7 were introduced into a genetic background expressing a high level of rate-reducing resistance, the effects of both types of gene (HR and non-HR) could be determined independently and consecutively (Niks & Kuiper, 1983).

Niks (1983a; 1983b) also studied the response of a non-host (wheat) to *P. hordei* and found that fungal abortion occurred just before haustoria were produced. This pre-haustorial abortion was not accompanied by host cell necrosis and resembled the situation in partially resistant genotyes of barley, except that no penetrations gave rise to successful colonization.

The expression of HR varies considerably between host-fungus combinations, and between different genotype-genotype combinations within any relationship. The speed with which HR occurs, the colouration of necrotic cells, the extent of necrosis in invaded and non-invaded cells and the extent of parasite development beyond the necrotic area all vary (Heath, 1974; Littlefield & Aronson, 1969; Littlefield, 1973; Skipp & Deverall, 1972; Coffey & Allen, 1983; Coffey, 1976; Ellingboe, 1972; Maclean & Tommerup, 1979). Ultrastructural studies emphasise that more than one sequence of events may lead to the eventual collapse of host cells.

Particular host resistance genes condition characteristic responses which may be recognised microscopically or macroscopically. However, the parasite isolate involved in the incompatible combination, and the host genetic background in which the resistance gene is located can have marked effects on expression of HR. (Ward & Buzzell, 1983; Crute & Norwood, 1978).

The physiology and biochemistry of HR in the association between resistant potato cultivars and avirulent isolates of *Phytophthora infestans* has been studied in detail (Tomiyama, 1971; 1976; Doke & Tomiyama, 1975; Nozue *et al.*, 1977a; 1977b; 1980; Tomiyama *et al.*, 1979; 1983; Kitazawa & Tomiyama, 1970). During the process of HR, host membrane potential decreased, cytoplasm aggregated around the invading fungus and cells lost their plasmolyzing ability. Experiments with inhibitors showed that *de novo* protein synthesis was required to enable cells to react hypersensitively, but that HR itself did not appear to require *de novo* synthesis. The suggestion is that HR requires the tricarboxylic acid cycle to be operative, and that this may need to be activated in some tissues. HR does require energy: the process can be inhibited by mitochondrial respiratory inhibitors, and addition of ATP can compensate for this. The involvement of the host plasmalemma in the recognition process which triggers HR is suggested by the fact that SH-reagents inhibit HR. In addition, D-chitobiose inhibited HR, apparently by interfering with the binding between the plasmalemma and the fungal cell wall.

Recently, studies with intact potato cells and protoplasts have shown that the superoxide anion (O_2^-) is generated and NADPH oxidation occurs at the plasmalemma during the determination of incompatibility (Doke, 1983a; 1983b). Superoxides are toxic to living cells. Both SH-reagents and $NADP^+$ inhibit the generation of O_2^- whereas respiratory inhibitors do not suggesting no association with mitochondrial metabolism. The O_2^- generating system appeared to be activated within minutes of the host cell membrane coming into contact with fungal wall components. It was suggested that the generation of O_2^- could be the cause of HR and the trigger for phytoalexin production. In compatible cultivar-isolate combinations there is evidence that water-soluble glucans from the fungus suppressed NADPH-dependent generation of O_2^-. Evidence for the occurrence of race-specific suppressors of HR is discussed further in sections 7.4.6 and 7.6.4.

These findings are particularly exciting since generation of O_2^- is known to occur in animal cells following recognition of alien substances (Babior *et al*., 1973; Gabig & Babior, 1979).

The relationship between HR and resistance. Suggested explanations of the occurrence of HR and its relationship with resistance include:

a) HR precedes inhibition of parasite growth and directly contributes to resistance;

b) HR precedes inhibition of parasite growth but cell death is not in itself reponsible for resistance;

c) HR occurs as a consequence of parasite growth having been inhibited by some other mechanism, and is induced by metabolites from the inhibited parasite;

d) HR is an inevitable consequence of an incompatible host-parasite combination but plays no part in the determination of incompatibility;

e) HR results from a general stress condition which occurs during a resistance response but does not contribute to resistance.

Because of the frequent association of HR with resistance to microbial parasites, it has been assumed that cell necrosis has an integral role in the resistance mechanism. However, apart from the endogenous elicitation of phytoalexins (section 7.4.2), there is little indication of what this role might be. On the other hand, there are several pieces of evidence suggesting that there is no causal relationship between HR and resistance.

Firstly, the occurrence of HR does not necessarily lead to death of the parasite; growth can resume if the parasite, along with the necrotic tissue, is transferred to a compatible host cultivar (Chakravarti, 1966). From detailed microscopical evidence, Tomiyama (1971) concluded that P. infestans is not restricted to necrotic cells but that HR nevertheless contributes in some way to the establishment of a chemically inhibitory environment.

Secondly, there are cases where host cell necrosis and the extent of parasite development appear to be unrelated. Brown et al. (1966) and Ogle & Brown (1971) observed that in some wheat varieties with partial resistance to Puccinia graminis, there was no relationship between host necrosis and the extent of colony development. Investigations with the temperature sensitive Sr6 gene in this host-parasite relationship led to similar conclusions (Harder et al., 1979a; 1979b), in that host cell necrosis and reduced fungal colonization were not invariably associated. Previous studies (Mayama et al., 1975) with lines carrying the same gene also showed that cellular response, as indicated by autofluorescence, varied independently from fungal growth when this was assessed using an assay for glucosamine. Koga et al. (1983) found no significant correlation between collapsed cells and colony size in barley cultivars resistant to Erisyphe graminis. Resistance correlated best with the occurrence of host cytoplasmic aggregation at the sites of attempted penetration.

Thirdly, Kiraly et al, (1972) reproduced the symptoms of HR in a compatible cultivar-isolate combination of potato and Phytophthora infestans by inhibiting fungal growth with chloramphenicol. Both differential heat treatments and fungicides can produce similar effects (Crute et al., 1977; Prusky et al., 1981) in different host-parasite relationships. The fungicide induced 'pseudo-HR' can also be accompanied by phytoalexin accumulation in the same way as genetically controlled HR (Borner et al., 1983). These observations indicate that host cells can die when the fungus within them is killed by various means. Whether parasite metabolites are responsible is unproved. Neither do these observations prove that the apparent localization of fungus in necrotic cells always results from inhibition prior to host-cell death.

Taking the evidence overall, probably the only conclusion that can be made concerning the role of HR in resistance is that reached by Ingram (1978): 'one is forced to note that 60 years after the publication of Stakman's 1915 paper we still do not know for certain whether cell death has an integral part to play in determining the resistance of any host to any biotrophic parasite'.

7.3.4. The accumulation of phytoalexins. Phytoalexins are generally defined as low molecular weight antimicrobial compounds that are synthesized by and accumulated in plants after exposure to micro-organisms (Paxton, 1982). HR and accumulation of phytoalexins very often coincide: this and their antimicrobial activity made phytoalexins clear candidates for a role in plant resistance to microbial parasites.

In a recent review, Keen (1982) states: 'There is substantial evidence from several gene-for-gene systems that phytoalexins accumulate to high quantity after inoculation of plants with avirulent but not virulent parasite races', and cites sixteen examples. Further: 'The consistent association of high phytoalexin production only with incompatible host reactions in these systems strongly supports their role as mediators to disease resistance'.

Fungal parasites. Mansfield (1982) has recently addressed the problem of whether phytoalexins are directly responsible for resistance to fungal parasites. He argues that: 'In order to prove whether or not inhibition of hyphal growth at the infection court is caused by phytoalexins, it would be necessary to measure their concentration near the hyphal tips at the time they stop growing. To obtain such direct proof it is necessary to carry out coordinated biochemical and microscopical studies'. Mansfield (1982) reviewed the role of phytoalexins in four host-parasite associations: *Botrytis* spp. - *Vicia faba*; *Colletotrichum lindemuthianum* - *Phaseolus vulgaris*; *Phytophthora infestans* - *Solanum tuberosum* and *Phytophthora megasperma* f. sp. *glycinea* - *Glycine max*. Another fifteen host-parasite combinations are also mentioned where the accumulation of phytoalexins has been examined in an appropriately critical manner.

In the association between *Botrytis* spp. and *Vicia faba* (broad bean), *Botrytis fabae* (a necrotrophic pathogen) induces significantly less phytoalexin accumulation than non-pathogenic species of *Botrytis*. A further difference between *B. fabae* and species non-pathogenic to *V. faba* is the greater tolerance of *B. fabae* to the phytoalexins (wyerone and wyeronic acid) produced by *V. faba*. In addition, *B. fabae* kills many cells during the early stages of infection thus resulting in a reduction in the synthetic activity of infected tissue. All species of *Botrytis* were shown to metabolize *V. faba* phytoalexins.

The association between *Botrytis* and *V. faba* can be envisaged as a balance between phytoalexin production and breakdown. *B. fabae* keeps the amount of phytoalexins low so a spreading lesion results. Other species of *Botrytis* kill fewer cells, induce the accumulation of high amounts of phytoalexin, and are very sensitive to them. The result is limited lesions development, and thus host resistance.

However, in the association between *Botrytis* spp. and *V. faba* there is no quadratic or reciprocal check (c.f. Figures 7.3 and 7.4) which makes it difficult to draw conclusions about the role of phytoalexins in the mechanism of disease resistance.

A different situation exists in the association between *C. lindemuthianum* and *P. vulgaris*. Many races of the fungus carry (putative) virulence genes which match the different resistance genes located in a range of bean cultivars; a reciprocal check exists. In compatible isolate-cultivar combinations, the fungus has a biotrophic phase followed by a necrotrophic phase which is characterized by extensive cell death, tissue collapse and the development of brown lesions. During the necrotrophic phase some fungal growth may still occur. In incompatible combinations, fungal growth is restricted to the first few penetrated cells, which express a hypersensitive response.

In this system, no phytoalexins were detectable in compatible combinations while the fungus was growing biotrophically. During the necrotrophic phase, phytoalexins (phaseollin and phaseollinisoflavan) accumulated but at significantly lower concentrations than in hypersensitively responding lesions in incompatible combinations. Phytoalexins are thought to be synthesized in living tissue around necrotic cells (Hargreaves & Bailey, 1978). Bailey *et al.* (1980) incubated compatible isolate-cultivar combinations at 16°C, which allowed the biotrophic relationship to become established. Subsequent transfer to 25°C induced necrosis; this facilitated a study of the timing of fungal growth inhibition in detail. The results indicated that infected cells died several hours before phytoalexins could be detected, and that fungal growth inhibition occurred shortly after phytoalexins began to accumulate. The evidence therefore suggests that phytoalexin accumulation is responsible for the cessation of growth of C. lindemuthianum.

The phytoalexin concept was first postulated for the extensively studied association between Phytophthora infestans and Solanum tuberosum (Müller & Börger, 1940). P. infestans has both biotrophic and nectrotrophic phases, and a gene-for-gene relationship exists. Where it has been examined, resistance is associated with HR in both leaves and tubers. However, the sesquiterpenoid phytoalexins (mainly rishitin and lubimin) accumulate to significant amounts in incompatible combinations only in tubers and tuber sprouts, but not in leaves. Sato et al. (1971) measured mycelial growth and phytoalexin accumulation in infected tubers. HR occurred approximately 2-3 h after inoculation with an avirulent race, and hyphae were mostly confined to the first few penetrated cells. Accumulation of rishitin was first detected 7 h after inoculation and preceded inhibition of fungal growth. Rishitin was localized in the hypersensitively responding cells, and reached local concentrations which were strongly inhibitory to P. infestans in vitro. Labelling studies have shown that rishitin may be present in incompatible race-cultivar combinations prior to the occurrence of HR (Tomiyama et al., 1979).

The accumulation of sesquiterpenoid phytoalexins in potato tubers is markedly affected by age and storage conditions. Bostock et al. (1983) found that young potato tubers (harvested in July, August and September) accumulated low levels of rishitin and lubimin regardless of whether they were inoculated with virulent or avirulent isolates. After storage at 4°C for several months the response changed; potato tubers of resistant cultivars accumulated high amounts of sesquiterpenes after inoculation with an avirulent race. In addition, avirulent races induced significantly greater accumulation of phytoalexins in resistant tubers than did virulent ones. The growth regulator abscisic acid (ABA) could either stimulate (in young tubers) or inhibit (in old, stored tubers) the accumulation of sesquiterpenes induced by an avirulent race. In cold-stored tubers, addition of ABA allowed an avirulent race to grow on tubers of a resistant cultivar. Interestingly, HR occurred at the same rate in young and cold-stored tubers, and young tubers were as resistant to avirulent races as were cold-stored tubers.

In the field, tubers become infected at the end of the growing season and therefore, in the epidemiology of P. infestans, the responses occurring in young tubers are of greater importance than those of stored tubers.

The differences between young and old tubers in accumulation of phytoalexins raises questions about their role in resistance. Clearly, phytoalexins cannot be the sole determinants of resistance in tubers. In potato leaves, they appear to play no part in resistance, and other defence mechanisms such as suberization, lignification or haustorial encasement may be more important (see sections 7.3.1, 7.3.5 and 7.3.7).

Yoshikawa _et al._ (1978) investigated fungal colonization in relation to phytoalexin accumulation in soybean infected with _P. megasperma_ f. sp. _glycinea_. Many races of this fungus can be identified by the response of cultivars carrying different genes for resistance. A gene-for-gene relationship is suspected although unproved. The fungus has a short biotrophic phase during the early stages of colonization. Yoshikawa _et al._ (1978) determined concentrations of the phytoalexin glyceollin in 0.25 mm sections around invading hyphae, in both compatible and incompatible combinations. In the latter, concentrations of glyceollin sufficient to inhibit mycelial growth by 90% _in vitro_ (ED_{90}) were present near the hyphal tips by 8 h after inoculation, at which time glyceollin could hardly be detected in compatible interactions. In these, ED_{90} values were recorded 24 h after inoculation but the hyphal tips had grown beyond the area of high glyceollin concentration. The accumulation of phytoalexins could be inhibited by blasticidin S (BcS). BcS also allowed an avirulent isolate to grow on a resistant cultivar as readily as a virulent isolate. This work provided strong evidence that glyceollin inhibited fungal growth in incompatible combinations, and suggested that _de novo_ protein synthesis was required for glyceollin production.

Ward _et al._ (1981) found a significant age-related effect on the expression of race-specific resistance in soybean hypocotyls to _P. megasperma_ f. sp. _glycinea_. The youngest tissue at the tops of intact 6-day-old soybean cotylodons displayed typical differential compatible or incompatible responses to inoculation with zoospores of virulent and avirulent isolates. However, the tissue became increasingly resistant with age; tissue at the bottom of the hypocotyl was resistant to both virulent and avirulent isolates. Recently, it was shown that etiolated hypocotyls could show resistant or susceptible reactions to a particular isolate of the parasite, but hypocotyls exposed to light all expressed resistance. In addition, some R-genes were clearly temperature-sensitive and when incompatibility was not expressed at high temperatures, glyceollin concentrations were as low as in compatible combinations at lower temperatures (Ward & Buzzell, 1983). The responses in race-specific resistance, age-related resistance and light-induced resistance were similar in that they were associated with host cell necrosis and accumulation of glyceollin.

Bacterial parasites. Phytoalexin accumulation has been detected during HR in several associations between bacteria and host plants (Keen & Kennedy, 1974; Lyon & Wood, 1975; Essenberg *et al.*, 1982). In detailed studies, Holliday & Keen (1982) observed that HR in an incompatible association between soybean and *Pseudomonas syringae* pv. *glycinea* was first visible 9 h after inoculation and that glyceollin was first detectable in leaf extracts at 16 h. Inhibition of bacterial multiplication was apparent by 30 h. When the inoculum concentration was relatively low, they found that HR did not occur until 24 h after inoculation; it appeared that the bacteria had to multiply sufficiently before HR was obvious.

The relationship between HR and the accumulation of phytoalexins is not always clear. Holliday & Keen (1982) separated the occurrence of HR and the accumulation of phytoalexins in experiments with the herbicide glyphosate, which inhibits the formation of chorismate from shikimate in the glyceollin synthesis pathway. Glyphosate did not inhibit HR but did significantly inhibit glyceollin accumulation. Bacterial populations in glyphosate-treated leaves of a resistant cultivar were twice as high as in untreated resistant controls, but still eight times lower than in a susceptible cultivar. Thus, inhibition of the accumulation of glyceollin by glyphosate only partly prevented the expression of resistance, suggesting that the accumulation of glyceollin is a component but not the only mechanism for resistance in this system.

Detoxification of phytoalexins and parasite sensitivity. Isoflavonoid, flavonoid and furanoacetylenic phytoalexins are selectively toxic to Gram-positive bacteria (Gnanamanickam & Smith, 1980; Gnanamanickam & Mansfield, 1981). However, most plant pathogenic bacteria are Gram-negative. Rishitin exhibited antimicrobial action against both Gram-negative and Gram-positive bacteria (Gnanamanickam & Mansfield, 1981). Structurally-related isoflavonoids showed differential antibacterial activity (Wyman & Van Etten, 1978; Pankhurst & Biggs, 1980). In general, the antibacterial activity of phytoalexins is less than their antifungal activity (Smith, 1982). There are no known examples of bacteria capable of detoxifying phytoalexins.

Some fungi are less sensitive to phytoalexins than others. Many examples of fungi capable of degrading phytoalexins have been reported (Van Etten *et al.*, 1982) and it has been suggested that the ability to metabolize phytoalexins might be related to pathogenicity. However, the ability to metabolize the phytoalexins of a particular plant species is not restricted to parasites of that species. In addition, some parasites are apparently unable to degrade phytoalexins produced on hosts to which they are virulent (Pueppke & Van Etten, 1976).

Van Etten et al. (1980) and Tegtmeier & Van Etten (1982) investigated the importance of pisatin metabolism for the pathogenicity of Nectria haematococca on peas. The pathogenicity of naturally occurring isolates of the fungus was correlated with ability to de-methylate pisatin. Isolates that were non-pathogenic or of low virulence did not have this facility. An isolate of low virulence unable to metabolize pisatin was crossed with an isolate of high virulence which could. Single ascospore progeny segregated for those which were unable to metabolize pisatin and exhibited low virulence, and those which were of high virulence and metabolized pisatin. However, there was also a class of progeny with low virulence but which were able to metabolize pisatin. This class indicates that the ability to metabolize pisatin is alone not sufficient for high virulence and that virulence characters segregating independently to this character are also required.

There is evidence for another mechanism of insensitivity to phytoalexin which does not depend on degradation. Denny & Van Etten (1983a; 1983b) found three groups of isolates of Nectria haematococca which differed in their pisatin degrading abilities. The first group degraded pisatin within 15 h, the second group needed more than 24 h, and the third group did not degrade. Isolates of all three groups grew equally well on pisatin-containing media, suggesting that detoxification was not required to allow growth. The two groups of isolates which degraded pisatin also had non-degradative tolerance, since these isolates could grow (i) before pisatin was degraded to non-inhibitory levels and (ii) when pisatin degradation was inhibited by 2% ethanol. After pretreatment with pisatin, the two groups of isolates capable of pisatin degradation grew better than those isolates incapable of degradation. This was because degradative and non-degradative tolerances were stimulated in the first two groups, which indicated an additive effect. In the third group only non-degradative tolerance was stimulated. Pretreatment of isolates with pisatin also induced non-degradative tolerance to other isoflavoid phytoalexins. Denny & Van Etten (1983b) suggested that during pretreatment with pisatin the plasma membrane was modified and that this might be associated with non-degradative insensitivity.

A comparable mechanism of inducible, non-degradative tolerance to the fungicide fenarimol (an inhibitor of ergosterol biosynthesis) has been reported in wild type strains of *Aspergillus nidulans* (De Waard & Van Nistelrooy, 1981). Genetically-determined tolerance to the fungicide in resistant mutants is based on a constitutive, energy-dependent fungicide efflux. Fenarimol efflux could also be activated by pretreatment with several ergosterol biosynthesis inhibitors and by unrelated chemicals such as pisatin (Fuchs *et al.*, 1983).

Degradative and non-degradative tolerance to phytoalexins might be important in host-parasite interactions where large quantities of phytoalexin are produced in compatible combinations. In cases where this occurs, the parasite involved is usually necrotrophic. With biotrophic parasites, phytoalexins are not normally detected in compatible combinations (Keen, 1982). At later stages of infection when these parasites become necrotrophic, phytoalexin can start to accumulate and insensitivity may be important. Most parasites which have a gene-for-gene relationship with their hosts have at least a short biotrophic phase and induce low, if any, amounts of phytoalexin. In these cases, virulence is not related to their ability to metabolize phytoalexins, but rather to their ability to avoid the induction of phytoalexins. With the exception of *C. lindemuthianum* on beans (Van den Heuvel & Glazener, 1975) there is little evidence that avirulent fungal isolates metabolize phytoalexins once they have induced them to form in associations controlled by a gene-for-gene relationship (Keen, 1982). However, metabolism of phytoalexins in this type of relationship has not been extensively studied.

It is also possible that the host may metabolize phytoalexins once they have accumulated. For instance, the sesquiterpenes reach a peak of accumulation 96-100 h after inoculation with a fungus or treatment with an elicitor and their levels decrease 7 to 10 days after treatment; Ward *et al.* (1977) reported that capsidiol and rishitin could be metabolized by plant cells. In addition, cell suspension cultures of kidney bean and mungbean have been shown to metabolize phaseollin (Glazener & Van Etten, 1978).

7.3.5. *Lignification*. Lignin is a phenolic polymer that occurs in enormous quantities in nature, especially in woody plants. It is formed mainly from hydroxycinnamyl alcohol precursors (p-coumaryl, coniferyl and sinapyl alcohol) by free radical condensation (see Figure 6.4). Lignin is deposited as a network in cell walls and forms covalent bonds with cellulose, hemicellulose, pectates and structural proteins. It can also form ester linkages with fatty acid polyesters to yield suberin.

Lignification has been associated with resistance in many host-parasite associations (reviewed by Vance et al., 1980). Histochemical studies show that lignification often occurs before penetration, as well as during colonization, in either the epidermis or the internal cells of many plant organs. Both biotrophic and necrotrophic organisms, as well as virulent and avirulent isolates of a parasite, may induce lignification. In some other associations, no lignification can be detected.

The association between potato and Phytophthora infestans has been studied in detail (Henderson & Friend, 1979). Three days afer inoculation, significantly more lignin-like material was deposited in host cell walls in incompatible cultivar-isolate combinations than in compatible ones. This response, however, occurred a good deal later than HR or the inhibition of fungal growth.

Bird & Ride (1981) have studied the role of lignification in four wheat varieties differing in their resistance to Septoria nodorum. Inoculation of wounded leaves of all varieties resulted in the same amount of lignification, although differences in resistance to the fungus were still apparent. Germ tube growth was slower on the leaves of more resistant varieties than on susceptible ones. On all varieties, the majority of appressoria failed to initiate successful penetrations. This failure was associated with the production of lignified papillae and 'haloes', i.e. rings of lignified tissue. The number of sucessful penetrations was highest on the least resistant variety, and those epidermal cells which were penetrated collapsed and became lignified. Hyphae which reached the mesophyll spread through the intercellular spaces and lignin continued to be deposited in this tissue. The rate of lignification did not explain cultivar resistance and the amount of lignification reflected the extent of fungal colonization rather than the degree of resistance. It was concluded that regardless of cultivar, lignification reduced penetration and the rate of fungal colonization, but that other mechanisms must further reduce fungal growth in the more resistant varieties. Recently, Beardmore et al. (1983) have presented evidence for cellular lignification in the hypersensitive resistance of wheat to Puccinia graminis (stem rust).

Lignification is by no means a universal resistance response and could not be detected in papillae formed by barley in response to invasion by Erysiphe graminis (powdery mildew) (Mayama & Shishiyama, 1978; Johnson et al., 1979).

In common with HR, the occurrence of lignification in response to a parasite may be a cause of resistance, a consequence of resistance, or in no way related to resistance. At present, no conclusion can be reached as to which is the most likely. Because lignin is a complex substance, it cannot be readily extracted in pure form, and quantification is therefore by necessity indirect. Furthermore, any biological activity of lignin cannot be tested _in vitro_. Thus the ability to inhibit fungal growth which is ascribed to lignin is a matter of speculation. Ride (1978; 1980) suggested five ways in which lignification might inhibit fungal growth:

a) lignin could make host cell walls more resistant to mechanical penetration;

b) lignification of plant cell walls at the point of attack might render them resistant to dissolution by fungal enzymes;

c) lignification of plant cell walls might restrict diffusion of enzymes and toxins from the fungus to the host, and of water and nutrients from the host to the fungus;

d) low molecular weight phenolic precursors of lignin and free radicals produced during polymerization might inactivate fungal membranes, enzymes, toxins and elictors, and

e) the hyphal tip might become lignified and plasticity necessary for growth might be reduced.

7.3.6. _The accumulation of hydroxyproline-rich glycoproteins_. Hydroxyproline-rich glycoproteins (HPRGP) in plant cells were first reported by Lamport & Northcote (1960), who called them 'extensin'. After infection by fungi or treatment with ethylene, plant cell walls become enriched in HPRGP (Esquerré-Tugayé & Lamport, 1979; Clarke _et al._, 1981). Esquerré-Tugayé _et al._ (1979); Hammerschmidt _et al._ (1982) and Fincher _et al._ (1983) have suggested a role for these glycoproteins in disease resistance. Esquerré-Tugayé _et al._ (1979) found that colonization of muskmelon seedlings by _Colletotrichum lagenarum_ was negatively correlated with the amount of HGRGP in cell walls. Pre-treatment with ethylene increased HGRGP in cell walls and hydroxyproline suppressed their accumulation, with corresponding effects on colonization. However, ethylene and hydroxyproline might have had effects on other processes influencing colonization. Clarke _et al._ (1981) concluded that enhancement of hydroxyproline was a necessary primary event for the successful development of _Erysiphe graminis_ on powdery mildew on wheat. With so little data available, however, the role of HGRGP in disease resistance is not clear.

7.3.7. <u>The accumulation of phenolic compounds (tannins, melanins and suberins)</u>. Phenolic compounds with antimicrobial activity often accumulate in plant cell walls as a result of infection or injury. Such compounds are mainly synthesized via the shikimic acid pathway (Kosuge, 1969; Friend, 1979). Glucosylation of various dihydroxyphenols gives rise to a range of tannins. Oxidation of ortho-dihydroxyphenolics and tannins by polyphenol oxidase (PPO) and perixodase (PO) gives rise to coloured quinone residues which can condense to form the brown or black pigments called melanins.

Very often, the HR associated with incompatible isolate-cultivar combinations is characterized by formation of melanins throughout the walls of the dead cells. This could indicate that melanins or their precursors contribute to resistance, perhaps by rendering cell walls resistant to degradation by microbial enzymes. In this respect, melanization may have effects similar to lignification.

Polyphenols and their oxidation products can result in crosslinkages between proteins including microbial enzymes, leading to their inactivation (Leatham <u>et al</u>., 1980).

Attempts to evaluate the contribution of melanins and related compounds to plant resistance have met with difficulties since reliable quantification is impossible. Many of the enzymes which are involved in their synthesis are also used for the synthesis of phytoalexins and lignin (Kosuge, 1969; Friend, 1979; Bell, 1981) so that measurement of enzyme activities is not a reliable route to quantification.

Suberin is a biopolymer containing aromatics, similar to those occurring in lignin, which are esterified to fatty acids (Kolattukudy, 1980). Suberin occurs as a normal constituent in healthy plants and functions as an impermeable barrier, particularly in roots and tubers (e.g. in the Casparian band of the endodermis). However, fungal infection and wounding can trigger the suberization of cell walls. This response could help to prevent the spread of an invading parasite in the host since suberized cells are rarely penetrated by fungi.

Kolattukudy (1980) has pointed out that many workers refer to suberin or lignin deposition without reliable information on the actual composition of the material they have been studying. Because few studies involving the accumulation of polyphenolics and related compounds have been carried out in systems involving quadratic (Figure 7.3) or reciprocal (Figure 7.4) checks their role in resistance is uncertain.

7.4. INDUCTION OF ACTIVE RESISTANCE MECHANISMS

In the previous section the host responses associated with expression of active resistance were described. This section discusses how these responses may be induced.

In gene-for-gene systems, active defence mechanisms are induced by avirulent but not by virulent races of a parasite. The reason why active defence responses do not result from compatible race-cultivar combinations or are ineffective is of considerable interest. Daly (1972) and Heath (1982), among others, have advanced the hypothesis that compatibility could be the result of a suppression of host resistance responses by a virulent parasite (induced susceptibility). If such induced susceptibility was an active plant response rather than a constitutive failure of recognition, metabolic inhibitors might be expected to turn a compatible combination into an incompatible one. This has usually not been found to occur. However, there are many reports of metabolic inhibitors turning an incompatible combination into a compatible one. This indicates that resistance needs to be actively induced (Keen et al., 1981; Yoshikawa et al., 1978; Keen et al., 1982), although it does not preclude the possibility that susceptibility results from a constitutive absence of the induction stimulus in the parasite.

Genetic analyses of gene-for-gene systems support the contention that incompatibility as opposed to compatibility is the specific event (Ellingboe, 1981, 1982), since the former is uniquely conditioned by the combination of resistance and avirulence alleles. This in turn implies that avirulent races of a parasite should specifically induce defence mechanisms. Much research has been directed to examination of this prediction, but investigations of specific induction have concentrated on only a few mechanisms, notably HR, phytoalexin accumulation and lignification.

Instances where resistance can be attributed solely to a single defence mechanism are very few; several responses may occur simultaneously or consecutively. It is likely therefore that more than one biochemical mechanism may be responsible for the restricted growth of a pathogen. For example, HR is very often associated with the accumulation of phytoalexins (Bailey, 1982) or with lignification (Vance et al., 1980), although very often only the induction of one response has been studied.

Induction of resistance mechanisms has been achieved with molecules that have been isolated from the cell walls, culture filtrates and cytoplasm of various parasitic as well as non-parasitic bacteria and fungi. These molecules have come to be called elicitors (Keen, 1975).

7.4.1. Elicitors of the hypersensitive response (HR).

Bacterial parasites. There is little information on the molecules responsible for the elicitation of HR by bacterial parasites (Keen & Holliday, 1982). Klement (1982) concluded that no compounds had been isolated from bacteria which are able to induce HR, this being a characteristic solely of living bacteria. It is possible that inducers of HR only exist *in vivo* and are highly unstable, so that their HR-inducing activity is lost during isolation and purification. There is a report that pectic enzymes produced by bacteria induce HR (Gardner & Kado, 1976), but whether the observed symptoms were equivalent to HR has been disputed. Bashan *et al.* (1982) reported that a substance produced by *Pseudomonas syringae* pv. *tomato*, which appeared to be a glycoprotein, induced necrosis on leaves of non-host species such as bean, pepper and cucumber, but not in the host species tomato. In addition, Bashan *et al.* (1982) and Whatley *et al.* (1980) have presented data to show that lipopolysaccharides (LPS) from avirulent strains of *Pseudomonas solanacearum* induce HR in tobacco, while LPS from virulent strains do not.

In incompatible combinations, bacteria often become enveloped (section 7.3.1) by fibrillar material. This may play a role in the HR by ensuring cell to cell contact thus facilitating an interaction between bacterial elicitor and host receptor molecules (Stall & Cook, 1979). It is possible that virulent bacteria have evolved a mechanism for preventing envelopment and thus direct host contact. It has been suggested that the production of extracellular polysaccharides (EPS) could prevent recognition of elicitors or induce a water-soaking condition necessary for bacterial multiplication.

Fungal parasites. Elicitors of HR produced by fungi have usually been studied in association with the accumulation of phytoalexins (Bailey, 1982). These studies are discussed in the following section (7.4.2) and investigations related solely to HR are discussed here.

Jones & Deverall (1978) found indirect evidence for the existence of an *Lr20* gene-specific 'toxin' produced by avirulent races of *Puccinia recondita* f. sp. *tritici*. This 'toxin' induced HR when it came into contact with mesophyll cells of resistant cultivars carrying the *Lr20* gene. Similar studies have been discussed in section 7.2.2. De Wit & Spikman (1982) and De Wit *et al.* (1984) found elicitors of chlorosis and necrosis in the intercellular fluids extracted from compatible combinations of *Cladosporium fulvum* and tomato. These elicitors were race-specific, probably of fungal origin, and induced chlorosis and necrosis only in cultivars resistant to the race concerned. Elicitor specificity was not related to the gene(s) for resistance carried by the susceptible cultivar from which the intercellular fluids were obtained, but was determined by the genes for virulence carried by the particular race in question. The chlorosis and necrosis induced by these specific elicitors reflected the HR which is typical for incompatible interactions between races of _C. fulvum_ and tomato cultivars (Lazarovits & Higgins, 1976; De Wit, 1977).

7.4.2. _Elicitors of phytoalexin accumulation_. Various types of molecule (glucans, proteins, glycoproteins and unsaturated fatty acids) will elicit phytoalexins. As mentioned above, HR and the accumulation of phytoalexins are often associated. Bailey (1982) obtained evidence that phytoalexin formation occurs after infected cells die, suggesting that phytoalexins accumulate as a consequence of cell death (HR). However, in this section, only the phytoalexin-inducing activity of elicitors will be emphasized although it is true that they also often induce necrosis or cell death. The specificity of these elicitors and their role in the expression of cultivar resistance are discussed in more detail in section 7.4.5.

Glucan elicitors. Glucans have been extensively studied in the relationship between _Phytophthora megasperma_ f. sp. _glycinea_ and soybean (Albersheim & Valent, 1978). The elicitors which have been isolated from culture filtrates and cell walls of the fungus are β-1,3-, β-1,6-glucans. Partial acid hydrolysis has led to isolation of the smallest cell wall component able to elicit the accumulation of glyceollin. This compound is a nonamer with two 3-linked, two 6-linked and three terminal glycosyl residues. Similar glucan elicitors have also been isolated from _Saccharomyces cerevisiae_.

Keen _et al_. (1983) isolated glucomannans from the cell wall of _P. megasperma_ f. sp. _glycinea_ with a soybean β-1,3-glucanase. These glucomannans are reportedly ten times more active as elicitors of glyceollin in soybean cotyledons than the glucan mentioned above. They also showed race-specificity in that they elicited more glyceollin in cultivars with which the isolate concerned is incompatible.

An elicitor from _Colletotrichum lindemuthianum_ found in culture filtrates and cell walls of the fungus was presumed to be a 3- and 4-linked glucan. This compound induces the accumulation of phaseollin, hydroxyphaseollin and other phytoalexins in bean (Anderson, 1978; 1980). Hoppe _et al_. (1980) have also isolated a glucan elicitor from *Uromyces phaseoli* that induces phytoalexins in bean. Hadwiger & Beckman (1980) and Hadwiger *et al.* (1981) found that chitosan, a 1,4-linked glucosamine isolated from cell walls of *Fusarium solani*, induced the accumulation of pisatin in pea.

Protein and glycoprotein elicitors. Cruickshank & Perrin (1968) isolated a protein from mycelium of *Monilinia fructicola*, monilicolin A, that induced the accumulation of phaseollin in bean.

Glycoproteins which are potent inducers of phytoalexins have been isolated from a number of fungi. Keen & Legrand (1980) found glycoproteins from *P. megasperma* f. sp. *glycinea* at the mycelial wall surface and which induced glyceollin in soybean cotyledons. However, the concentration of glycoproteins required was considerably higher than with the glucomannan and glucan elicitors from the same fungus discussed previously. Stekoll & West (1978) found a glycoprotein elicitor in culture filtrates of *Rhizopus stolonifer* that induced the accumulation of casbene in castor bean. It also showed an endo-polygalacturonase activity (Lee & West, 1981a; 1981b). De Wit & Roseboom (1980) and De Wit & Kodde (1981) found glycoproteins in culture filtrates and cell walls of *Cladosporium fulvum* (syn. *Fulvia fulva*) which induced rishitin accumulation in tomato fruits. These same elicitors from *C. fulvum* also induced pisatin in pea pods and glyceollin in soybean cotyledons.

Fatty acids. Bostock *et al.* (1981) found that arachidonic and eicosapentaenoic acid were effective inducers of sesquiterpenoid accumulation in potato tubers. These acids were released from the lipophylic materials in cell wall preparations of *Phytophthora infestans*. According to Bostock *et al.* (1981), the lipids alone were sufficient to induce phytoalexin accumulation, but Kurantz & Zacharius (1981) only found significant accumulation of sesquiterpenes when carbohydrate and lipid fractions from *P. infestans* were applied to potato tuber tissue in combination. More recently, Bostock *et al.* (1982) reported that the elicitor activity of these fatty acids could be greatly enhanced by addition of β-1,3- and β-2,6-glucans. Interestingly, combining the carbohydrate fractions with other unsaturated C_{20} and C_{18} fatty acids also led to sesquiterpenoid accumulation (Preisig & Kuć, 1983). Arachidonic and eicosapentaenoic acid also induced capsidiol accumulation in pepper but not, however, rishitin in tomato fruits (Bloch *et al.*, 1984). Apparently, species of the genus *Phytophthora* produce at least three types of elicitors: glucans, glycoproteins and unsaturated fatty acids. The first two induce glyceollin accumulation in soybean cotyledons, but not rishitin in potato tubers, while the latter induces sesquiterpene accumulation in potato tubers, capsidiol in pepper fruits but not rishitin in tomato fruits.

Abiotic elicitors. Many abiotic factors can cause the accumulation of phytoalexins in a range of plant species (Bailey, 1982). However, since host tissues are never exposed to these conditions or substances in nature, they will not be discussed further here, other than to point out that abiotic elicitors indicate the non-specific nature of the phytoalexin response.

Constitutive (endogenous) elicitors. It is suggested that these elicitors, of host origin, are released as a consequence of damage to the plant (Hargreaves & Bailey, 1978; Hahn et al., 1981; Lyon & Albersheim, 1982), for example during HR or the necrotrophic phase of parasitism. Hahn et al. (1981) suggested that endogenous elicitors were fragments of cell wall polysaccharides. Bruce & West (1982) found that pectic fragments released by fungal endo-polygalacturonase could also act as endogenous elicitors. Recently, Cooksey et al. (1983) have found that sucrose induces the accumulation of three phytoalexins in wounded leaves of pigeon pea (Cajanus cajan.

Bailey (1982) has proposed a hypothesis to explain the mechanisms of phytoalexin synthesis and accumulation. The hypothesis is that cell death gives rise to endogenous elicitors which in turn induce phytoalexin accumulation. This takes account of the facts that: (i) phytoalexins are not induced in compatible interactions while the fungus is in its biotrophic phase; (ii) at the onset of the necrotrophic phase, phytoalexin accumulation has often been reported in susceptible cultivar-isolate combinations, and (iii) in incompatible interactions of the HR-type, phytoalexins accumulate immediately. The hypothesis also suggests that necrotrophic parasites inevitably induce phytoalexin accumulation but may have evolved a way of escaping their deleterious effects (see section 7.3.4). Alternatively, necrotrophic parasites might suppress the accumulation of phytoalexins by killing host cells so quickly that the host synthetic capability is impaired.

7.4.3. Elicitors of lignification and papilla formation. There have been few studies of the elicitation of lignification or papilla formation; most have been conducted with living organisms (Vance et al., 1980). The induction of lignin in wheat appears to be almost specifically a feature of filamentous fungi (saprophytes or parasites) (Pearce & Ride, 1980). Yeasts and bacteria are poor elicitors of lignification. Chitin, chitosan and ethylene glycol induce lignification in wounded wheat leaves (Pearce & Ride, 1982), the first of these being of interest since it is a major constituent of the cell walls of many fungal parasites. As chitin itself is highly insoluble and may be unlikely to come in contact with plant cell membranes during the infection process, degradation products of chitin (N-acetylglucosamine, small chitin oligomers and glucosamine) have also been examined. However, all these substances appeared to be inactive as inducers of the lignification. In addition, purified cell walls from Botrytis cinerea and Agaricus bisporus, which do not contain chitin, did nevertheless elicit lignification. Thus it appears that cell wall components other than chitin can elicit the lignification response. There are no reports of species or race-specific induction of lignification.

7.4.4. Elicitors of the hydroxyproline-rich glycoproteins. Ethylene induces the accumulation of hydroxyproline-rich glycoproteins (Toppan et al., 1982), and during fungal infection ethylene production may be stimulated (Pegg, 1976). This may in turn induce the accumulation of hydroxyproline-rich glycoproteins.

7.4.5. Elicitors and their specificity. HR and the associated accumulation of phytoalexins are frequently associated with incompatibility in gene-for-gene systems. For this reason a great deal of effort has been put into looking for molecules produced by avirulent races of a pathogen which specifically induce HR and the accumulation of phytoalexins: the so-called specific elicitors.

Most of the elicitors that have been isolated so far do not exhibit race-specificity; many do not even have species-specificity (De Wit & Roseboom, 1980). Often, specificity would not be expected since no differential relationship between races of the fungus and cultivars of the host plant occurs. The relationships that have been most studied are those for which a reciprocal check arrangement is evident. These are the associations between: potato and Phytophthora infestans, soybean and P. megasperma f. sp. glycinea, and bean and C. lindemuthianum. The conclusions from these studies are contradictory and undoubtedly the procedures used to isolate elicitors are crucial.

Elicitors obtained from cell walls of different fungal races by using crude homogenization procedures appear to be non-specific. These elicitors in addition seem to be rather diverse (unsaturated fatty acids from P. infestans; β-1,3- and β-1,6- branched glucans from P. megasperma f. sp. glycinea; and a glycoprotein from C. fulvum (see section 7.4.2).

When milder extraction methods have been used, some elicitors obtained have indeed appeared to be race-specific (Bruegger & Keen, 1979; Keen & Legrand. 1980; Anderson, 1980). Keen et al. (1983) and Keen & Yoshikawa (1983) found that a β-1,3-endoglucanase from soybean released potent race-specific elicitors from P. megasperma f. sp. glycinea; they appeared to be glucomannans and were different from the compounds previously reported for this fungus. Wade & Albersheim (1979) also obtained race-specific molecules from avirulent races of P. megasperma f. sp. glycinea that could protect soybeans against virulent races of the same pathogen. However, these molecules were not potent elicitors of glyceollin accumulation and probably induced an unknown defence mechanism. Desjardins et al. (1982) have been unable consistently to confirm the results obtained by Wade & Albersheim (1979). They found that compounds produced by virulent as well as avirulent races gave protection against subsequent inoculation.

De Wit & Spikman (1982) have obtained race-specific elicitors of chlorosis and necrosis from in vivo intercellular fluids extracted from compatible combinations of C. fulvum and tomato. These elicitors are probably of fungal origin (De Wit et al., 1984) and are different from the non-specific glycoproteins isolated from the fungus grown in vitro (De Wit & Kodde, 1981). In one cultivar, chlorosis and necrosis induced by these elicitors could be seen as early as 7 h after injection into tomato leaves. The elicitor of necrosis appears to be a small peptide (De Wit, unpublished results).

7.4.6. Suppressors and their specificity. In some studies, no evidence for the existence of specific elicitors has been obtained and a contrary hypothesis has emerged. It is suggested that the observed specificity in vivo could be explained by the existence of specific suppressor molecules which eliminate or decrease the effect of non-specific elicitors in compatible interactions. In other words, non-specific elicitors and specific suppressors produced by virulent races only, would work in concert resulting in the observed in vivo specificity. Garas et al. (1979), Doke et al. (1979; 1980) and Doke & Tomiyama (1980) have isolated high molecular weight phytoalexin elicitors from mycelium of P. infestans. In addition, however, low-molecular-weight glucan molecules from homogenates and germination fluids of the same pathogen appeared to function as race-specific suppressors of HR and phytoalexin accumulation. These workers have proposed that race-specific suppressors confer gene-for-gene specificity. These suppressor molecules are branched β-1,3 glucans of 17-23 glucose units. However they did not suppress, but indeed enhanced the phytoalexin-inducing activity of arachidonic and eicosapentaenoic acids (the purified elicitors from P. infestans; Bostock et al., 1983). This indicates that in vivo these acids are not free but must be esterified or otherwise linked to other macromolecules.

Ziegler & Pontzen (1982) have found specific inhibition of glucan-elicited glyceollin accumulation in soybeans by an extracellular mannan-glycoprotein obtained from virulent races of P. megasperma f. sp. glycinea. The mannan-glycoprotein appeared to be an extracellular invertase. From these results, it was concluded that a susceptible host response (suppression of glyceollin accumulation) and not a resistance response (high glyceollin accumulation) is specifically induced by the fungus.

7.5. RECEPTORS FOR ELICITORS OF ACTIVE RESISTANCE MECHANISMS

Elicitors of active defence responses have been studied more extensively than their receptors in host plants. Keen & Bruegger (1977) distinguished distinct determinative (recognition) and expressive phases in their model to explain gene-for-gene complementation. Recognition was envisaged as a static process while expression was thought of as dynamic. The contention is that recognition of an elicitor confers resistance by setting in motion a series of steps from, for example, de novo mRNA synthesis, through protein synthesis to the accumulation of phytoaloxins (or any other active resistance response).

Recognition factors in plants are envisaged as constitutively produced surface molecules, as are elicitors from the parasite. Their occurrence at host and parasite cell surfaces would facilitate contact during the early stages of host-parasite relationship. The recognition factors in plants that have attracted most attention are the lectins. These are proteins or glycoproteins that have the ability to bind certain carbohydrate structures. Lectins are also known as haemagglutinins or phytohaemagglutinins due to their ability to agglutinate red blood cells. Some lectins also agglutinate bacteria and fungal spores. Their hapten specificity makes lectins plausible candidates for the receptors of elicitors, which are frequently carbohydrates or glycoproteins. Recognition events between two organisms through complementary surface-macromolecules including lectins have been the subject of intensive study and several recent reviews have appeared (Callow, 1977; Albersheim & Valent, 1978; Sequeira, 1978; 1982; Bauer, 1981; Dazzo & Truchet, 1983).

7.5.1. The role of lectins.

Rhizobium. Much is now known about interactions between lectins and saccharides in relationships between bacteria and plants, especially in the Rhizobium-legume symbiosis. Rhizobium appears not to have evolved the same degree of specificity with respect to its host plants as has occurred with parasitic bacteria and fungi. One reason for this may be that root nodulation and infection by Rhizobium is advantageous to the host. From a genetic standpoint there are also differences between this mutualistic symbiosis and host-parasite relationships. 'Resistance' to nodulation is generally inherited as a recessive character (Nutman, 1969) while resistance to a fungal or bacterial parasite is usually dominant (Ellingboe, 1981). In other words, recognition in the Rhizobium-legume symbiosis leads to nodulation, while recognition in a parasitic system is thought to lead to a resistance response. Some of the most important aspects of the role of lectins in recognition processes in the Rhizobium-legume symbiosis (reviewed by Bauer, 1981 and Dazzo & Truchet, 1983) may have analogy to recognition of plant parasitic bacteria, and are summarized here.

Infection of legume root hairs by the bacterial symbiont involves many steps from initial recognition to the formation of a root nodule. These steps are: attachment, curling and branching of the root hair, penetration of the root hair cell wall, formation of an infection thread, host cell proliferation in the inner cortex in front of the advancing infection thread, development of bacteroids and so forth (Dazzo & Truchet, 1983). Although attachment of infective rhizobia to root hairs is a prerequisite, several observations indicate that other, undefined events must occur to intiate root hair infection. Bohlool & Schmidt (1974) and Dazzo & Hubell (1975) proposed that complementary lectin-polysaccharide interactions serve as a basis for host-specificity in the Rhizobium-legume symbiosis. During the early stages of infection, bacteria attach to the host by a hapten-reversible interaction. Subsequently, bacterial cells become irreversibly anchored to the host cell. The ability of bacteria to attach to root hairs is controlled by Roa (root hair attachment)-genes. Despite some reported exceptions (Chen & Phillips, 1976) there seems to be a good correlation between the ability of Rhizobium japonicum and R. trifolii to nodulate their respective hosts and also to bind to host lectins. The exceptions include nodulating strains of R. japonicum which do not bind to soybean lectin but this could be explained by the observation that lectin-binding sites on the surface of rhizobia are transitory (Bhuvaneswari & Bauer, 1978). For example, certain strains of R. japonicum only possessed lectin-binding sites when they were grown in media containing soybean root exudates. Lectin receptors on *R. trifolii* also appear to be transient since they were exposed for only short periods during culture (Dazzo *et al.*, 1979). The exposure of lectins on the root hair surface is also apparently transient (Bhuvaneswari *et al.*, 1980). However the observation that *R. japonicum* still infected soybean lines which appeared to lack the soybean lectin and that root nodulation in these hosts remained specific for soybean rhizobia requires explanation (Dazzo & Truchet, 1983). It is conceivable that the root hairs of these lines did carry the soybean lectin or other lectins which recognize the rhizobia of soybean, since the absence of lectin was only established for seeds.

The identities of the lectin receptors on different strains of *Rhizobium* are not yet known, but the evidence is that there are multiple receptors. A major controversy is whether the lectin receptor is the capsular polysaccharide (CPS), lipopolysaccharide (LPS) or extracellular polysaccharide (EPS). It seems that *Rhizobium* species may have different receptors: *R. japonicum* (CPS and EPS); *R. trifolii* and *R. leguminosarum* (CPS, EPS and LPS), and *R. meliloti* (LPS) (Dazzo & Truchet, 1983). The presence of multiple lectin receptors in *Rhizobium* species suggests that each one could have a different role in the process of root hair infection.

In the association between *Agrobacterium tumefaciens* (a member of the *Rhizobiaceae*) and its hosts, recognition leads to tumor induction and proliferation of the bacterium in the tumor (i.e. as with *Rhizobium* recognition leads to compatibility). However, in this relationship no lectin-mediated recognition has been reported (Pueppke *et al.*, 1982). Here, pectic substances in the cell walls of dicot plants appear to be the receptors which mediate bacterial attachment (Lippincott & Lippincott, 1977).

Bacterial parasites. It has been suggested that lectins present on higher plant cell walls may function as recognition factors for avirulent strains of plant parasitic bacteria (Sequeira, 1978; 1981; 1982). In this situation, it is envisaged that recognition results in incompatibility associated with HR and sometimes phytoalexin accumulation. Sequeira & Graham (1977) proposed that in incompatible combinations between tobacco and avirulent strains of *P. solanacearum*, the LPS portion of the bacterial wall bound with a lectin (potato lectin) from the plant cell wall (Leach *et al.*, 1982). It was suggested that virulent strains avoided this binding by their ability to secrete soluble EPS which became bound instead, thus saturating the lectin binding sites. This would prevent the attachment of virulent bacterial cells and allow them to multiply freely in the intercellular spaces. However, contrary to the hypothesis, injection of EPS prior to inoculation with avirulent bacteria did not result in the prevention of HR. Since binding of EPS to the lectin was affected by ionic strength it is questionable whether EPS plays an important role in preventing attachment of bacteria to plant cell walls in vivo. In addition, the hypothesis does not explain why certain isolates of P. solanacearum which produce copious amounts of EPS in culture also induce HR in tobacco (Sequeira, 1982). Furthermore, certain avirulent strains of P. solanacearum are neither agglutinated by the lectin nor do they induce HR.

Whatley et al. (1980) have suggested that differences in structure of LPS from virulent and avirulent strains of P. solanacearum are correlated with the induction of HR. Most HR-inducing strains were shown to have low molecular-weight (rough) LPS while non-HR-inducing strains mostly had high molecular-weight (smooth) LPS. The hypothesis was that initial recognition, a prerequisite for induction of HR, required the presence of rough LPS. It was argued that smooth LPS-producing strains may not be recognized by the host because the portion of the LPS involved in recognition may be masked by the specific O-antigen of the LPS molecule (somatic or non-flagella antigen). In rough LPS producing strains, part or all of the O-antigen of the LPS is absent. It is now believed, however, that other properties of EPS, such as its ability to cause water-soaking, might be more important than its lectin-binding properties in preventing attachment of virulent bacteria to cell walls.

Slusarenko & Wood (1981) have extracted factors from green bean leaves that specifically agglutinate cells of Pseudomonas syringae pv. phaseolicola in vitro. These extracts particularly agglutinated cells of an avirulent strain but agglutination of a virulent strain occurred to a lesser extent. The purified and isolated agglutinins appeared not to be lectins but pectic polysaccharides (Slusarenko & Wood, 1983). Interestingly, the agglutination phenomenon was studied in a quadratic check arrangement (Figure 7.3) and it was found that pectic polysaccharides isolated from cv. Red Mexican agglutinated race 1 to a greater degree than race 2 (race 1 is virulent on cv. Canadian Wonder and avirulent on cv. Red Mexican; race 2 is virulent on both cultivars). However, in spite of differential activity in vitro the active pectic polysaccharide was obtained from both cultivars and the chemical composition was shown to be similar. This indicated that differential attachment of bacterial cells to pectic polysaccharides in plant cell walls was not responsible for the specific resistance of cv. Red Mexican to race 1.

In the association between soybean and Pseudomonas syringae pv. glycinea, no specific attachment of avirulent strains to host cells was observed in incompatible interactions (Fett & Jones, 1982). It was therefore concluded that attachment of bacterial cells was not a prerequisite for HR and phytoalexin accumulation to occur. However, Ersek et al. (1981) have extracted factors from the intercellular spaces of a cultivar resistant to strain 1 of the bacterium but susceptible to strain 2, which specifically agglutinate the avirulent strain. These agglutination factors were unlike soybean lectin in that agglutination was not inhibited by N-acetyl-galactosamine, a specific hapten for the lectin.

Fett & Sequeira (1980a; 1980b) have found a bacterial agglutinin in soybean that agglutinated several strains of Xanthomonas phaseoli var. sojensis but did not agglutinate P. syringae pv. glycinea. There was, however, no correlation between agglutination and pathogenicity.

In the association between Erwinia amylovora and apple, the production of EPS appears to be required for virulence. Bacteria lacking EPS were agglutinated in the xylem vessels of apple petioles (Hseu & Goodman, 1978; Goodman, 1980). One suggestion is that EPS blocks host cell recognition of bacterial LPS thus preventing the induction of host defence responses. Romeiro et al. (1981a; 1981b) have also isolated a low molecular-weight, heat-stable protein from apple which agglutinates avirulent strains of E. amylovora to a greater extent than virulent strains. Evidence was also obtained for binding of bacterial LPS to this apple agglutinin. However, since no quadratic or reciprocal check exists in this association, no insight into cultivar specificity is provided.

There is therefore some evidence that lectins may be receptors for parasite 'elicitors' but this is by no means certain and no other candidates have been investigated. In host plants from which lectins have been obtained there is some evidence for specific binding at the host-bacterial species level, but there is no good evidence for race-specific binding by lectins. The role for lectins as receptors for elicitors has not so far been examined in associations where a quadratic or reciprocal check arrangement occurs. The host-parasite relationships studied conform to the patterns shown in Figures 7.1 and 7.2.

Fungal parasites. There are a few indications that lectins could also play

	Host lines	
	A	B
Parasite	C	I

FIGURE 7.1. The possible combinations between one parasite and two host lines (A and B) which vary in resistance. C = compatible or susceptible; I = incompatible or resistant.

	Parasite isolates	
	X	Y
Host	C	I

FIGURE 7.2. The possible combinations between two parasite isolates (X and Y) which differ in parasitic capability, and one host line. C and I as above.

	Genotypes of host lines	
Genotypes of parasite isolates	R_1R_1	r_1r_1
A_1	I	C
a_1	C	C

FIGURE 7.3. The 'quadratic check'. Possible combinations between two host lines and two haploid parasite isolates differing in resistance and parasitic capability respectively. R_1 and A_1 are alleles for resistance and avirulence respectively, and r_1 and a_1 are the opposite alleles for susceptibility and virulence. C and I as above.

	Genotypes of host lines	
Genotypes of parasite isolates	$R_1R_1r_2r_2$	$r_1r_1R_2R_2$
A_1a_2	I	C
a_1A_2	C	I

FIGURE 7.4. The 'double reciprocal check'. Possible combinations between two host lines and two haploid parasite isolates differing in resistance and parasitic capability respectively. R_1 and R_2 are different alleles for resistance and r_1 and r_2 are the opposite alleles for susceptibility. A_1 and A_2 are different alleles for specific avirulence on R_1 and R_2 respectively; a_1 and a_2 are the opposite alleles for virulence. C and I as above.

a role in relationships between fungal parasites and their hosts. Furuichi et al. (1980) suggested that potato lectin is involved in binding the cell wall surfaces of Phytophthora infestans to cell membranes. However, no race-specificity was found. Garas & Kuć (1981) found that potato lectin precipitated elicitors of terpenoid accumulation which had been extracted from the same fungus, while Nozue et al. (1980) found that the induction of HR in incompatible race-cultivar combinations could be inhibited by adding NN'-diacetylchitobiose, a specific hapten for the potato lectin.

Kojima et al. (1982) reported that a lectin-like agglutination factor from sweet potato roots can agglutinate non-germinated spores of seven strains of Ceratocystis fimbriata including one parasitic on sweet potato. This factor, however, also showed agglutinating activity with germinated spores of five strains of the fungus parasitic on hosts other than sweet potato, while germinated spores of strains parasitic on sweet potato and almond were not agglutinated. The agglutination factor in this case was thought to be a high molecular-weight polysaccharide. Yoshikawa et al. (1983) have recently reported a receptor on soybean membranes for a non-specific fungal elicitor (mycolaminarin). This receptor appears to be a protein or glycoprotein. However, binding of ^{14}C-mycolaminarin was irreversible and could not be displaced with an excess of unlabelled ligand. This finding suggests that the receptor was not likely to be a lectin.

There are other reports of lectins binding fungal elicitors, but often the lectins concerned are unrelated to those found in the host of the parasite in question. Interestingly, Hinch & Clarke (1980) reported an inverse binding phenomenon: a fungal lectin interacts with a carbohydrate receptor from the host, but the significance of this is unclear. Little experimental evidence has therefore been accumulated to suggest a role for lectins in the recognition of fungal parasites.

7.6. PHYSIOLOGICAL MODELS TO EXPLAIN GENE-FOR-GENE RELATIONSHIPS

Although speculative, a widely held opinion is that basic compatibility between host and parasite evolved first and that the gene-for-gene relationship has become superimposed upon this. (Ellingboe, 1976). The theory that plant parasites evolved from saprophytes (Lewis, 1973) raises the question of how many different attributes are required for an organism to be a successful parasite. Clearly, the evolutionary changes required for a saprophyte to become a parasite are greater than for a race of a parasitic species avirulent on a particular cultivar to become virulent to it. Most characteristics conferring parasitic ability are envisaged as positive functions enabling successful host colonization. Some of these functions, for example, could be the production of toxins; pectic enzymes; enzymes that degrade preformed antimicrobial compounds or phytoalexins; substances that mask recognition factors at parasite surfaces and substances that suppress the expression of HR (Heath, 1981a; Keen, 1982). Undoubtedly, many genes must be involved in the successful development of a parasite in its host and the relationship with host gene functions could be complex.

Gabriel *et al.* (1979) attempted to obtain temperature-sensitive parasite mutants of *Phyllosticta maydis* to probe which characters were crucial for basic compatibility. A class of mutants was obtained which was temperature sensitive *in vivo* but not *in vitro*. The conclusion was therefore that these mutations affected genes crucial to parasitism. If large numbers of this type of mutant could be obtained, it would be possible both to map the loci concerned and estimate how many genes are necessary for the establishment of basic compatibility. Perhaps more important than the numbers of genes would be knowledge of the relationship between genes controlling parasitic capability and host genes governing susceptibility. Is there a relationship analagous with the gene-for-gene system for race-specific resistance? Since almost nothing is known about the genetics of basic compatibility only speculation is possible. Lack of knowledge is not surprising since such an investigation is equivalent to unravelling the course of the presumed evolution.

Given the existence of basic compatibility, there is likely to be a selection pressure on the plant towards resistance. In the unnatural circumstances of crop species, a plant breeder's role in the selection of resistant cultivars accelerates this process. Once resistant genotypes emerge either naturally or under man's influence, selection pressure is again imposed on the parasite to respond by overcoming the resistance. It has been suggested that this is how the gene-for-gene relationship comes into operation. Only after basic compatibility has been established can specificity at the race-cultivar level develop. Heath (1981a) has pointed out that mechanisms controlling basic compatibility and race-specific resistance may operate simultaneously and this has a bearing on the interpretation of biochemical and physiological data as well as the design of models to explain gene-for-gene relationships. However, race-specific resistance has generally been studied with little regard to the processes responsible for basic compatibility. Many of the apparent contradictions between studies may be a reflection of this, and paying too little regard to basic compatibility processes may be one reason why mechanisms of race-specific resistance are still poorly understood.

7.6.1. Genetical implications. Ellingboe (1981) has summarized the development of research on the physiology of host resistance since the 1950s. Much of the early work was conducted using two host lines, one susceptible and one resistant to a particular isolate of a parasite (Figure 7.1). In most studies the different host lines may have differed greatly for many characters other than resistance. Thus, it is inevitable that any difference between the two lines is absolutely correlated with resistance and susceptibility, and the same argument holds for investigations where a single cultivar or a line of a plant species is inoculated with virulent or avirulent isolates of a parasite (Figure 7.2). Any difference between two isolates is absolutely correlated with virulence and avirulence. But establishing 'cause and effect' is almost impossible from this type of study. Rowell *et al.* (1963) influenced thinking by introducing the concept of the quadratic check as a genetical model for physiological studies of resistance to *Puccinia graminis* (stem rust) in wheat (Figure 7.3). Examination of the quadratic check leads to the conclusion that three different genotypical combinations (A_1/r_1; a_1/r_1; a_1/R_1) all lead to a compatible phenotype while only one unique genotype combination (A_1/R_1) leads to incompatibility. This can be interpreted to mean that specificity is controlled by the interaction between the gene products of the dominant A and R alleles. In all other cases, it is envisaged that there is no recognition or interaction between gene products and as a consequence compatibility results. However, Martin & Ellingboe (1976) have found small differences between the phenotypes of the three allele combinations leading to compatibility (see Chapter 5.5).

Loegering & Harmon (1969) introduced the use of near-isogenic lines, more or less identical to one another except for their resistance genes. This made it possible to study differences between resistant and susceptible lines that were more likely to be related to genes for resistance. The arrangement between isolates and cultivars employed in this study was called the (double) reciprocal check (Figure 7.4). This arrangement can be interpreted to mean that A_1 interacts specifically with R_1 and not with R_2 and that A_2 interacts specifically with R_2 and not with R_1. There is a strict gene-for-gene relationship. In addition, this arrangement implies that incompatibility is epistatic to compatibility.

Ellingboe (1982) and Keen (1982) have individually reviewed a number of models which have been proposed to explain the gene-for-gene relationship on a mechanistic basis. These models are briefly elaborated and commented upon here.

7.6.2. Constitutive models. One model has been advanced by Ellingboe (1982) and is derived from genetic studies of the gene-for-gene relationship. It is suggested that the primary product of the dominant allele for avirulence (A_1) interacts directly with the primary product of the dominant gene for resistance (R_1) and that the interaction itself is responsible for incompatibility (Figure 7.5). It is suggested that a structural dimer is directly responsible for inhibiting parasite growth.

Lectins might contribute to inhibition of growth on the basis of this model, but there are no reports of lectins with race-specific binding properties, nor are there indications that cultivar or R-gene-specific lectins exist. Plant lectins can inhibit fungal growth (Mirelman et al., 1975; Gibson et al., 1982) but this is probably a non-specific phenomenon. A further objection that has been raised to this model is that compatibility can be induced with metabolic inhibitors, thus suggesting that resistance is not a constitutive character and that the expression of incompatibility requires active processes.

Vanderplank (1978; 1982) has also advanced an intriguing although highly speculative constitutive model to explain gene-for-gene relationships. This model is almost the antithesis of the one above, although both propose the direct involvement of primary gene products. It is argued that only nucleic acids, proteins or polysaccharides can store the degree of qualitative variation required to explain the extreme variability and specificity observed in gene-for-gene relationships. Polysaccharides are ruled out because the necessary variation could only result from the action of different enzymes which would all need to be catalytically distinct. It is not considered likely that catalytically distinct enzymes would be coded for by alleles or pseudoalleles at an R-gene locus. The gene products of these alleles are thought more likely to be isoenzymes with similar catalytic function. Nucleic acids are also considered unlikely candidates because of the lack of evidence for direct associations between this type of molecule in host-parasite combinations exhibiting a gene-for-gene relationship. A protein-for-protein hypothesis is therefore advanced.

It is envisaged that the process of protein polymerization is the basis for

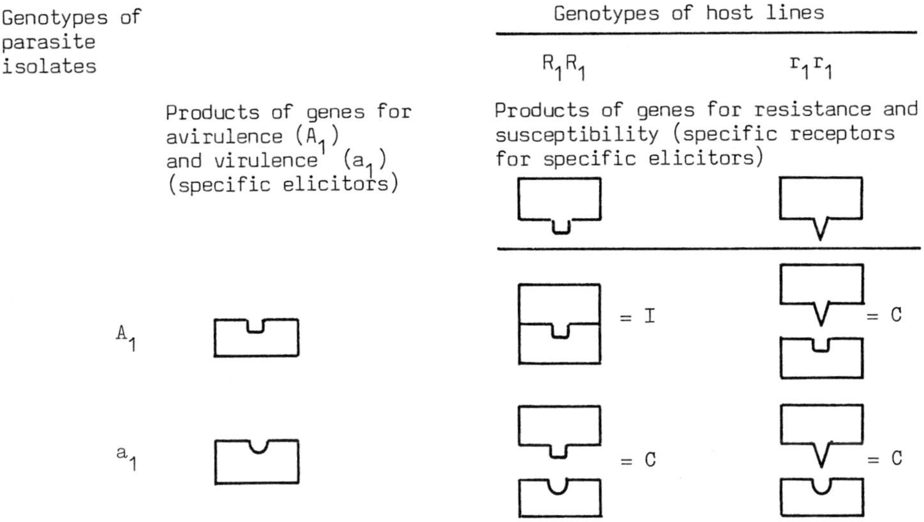

FIGURE 7.5. A physiological model for expression of incompatibility after interactions between the product of the gene for resistance (R_1) and the product of the gene for avirulence (A_1). I = incompatible; C = compatible.

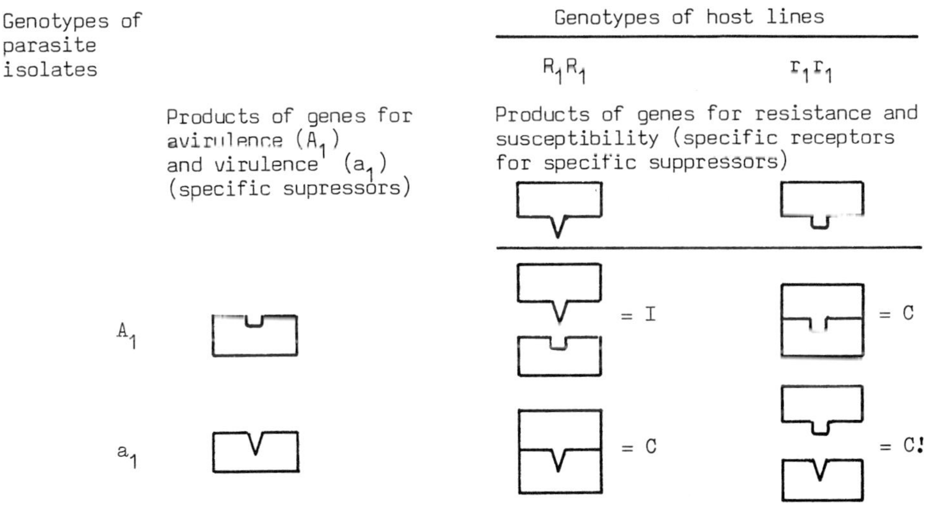

FIGURE 7.6. A physiological model for expression of compatibility after interaction between products of genes for resistance (R_1) and susceptibility (r_1) and products of genes for avirulence (A_1) and virulence (r_1). Note that products of a_1 and r_1 do not interact yet a compatible association is observed in practice. C and I as above.

gene-for-gene recognition. In contrast to Ellingboe's (1981) model, Vanderplank (1978) proposed that in a compatible combination, parasite proteins polymerize each with their partner host proteins. This polymerization it is suggested, interferes with host gene regulation, causing the production of more host protein from which the parasite benefits nutritionally. Any failure of polymerization of a parasite protein leads to incompatibility as a consequence of its being enzymically active. Any unpolymerized protein is therefore the product of an avirulence allele and behaves as an 'elicitor' or 'toxin'. On the basis of this hypothesis any parasite gene may be either a gene for virulence or avirulence depending on whether the host carries a gene capable for coding for a protein with which it can polymerize.

Protein polymerization is an endothermic reaction and the circumstantial evidence to support the protein-for-protein hypothesis comes from the observations that the outcome of some host-parasite combinations can be temperature sensitive. The hypothesis dictates that if temperature sensitivity is observed, it should be such that combinations which are incompatible at low temperatures become more compatible at high temperature and this is often observed (Watson & Luig, 1968; Johnson & Schafer, 1965) (see Chapter 5.5). The opposite effect i.e. increased incompatibility with higher temperature cannot be accommodated by the hypothesis, yet has been recorded (Rajaram *et al.*, 1971). This discrepancy has been explained by suggesting that above certain temperatures, proteins begin to denature and therefore polymers dissociate thus restoring incompatibility. However, a change of incubation temperature from 15° to 27°C caused some wheat cultivars to become susceptible to avirulent isolates of *Puccinia recondita* (Johnson & Schafer, 1965) while others became resistant to compatible isolates (Rajaram *et al.*, 1971).

Temperature has also been observed to result in changes in dominance relationships in accordance with the predictions of the hypothesis (Luig & Watson, 1965). Both resistance and avirulence are expected to become less dominant at higher temperatures.

The protein polymerization hypothesis also offers an explanation for the observation that compatible hosts and parasites share common antigens (Doubly *et al.*, 1960; DeVay & Alder, 1976). It is envisaged that the polymerizing surfaces of the host and parasite proteins may also be antigenic determinants (see Chapter 5.5).

The protein-for-protein model accommodates the double reciprocal check arrangement (Figure 7.4) based on two interacting gene pairs in the host and parasite. The model also explains the fact that incompatibility is epistatic to compatibility. However, this hypothesis only explains how the r/A, r/a and R/a combinations of the quadratic check (Figure 7.3) all result in compatibility if the a allele does not code for a protein with an enzymatic capability.

A different type of constitutive model has been discussed previously in relation to the action of host-selective toxins (section 7.2.2). This model is based on the supposition that a parasite gene controls the production of a toxin which is detected by the product of a host gene for sensitivity to that toxin. Sensitivity to the toxin leads the host to be susceptible to the parasite, for example in the case of some toxin-producing <u>Helminthosporium</u> species. Since there is no double reciprocal check arrangement in diseases involving host-selective toxins, evidence for a gene-for-gene relationship is lacking. However, in other host-parasite combinations which are demonstrably gene-for-gene controlled, it has also been suggested that HR may be a specific response to a selective toxin, the production of which is controlled by an avirulence gene. The hypothetical toxin is thought to cause cellular disruption and the release of pre-formed anti-microbial metabolites (Kaars Sijpesteijn, 1976).

7.6.3. <u>The double induction model</u>. This model proposes that there is a host mediated release of avirulence gene product (specific elicitor) when host-parasite contact is established, which in turn initiates the host defence response (Cruickshank, 1980). Although this model is hypothetical, some supporting experimental data have recently been forthcoming. Plant enzymes like β-1,3 glucanases and chitinases can rapidly liberate soluble phytoalexin elicitors. Keen & Yoshikawa (1983) and Keen <u>et al</u>. (1983) found that a β-1,3 glucanase from soybean released glucomannans with elicitor activity from cell walls of the fungus <u>P</u>. <u>megasperma</u> f. sp. <u>glycinea</u>. These elicitors appeared to be race-specific. Alcu, De Wit & Spikman (1982) found host-mediated release (or production) of race-specific elicitors of necrosis in intercellular fluids of compatible combinations between <u>Cladosporium fulvum</u> and tomato. The factor mediating the production of these elicitors in tomato is unknown but it is not dependent upon the R-gene carried by the host in which the fungus is growing (De Wit <u>et al</u>., 1984). This model is in agreement with genetic studies of gene-for-gene relationships and its implications are discussed in more detail in section 7.6.5.

7.6.4. <u>The non-specific elicitor and specific suppressor model</u>. This model is based on the premise that non-specific elicitors in concert with race-specific suppressors confer specificity in host-pathogen associations. It is envisaged that race-specific suppressors produced by virulence genes specifically bind to hosts receptors (products of genes for resistance). In this way, they either prevent the binding of non-specific elicitors, thus causing the suppression of a resistance response, or alternatively prevent the expression of resistance in another way. Doke et al. (1979; 1980), Garas et al. (1979), Doke & Tomiyama (1980) and Doke (1983a; 1983b) found experimental evidence to support this hypothesis from studies with potato and Phytophthora infestans. Ziegler & Pontzen (1982) obtained data from studies with soybean and P. megasperma f. sp. glycinea which could also be interpreted on the basis of this hypothesis. The suppressors identified by Doke et al. (1979, 1980) and Garas et al. (1979) are active in crude preparations of non-specific elicitors from P. infestans but unaccountably, they enhanced the activity of purified elicitors (Kurantz & Zacharius, 1981; Kurantz & Osman, 1983; Bostock et al., 1982; Preisig & Kuć, 1983). Hence the role of race-specific suppressors is unclear.

Apart from the confusing experimental evidence, this model is not entirely in agreement with genetic studies of gene-for-gene relationships. In the model, compatibility is thought to be the specific event, yet the combinations A_1/r_1, a_1/R_1 and a_1/r_1 (Figure 7.3) all lead to compatibility. As represented in Figure 7.6 it can be seen that specific recognition between products of A_1 and r_1, and a_1 and R_1 could occur, but not between a_1 and r_1. It seems to be more feasible to consider suppressors to be one of the many characters needed by a parasite to establish basic compatibility.

In this context, Davidse & Boekeloo (1984) have found evidence for the production of race non-specific suppressors by P. infestans. These suppressors inhibited the induction of necrosis in potato leaves caused by non-specific elicitors from P. infestans. Heath (1980b; 1981b) also presented evidence for general suppressors and has suggested (Heath, 1981a) that they may act below the level of race-cultivar specificity. Bushnell & Rowell (1981) have proposed a model in which race-specific suppressors may have evolved from general species-specific suppressors. This implies that one macromolecule possesses general as well as race-specific suppressor activity, and that basic compatibility in addition to race-specific compatibility could be determined by one macromolecule. According to this model, both basic compatibility and race-specific compatibility would be inherited as dominant characters. But whilst almost nothing is known about the genetics of basic compatibility, race-specific compatibility in host-parasite combinations involving diploid biotrophic fungi usually involves combinations of recessive alleles (Flor, 1956).

7.6.5. The specific elicitor-specific receptor model. This model is similar to the first constitutive model described above (section 7.6.2) (Figure 7.5). However, it differs in that initial recognition leads to a sequence of events culminating in (locally) induced resistance (HR, accumulation of phytoalexins, lignification etc.). Incompatibility is the specific event. The suggestion is that the primary product of a gene for avirulence (A_1) interacts with the primary product of a gene for resistance (R_1). As a consequence, a second messenger-like substance transfers information to the plant cell nucleus, where defence mechanisms are initiated. These second messengers could be the endogenous elicitors mentioned earlier (section 7.4.2).

Jones & Deverall (1978), Bruegger & Keen (1979), Wade & Albersheim (1979), Anderson (1980), Keen & Legrand (1980), De Wit & Spikman (1982), Keen & Yoshikawa (1983), Keen et al. (1983) and De Wit et al. (1984) have all found evidence for race-specific elicitors of defence responses. The specific elicitor-specific receptor model does not conflict with the idea of there being a basic compatibility since the gene-for-gene system is envisaged as superimposed upon it. The implication is that the general suppressors discussed earlier (section 7.6.4) (Heath, 1980b; 1981a; 1981b; 1982; Davidse & Boekeloo, 1984) are a component of basic compatibility.

Ellingboe (1982) has argued that if responses such as phytoalexin accumulation are important in gene-for-gene controlled resistance, one gene-for-one-gene relationships would not be expected. The argument is that many enzymes and hence genes are responsible for phytoalexin synthesis, and that mutations at any of these loci should impair resistance. Similarly, since glycoproteins (as specific elicitors) are also synthesized by many enzymes, it is argued that their production is unlikely to relate to a single avirulence gene. Ellingboe (1982) is of the opinion that if this model were to operate, specific resistance and virulence would both be polygenic traits. With respect to a glycoprotein, the glycosyl transferase which provides the final structure is the most important in terms of potential specificity, and this could be coded by the avirulence locus. Mutants lacking enzymes, or having less efficient enzymes for synthesis of phytoalexins have not been studied and may simply be lethal.

7.6.6. Does recognition in the pistil-pollen system provide a model for host-pathogen recognition? Bushnell (1979) has drawn attention to the similarities that exist between the determination of plant-parasite compatibility and pistil-pollen compatibility.

Hogenboon (1975; 1983) distinguishes between incongruity and incompatibility in pistil-pollen relationships. Incongruity largely refers to what has also been described as interspecies incompatibility. Hogenboon (1975) restricts the use of incompatibility to intra-species and self-incompatibility. Conceptually, incongruity can be equated with non-host resistance or the lack of basic compatibility between host and parasite. In addition some forms of cultivar resistance may result from mechanisms equivalent to incongruity.

In contrast, incompatibility can be equated with gene-for-gene controlled resistance. Bushnell (1979) has shown how the double reciprocal check arrangement (Figure 7.4) is common to both gene-for-gene controlled host-parasite relationships and S-allele controlled pistil-pollen relationships.

Incongruity results from a 'lack of genetic information' in one partner about some relevant character of the other. An incomplete matching of genetic systems is envisaged and the relationship is non-functioning. The absence of basic compatibility may be considered in the same way. At one of many stages of development of a parasite in a non-host plant, the relationship may become non-functional because either the host lacks something which the parasite requires or the parasite lacks some character essential for its continued development. In this context, resistance and susceptibility controlled by host-selective toxins may be an example of host-parasite incongruity.

Incompatibility on the other hand 'prevents or disturbs' a relationship where each partner contains the genetic information to allow a functioning relationship. This is the superimposed S-allele or gene-for-gene system.

In the context of this Chapter, many of the resistance mechanisms discussed where host-parasite contact is made but where there is no apparent gene-for-gene relationship appear to be analogous to incongruity. Are there any indications that the mechanisms which control self-incompatibility and gene-for-gene incompatibility are similar?

First, there is one difference between the two types of relationship. In pistil-pollen relationships, possession of the same S-allele or products of that allele (in the case of sporophytic systems) in both partners results in incompatibility. There are no indications that gene-for-gene incompatibility is based upon shared genes between host and parasite. Complete correspondence between these two types of relationship would imply that a resistance allele behaves as it does because it codes for the same protein as an allele located in the parasite which thus becomes identifiable as an avirulence allele.

In the case of sporophytic systems, incompatibility can be determined within minutes of the pollen arriving at the stigmatic surface (Heslop-Harrison, 1975). Often, incompatible pollen grains fail to germinate, or if they do, they fail to penetrate the cuticle. When penetration does occur, callose-like material may be rapidly deposited around incompatible pollen tubes (Heslop-Harrison et al., 1974). This response is at least superficially similar to encasement of haustoria and papilla formation in host resistance responses to fungal parasites (section 7.3.1).

Compatibility may be achieved in pistil-pollen combinations which would normally be incompatible, when immature stigmas are pollinated or when diffusates from compatible pollen are applied to the stigma (Howlett et al., 1973). Pollination at high temperatures may cause loss of incompatibility.

In gametophytic systems, since the outer layers of the pollen grain play no part, incompatibility is usually only expressed after penetration of the stigmatic surface. The pollen tube may then grow abnormally, degenerate or cease growth. Again, callose-like deposits are commonly associated with incompatibility. Evidence that incompatibility may require active induction comes from the finding that inhibitors of RNA synthesis can promote compatibility (Cambell & Ascher, 1976).

S-allele controlled incompatibility is commonly supposed to result from a protein-protein recognition event and a number of different mechanistic models have been proposed (Nasrallah et al., 1970; Lewis, 1965; Burret, 1971; Abdalla & Hermsen, 1972). Events following recognition which result in inhibited pollen-germination or germ tube growth are largely unknown but it is intriguing that 'phytoalexin-like' inhibitors of pollen germination have recently been described (Hodgkin & Lyon, 1983).

7.6.7. Further considerations. Models such as those discussed above are important in providing a stimulus for further experimentation, but there is a danger that they channel thinking in a single direction. Allegiance to a particular model is soon likely to inhibit genuine testing of hypotheses.

There seems to be one major point at issue concerning the mechanistic basis of gene-for-gene relationships. To explain specific resistance, is there a need to invoke expressions of incompatibility such as HR, phytoalexin accumulation and lignification, or are these merely symptoms that inevitably follow incompatibility resulting from some prior event? This is the distinction between constitutive models and those which explain resistance in terms of some actively induced host response.

It cannot be denied that these host responses occur and that they are strongly associated with incompatibility in gene-for-gene relationships. There is also persuasive evidence, discussed previously, which points to their direct involvement in the cessation of parasite development in some cases. Certainly, the onus is on those sceptical that active defence mechanisms are involved in gene-for-gene resistance to demonstrate that these types of response are not required for resistance.

It is nevertheless possible to provide alternative explanations to account for the universal occurrence of some active reaction in plants associated with gene-for-gene incompatibility. For example, if cell death is a necessary consequence to post-penetration incompatibility (as envisaged in at least two of the constitutive models), its occurrence in the absence of some response could be potentially detrimental to the plant. All sorts of substances leak from dead cells, encouraging the growth of micro-organisms in their vicinity. The entry and establishment of unspecialized and damaging necrotrophic parasites would be facilitated, unless a 'wound-healing' process prevented this eventuality. Accumulation of phytoalexins or cell wall modification for example may provide essential non-specific protection against secondary invasion. It is probable that small lesions resulting from penetration by avirulent parasites or particularly those organisms to which the plant is a non-host, abound at plant surfaces under natural circumstances. Some mechanism to restrict access of micro-organisms to the plant via these wounds is to be expected.

7.7. CONCLUSION: PERSPECTIVES AND CHALLENGES

Plant pathology, like most fields of endeavour, has evolved by trial and error rather than by maturing to a plan. Inevitably, this has lead to great leaps forward, gaps and inconsistencies. The study of resistance calls on many disciplines: genetics; biochemistry; plant and microbial physiology; molecular biology, and analytical chemistry. Each has made valuable contributions to the whole, but clearly there is a limit to what can be achieved by any single approach. Thus physiological studies cannot alone explain the underlying genetic controls, and genetic analysis cannot fully explain biochemical mechanisms. A major challenge in studying resistance mechanisms today is to bring together these disparate but complementary disciplines. Another is to translate fundamental knowledge into measures for practical crop protection.

These are easier said than done; there are problems inherent in the systems chosen for study. Near-isogenic lines or races of hosts and pathogens are only rarely available. From earlier sections in this Chapter, it is clear that host-pathogen associations which are amenable to genetic study all too often involve obligate parasites such as rusts and mildews, where certain types of biochemical or physiological study become very difficult. On the other hand, those organisms which can be cultured easily and are more amenable to biochemical study are often <u>fungi imperfecti</u>, making genetic analysis difficult or impossible. Some organisms are renowned for having unstable genomes, making long-term experiments difficult. Pathogens grown in culture can lose their parasitic ability unless they are periodically 'recycled' through the plant.

There are also difficulties with plant material. The conditions of propagation can have a profound influence. The urge is to chose a 'model' host-pathogen system which is most convenient for the type of analysis involved, but even the best systems are prone to artifacts or other factors which confuse the issue. Tissue age and type are among the commonest offenders. Some plant tissues are easier to examine physiologically and biochemically than others, which may be more important in the pathogenic relationship. Basing general assumptions on experiments carried out on specific tissues is a constant temptation. Much work on elicitors of phytoalexin accumulation, for example, has been done with cotyledons, hypocotyls, cultured cells and various storage organs. How representative are they of stems and leaves, and what is the relevance of a result from a laboratory experiment to the crop in the field?

The concentration on model systems also produces a fragmentary picture. There is a temptation to extrapolate findings to other cases, but the enormous diversity of reaction types makes this a dangerous practice, and valid generalities are hard to come by, even for closely related systems.

Having emphasised the difficulties, it is appropriate to balance the picture by marking the progress that has been made. Fortunately, studies on disease resistance mechanisms have moved on from an early fixation with the isolation and characterization of phytoalexins. Effort is now being devoted to answering more important questions on the mechanisms that lead to their accumulation, and their role in defence. The scope is being broadened to include other active responses. Progress is being made in matching evidence from genetics, biochemistry and physiology, thus bringing an understanding of the mechanisms of race specificity into sight. Recognition of the concept of basic compatibility has recently given a timely persepective to fundamental plant pathology.

The emphasis on active defence is currently eclipsing the relevance of passive defences. Yet it is likely that the majority of potential parasites that plants have to contend with are repelled by passive barriers long before active defences are called upon. Perhaps some effort should be returned to understanding the mechanisms and functions of passive defences as a key to enduring horizontal resistance.

REFERENCES

Abdalla, M.M.F. & Hermsen, J.G.T. (1972). Unilateral incompatibility: hypotheses, debate and its implications for plant breeding. Euphytica 21, 32-47.

Abeles, F.B., Bosshard, R.P., Forrence, L.E. & Habig, W.H. (1971). Preparation and purification of glucanase and chitinase from bean leaves. Plant Physiology 47, 129-134.

Aist, J.R. (1976). Papillae and related wound plugs of plant cells. Annual Review of Phytopathology 14, 145-163.

Akai, S. (1959). Histology of defense in plants. In 'Plant Pathology, An Advanced Treatise. Vol. I. The Diseased Plant' (Horsfall, J.G. and Dimond, A.E., eds), pp. 391-434. Academic Press, New York.

Akai, S. & Fukutomi, M. (1980). Preformed internal physical defenses. In 'Plant Disease, An Advanced Treatise. Vol. V. How Plants Defend Themselves' (Horsfall, J.G. & Cowling, E., eds), pp. 139-159). Academic Press, New York.

Albersheim, P. & Anderson-Prouty, A. (1975). Carbohydrates, proteins, cell surfaces, and the biochemistry of pathogenesis. Annual Review of Plant Physiology 26, 31-52.

Albersheim, P., Jones, T.M. & English, P.D. (1969). Biochemistry of the cell wall in relation to infective processes. Annual Review of Phytopathology 7, 171-194.

Albersheim, P. & Valent, B.S. (1978). Host-pathogen interactions in plants. Plants when exposed to oligosaccharides of fungal origin defend themselves by accumulating antibiotics. Journal of Cell Biology 78, 627-643.

Al-Issa & Sigee, D.C. (1982). The hypersensitive reaction in tobacco leaf tissue infiltrated with Pseudomonas pisi. 1. Active growth and division of bacteria entrapped at the surface of mesophyll cells. Phytopathologische Zeitschrift 104, 104-114.

Allen, F.H.E. & Friend, J. (1983). Resistance of potato tubers to infection by Phytophthora infestans: a structural study of haustorial encasement. Physiology Plant Pathology 22, 285-292.

Anderson, A.J. (1978). Isolation from three species of Colletotrichum of glucan containing polysaccharides that elicit browning and phytoalexin production in bean. Phytopathology 68, 189-194.

Anderson, A.J. (1980). Studies on the structure and elicitor activity of fungal glucans. Canadian Journal of Botany 58, 2343-2348.

Anderson, A.J. & Albersheim, P. (1972). Host-pathogen interactions. V. Comparison of the abilities of proteins isolated from three varieties of Phaseolus vulgaris to inhibit endopolygalacturonases secreted by three races of Colletotrichum lindemuthianum. Physiological Plant Pathology 2, 339-346.

Angell, H.R., Walker, J.C. & Link, K.P. (1930). The relation of protocatechuic acid to disease resistance in the onion. Phytopathology 20, 431-438.

Appel, O. (1915). Disease resistance in plants. Science 41, 773.

Asher, M.J.C. & Thomas, C.E. (1983). The expression of partial resistance to Erysiphe graminis in spring barley. Plant Pathology 32, 79-89.

Atkinson, T.G., Neal, J.L. & Larson, R.I. (1974). Root rot reaction in wheat: resistance not mediated by rhizosphere or laimosphere antagonists. Phytopathology 64, 97-101.

Atkinson, T.G., Neal, J.L. & Larson, R.I. (1975). Genetic control of rhizosphere microflora of wheat. In 'Biology and Control of Soil-borne Plant Pathogens' (G. W. Bruehl, ed.), pp. 116-122. American Phytopathological Society, St Paul, Minnesota.

Babior, B.M., Kipnes, R.S. & Curnutte, J.T. (1973). Biological defense mechanisms. The production by leucocytes of superoxide, a potential bactericidal agent. Journal of Clinical Investigation 52, 741-744.

Bailey, J.A. (1982). Mechanisms of phytoalexin accumulation. In 'Phytoalexins' (Bailey, J.A. & Mansfield, eds), pp. 289-318, Blackie, Glasgow & London.

Bailey, J. A., Rowell, P.M. & Arnold, G.M. (1980). The temporal relationship between host cell death, phytoalexin accumulation and fungal inhibition during hypersensitive reactions of Phaseolus vulgaris to Colletotrichum lindemuthianum. Physiological Plant Pathology 17, 329-339.

Baker, R. (1968). Mechanisms of biological control of soil-borne pathogens. Annual Review of Phytopathology 6, 263-294.

Bantarri, E.E., Anderson, W.H. & Rasmusson, D.C. (1975). Helminthosporium headblight resistance in six-row spring barleys. Plant Disease Reporter 59, 274-277.

Bashan, Y., Okon, Y. & Henis, Y. (1982). Detection of a necrosis-inducing factor of non-host plant leaves produced by Pseudomonas syringae pv. tomato. Canadian Journal of Botany 60, 2453-2460.

Bateman, D.F. & Bashan, H.G. (1976). Degradation of plant cell walls and membranes by microbial enzymes. In 'Encyclopedia of Plant Physiology Vol. 4. Physiolgical Plant Pathology' (Heitefuss, R. & Williams, P.H. eds), pp. 316-355. Springer-Verlag, Berlin, Heidelberg & New York.

Bauer, W.D. (1981). Infection of legumes by Rhizobia. Annual Review of Plant Physiology 32, 407-449.

Beardmore, J., Ride, J.P. & Granger, J.W. (1983). Cellular lignification as a factor in hypersensitive resistnce of wheat to stem rust. Physiological Plant Pathology 22, 209-220.

Beckman, C.H. (1966). Cell irritability and localisation of vascular infection in plants. Phytopathology 56, 821-824.

Beckman, C.H. (1968). An evaluation of possible resistance mechanisms in broccoli, cotton, and tomato to vascular infection by Fusarium oxysporum. Phytopathology 50, 429-433.

Beckman, C.H. (1971). The plasticizing of plant cell walls and tylose formation - a model. Physiological Plant Pathology 1, 1-10.

Beckman, C.H. (1980). Defenses triggered by the invader: physical defenses. In 'Plant Pathology: an Advanced Treatise' Vol. V. (Horsfall, J.G. & Cowling, E.B., eds), pp. 225-245. Academic Press, New York.

Beckman, C.H., Elgersma, D.M. & McHardy, W.E. (1972). The localization of fusarial infections in the vascular tissue of single-dominant gene resistant tomatoes. Phytopathology 62, 1256-1260.

Beckman, C.H. & Halmos, S. (1962). Relation of vascular occluding reactions in banana roots to pathogenicity of root-invading fungi. Phytopathology 52, 893-897.

Beckman, C.H., Halmos, S. & Mace, M.E. (1962). The interaction of host, pathogen and soil temperature in relation to susceptibility to Fusarium wilt of bananas. Phytopathology 52, 134-140.

Bednarski, M.A., Izawa, S. & Scheffer, R.P. (1977). Reversible effects of toxins from Helminthosporium maydis race T on oxidative phosphorylation by mitochondria from maize. Plant Physiology 59, 540-545.

Bell, A.A. (1981). Biochemical mechanisms of disease resistance. Annual Review of Plant Physiology 32, 21-81.

Bell, A.A. & Mace, M.E. (1980). Biochemistry and physiology of resistance. In 'Fungal Wilt Diseases of Plants' (Mace, M.E., Bell, A.A. & Beckman, C.H., eds), pp. 431-521. Academic Press, New York, San Francisco & London.

Bergman, B.H.H. & Beijersbergen, J.C.M. (1968). A fungitoxic substance extracted from tulips and its possible role as a protectant against disease. Netherlands Journal of Plant Pathology 74, 157-162.

Berlin, J.D. & Bowen, C.C. (1964). The host-parasite interface of Albugo candida on Raphanus sativus. American Journal of Botany 51, 445-452.

Bhuvaneswari, T.V. & Bauer, W.D. (1978). Role of lectins in plant-microorganism interactions. 3. Influence of rhizosphere-rhizoplane culture conditions on soybean lectin-binding properties of Rhizobium. Plant Physiology 62, 71-74.

Bhuvaneswari, T.V., Turgeon, G. & Bauer, W.D. (1980). Early stages in the infection of soybean (Glycine max L. Merr) by Rhizobium japonicum. I. Localization of infectible root cells. Plant Physiology 66, 1027-1031.

Bird, P.M. & Ride, J.P. (1981). The resistance of wheat to Septoria nodorum: fungal development in relation to host lignification. Physiological Plant Pathology 19, 289-299.

Blackhurst, F.M. & Wood, R.K.S. (1963). Resistance of tomato plants to Verticillium albo-atrum. Transactions of the British Mycological Society 46, 385-392.

Blakeman, J.P. & Atkinson, P. (1976). Antimicrobial substances associated with the aerial surfaces of plants. In 'Microbial Ecology of the Phylloplane' (Blakeman, J.P. ed.), pp. 245-263. Academic Press, London, New York & San Francisco.

Blakeman, J.P. & Brodie, I.D.S. (1976). Inhibition of pathogens by epiphytic bacteria on aerial plant surfaces. In 'Microbiology of Aerial Plant Surfaces' (Dickinson, C.H. & Preece, T.F., eds), pp. 529-557. Academic Press, London, New York & San Francisco.

Blakeman, J.P. & Fokkema, N.J. (1982). Potential for biological control of plant diseases on the phylloplane. Annual Review of Phytopathology 20, 167-192.

Bloch, C.B. De Wit, P.J.G.M. & Kuć, J. (1984). Elicitation of phytoalexins by arachidonic and eicosapentaenoic acids: a host survey. Physiological Plant Pathology, in press.

Bohlool, B.B. & Schmidt, F.L. (1974). Lectins: a possible basis for specificity in the Rhizobium-legume root nodule symbiosis. Science 185, 269-271.

Boller, T., Gehri, A., Mauch, F. & Vögeli, U. (1983a). Chitinase in bean leaves: induction by ethylene, purification, properties and possible function. Planta 157, 22-31.

Boller, T., Gehri, A., Mauch, F. & Vögeli, U. (1983b). Ethylene-induced chitinase: is it a pathogenesis-related protein? Netherlands Journal of Plant Pathology 89, 318.

Booth, J.A. (1969). Gossypium hirsutum tolerance to Verticillium albo-atrum infection. I. Amino-acid exudation from aseptic roots of tolerant and susceptible cotton. Phytopathology 59, 43-46.

Booth, J.A. (1974). Effect of cotton root exudate constituents on growth and pectolytic enzyme production by Verticillium albo-atrum. Canadian Journal of Botany 52, 2219-2224.

Borner, H., Schatz, G. & Grisebach, H. (1983). Influence of the systemic fungicide metalaxyl on glyceollin accumulation in soybean infected with Phytophthora megasperma f. sp. glycinea. Physiological Plant Pathology 23, 145-152.

Bostock, R.M., Kuć, J.A. & Laine, R.A. (1981). Eicosapentaenoic and arachidonic acids from Phytophthora infestans elicit fungitoxic sesquiterpenes in the potato. Science 212, 67-69.

Bostock, R.M., Laine, R.A. & Kuć, J.A. (1982). Factors affecting the elicitation of sesquiterpenoid phytoalexin accumulation by eicosapentaenoic and arachidonic acids in potato. Plant Physiology 70, 1417-1424.

Bostock, R.M., Nuckles, E., Henfling, J.W.D.M. & Kuć, J.A. (1983). Effect of potato tuber age and storage on sesquiterpenoid stress metabolite accumulation, steroid glycoalkaloid accumulation and response to abcisic and arachidonic acids. Phytopathology 73, 435-438.

Bracker, C.E. & Littlefield, L.J. (1973). Structural concepts of host-pathogen interfaces. In 'Fungal Pathogenicity and the Plant's Response' (Byrde, R.J.W. & Cutting, C.V. eds), pp. 159-318. Academic Press, London & New York.

Braun, A.E. (1942). Resistance of watermelon to wilt disease. American Journal of Botany 29, 683-684.

Brown, J.F. & Shipton, W.A. (1964). Relationship of penetration to infection type when seedling wheat leaves are inoculated with Puccinia graminis tritici. Phytopathology 54, 89-91.

Brown, J.F., Shipton, W.A. & White, N.H. (1966). The relationship between hypersensitive tissue and resistance in wheat seedlings infected with Puccinia graminis tritici. Annals of Applied Biology 58, 279-290.

Bruce, R.J. & West, C.A. (1982). Elicitation of casbene synthetase activity in castor bean Ricinus communis. The role of pectic fragments of the plant cell wall in elicitation by a fungal endo-polygalacturonase. Plant Physiology 69, 1181-1188.

Bruegger, B.B. & Keen, N.T. (1979). Specific elicitors of glyceollin accumulation in the Pseudomonas glycinea-soybean host-parasite system. Physiological Plant Pathology 15, 43-51.

Bugbee, W.M. (1979). Vascular response of cotton to infection by Fusarium oxysporum f. sp. vasinfectum. Phytopathology 60, 121-123.

Burchill, R.T. & Cook, R.T.A. (1971). The interaction of urea and microorganisms in suppressing development of perithecia of Venturia inaequalis (Cke) Wint. In 'Ecology of Leaf-Surface Microorganisms' (Preece, T.F. & Dickinson, C.H. eds), pp. 471-483. Academic Press, London & New York.

Burret, F.M. (1971). 'Self-recognition' in colonial marine forms and flowering plants in relation to the evolution of immunity. Nature 232, 230-235.

Bushnell, W.R. (1972). Physiology of fungal haustoria. Annual Review of Phytopathology 10, 151-176.

Bushnell, W.R. (1979). The nature of basic compatibility: comparisons between pistil-pollen and host-parasite interaction. In 'Recognition and Specifity in Plant Host-Parasite Interactions' (Daly, J.M. & Uritani, I., eds), pp. 211-226. Japan Scientific Societies Press, Tokyo; University Park Press, Baltimore.

Bushnell, W.R. & Rowell, J.B. (1981). Suppressors of defence reactions: a model for roles of specificity. Phytopathology 71: 1012-1014.

Buxton, E.W. (1957a). Some effects of pea root exudates on physiologic races of Fusarium oxysporum Fr. f. pisi (Linf.) Snyder & Hansen. Transactions of the British Mycological Society 40, 145-154.

Buxton, E.W. (1957b). Differential rhizosphere effects of three pea cultivars on physiologic races of Fusarium oxysporum f. pisi. Transactions of the British Mycological Society 40, 305-317.

Buxton, E.W. (1962). Root exudates from banana and their relationship to strains of the Fusarium causing Panama wilt. Annals of Applied Biology 50, 269-282.

Callow, J.A. (1977). Recognition, resistance and the role of plant lectins in host-parasite interactions. Advances in Botanical Research 4, 1-149.

Camacho-Henriquez, A. & Sänger, H.L. (1982). Analysis of acid-extractable tomato leaf proteins after infection with a viroid, two viruses and a fungus and partial purification of the 'pathogenesis-related' protein p14. Archives of Virology 74, 181-196.

Campbell, R.J. & Ascher, P.D. (1976). Incorporation of label from radioactive uridine into the stylar nucleic acids of Lilium longiflorum Thunb. as affected by heat, 6-methylpurine and actinomycin D. Theoretical and Applied Genetics 47, 215-226.

Carlson, P.S. (1973). Methionine sulphoxamide-resistant mutants of tobacco. Science 180, 1366-1368.

Carson, E.T., Richardson, P.E., Essenberg, M.K., Brinkerhoff, L.A., Hohnson, W.M. & Venere, R.J. (1978). Ultrastructural cell wall alterations in immune cotton leaves inoculated with Xanthomonas malvacearum. Phytopathology 68, 1015-1021.

Carver, T.L.W. & Bushnell, W.R. (1983). The probable role of primary germ tubes in water uptake before infection by Erysiphe graminis. Physiological Plant Pathology 23, 229-240.

Chakravarti, B.P. (1966). Attempts to alter infection processes and aggressiveness of Puccinia

graminis var. tritici. Phytopathology 56, 223-229.
Chen, A.P. & Phillips, D.A. (1976). Attachment of Rhizobium to legume roots as the basis for specific interactions. Physiologia Plantarum 38, 83-88.
Cho, Y.S., Wilcoxson, R.D. & Frosheiser, F.I. (1973). Differences in anatomy, plant-extracts and movement of bacteria in plants of bacterial wilt resistant and susceptible varieties of alfalfa. Phytopathology 63, 760-765.
Clarke, A.E. & Stone, B.A. (1962). β-1,3-glucan hydrolases from the grape vine (Vitis vinifera) and other plants. Phytochemistry 1, 175-188.
Clarke, J.A., Lisker, N., Lamport, D.T.A. & Ellingboe, A.H. (1981). Hydroxyproline enhancement as a primary event in the successful development of Erysiphe graminis in wheat. Plant Physiology 67, 188-189.
Coffey, M.D. (1976). Flax rust resistance involving the K gene: an ultrastructural survey. Canadian Journal of Botany 54, 1443-1457.
Coffey, M.D. & Allen, F.H.E. (1983). A quantitative histological and ultrastructural analysis of interactions between flax rust and near-isogenic host lines varying in their degree of incompatibility. Canadian Journal of Botany 61, 1831-1850.
Coffey, M.D. & Wilson, U.E. (1983). An ultrastructural study of the late blight fungus Phytophthora infestans and its interaction with the foliage of two potato cultivars possessing different levels of general (field) resistance. Canadian Journal of Botany 61, 2669-2685.
Coley-Smith, J.R. & Holt, R.W. (1966). The effect of species of Allium on germination in soil of sclerotia of Sclerotium cepivorum Berk. Annals of Applied Biology 58, 273-278.
Coley-Smith, J.R., King, J.E., Dickinson, D.J. & Holt, R.W. (1967). Germination of sclerotia of Sclerotium cepivorum Berk. under aseptic conditions. Annals of Applied Biology 60, 109-115.
Collins, W.W. & Nielsen, L.W. (1976). Fusarium wilt resistance in sweet potatoes. Phytopathology 66, 489-493.
Comstock, J.C., Martinson, C.A. & Gengenbach, B.G. (1973). Host specificity of a toxin from Phyllosticta maydis for Texas cytoplasmically male-sterile maize. Phytopathology 63, 1357-1361.
Comstock, J.C. & Scheffer, R.P. (1973). Role of host-selective toxin in colonization of corn leaves by Helminthosporium carbonum. Phytopathology 63, 24-29.
Cooke, R.C. & Whipps, J.M. (1980). The evolution of modes of nutrition in fungi parasitic on terrestrial plants. Biological Reviews 55, 341-362.
Cooksey, C.J., Garrat, P.J., Dahya, J. & Strange, R. (1983). Sucrose, a constitutive elicitor of phytoalexin synthesis. Science 220, 1398-1400.
Cooper, R.M. & Wood, R.K.S. (1975). Regulation of synthesis of cell wall-degrading enzymes by Verticillium albo-atrum and Fusarium oxysporum f. sp. lycopersici. Physiological Plant Pathology 5, 135-156.
Cooper, R.M., Rankin, B. & Wood, R.K.S. (1978). Cell wall-degrading enzymes of vascular wilt fungi. II. Properties and modes of action of polysaccharides of Verticillium albo-atrum and Fusarium oxysporum f. sp. lycopersici. Physiological Plant Pathology 13, 101-134.
Cruickshank, I.A.M. (1980). Defenses triggered by the invader: chemical defenses. In 'Plant Disease: An Advanced Treatise. Vol. V.' (Horsfall, J.G. & Cowling, E.B., eds), pp. 247-267. Academic Press, New York, San Francisco & London.
Cruickshank, I.A.M. & Perrin, D.R. (1968). The isolation and partial characterization of monilicolin A, a polypeptide with phaseollin-inducing activity from Monilinia fructicola. Life Science 7, 449-458.
Cruickshank, I.A.M., Perrin, D.R. & Mandryk, M. (1977). Fungitoxicity of duvatrienediols associated with the cuticular wax of tobacco leaves. Phytopathologische Zeitschrift 90, 243-249.
Crute, I.R. (1985). The relationship between Plasmodiophora brassicae and its hosts - the application of concepts relating to variation in interorganismal associations. Advances in Plant Pathology (in press).
Crute, I.R. & Norwood, J.M. (1978). Incomplete specific resistance to Bremia lactucae in lettuce. Annals of Applied Biology 89, 467-474.

Crute, I.R., Wolfman, S.A. & Davis, A.A. (1977). A laboratory method of screening fungicides for systemic activity against Bremia lactucae. Annals of Applied Biology 85, 147-152.

Daly, J.M. (1972). The use of near-isogenic lines in biochemical studies of the resistance of wheat to stem rust. Phytopathology 62, 392-400.

Daly, J.M. (1981). Mechanisms of action. In 'Toxins in Plant Disease' (Durbin, R.D. ed.), pp. 331-394. Academic Press, New York, San Francisco & London.

Daub, M.E. & Hagedorn, D.J. (1980). Growth kinetics and interactions of Pseudomonas syringae with susceptible and resistant bean tissue. Phytopathology 70, 429-436.

Davidse, L.C. & Boekeloo, M. (1984). Elicitation and suppression of necrosis in potato leaves by culture filtrate compounds of Phytophthora infestans (Mont.) de Bary. Acta Botanica Neerlandica 33, 234).

Davis, D. (1964). Host fungitoxicity in selective pathogenicity of Fusarium oxysporum. Phytopathology 54, 290-293.

Davis, D. (1966). Cross infection in Fusarium wilt diseases. Phytopathology 56, 825-828.

Davison, E.M. (1968). Cytochemistry and ultrastructure of hyphae and haustoria of Peronospora parasitica (Pers. ex. Fr.) Fr. Annals of Botany 32, 613-621.

Dazzo, F.B. & Hubbel, D.H. (1975). Cross-reactive antigens and lectins as determinants of symbiotic specificity in the Rhizobium-clover association. Applied Microbiology 30: 1017-1033.

Dazzo, F.B. & Truchet, G.L. (1983). Interactions of lectins and their saccharide receptors in the Rhizobium-like symbiosis. Journal of Membrane Biology 73, 1-16.

Dazzo, F.B., Urbano, M.R. & Brill, W.J. (1979). Transient appearance of lectin receptors on Rhizobium trifolii. Current Microbiology 2, 15-20.

Défago, G. & Kern, H. (1983). Induction of Fusarium solani mutants insensitive to tomatine, their pathogenicity and aggressiveness to tomato fruits and pea plants. Physiological Plant Pathology 22, 29-37.

Défago, G., Kern, H. & Sedlar, L. (1983). Genetic analysis of tomatine insensitivity, sterol content and pathogenicity for green tomato fruits in mutants of Fusarium solani. Physiological Plant Pathology 22, 39-43.

Denny, T.P. & Van Etten, H.D. (1983a). Tolerance of Nectria haematococca MPVI to the phytoalexin pisatin in the absence of detoxification. Journal of General Microbiology 129, 2893-2901.

Denny, T.P. & Van Etten, H.D. (1983b). Characterization of an inducible, non-degradative tolerance of Nectria haematococca MPVI to phytoalexins. Journal of General Microbiology 129, 2903-2913.

DeSilva, R.L. & Wood, R.K.S. (1964). Infection of plants by Corticum solani and C. praticola - effect of plant exudates. Transactions of the British Mycological Society 47, 15-24.

Desjardins, A.E., Ross, L.M., Spellman, M.W., Darvill, A.C. & Albersheim, P. (1982). Host pathogen interactions XX. Biological variation in the protection of soybeans Glycinea max from infection by Phytophthora megasperma f. sp. glycinea. Plant Physiology 69, 1046-1050.

DeVay, J.F. & Alder, H.E. (1976). Antigens common to hosts and parasites. Annual Review of Microbiology 30, 147-168.

De Waard, M.A. & Van Nistelrooy, J.G.M. (1981). Induction of fenarimol-efflux activity in Aspergillus nidulans by fungicides inhibiting sterol biosynthesis. Journal of General Microbiology 126, 483-489.

De Wit, P.J.G.M. (1977). A light and scanning-electron microscope study of infection of tomato plants by virulent and avirulent races of Cladosporium fulvum. Netherlands Journal of Plant Pathology 83, 109-122.

De Wit, P.J.G.M. & Bakker, J. (1980). Differential changes in soluble tomato leaf proteins after inoculation with virulent and avirulent races of Cladosporium fulvum (syn. Fulvia fulvum). Physiological Plant Pathology 17, 121-130.

De Wit, P.J.G.M., Hofman, J.E. & Aarts, J.M.M.J.G. (1984). Origin of specific elicitors of chlorosis and necrosis occurring in intercellular fluids of compatible interactions of Cladosporium fulvum (syn. Fulvia fulva) and tomato. Physiological Plant Pathology 24, 17-23.

De Wit, P.G.J.M. & Kodde, E. (1981). Further characterization and cultivar specificity of glycoprotein elicitors from culture filtrates and cell walls of Cladosporium fulvum (syn.

Fulvia fulva). Physiological Plant Pathology 18, 297-314.

De Wit, P.J.G.M. & Roseboom, P.H.M. (1980). Isolation, partial characterization and specificity of glycoprotein elicitors from culture filtrates, mycelium and cell walls of Cladosporium fulvum (syn. Fulvia fulva). Physiological Plant Pathology 16, 391-408.

De Wit, P.J.G.M. & Spikman, G. (1982). Evidence for the occurrence of race and cultivar-specific elicitors of necrosis in intercellular fluids of compatible interactions of Cladosporium fulvum and tomato. Physiological Plant Pathology 21, 1-11.

Dickinson, S. (1960). The mechanical ability to breach the host barriers. In 'Plant Pathology: an Advanced Treatise' Vol. II (Horsfall, J.G. & Dimond, A.E., eds), pp. 203-232. Academic Press, New York.

Dixon, G.R. & Pegg, G.F. (1969). Hyphal lysis and tylose formation in tomato cultivars infected by Verticillium albo-atrum. Transactions of the British Mycological Society 53, 109-118.

Doke, N. (1982). A further study on the role of hypersensitivity in resistance of potato cultivars to infection by an incompatible race of Phytophthora infestans. Physiological Plant Pathology 21, 85-95.

Doke, N. (1983a). Involvement of superoxide anion generation in the hypersensitive response of potato tuber tissues to infections with an incompatible race of Phytophthora infestans and to the hyphal wall components. Physiological Plant Pathology 23, 345-357.

Doke, N. (1983b). Generation of superoxide anion by potato tuber protoplasts during the hypersensitive response to hyphal wall components of Phytophthora infestans and specific inhibition of the reaction by suppressors of hypersensitivity. Physiological Plant Pathology 23, 359-367.

Doke, N., Garas, N.A. & Kuc, J. (1979). Partial characterization and aspects of the mode of action of a hypersensitivity-inhibiting factor (HIF) isolated from Phytophthora infestans. Physiological Plant Pathology 15, 127-140.

Doke, N., Garas, N.A. & Kuc, J. (1980). Effect on host hypersensitivity of suppressors released during the germination of Phytophthora infestans cytospores. Phytopathology 70, 35-39.

Doke, N. & Tomiyama, K. (1975). Effect of blasticidin S on hypersensitive death of potato leaf petiole cells caused by infection with an incompatible race of Phytophthora infestans. Physiological Plant Pathology 6, 169-175.

Doke, N. & Tomiyama, K. (1980). Suppression of the hypersensitive response of potato tuber protoplasts to hyphal wall components by water soluble glucans isolated from Phytophthora infestans. Physiological Plant Pathology 16, 177-186.

Doubly, J.A., Flor, H.H. & Clagett, C.O. (1960). Relation of antigens of Melampsora lini and Linum usitatissimum to resistance and susceptibility. Science 131, 229.

Dow, J.M. & Callow, J.A. (1978). A possible role for .-tomatin in the varietal-specific resistance of tomato to Cladosporium fulvum. Phytopathologische Zeitschrift 92, 211-216.

Drysdale, R.B. & Langcake, P. (1973). Response of tomato to infection by Fusarium oxysporum f. lycopersici. In 'Fungal Pathogenicity and the Plant's Response' (Byrde, R.J.W. & Cutting, C.V. eds), pp. 423-433. Academic Press, London & New York.

Durbin, R.D. (1981). Toxins in Plant Disease. Academic Press, New York. 515 pp.

Earl, E.D., Gracen, V.E., Yoder, O.C. & Gemmill, K.P. (1978). Cytoplasmic specific effects of Helminthosporium maydis race T toxin on survival of corn mesophyll protoplasts. Plant Physiology 61, 420-424.

El-Banoby, F.E. & Rudolph, K. (1979). A polysaccharide from liquid cultures of Pseudomonas phaseolicola which specifically induces water-soaking in bean leaves. Phytopathologische Zeitschrift 95, 38-50.

El-Banoby, F.E. & Rudolph, K. (1980). Purification of extracellular polysaccharides from Pseudomonas phaseolicola which induce water-soaking in bean leaves. Physiological Plant Pathology 16, 425-437.

El-Banoby, F.E., Rudolph, K. & Huttermann, A. (1980). Biological and physical properties of an extracellular polysaccharide from Pseudomonas phaseolicola. Physiological Plant Pathology 17, 291-301.

El-Banoby, R.E., Rudolph, K. & Mendgen, K. (1981). The fate of extracellular polysaccharide from Pseudomonas phaseolicola in leaves and leaf extracts from halo-blight susceptible and resistant bean plants (Phaseolus vulgaris L.). Physiological Plant Pathology 18, 91-98.

Elgersma, D.M. (1970). Length and diameter of xylem vessels as a factor in resistance of elms to Ceratocystis ulmi. Netherlands Journal of Plant Pathology 76, 179-182.

Elgersma, D.M. (1973). Tylose formation in elms after inoculation with Ceratocystis ulmi, a possible resistance mechanism. Netherlands Journal of Plant Pathology 79, 218-220.

Elgersma, D.M. (1980). Accumulation of rishitin in susceptible and resistant tomato plants after inoculation with Verticillium albo-atrum. Physiological Plant Pathology 16, 149-153.

Elgersma, D.M., MacHardy, W.E. & Beckman, C.H. (1972). Growth and distribution of Fusarium oxysporum f. lycopersici in near-isogenic lines of tomato resistant or susceptible to wilt. Phytopathology 62, 1232-1237.

Ellingboe, A.H. (1972). Genetics and physiology of primary infection by Erysiphe graminis. Phytopathology 62, 401-406.

Ellingboe, A.H. (1976). Genetics of host-parasite interactions. In 'Encyclopedia of Plant Physiology' (Heitefuss, R. & Williams, P.H., eds), pp. 761-778. Springer Verlag, Berlin, Heidelberg & New York.

Ellingboe, A.H. (1981). Changing concepts in host-pathogen genetics. Annual Review of Phytopathology 19, 125-143.

Ellingboe, A.H. (1982). Genetic aspects of active defense. In 'Active Defense Mechanisms in Plants' (Wood, R.K.S., ed.), pp. 179-192. Plenum Press, New York & London.

English, P.D. & Albersheim, P. (1969). Host-pathogen interactions I. A correlation between α-galactosidase production and virulence. Plant Physiology 44, 217-224.

Ersek, T., Sarham, A.R.T. & Pongar, S. (1981). Isolation of an agglutinating factor that may determine race-specific resistance of soybean leaves to Pseudomonas glycinea. Acta Phytopathologia Academiae Scientiarum Hungaricae 16, 137-138.

Esquerré-Tugayé, M.T. & Lamport, D.T.A. (1979). Cell surfaces in plant-microorganism interactions. I. A structural investigation of cell wall hydroxyproline-rich glycoproteins which accumulate in fungus-infected plants. Plant Physiology 64, 314-319.

Esquerré-Tugayé, M.T., Lafitte, D., Mazau, D., Toppan, A. & Touze, A. (1979). Cell surfaces in plant-microorganism interactions. II. Evidence for the accumulation of hydroxyproline-rich glycoproteins in the cell wall of diseased plants as a defense mechanism. Plant Physiology 64, 320-326.

Essenberg, M., Cason, E.T., Hamilton, B., Brinkerhoff, L.A., Gholson, R.K. & Richardson, P.E. (1979). Single cell colonies of Xanthomonas malvacearum in susceptible and immune cotton leaves and the local resistant response to colonies in immune leaves. Physiological Plant Pathology 15, 53-68.

Essenberg, M., Doherty, M.D., Hamilton, B.K., Henning, V.T., Cover, E.C., McFaul, S.J. & Johnson, W.M. (1982). Identification and effects on Xanthomonas campestris pv. malvacearum of two phytoalexins from leaves and cotyledons of resistant cotton. Phytopathology 72, 1349-1356.

Fett, W.F. & Jones, S.B. (1982). Role of bacterial immobilization in race-specific resistance of soybean to Pseudomonas syringae pv. glycinea. Phytopathology 72, 488-492.

Fett, W.F. & Sequeira, L. (1980a). A new bacterial agglutinin from soybean. I. Isolation, partial purification and characterization. Plant Physiology 66, 847-852.

Fett, W.F. & Sequeira, L. (1980b). A new bacterial agglutinin from soybean. II. Evidence against a role in determining pathogen specificity. Plant Physiology 66, 853-858.

Fincher, G.B., Stone, B.A. & Clarke, A.E. (1983). Arabinogalactan-protein: structure, biosynthesis, and function. Annual Review of Plant Physiology 34, 47-70.

Fisher, M.L., Anderson, A.J. & Albersheim, P. (1973). Host-pathogen interactions VI. A single plant protein efficiently inhibits endopolygalacturonases secreted by Colletotrichum lindemuthianum and Aspergillus niger. Plant Physiology 51, 489-491.

Flentje, N.T., Dodman, R.L. & Kerr, A. (1963). The mechanism of host penetration by Thanatephorus cucumeris. Australian Journal of Biological Science 16, 784-799.

Flor, H.H. (1956). The complementary genetic systems in flax and flax rust. Advances in Genetics 8, 29-54.

Fokkema, N.J. (1976). Antagonism between fungal saprophytes and pathogens on aerial plant surfaces. In 'Microbiology of Aerial Plant Surfaces' (Dickinson, C.H. & Preece, T.F., eds),

pp. 487-506. Academic Press, London, New York & San Francisco.
Friend, J. (1979). Phenolic substances and plant disease. In 'Recent Advances in Phytochemistry: Biochemistry of Plant Phenolics' (Swain, T., Harborne, J.B. & Van Sumere, C. eds), Vol. 12, pp. 557-588. Plenum Press, New York & London.
Fry, W.E. & Millar, R.L. (1971a). Cyanide tolerance in Stemphylium loti. Phytopathology 61, 494-500.
Fry, W.E. & Millar, R.L. (1971b). Development of cyanide tolerance in Stemphylium loti. Phytopathology 61, 501-506.
Fuchs, A., Davidse, L.C., De Waard, M.A. & De Wit, P.J.G.M. (1983). Contemplations and speculations on novel approaches in the control of fungal plant diseases. Pesticide Science 14, 272-293.
Furuichi, N., Tomiyama, K. & Doke, N. (1980). The role of potato lectin in binding of germ tubes of Phytophthora infestans to potato cell membrane. Physiological Plant Pathology 16, 249-256.
Gabe, H.L. (1975). Standardization of nomenclature for pathogenic races of Fusarium oxysporum f. sp. lycopersici. Transactions of the British Mycological Society 64, 156-159.
Gabig, T.G. & Babior, B.M. (1979). The O_2^--forming oxidase responsible for the respiratory burst in human neutrophiles. Properties of the solubilized enzyme. Journal of Biological Chemistry 254, 9070-9074.
Gabriel, D.W., Ellingboe, A.H. & Rossman, E.C. (1979). Mutations affecting virulence in Phyllosticta maydis. Canadian Journal of Botany 57, 2639-2643.
Garas, N.A., Doke, N. & Kuć, J. (1979). Suppression of the hypersensitive reaction in potato tubers by mycelial components from Phytophthora infestans. Physiological Plant Pathology 15, 117-126.
Garas, N.A. & Kuć, J. (1981). Potato lectin lyses zoospores of Phytophthora infestans and precipitates elicitors of terpenoid accumulation produced by the fungus. Physiological Plant Pathology 18, 227-237.
Garber, E.D. (1956). A nutrition-inhibition hypothesis of pathogenicity. The American Naturalist 90, 183-194.
Garber, E.D. & Shaeffer, S.G. (1957). Free histidine content of turnip varieties and their resistance to histidine-requiring mutants of Erwinia aroideae. Journal of Bacteriology 74, 312-395.
Garber, E.D. & Heggestad, H.E. (1958). Observations on the pathogenicity of biochemical mutants of Pseudomonas tabaci. Phytopathology 48, 535-537.
Garber, E.D., Shaeffer, S.G. & Goldman, M. (1956). The virulence of biochemical mutants of Erwinia aroideae for varieties of radish and turnip. Journal of General Microbiology 14, 261-267.
Gardner, J.M., Felman, A.W. & Stamper, D.H. (1983). Role and fate of bacteria in vascular occlusions of citrus. Physiological Plant Pathology 23, 295-309.
Gardner, J.M. & Kado, C.I. (1976). Polygalacturonic acid trans-eliminase in the osmotic shock fluid of Erwinia rubri faciens: characterization of the purified enzyme and its effect on plant cells. Journal of Bacteriology 127, 451-460.
Gengenbach, B.G., Green, C.E. & Donovan, C.M. (1977). Inheritance of selected pathotoxin resistance in maize plants regenerated from cell cultures. Proceedings of the National Academy of Sciences, USA 74, 5113-5117.
Gessler, C. & Kuć, J. (1982). Appearance of a host protein in cucumber plants infected with viruses, bacteria and fungi. Journal of Experimental Botany 33, 58-66.
Gianinazzi, S., Ahl, P., Cornu, A., Scalla, R. & Cassini, R. (1980). First report of host b-protein appearance in response to a fungal infection in tobacco. Physiological Plant Pathology 16, 337-342.
Gibson, D.M., Stack, S., Krell, K. & House, J. (1982). A comparison of soybean agglutinin in cultivars resistant and susceptible to Phytophthora megasperma var. sojae (race 1). Plant Physiology 70, 560-566.
Gilchrist, D.G. & Grogan, R.G. (1976). Production and nature of a host-specific toxin from Alternaria alternata f. sp. lycopersici. Phytopathology 66, 165-171.

phaseollin by cell suspension cultures of Phaseolus vulgaris. Phytopathology 68, 111-117.

Gnanamanickam, S.S. & Mansfield, J.W. (1981). Selective toxicity of wyerone and other phytoalexins to Gram-positive bacteria. Phytochemistry 20, 997-1000.

Gnanamanickam, S. S. & Patil, S.S. (1977). Phaseotoxin suppresses bacterially induced hypersensitive reaction and phytoalexin synthesis in bean cultivars. Physiological Plant Pathology 10, 169-179.

Gnanamanickam, S.S. & Smith, D.A. (1980). Selective toxicity of isoflavonoid phytoalexin to Gram-positive bacteria. Phytopathology 70, 894-896.

Godfrey, B.E.S. (1976). Leachates from aerial parts of plants and their relation to plant surface microbial populations. In 'Microbiology of Aerial Plant Surfaces' (Dickinson, C.H. & Preece, T.F., eds), pp. 433-440. Academic Press, London, New York & San Francisco.

Goodman, R.N. (1980). Defenses triggered by previous invaders: Bacteria. In 'Plant Disease: a Comprehensive Treatise, Vol. V' (Horsfall, J.G. & Cowling, E.B., eds), pp. 305-317. Academic Press, New York, San Francisco & London.

Goodman, R.N., Huang, J.S. & Huang, P. (1975). Host-specific phytotoxic polysaccharide from apple tissue infected by Erwinia amylovora. Science 183, 1081-1082.

Goodman, R.N., Huang, P. & White, J.A. (1976). Ultrastructural evidence for immobilization of an incompatible bacterium, Pseudomonas pisi in tobacco leaf tissue. Phytopathology 66, 754-764.

Gracen, V.E., Forster, M.J., Sayre, K.D. & Grogan, C.O. (1971). Rapid method of selecting resistant plants for control of southern corn leaf blight. Plant Disease Reporter 55, 469-470.

Gray, L.E. & Chamberlain, D.W. (1975). Evidence for toxin production by a strain of Cephalosporium gregatum. Phytopathology 65, 89-90.

Greenhalgh, J.R. & Mitchell, N.D. (1976). The involvement of flavour volatiles in the resistance to downy mildew of wild and cultivated forms of Brassica oleracea. New Phytologist 77, 391-398.

Gregory, P., Earle, E.D. & Gracen, V.E. (1980). Effects of purified Helminthosporium maydis race T toxin on the structure and function of corn mitochondria and protoplasts. Plant Physiology 66, 477-481.

Hadwiger, L.A. & Beckman, J.M. (1980). Chitosan as a component of pea-Fusarium solani interactions. Plant Physiology 66, 205-211.

Hadwiger, L.A., Beckman, J.M. & Adams, M.J. (1981). Localization of fungal compounds in the pea-Fusarium interaction detected immunochemically with anti-chitosan and anti-fungal cell wall antisera. Plant Physiology 67, 170-175.

Hafiz, A. (1952). Basis for resistance in gram to Mycosphaerella blight. Phytopathology 42, 422-424.

Hahn, M.G., Darvill, A.G. & Albersheim, P. (1981). Host-pathogen interactions. XIX. The endogenous elicitor, a fragment of a plant cell wall polysaccharide that elicits phytoalexin accumulation in soybeans. Plant Physiology 68, 1161-1169.

Hammerschlag, F. & Mace, M.E. (1975). Antifungal activity of extracts from Fusarium wilt-susceptible and -resistant tomato plants. Phytopathology 65, 93-94.

Hammerschmidt, R., Lamport, D.T.A. & Muldoon, E.P. (1982). Hydroxyproline enhancement as an early event in the resistance of cucumber to Cladosporium cucumerinum. Phytopathology 72, 981.

Harder, D.E., Rohringer, R., Samborski, D.J., Rimmer, S.R., Kim, W.K. & Chong, J. (1979a). Electron microscopy of susceptible and resistant near-isogenic (sr6/Sr6) lines of wheat infected by Puccinia graminis tritici II. Expression of incompatibility in mesophyll and epidermal cells and the effect of temperature on host-parasite interactions in these cells. Canadian Journal of Botany 57, 2617-2625.

Harder, D.E., Samborski, R., Rohringer, R., Rimmer, S.R., Kim, W.K. & Chung, J. (1979b). Electron microscopy of susceptible and resistant near-isogenic (sr6/Sr6) lines of wheat infected by Puccinia graminis tritici. III. Ultrastructure of incompatible interactions. Canadian Journal of Botany 57, 2626-2634.

Hardwick, N.V., Greenwood, A.D. & Wood, R.K.S. (1971). The fine structure of the haustorium of

Hargreaves, J.A. & Bailey, J.A. (1978). Phytoalexin production by hypocotyls of Phaseolus vulgaris in response to constitutive metabolites released by damaged cells. Physiological Plant Pathology 13, 89-100.

Harper, J.L. (1950). Studies on the resistance of certain varieties of banana to Panama disease. II. The rhizosphere. Plant and Soil 2, 383-394.

Harrison, M.A. & Beckman, C.H. (1982). Time/space relationship of colonization and host response in wilt-resistant and wilt-susceptible cotton (Gossypium) cultivars inoculated with Verticillium dahliae and Fusarium oxysporum f. sp. vasinfectum. Physiological Plant Pathology 21, 193-207.

Harrison, M., Payne, M.G. & Gaskill, J.O. (1961). Some chemical aspects of resistance to Cercospora leaf spot in sugar beets. Journal of the American Society for Sugar Beet Technology 11, 457-468.

Hart, H. (1931). Morphological and physiologic studies on stem rust resistance in cereals. USDA Technical Bulletin 266, 1-75.

Heath, M.C. (1971). Haustorial sheath formation in cowpea leaves immune to rust infection. Phytopathology 61, 383-388.

Heath, M.C. (1974). Light and electron microscope studies of the interactions of host and non-host plants with cowpea rust, Uromyces phaseoli var. vignae. Physiological Plant Pathology 4, 403-414.

Heath, M.C. (1980a). Reactions of nonsuscepts to fungal pathogens. Annual Review of Phytopathology 18, 211-236.

Heath, M.C. (1980b). Effects of infection by compatible species or injection of tissue extracts on the susceptibility of nonhost plants to rust fungi. Phytopathology 70, 356-360.

Heath, M.C. (1981a). A generalized concept of host-parasite specificity. Phytopathology 71, 1121-1123.

Heath, M.C. (1981b). The suppression of the development of silicon containing deposits in French bean leaves by exudates of the bean rust fungus and extracts from bean rust-infected tissue. Physiological Plant Pathology 18, 149-155.

Heath, M.C. (1982). Absence of active defense mechanisms in compatible host pathogen interactions. In 'Active Defense Mechanisms in Plants' (Wood, R.K.S., ed.), pp. 143-156. Plenum Press, New York & London.

Heath, M.C. & Heath, I.B. (1971). Ultrastructure of an immune and a susceptible reaction of cowpea leaves to rust infection. Physiological Plant Pathology 1, 277-287.

Henderson, S.J. & Friend, J. (1979). Increase in PAL and lignin-like compounds as race-specific responses of potato tubers to Phytophthora infestans. Phytopathologische Zeitschrift 94, 323-334.

Heslop-Harrison, J. (1975). Incompatibility and the pollen-stigma interaction. Annual Review of Plant Physiology 26, 403-425.

Heslop-Harrison, J., Knox, R.B. & Heslop-Harrison, Y. (1974). Pollen-wall proteins: exine-held fractions associated with the incompatible response in Cruciferae. Theoretical and Applied Genetics 44, 133-137.

Hickey, E.L. & Coffey, M.D. (1977). A fine-structural study of the pea downy mildew fungus Peronospora pisi in its host Pisum sativum. Canadian Journal of Botany 55, 2845-2858.

Hickey, E.L. & Coffey, M.D. (1980). The effects of Ridomil on Peronospora pisi parasitizing Pisum sativum: an ultrastructural investigation. Physiological Plant Pathology 17, 199-204.

Hildebrand, D.C., Alosic, M.C. & Schroth, M.N. (1980). Physical entrapment of pseudomonads in bean leaves by films formed at air-water interfaces. Phytopathology 70, 98-109.

Hildebrand, D.C., Powell, C.C. & Schroth, M.N. (1969). Fire blight resistance in Pyrus: localization of arbutin and β-glucosidase. Phytopathology 59, 1534-1539.

Hildebrand, D.C. & Schroth, M.N. (1964). Arbutin-hydroxyquinone complex in pear as a factor in fire blight development. Phytopathology 54, 640-645.

Hill, K., Coyne, D.P. & Schuster, M.L. (1972). Leaf, pod and systemic chlorosis reactions in Phaseolus vulgaris to halo-blight controlled by different genes. Journal of the American Society of Horticultural Science 97, 494-498.

carbohydrate determinants of the root slime. Physiological Plant Pathology 16, 303-307.

Hodgkin, T. & Lyon, G.D. (1983). Detection of pollen germination inhibitors in *Brassica oleracea* tissue extracts. Annals of Botany 52, 781-789.

Hogenboom, N.G. (1975). Incompatibility and incongruity: two different mechanisms for the non-functioning of intimate partner relationships. Proceedings of the Royal Society B, 188, 361-375.

Hogenboom, N.G. (1983). Bridging a gap between related fields of research: pistil-pollen relationships and the distinction between incompatibility and incongruity in nonfunctioning host-parasite relationships. Phytopathology 73, 381-383.

Hohl, H.R. & Stössel, P. (1976). Host-parasite interfaces in a resistant and a susceptible cultivar of *Solanum tuberosum* inoculated with *Phytophthora infestans*: tuber tissue. Canadian Journal of Botany 54, 900-912.

Hohl, H.R., Stössel, P. & Hachler, H. (1980). Papilla formation and partial inhibition of fungal glucanases by phytoalexins in the *Phytophthora infestans* - *Solanum tuberosum* system. Annales de Phytopathologie 12, 353-362.

Hohl, H.R. & Suter, E. (1976). Host-parasite interfaces in a resistant and a susceptible cultivar of *Solanum tuberosum* inoculated with *Phytophthora infestans*: leaf tissue. Canadian Journal of Botany 54, 1956-1970.

Holliday, M.J. & Keen, N.T. (1982). The role of phytoalexins in the resistance of soybean leaves to bacteria: effect of glyphosate on glyceollin accumulation. Phytopathology 72, 1470-1474.

Holliday, M.J., Long, M. & Keen, N.T. (1981). Manipulation of the temperature-sensitive interaction between soybean leaves and *Pseudomonas syringae* pv. *glycinea* - implications on the nature of determinative events modulating hypersensitive resistance. Physiological Plant Pathology 19, 209-216.

Hoppe, H.H., Humme, B. & Heitefuss, R. (1980). Elicitor induced accumulation of phytoalexins in healthy and rust infected leaves of *Phaseolus vulgaris*. Phytopathologische Zeitschrift 97, 85-88.

Howlett, B.J., Knox, R.B., Paxton, J.D. & Heslop-Harrison, J. (1975). Pollen wall proteins: physicochemical characterization and role in self-incompatibility in *Cosmos bipinnatus*. Proceedings of the Royal Society B 188, 167-182.

Hseu, S.T. & Goodman, R.N. (1978). Agglutination activity in apple cell suspension cultures inoculated with a virulent strain of *Erwinia amylovora*. Phytopathology 68, 355-360.

Huang, P.Y., Huang, J.S. & Goodman, R.M. (1975). Resistance mechanisms of apple shoots to an avirulent strain of *Erwinia amylovora*. Physiological Plant Pathology 6, 283-287.

Hutson, R.A. & Smith, I.M. (1980). Phytoalexins and tyloses in tomato cultivars infected with *Fusarium oxysporum* f. sp. *lycopersici* or *Verticillium albo-atrum*. Physiological Plant Pathology 17, 245-257.

Ingham, J.L. (1973). Disease resistance in higher plants. The concept of pre-infectional and post-infectional resistance. Phytopathologische Zeitschrift 78, 314-355.

Ingram, D.S. (1978). Cell death and resistance to biotrophs. Annals of Applied Biology 89, 291-295.

Jacobi, W.R. & MacDonald, W.L. (1980). Colonization of resistant and susceptible oaks by *Ceratocystis fagacearum*. Phytopathology 70, 618-623.

Jennings, D.L. (1962). Some evidence on the influence of the morphology of raspberry canes upon their liability to be attacked by certain fungi. Horticultural Research 1, 100-111.

Jennings, P.R. & Ullstrup, A.J. (1959). A histological study of three *Helminthosporium* leaf blights of corn. Phytopathology 47, 707-714.

Jensen, N.F. & Driscol, C.J. (1962). Inheritance of the waxless character in wheat. Crop Science 2, 504-505.

Jhooty, J.S. & McKeen, W.C. (1965). Studies on powdery mildew of strawberry caused by *Sphaerotheca macularis*. Phytopathology 55, 281-285.

Johnson, G. & Schaal, L.A. (1957). Chlorogenic acid and other ortho-dihydrophenols in scab-resistant Russet Burbank and scap-susceptible Triumph potato tubers of different maturities. Phytopathology 47, 253-255.

by differential temperature effects. Plant Disease Reporter 49, 222-224.

Johnson, L.E.B., Bushnell, W.R. & Zeyen, R.J. (1979). Binary pathways for analysis of primary infection and host responses in populations of powdery mildew fungi. Canadian Journal of Botany 57, 497-511.

Johnson, R. (1976), Development and use of some genetically controlled lines for studies of host-parasite interactions. In 'Biochemical Aspects of Plant Parasite Relationships' (Friend, J. & Threlfall, D.R., eds), pp. 25-41. Academic Press, London, New York & San Francisco.

Jones, D.R. & Deverall, B.J. (1978). The use of leaf transplants to study the cause of hypersensitivity to leaf rust, Puccinia recondita in wheat carrying the Lr20 gene. Physiological Plant Pathology 12, 311-319.

Jones, T.M., Anderson, A.J. & Albersheim, P. (1972). Host-pathogen interactions. IV. Studies on the polysaccharide-degrading enzymes secreted by Fusarium oxysporum f. sp. lycopersici. Physiological Plant Pathology 2, 153-166.

Jørgensen, H.J. & Mortensen, K. (1977). Primary infection by Erysiphe graminis f. sp. hordei of barley mutants with resistance genes in the ml-o locus. Phytopathology 67, 678-685.

Kaars Sijpesteijn, A. & Van Dijkman, A. (1973). The host-parasite interactions in resistance of tomatoes to Cladosporium fulvum. In 'Fungal Pathogenicity and the Plant's Response' (Byrde, R.J.W. & Cutting, C.V., eds), pp. 437-445. Academic Press, London & New York.

Kaars Sijpesteijn, A. (1976). Induction of hypersensitive responses to fungal pathogens. In 'Specificity in Plant Diseases' (Wood, R.K.S. & Graniti, A., eds), pp. 273-283. Plenum Press, New York & London.

Keen, N.T. (1975). Specific elicitors of plant phytoalexin production: determinants of race specificity in pathogens. Science 187, 74-75.

Keen, N.T. (1982). Specific recognition in gene-for-gene host parasite systems. Advances in Plant Pathology 1, 35-82.

Keen, N.T. & Bruegger, B.B. (1977). Phytoalexins and chemicals that elicit their production in plants. In 'Host Plant Resistance to Pests' (Hedin, P., ed.), pp. 1-26. American Chemical Society Symposium Series 62.

Keen, N.T., Ersek, T., Long, M., Bruegger, B. & Holliday, M.J. (1981). Inhibition of the hypersensitive reaction of soybean leaves to incompatible Pseudomonas spp. by blasticidin S, streptomycin or elevated temperature. Physiological Plant Pathology 18, 325-337.

Keen, N.T. & Holliday, M.J. (1982). Recognition of bacterial pathogens by plants. In 'Phytopathogenic Prokaryotes Vol. 2' (Mount, M.S. & Lacy, G.H., eds), pp. 179-217. Academic Press, New York & London.

Keen, N.T., Holliday, M.J. & Yoshikawa, M. (1982). Effects of glyphosphate on glyceollin production and the expression of resistance to Phytophthora megasperma f. sp. glycinea in soybean. Phytopathology 72, 1467-1470.

Keen, N.T. & Kennedy, B.W. (1974). Hydroxyphaseollin and related isoflavonoids in the hypersensitive resistance response of soybean against Pseudomonas glycinea. Physiological Plant Pathology 4, 173-185.

Keen, N.T. & Legrand, M. (1980). Surface glycoproteins: evidence that they may function as the race specific phytoalexin elicitors of Phytophthora megasperma f. sp. glycinea. Physiological Plant Pathology 17, 175-192.

Keen, N.T. & Yoshikawa, M. (1983). β-1,3-endoglucanase from soybean releases elicitor-active carbohydrates from fungal cell walls. Plant Physiology 71, 460-465.

Keen, N.T., Yoshikawa, M. & Wang, M.C. (1983). Phytoalexin elicitor activity of carbohydrates from Phytophthora megasperma f. sp. glycinea and other sources. Plant Physiology 71, 466-471.

Kerr, A. & Flentje, N.T. (1957). Host infection in Pellicularia filamentosa controlled by chemical stimuli. Nature 179, 204-205.

Keyworth, W.G. (1963). The reaction of monogenic resistant and susceptible varieties of tomato to inoculation with Fusarium oxysporum f.sp. lycopersici into stems or through Bonny Best rootstocks. Annals of Applied Biology 52, 257-270.

in Bonny Best rootstocks invaded by Fusarium oxysporum f. sp. lycopersici. Annals of Applied Biology 54, 99-105.

Kiraly, Z., Barna, B. & Ersek, T. (1972). Hypersensitivity as a consequence, not a cause, of plant resistance to infection. Nature 239, 456-458.

Kitazawa, K. & Tomiyama, K. (1969). Microscopic observations of infection of potato cells by compatible and incompatible races of Phytophthora infestans. Phytopathologische Zeitschrift 66, 317-324.

Kitazawa, K. & Tomiyama, K. (1970). Cinephotomicrographic observation on the dynamic response of a potato cell to infection by Phytophthora infestans. Phytopathologische Zeitschrift 76, 80-86.

Klement, Z. (1982). Hypersensitivity. In 'Phytopathogenic Prokaryotes Vol. 2' (Mount, M. S. & Lacey, G.H., eds), pp. 149-177. Academic Press, New York & London.

Kline, D.M., Boone, D.M. & Keitte, G.W. (1957). Venturia inaequalis (Cke.) Wint. XIV Nutritional control of pathogenicity of certain induced biochemical mutants. American Journal of Botany 44, 797-803.

Kloepper, J.W., Leong, J., Teintze, M. & Schroth, M.N. (1980). Enhanced plant growth by siderophores produced by plant growth-promoting bacteria. Nature 286, 885-886.

Kochman, J.K. & Brown, J.F. (1975). Development of the stem and crown rust fungi on leaves, sheaths and peduncles of oats. Phytopathology 65, 1404-1408.

Koga, H., Mayama, S. & Shishiyama, J. (1980). Correlation between the deposition of fluorescent compounds in papillae and resistance in barley against Erysiphe gramminis f. sp. hordei. Canadian Journal of Botany 58, 536-541.

Koga, H., Toyoda, H., Mayama, S., Shishiyama, J. & Hiura, U. (1983). Characteristic cellular responses as expressions of genes for resistance to Erisyphe graminis f. sp. hordei in barley. Phytopathology 73, 907-910.

Kojima, M., Kawakita, K. & Uritani, I. (1982). Studies on a factor in sweet potato roots which aggutinates spores of Ceratocystis fimbriata black rot fungus. Plant Physiology 69, 474-478.

Kolattukudy, P.E. (1980). Biopolyester membranes of plants: cutin and suberin. Science 208, 990-1000.

Kommendahl, T. (1966). Relation of exudates of pea roots to germination of spores in races of Fusarium oxysporum f.sp. pisi. Phytopathology 56, 721-722.

Kono, Y. & Daly, J.M. (1979). Characterization of the host-specific pathotoxin produced by Helminthosporium maydis, Race T, affecting corn with Texas male-sterile cytoplasm. Biorganic Chemistry 8, 391-397.

Kosuge, T. (1969). The role of phenolics in host response to infection. Annual Review of Phytopathology 7, 195-222.

Kuo, M. & Scheffer, R.P. (1970). Comparative effects of host specific toxins and Helminthosporium infections on respiration and carboxylation by host tissue. Phytopathology 60, 1391-1394.

Kuo, M., Yoder, O.C. & Scheffer, R.P. (1970). Comparative specificity of the toxins of Helminthosporium victoriae. Phytopathology 60, 365-368.

Kurantz, M.J. & Osman, S.F. (1983). Class distribution, fatty acid composition and elicitor activity of Phytophthora infestans mycelial lipids. Physiological Plant Pathology 22, 363-370.

Kurantz, M.J. & Zacharius, R.M. (1981). Hypersensitive response in potato tuber: elicitation by combination of non-eliciting components from Phytophthora infestans. Physiological Plant Pathology 18, 67-77.

Lacey, G.H., Hirano, S.S., Victoria, J.I., Kelman, A. & Upper, C.D. (1969). Inhibition of soft-rotting Erwinia spp. stains by 2,4-dihydroxy-7-methoxy-2H-1,4-benzoxazin-3(4H)-one in relation to their pathogenicity on Zea mays. Phytopathology 69, 757-763.

Lacey, M.L. & Horner, C.E. (1966). Behaviour of Verticillium dahliae in the rhizosphere and on roots of plants susceptible, resistant, and immune to wilt. Phytopathology 56, 427-430.

Lampard, J.F. & Carter, G.A. (1973). Chemical investigations on resistance to coffee berry disease in Coffea arabica. An antifungal compound in coffee cuticular wax. Annals of

Lamport, D.T.A. & Northcote, D.H. (1960). Hydroxyproline in primary cell walls of higher plants. Nature 188, 665-666.

Larson, R.I. & Atkinson, T.G. (1970). A cytogenetic analysis of reaction to common root rot in some hard red spring wheats. Canadian Journal of Botany 48, 2059-2067.

Lazarovits, G. & Higgins, V.J. (1976). Histological comparison of Cladosporium fulvum race 1 on immune, resistant and susceptible tomato varieties. Canadian Journal of Botany 54, 224-234.

Leach, J.E., Cantrell, M.A. & Sequeira, L. (1982). A hydroxyproline-rich bacterial agglutinin from potato: its localization by immunofluorescence. Physiological Plant Pathology 21, 319-325.

Leach, J.E., Sherwood, J., Fulton, R.W. & Sequeira, L. (1983). Comparison of soluble proteins associated with disease resistance induced by bacterial lipopolysaccharide and by viral necrosis. Physiological Plant Pathology 23, 377-385.

Leatham, G.F., King, V. & Stahmann, M.A. (1980). In vitro protein polymerization by quinones or free radicals generated by plant or fungal oxidative enzymes. Phytopathology 70, 1134-1140.

Lee, S.C. & West, C.A. (1981a). Polygalacturonase from Rhizopus stolonifer, an elicitor of casbene synthetase activity in castorbean Ricinus communis seedlings. Plant Physiology 67, 633-639.

Lee, S.C. & West, C.A. (1981b). Properties of Rhizopus-stolonifer polygalacturonase an elicitor of casbene synthetase activity in castorbean Ricinus communis seedlings. Plant Physiology 67, 640-645.

Lewis, D. (1965). A protein dimer hypothesis on incompatibility. Genetics Today 3, 657-663.

Lewis, D.H. (1973). Concepts in fungal nutrition and the origin of biotrophy. Biological Reviews 48, 261-278.

Lewis, D.H. (1974). Micro-organisms and plants: the evolution of parasitism and mutualism. Symposia of the Society for General Microbiology 24, 367-392.

Link, P.K., Dickson, A.D. & Walker, J.C. (1929). Further observations on the occurrence of protocatechuic acid in pigmented onion scales and its relation to disease resistance in the onion. Journal of Biological Chemistry 84, 719-725.

Link, P.K. & Walker, J.C. (1933). The isolation of catechol from pigmented onion scales and its significance in relation to disease resistance in onions. Journal of Biological Chemistry 100, 379-383.

Lippincott, J.A. & Lippincott, B.B. (1977). Nature and specificity of the bacterium-host attachment in Agrobacterium infection. In 'Cell Wall Biochemistry Related to Specificity in Host Pathogen Interactions' (Solheim, B. & Raa, J., eds), pp. 439-451. Universitetsforlaget, Oslo, Norway.

Littlefield, L.J. (1973). Histological evidence for diverse mechanisms of resistance to flax rust, Melampsora lini (Ehrenb.) Lev. Physiological Plant Pathology 3, 241-247.

Littlefield, L.J. & Aronson, S.J. (1969). Histological studies of Melampsora lini resistance in flax. Canadian Journal of Botany 47, 1713-1717.

Lizenberger, S.C. (1949). Nature of susceptibility to Helminthosporium victoriae and resistance to Puccinia coronata in Victoria oats. Phytopathology 39, 300-318.

Loegering, W.Q. & Harmon, D.L. (1969). Wheat lines near-isogenic for reaction to Puccinia graminis tritici. Phytopathology 59, 456-459.

Ludtke, M. & Hahn, H. (1953). Über den Limanaringehalt gesunder und von Colletotrichum lini befallener junger Leinpflanzen. Biochemische Zeitschrift 325, 433-442.

Luig, N.H. & Watson, I.A. (1965). Studies on the genetic nature of resistance to Puccinia graminis var. tritici in six varieties of common wheat. Proceedings of the Linnean Society of New South Wales 90, 299-327.

Luke, H.H. & Wheeler, H. (1964). An intermediate reaction to victorin. Phytopathology 54, 1492-1493.

Lukezic, F.L. & DeVay, J.E. (1964). Effect of myo-inositol in host tissues on the parasitism of Prunus domestica var. President by Rhodosticta quercina. Phytopathology 54, 697-700.

Lyon, G.D. & Albersheim, P. (1982). Host-pathogen interactions. XXI. Extraction of a heat-labile elicitor of phytoalexin accumulation from frozen soybean stems. Plant Physiology

Lyon, F.M. & Wood, R.K.S. (1975). Production of phaseollin, coumestrol and related compounds in bean leaves inoculated with Pseudomonas spp. Physiological Plant Pathology 2, 411-416.

Mace, M.E. (1978). Contributions of tyloses and terpenoid aldehyde phytoalexins to Verticillium wilt resistance in cotton. Physiological Plant Pathology 12, 1-11.

Maclean, D.J. & Tommerup, I.C. (1979). Histology and physiology of compatibility and incompatibility between lettuce and the downy mildew fungus Bremia lactucae Regel. Physiological Plant Pathology 14, 291-312.

Maiti, I.B. & Kolattukudy, P.E. (1979). Prevention of fungal infection of plants by specific inhibition of cutinase. Science 205, 507-508.

Malik, M.M.S. & Batts, C.C.V. (1960). The determination of the reaction of barley varieties to loose smut. Annals of Applied Biology 48, 39-50.

Mansfield, J.W. (1982). The role of phytoalexins in disease resistance. In 'Phytoalexins' (Bailey, J.A. & Mansfield, J.W., eds), pp. 253-288. Blackie, Glasgow & London.

Marryat, D. (1907). Notes on the infection and histology of two wheats immune to the attacks of Puccinea glumarum, yellow rust. Journal of Agricultural Science, Cambridge 2, 129-138.

Martin, T.J. & Ellingboe, A.H. (1976). Differences between compatible parasite/host genotypes involving the Pm4 locus of wheat and the corresponding genes in Erysiphe graminis f. sp. tritici. Phytopathology 66, 1435-1438.

Masri, S.S. & Ellingboe, A.H. (1966). Primary infections of wheat and barley by Erysiphe graminis. Phytopathology 56, 389-395.

Matthews, D.E., Gregory, P. & Gracen, V.E. (1979). Helminthosporium maydis race T toxin induces leakage of NAD^+ from T cytoplasm corn mitochondria. Plant Physiology 63, 1149-1153.

Mayama, S., Daly, J.M., Rehfeld, D.W. & Daly, C.R. (1975). Hypersensitive response of near-isogenic wheat carrying the temperature-sensitive Sr6 allele for resistance to stem rust. Physiological Plant Pathology 7, 35-47.

Mayama, S. & Shishiyama, J. (1978). Localized accumulation of fluorescent and UV-absorbing compounds at penetration sites in barley leaves infected with Erysiphe graminis hordei. Physiological Plant Pathology 13, 347-354.

Mazzuchi, U., Bazzi, C. & Bonatti, P.M. (1982). Encapsulation of Pseudomonas syringae pv. tabaci in relation to its growth in tobacco leaves both pretreated and not pretreated with protein-lipopolysaccharide complexes. Physiological Plant Pathology 21, 105-112.

McNabb, H.S., Heybroek, H.V. & Macdonald, W.S. (1970). Anatomical factors in resistance to Dutch elm disease. Netherlands Journal of Plant Pathology 76, 196-204.

McLean, F.T. (1921). A study of the structures of the stomata of two species of citrus in relation to citrus canker. Bulletin of the Torrey Botanical Club 48, 101-106.

McLean, F.T. & Lee, H.A. (1922). Pressure required to cause stomatal infections with citrus canker organism. Phillipine Journal of Science 20, 309-321.

McLean, J.G., Le Tourneau, D.J. & Guthrie, J.W. (1961). Relation of histochemical tests for phenols to Verticillium wilt resistance of potatoes. Phytopathology 51, 84-89.

Meier, H., Buchs, L., Buchala, A.J. & Homewood, T. (1981). (1,3)-β-D-glucan (callose) is a probable intermediate in biosynthesis of cellulose in cotton fibres. Nature 289, 821-822.

Mendgen, K. (1977). Reduced lysine uptake by bean rust haustoria in a resistant reaction. Naturwissenschaften 64, 438-439.

Mendgen, K. (1981). Nutrient uptake in rust fungi. Phytopathology 71, 983-989.

Mercer, P.C., Wood, R.K.S. & Greenwood, A.D. (1974). Resistance to anthracnose of French beans. Physiological Plant Pathology 4, 291-306.

Miller, R.J. & Koeppe, D.E. (1971). Southern corn leaf blight: susceptible and resistant mitochondria. Science 173, 67-69.

Mirelman, D., Galun, E., Sharon, N. & Lotan, R. (1975). Inhibition of fungal growth by wheat germ agglutinin. Nature 156, 414-416.

Misaghi, I.J. (1982). Physiology and Biochemistry of Plant-Pathogen Interactions. Plenum Press, New York and London. 287 pp.

Misaghi, I.J., Stowell, L.J., Grogan, R.G. & Spearman, L.C. (1982). Fungistatic activity of water-soluble fluorescent pigments of fluorescent pseudomonads. Phytopathology 72, 33-36.

bacterial wilt resistance in tomato. Phytopathology 59, 14.

Mollenhauer, H.H. & Hopkins, D.L. (1979). Xylem morphology of Pierce's disease-infected grapevines with different levels of tolerance. Physiological Plant Pathology 9, 95-100.

Morgan, W. (1983). Viability of Bremia lactucae oospores and stimulation of their germination by lettuce seedlings. Transactions of the British Mycological Society 80, 403-408.

Müller, K.O. (1959). Hypersensitivity. In 'Plant Pathology: an Advanced Treatise'. Vol. 1. (Horsfall, J.G. & Dimond, A.E., eds), pp. 469-519. Academic Press, New York.

Müller, K.O. & Börger, H. (1940). Experimentelle Untersuchungen über die Phytophthora Resistenz der Kartoffel. Arbeiten aus dem Biologischen Reichsanstalt (Berl.) 23, 189-231.

Murray, T.D. & Bruehl, G.W. (1981). Some anatomical characteristics of wheat culms associated with resistance to Pseudocercosporella herpatrichoides. Phytopathology 71, 107.

Murray, T.D. & Bruehl, G.W. (1982). Association between crude fibre, cellulose and lignin in stems of winter wheat and resistance to Pseudocercosporella herpatrichoides. Phytopathology 72, 1002.

Mussell, H.W. & Green, R.J. (1970). Host colonisation and polygalacturonase production by two tracheonytic fungi. Phytopathology 60, 192-195.

Nagdy, G.A. & Boyd, A.E.W. (1965). Susceptibility of potato varieties to skin spot (Oospora pustulans) in relation to the structure of the skin and eye. European Potato Journal 8, 200-214.

Nasrallah, M.E., Barber, J.T. & Wallace, D.H. (1970). Self-incompatibility proteins in plants: detection, genetics and possible mode of action. Heredity 25, 23-27.

Neal, J.L., Atkinson, T.G. & Larson, R.I. (1970). Changes in rhizosphere microflora in spring wheat induced by disomic substitution of a chromosome. Canadian Journal of Microbiology 16, 153-158.

Neal, J.L., Larson, R.I. & Atkinson, T.G. (1973). Changes in rhizosphere populations of selected physiological groups of bacteria related to substitution of specific pairs of chromosomes in spring wheat. Plant and Soil 39, 209-212.

Nelson, O.E. & Ullstrup, A.J. (1964). Resistance to leaf spot in maize. Journal of Heredity 55, 195-199.

Nelson, R.R., Scheffer, R.P. & Pringle, R.B. (1963). Genetic control of toxin production in Helminthosporium victoriae. Phytopathology 53, 385-387.

Netzer, D., Kritzman, G. & Chet, I. (1979). β-(1,3) glucanase activity and quantity of fungus in relation to Fusarium wilt in resistant and susceptible near-isogenic lines of muskmelon. Physiological Plant Pathology 14, 47-55.

Nichols, E.J., Beckman, J.M. & Hadwiger, L.A. (1980). Glycosidic enzyme activity in pea tissue and pea - Fusarium solani interactions. Plant Physiology 66, 199-204.

Niks, R.E. (1983a). Comparative histology of partial resistance and the nonhost reaction to leaf rust pathogens in barley and wheat seedlings. Phytopathology 73, 60-64.

Niks, R.E. (1983b). Haustorium formation by Puccinia hordei in leaves of hypersensitive, partly resistant and nonhost plant genotypes. Phytopathology 73, 64-66.

Niks, R.E. & Kuiper, H.J. (1983). Histology of the relation between minor and major genes for resistance of barley to leaf rust. Phytopathology 73, 55-59.

Nims, R.C., Halliwell, R.S. & Rosberg, D.W. (1967). Wound healing in cultured tobacco cells following microinjection. Protoplasma 64, 305-314.

Nishimura, S., Kohmoto, K., & Otami, H. (1979). The role of host-specific toxins in saprophytic pathogens. In 'Recognition and Specificity in Plant Host-Parasite Interactions' (Daly, J.M. & Uritani, I., eds), pp. 133-146. Japan Scientific Societies Press, Tokyo; University Park Press, Baltimore.

Nishimura, S., Kohmoto, K., Otami, H., Ramachandran, P. & Tamora, F. (1982). Pathological and epidemiological aspects of Alternaria alternata infection depending on host-specific toxin. In 'Plant Infection: the Physiological and Biochemical Basis' (Asada, Y., Bushnell, W.R., Ouchi, S. & Vance, C.P., eds), pp. 199-213. Japan Scientific Societies Press, Tokyo; Springer Verlag, Berlin, Heidelberg & New York.

Nitzsch, W. (1983). Chitinase as a possible resistance factor for higher plants. Theoretical

Noveroske, R.L., Kuc, J. & Williams, E.B. (1964). Oxidation of phloridzin and phloretin related to resistance of Malus to Venturia inaequalis. Phytopathology 54, 92-97.

Noyes, R.D. & Hancock, J.G. (1981). Role of oxalic acid in the Sclerotinia wilt of sunflower. Physiological Plant Pathology 18, 123-132.

Nozue, M., Tomiyama, K. & Doke, M. (1977a). Effect of adenosine 5'-triphosphate on hypersensitive death of potato cells infected by Phytophthora infestans. Phytopathology 68, 873-876.

Nozue, M., Tomiyama, K. & Doke, M. (1977b). Effect of blasticidin S on development of potential of potato tuber cells to react hypersensitively to infection by Phytophthora infestans. Physiological Plant Pathology 10, 181-189.

Nozue, M., Tomiyama, K. & Doke, N. (1980). Effect of N,N'-diacetyl-D-chitobiose, the potato lectin hapten and other sugars on hypersensitive reaction of potato tuber cells infected by incompatible and compatible races of Phytophthora infestans. Physiological Plant Pathology 17, 221-227.

Nutman, F.J. & Roberts, F.M. (1960). Investigations on a disease of Coffea arabica caused by a form of Colletotrichum coffeanum Noack. I. Some factors affecting infection by the pathogen. Transactions of the British Mycological Society 43, 489-505.

Nutman, P.S. (1969). Genetics of symbiosis and nitrogen fixation in legumes. Proceedings of the Royal Society London B 172, 417-437.

Odvody, G.M., Dunkle, L.D. & Edmunds, L.K. (1977). Characterization of the Periconia circinata population in a Milo disease nursery. Phytopathology 67, 1485-1489.

Ogle, H. & Brown, J.F. (1971). Quantitative studies of the post penetration phase of infection by Puccinia graminis tritici. Annals of Applied Biology 67, 309-319.

Onesirosan, P., Mabuni, C.T., Durbin, R.B., Murin, D.H., Rich, D.H. & Arny, D.C. (1975). Toxin production by Corynespora cassiicola. Physiological Plant Pathology 5, 289-295.

Pankhurst, C.E. & Briggs, D.R. (1980). Sensitivity of Rhizobium to selected isoflavonoids. Canadian Journal of Microbiology 26, 542-545.

Parlevliet, J.E. (1977). Plant pathosystems: an attempt to elucidate horizontal resistance. Euphytica 26, 553-556.

Parlevliet, J.E. (1979). The co-evolution of host-parasite systems. Symbolae Botanicae Upsaliensis 22, 39-45.

Parlevliet, J.E. & Zadoks, J.C. (1977). The integrated concept of disease resistance; a new view including horizontal and vertical resistance in plants. Euphytica 26, 5-21.

Patil, S.S. (1974). Toxins produced by phytopathogenic bacteria. Annual Review of Phytopathology 12, 259-279.

Patil, S.S. (1980). Defenses triggered by the invader: detoxifying the toxins. In 'Plant Disease, An Advanced Treatise Vol. V' (Horsfall, J.G. & Cowling, E.B., eds), pp. 289-277. Academic Press, London & New York.

Patil, S.S. & Dimond, A.E. (1967). Inhibition of Verticillium polygalacturonase by oxydation products of polyphenols. Phytopathology 57, 492-496.

Patil, S.S., Hayward, A.C. & Emmons, R. (1974). An ultra-violet-induced non toxigenic mutant of Pseudomonas phaseolicola of altered pathogenicity. Phytopathology 64, 590-595.

Paxton, J.D. (1982). Phytoalexins. In 'Active Defense Mechanisms in Plants' (Wood, R.K.S., ed.), pp. 344-346. Plenum Press, New York & London.

Payne, G.A. & Yoder, O.C. (1978). Effect of the nuclear genome of corn on sensitivity to Helminthosporium maydis race T toxin and on susceptibility to H. maydis race T. Phytopathology 68, 331-337.

Payne, G.A., Kono, Y. & Daly, J.M. (1980). A comparison of purified host specific toxin from Helminthosporium maydis, race T, and its acetate derivative on oxidation by mitochondria from susceptible and resistant plants. Plant Physiology 65, 785-791.

Pearce, R.B. & Ride, J.P. (1980). Specificity of induction of the lignification response in wounded wheat leaves. Physiological Plant Pathology 16, 197-204.

Pearce, R.B. & Ride, J.P. (1982). Chitin and related compounds as elicitors of the lignification in wounded wheat leaves. Physiological Plant Pathology 20, 119-123.

resistant and susceptible tomato plants. Journal of Experimental Botany 27, 1093-1101.
Pegg, G.F. (1976b). The involvement of ethylene in plant pathogenesis. In 'Encyclopaedia of Plant Physiology Vol. 4. Physiological Plant Pathology' (Heitefuss, R. & Williams, P.H., eds), pp. 582-591. Springer Verlag, Berlin, Heidelberg & New York.
Pegg, G.F. & Dixon, G.R. (1969). The reaction of susceptible and resistant tomato cultivars to strains of Verticillium albo-atrum. Annals of Applied Biology 63, 389-400.
Pegg, G.F. & Vessey, J.C. (1973). Chitinase activity in Lycopersicon esculentum and its relationship to the in vivo lysis of Verticillium albo-atrum mycelium. Physiological Plant Pathology 3, 207-222.
Pegg, G.F. & Young, D.H. (1981). Changes in glycosidase activity and their relationship to fungal colonisation during infection of tomato by Verticillium albo-atrum. Physiological Plant Pathology 19, 371-382.
Pegg, G.F. & Young, D.H. (1982). Purification and characterisation of chitinase from healthy and Verticillium albo-atrum - infected tomato plants and from V. albo-atrum. Physiological Plant Pathology 21, 389-409.
Pelcher, L.E., Kao, K.H., Gamborg, O.L., Yoder, O.C., Gracen, V.E. (1975). Effects of Helminthosporium maydis race T toxin on protoplasts of resistant and susceptible corn (Zea mays L.). Canadian Journal of Botany 53, 427-431.
Peries, O.S. (1962). Studies on strawberry mildew caused by Sphaerotheca macularis (Wallr. ex Fries) Jaczewski. II. Host-parasite relationships on foliage of strawberry varieties. Annals of Applied Biology 50, 225-233.
Perkins, J.M. & Hooker, A.L. (1981). Reactions of eighty-four sources of chlorotic lesion resistance in corn to three biotypes of Helminthosporium turcicum. Plant Disease 65, 502-504.
Person, C. (1959). Gene-for-gene relationships in host:parasite systems. Canadian Journal of Botany 37, 1101-1130.
Powers, H.R. (1954). The mechanism of wilting in tobacco plants affected by black shank. Phytopathology 44, 513-521.
Preisig, C. & Kuć, J. (1983). Regulation of the sesquiterpenoid-eliciting activity of C_{20} fatty acids in potato by carbohydrate isolated from Phytophthora infestans. Phytopathology 73, 831.
Pring, R.J. & Richmond, D.V. (1976). An ultrastructural study of the effect of oxycarboxin on Uromyces phaseoli infecting leaves of Phaseolus vulgaris. Physiological Plant Pathology 8, 155-162.
Pringle, R.B. & Scheffer, R.P. (1964). Host specific plant toxins. Annual Review of Phytopathology 2, 133-156.
Prusky, D., Dinoor, A. & Jacoby, B. (1981). The fungicide or heat induced hypersensitive reaction of oats to crown rust: relations between various treatments and infection type. Physiological Plant Pathology 18, 181-186.
Pueppke, S.G. & Van Etten, H.D. (1976). The relation between pisatin and the development of Aphanomyces euteiches in diseased Pisum sativum. Phytopathology 66, 1174-1185.
Pueppke, S.G., Kluepfel, D.A. & Anand, V.K. (1982). Interaction of Agrobacterium tumefaciens with potato lectin and concanavalin A and its effect on tumor induction in potato. Physiological Plant Pathology 20, 35-42.
Puranik, S.B. & Mathre, D.E. (1971). Biology and control of ergot on male sterile wheat and barley. Phytopathology 61, 1075-1080.
Raa, J. & Overeem, J.C. (1968). Transformation reactions of phloridzin in the presence of apple leaf enzymes. Phytochemistry 7, 721-731.
Raa, J. & Kaars Sijpesteijn, A. (1968). A biochemical mechanism of natural resistance of apple to Venturia inaequalis. Netherlands Journal of Plant Pathology 74, 229-331.
Rabenantoandro, Y., Auriol, P. & Touze, A. (1976). Implication of β-(1,3) glucanase in melon anthracnose. Physiological Plant Pathology 8, 313-324.
Rajaram, S., Luig, N.H. & Watson, I.A. (1971). Inheritance of leaf rust resistance in four varieties of common wheat. Euphytica 20, 574-585.

245-251.

Ride, J.P. (1978). The role of cell wall alterations in resistance to fungi. Annals of Applied Biology 89, 302-306.

Ride, J.P. (1980). The effect of induced lignification on the resistance of wheat cell walls to fungal degradation. Physiological Plant Pathology 16, 187-196.

Robb, J. & Busch, L.V. (1983). Structure and significance of 'blocked veins' in Verticillium-infected chrysanthemum. Physiological Plant Pathology 23, 35-53.

Rockwood, D.L. (1974). Cortical monoterpene and fusiform rust resistance relationships in slash pines. Phytopathology 64, 976-979.

Roebuck, P., Sexton, R. & Mansfield, J.W. (1978). Ultrastructural observations on the development of the hypersensitive reaction in leaves of Phaseolus vulgaris cv. Red Mexican inoculated with Pseudomonas phaseolicola (race 1). Physiological Plant Pathology 12, 151-157.

Rohringer, R., Kim, W.K. & Samborski, D.J. (1979). A histological study of interactions between avirulent races of stem rust and wheat containing resistance genes Sr_5, Sr_6, Sr_8 and Sr_{22}. Canadian Journal of Botany 57, 324-331.

Rombouts, J.E. (1953). The microorganisms in the rhizosphere of banana plants in relation to susceptibility or resistance to Panama disease. Plant & Soil 4, 276-288.

Romeiro, R., Karr, A. & Goodman, R.N. (1981a). Isolation of a factor from apple that agglutinates Erwinia amylovora. Plant Physiology 69, 772-777.

Romeiro, R., Karr, A.L. & Goodman, R.N. (1981b). Erwinia amylovora cell wall receptor for apple agglutinin. Physiology Plant Pathology 19, 383-390.

Romig, R.W. & Caldwell, R.M. (1964). Stomatal exclusion of Puccinia recondita by wheat peduncles and sheaths. Phytopathology 54, 214-218.

Rowell, J.B., Loegering, W.Q. & Powers, H.R. (1963). Genetic model for physiologic studies of mechanisms governing development of infection type in wheat stem rust. Phytopathology 53, 932-937.

Royle, D.J. (1976). Structural features of resistance to plant diseases. In 'Biochemical Aspects of Plant-Parasite Relationships' (Friend, J. & Threlfall, D.R., eds), pp. 161-193. Academic Press, London, New York & San Francisco.

Rudolph, K. (1972). The halo-blight toxin of Pseudomonas phaseolicola: influence on host parasite relationships and counter effects of metabolites. In 'Phytotoxins in Plant Disease' (Wood, R.K.S., Ballio, A. & Graniti, A., eds), pp. 373-375. Academic Press, London & New York.

Rudolph, K. (1976). Non-specific toxins. In 'Physiological Plant Pathology, Encyclopaedia of Plant Physiology' Vol. 4 (Heitefuss, R. & Williams, P.H., eds), pp. 270-315. Springer-Verlag, Berlin, Heidelberg & New York.

Russell, G.E. (1975). Deposition of Erisyphe graminis f. sp. hordei conidia on barley varieties of different growth habit. Phytopathologische Zeitschrift 84, 316-321.

Samaddar, K.R. & Scheffer, R.P. (1968). Effect of the specific toxin of Helminthosporium victoriae on host cell membranes. Plant Physiology 43, 21-28.

Samaddar, K.R. & Scheffer, R.P. (1970). Effects of Helminthosporium victoriae toxin on germination and aleurone secretion by resistant and susceptible seeds. Plant Physiology 45, 586-590.

Samborski, D.J., Kim, W.K., Rohringer, R., Howes, N.K. & Baker, R.J. (1977). Histological studies on host-cell necrosis conditioned by the Sr6 gene for resistance to stem rust. Canadian Journal of Botany 55, 1445-1452.

Sato, N., Kitazawa, K. & Tomiyama, K. (1971). The role of rishitin in localizing the invading hyphae of Phytophthora infestans in infection sites at the cut surface of potato tubers. Physiological Plant Pathology 1, 289-295.

Schaal, L.A. & Johnson, G. (1955). The inhibitory effects of phenol compounds on the growth of Streptomyces scabies as related to the mechanism of scab resistance. Phytopathology 45, 626-628.

Scharvelle, E.G. (1936). The nature of resistance of flax to Melampsora lini. Journal of

Scheffer, R.P. (1976). Host specific toxins in relation to pathogenesis and disease resistance. In 'Encyclopedia of Plant Physiology Vol. 4 Physiological Plant Pathology' (Heitefuss, R. & Williams, P.H., eds), pp. 247-269. Springer Verlag, Berlin, Heidelberg & New York.

Scheffer, R.P. & Livingston, R.S. (1980). Sensitivity of sugarcane clones to toxin from Helminthosporium sacchari as determined by electrolyte leakage. Phytopathology 70, 400-404.

Scheffer, R.P., Nelson, R.R. & Ullstrup, A.J. (1967). Inheritance of toxin production and pathogenicity in Cochliobolus carbonum and Cochliobolus victoriae. Phytopathology 57, 1288-1291.

Scheffer, R.P. & Pringle, R.B. (1961). A selective toxin produced by Periconia circinata. Nature 191, 912-913.

Scheffer, R.P. & Pringle, R.B. (1967). Pathogen-produced determinants of disease and their effects on host plants. In 'The Dynamic Role of Molecular Constituents in Plant-Parasite Interaction' (Mirocha, C.J. & Uritani, I., eds), pp. 217-236. Bruce Publishing Co., St Paul, Minnesota.

Sheffer, R.P. & Ullstrup, A.J. (1965). A host-specific toxic metabolite from Helminthosporium carbonum. Phytopathology 55, 1037-1038.

Scheffer, R.P. & Yoder, O.C. (1972). Host-specific toxins and selective toxicity. In 'Phytotoxins in Plant Diseases' (Wood, R.K.S., Ballio, A. & Graniti, A., eds), pp. 251-272. Academic Press, London & New York.

Schertz, K.F. & Tai, Y.P. (1969). Inheritance of reaction of Sorghum bicolor to toxin produced by Periconia circinata. Crop Science 9, 621-624.

Schlösser, E. (1975). Role of saponins in antifungal resistance III. Tomatin dependent development of fruit rot organisms on tomato fruits. Zeitschrift für Pflanzenkrankheiten und Pflanzenschutz 82, 476-484.

Schlösser, E.W. (1980). Preformed internal chemical defenses. In 'Plant Disease: an Advanced Treatise. Vol. V How Plants Defend Themselves' (Horsfall, J.G. & Cowling, E.B., eds), pp. 161-177. Academic Press, New York, San Francisco & London.

Schnathorst, W.C. (1959). Resistance in lettuce to powdery mildew related to osmotic value. Phytopathology 49, 562-571.

Schönbeck, F. & Schroeder, C. (1972). Role of antimicrobial substances (tuliposides) in tulips attacked by Botrytis spp. Physiological Plant Pathology 2, 91-99.

Schönbeck, F. & Schlösser, E. (1976). Preformed substances as potential protectants. In 'Encyclopedia of Plant Physiology Vol. 4 Physiological Plant Pathology' (Heitefuss, R. & Williams, P.H., eds), pp. 653-678. Springer Verlag, Berlin, Heidelberg & New York.

Schroth, M.N. & Cook, R.J. (1964). Seed exudation and its influence on pre-emergence damping-off of bean. Phytopathology 54, 670-673.

Schroth, M.N. & Hildebrand, D.C. (1964). Influence of plant exudates on root-infecting fungi. Annual Review of Phytopathology 2, 101-132.

Scott, P.R. (1973). Incidence and effects of Septoria nodorum on wheat cultivars. Annals of Applied Biology 75, 321-329.

Scott, P.R., Benedikz, P.W. & Cox, C.J. (1982). A genetic study of the relationship between height, time of ear emergence and resistance to Septoria nodorum in wheat. Plant Pathology 31, 45-60.

Sequeira, L. (1978). Lectins and their role in host-pathogen specificity. Annual Review of Phytopathology 16, 453-481.

Sequeira, L. (1981). Induction of host physical responses to bacterial infection: a recognition phenomenon. In 'Plant Disease Control' (Staples, R.C. & Toenniesen, eds), pp. 143-153. J. Wiley and Sons, New York.

Sequeira, L. (1982). Determinants of plant response to bacterial infection. In 'Active Defence Mechanisms in Plants' (Wood, R.K.S., ed.), pp. 85-102. Plenum Press, New York & London.

Sequeira, L., Gaard, G. & De Zoeten, G.A. (1977). Interaction of bacteria and host cell walls: its relation to mechanisms of induced resistance. Physiological Plant Pathology 10, 43-50.

Sequeira, L. & Graham, T.L. (1977). Agglutination of avirulent strains of Pseudomonas solanacearum by potato lectin. Physiological Plant Pathology 11, 43-54.

susceptibility and maturity on the germination of conidia of Colletotrichum graminicola (Ces.) Wilson. In 'Ecology of Leaf-Surface Micro-organisms' (Preece, T.F. & Dickinson, C.H., eds), pp. 597-601. Academic Press, London & New York.

Shaykh, M., Soliday, C. & Kolattukudy, P.E. (1977). Proof for the production of cutinase by Fusarium solani f.sp. pisi during penetration into its host, Pisum sativum. Plant Physiology 60, 170-172.

Sheppard, J.W. & Peterson, J.F. (1976). Chlorogenic acid and Verticillium wilt of tobacco. Canadian Journal of Plant Science 56, 157-160.

Sherwood, R.T. & Vance, C.P. (1980). Resistance to fungal penetration. Phytopathology 70, 273-278.

Shimony, C. & Friend, J. (1975). Ultrastructure of the interaction between Phytophthora infestans and leaves of two cultivars of potato (Solanum tuberosum L.), Orion and Majestic. New Phytologist 74, 59-65.

Sidhu, G.S. & Webster, J.M. (1979). Genetics of tomato resistance to the Fusarium - Verticillium complex. Physiological Plant Pathology 15, 93-98.

Sidhu, G.S. & Webster, J.M. (1983). Horizontal resistance in tomato against the Meloidogyne - Fusarium complex: an artefact of parasitic epistasis. Crop Protection 2, 205-210.

Sigee, D.C. & Epton, H.A.S. (1976). Ultrastructural changes in resistant and susceptible varieties of Phaseolus vulgaris following artificial inoculation with Pseudomonas phaseolicola. Physiological Plant Pathology 9, 1-8.

Sinclair, W.A., Zahand, J.P. & Melching, J.B. (1975). Localization of infection in American elm resistant to Ceratocystis ulmi. Phytopathology 65, 129-133.

Sing, V. O. & Schroth, M.N. (1977). Bacteria-plant cell surface interactions: active immobilization of saprophytic bacteria in plant leaves. Science 197, 759-761.

Sinha, A.K. & Wood, R.K.S. (1967). An analysis of responses of resistant and of susceptible tomato plants to Verticillium infection. Annals of Applied Biology 59, 143-154.

Skipp, R.A. & Deverall, B.J. (1972). Relationship between fungal growth and host changes visible by light microscopy during infection of bean hypocotyls (Phaseolus vulgaris) susceptible and resistant to physiological races of Colletotrichum lindemuthianum. Physiological Plant Pathology 2, 357-374.

Skou, J.P. (1982). Callose formation responsible for the powdery mildew Erysiphe graminis f. sp. hordei resistance in barley Hordeum vulgare with genes in the Ml-o locus. Phytopathologische Zeitschrift 104, 90-95.

Skovmand, B., Wilcoxon, R.D. & Heiner, R.F. (1977). Genetic and environmental variability in wheat characteristics reported to be involved in morphological resistance to wheat stem rust. Euphytica 26, 123-128.

Slesinski, R.S. & Ellingboe, A.H. (1969). The genetic control of primary infection of wheat by Erysiphe graminis f. sp. tritici. Phytopathology 59, 1833-1837.

Slusarenko, A.J. & Wood, R.K.S. (1981). Differential agglutination of races 1 and 2 of Pseudomonas phaseolicola by a fraction from cotyledons of Phaseolus vulgaris cv. Red Mexican. Physiological Plant Pathology 18, 187-193.

Slusarenko, A.J. & Wood, R.K.S. (1983). Agglutination of Pseudomonas phaseolicola by pectic polysaccharide from leaves of Phaseolus vulgaris. Physiological Plant Pathology 23, 217-227.

Smale, B.C. & Keil, H.L. (1966). A biochemical study of the intervarietal resistance of Pyrus communis to fire blight. Phytochemistry 5, 1113-1120.

Smith, D.A. (1982). Toxicity of phytoalexins. In 'Phytoalexins' (Bailey, J.A. & Mansfield, J.W., eds), pp. 218-255. Blackie, Glasgow & London.

Smith, F.G., Walker, J.C. & Hooker, W.J. (1946). Effect of hydrogen ion concentration on the toxicity to Colletotrichum circinans (Berk.) Vogl. of some carboxylic acids, phenols and crucifer extracts. American Journal of Botany 33, 351-356.

Smith, J.J. & Mansfield, J.W. (1981). Interactions between pseudomonads and leaves of oats, wheat and barley. Physiological Plant Pathology 18, 345-356.

Smith, J.J. & Mansfield, J.W. (1982). Ultrastructure of interactions between pseudomonads and oat leaves. Physiological Plant Pathology 21, 259-266.

attack. Journal of Agricultural Research 4, 193-199.
Stall, R.E. & Cook, A.A. (1979). Evidence that bacterial contact with the plant cell is necessary for the hypersensitive reaction but not the susceptible reaction. Physiological Plant Pathology 14, 77-84.
Steiner, G.W. & Byther, R.S. (1971). Partial characterization and use of a host-specific toxin from Helminthosporium sacchari on sugarcane. Phytopathology 61, 691-695.
Steiner, G.W. & Strobel, G.A. (1971). Helminthosporoside, a host-specific toxin from Helminthosporium sacchari. Journal of Biological Chemistry 246, 4350-4357.
Stekoll, M. & West, C.A. (1978). Purification and properties of an elicitor of castor bean phytoalexin from culture filtrates of the fungus Rhizopus stolonifer. Plant Physiology 61, 38-45.
Strobel, G.A. (1973a). The helminthosporoside-binding protein of sugarcane: its properties and relationship to susceptibility to the eye-spot disease. Journal of Biological Chemistry 248, 1321-1328.
Strobel, G.A. (1973b). Biochemical basis of the resistance of sugarcane to eyespot disease. Proceedings of the National Academy of Sciences USA 70, 1693-1696.
Strobel, G.A. (1974a). Phytotoxins produced by plant parasites. Annual Review of Plant Physiology 25, 541-566.
Strobel, G.A. (1974b). The toxin-binding protein of sugarcane, its role in the plant and in disease development. Proceedings of the National Academy of Sciences USA 71, 4232-4236.
Strobel, G.A. (1975). A mechanism of disease resistance in plants. Scientific American 232, 80-88.
Strobel, G.A. (1976). Toxins of plant pathogenic bacteria and fungi. In 'Biochemical Aspects of Plant Parasite Relationships' (Friend, J.E. & Threlfall, D.R., eds), pp. 135-159. Academic Press, London, New York & San Francisco.
Strobel, G.A. (1977). Bacterial phytotoxins. Annual Review of Microbiology 31, 205-224.
Strobel, G.A. & Steiner, G.W. (1972). Runner lesion formation in relation to helminthosporoside in sugarcane leaves infected by Helminthosporium sacchari. Physiological Plant Pathology 2, 129-132.
Stromberg, E.L. & Corden, M.E. (1980). Scanning electron microscopy of Fusarium oxysporum f. sp. lycopersici in xylem vessels of wilt-resistant and susceptible tomato plants. Canadian Journal of Botany 58, 2360-2366.
Struckmeyer, E.B., Beckman, C.H., Kuntz, J.E. & Riker, A.J. (1954). Plugging of vessels by tyloses and gums in wilting oaks. Phytopathology 44, 148-153.
Stuehling, B.A. & Nelson, P.E. (1981). Anatomy of a tolerant chrysanthemum cultivar infected with Fusarium oxysporum f. sp. chrysanthemi. Phytopathology 71, 1162-1168.
Subba-Rao, N.S. & Bailey, D.L. (1961). Rhizosphere studies in relation to varietal resistance or susceptibility of tomato to Verticillium wilt. Canadian Journal of Botany 31, 1747-1758.
Swain, T. (1977). Secondary compounds as protective agents. Annual Review of Plant Physiology 28, 479-501.
Swinburne, T.R. (1981). Iron and iron chelating agents as factors in germination, infection and aggression of fungal pathogens. In 'Microbial Ecology of the Phylloplane' (Blakeman, J.P., ed.), pp. 227-243. Academic Press, London & New York.
Talboys, P.W. (1972). Resistance to vascular wilt fungi. Proceedings of the Royal Society B 181, 319-332.
Teakle, D.S., Smith, P.M. & Steindl, D.R.L. (1975). Ratoon stunting disease of sugarcane: possible correlation of resistance with vascular anatomy. Phytopathology 65, 138-141.
Tegtmeier, K.J. & Van Etten, H.D. (1982). The role of pisatin tolerance and degradation in the virulence of Nectria haematococca. A genetic analysis. Phytopathology 72, 608-612.
Temmen, K.H., Kruppe, W. & Schlösser, E. (1980). Investigations on the resistance of plants to powdery mildew III. Basis for the horizontal resistance of Ribes cv. to Sphaerotheca mors-uvae. Zeitschrift für Pflanzenkrankheiten und Pflanzenschutz 87, 129-136.
Timonin, M.I. (1941). The interaction of higher plants and soil microorganisms. III. Effects of by-products of plant growth on activity of fungi and actinomycetes. Soil Science 52,

Tinline, R.D. (1963). *Cochliobolus sativus* VII. Nutritional control of the pathogenicity of some auxotrophs to wheat seedlings. Canadian Journal of Botany 41, 489-497.

Tjamos, E.C. & Smith, I.M. (1974). The role of phytoalexins in the resistance of tomato to *Verticillium* wilt. Physiological Plant Pathology 4, 249-260.

Tjamos, E.C. & Smith, I.M. (1975). The expression of resistance to *Verticillium albo-atrum* in monogenically resistant tomato varieties. Physiological Plant Pathology 6, 215-225.

Tomiyama, K. (1971). Cytological and biochemical studies of the hypersensitive reaction of potato cells to *Phytophthora infestans*. In 'Morphological and Biochemical Events in Plant Parasite Interaction' (Akai, S. & Ouchi, S., eds), pp. 387-401. Phytopathological Society of Japan, Tokyo.

Tomiyama, K. (1976). Further observation on the time requirement for hypersensitive cell death of potatoes infected by *Phytophthora infestans* and its relation to metabolic activity. Phytopathologische Zeitschrift 58, 367-373.

Tomiyama, K., Doke, N., Nozue, M. & Ishiguri, Y. (1979). The hypersensitive response of resistant plants. In 'Recognition and Specificity in Plant Host-Parasite Interactions' (Daly, J.M. & Uritani, I., eds), pp. 69-84. Japan Scientific Societies Press, Tokyo; University Park Press, Baltimore.

Tomiyama, K., Okamoto, H. & Katou, K. (1983). Effect of infection by *Phytophthora infestans* on membrane potential of potato cells. Physiological Plant Pathology 22, 233-243.

Toppan, A., Roby, D. & Esquerré-Tugayé, M.T. (1982). Cell surfaces in plant-microorganism interactions III. *In vivo* effect of ethylene on hydroxyproline-rich glycoprotein accumulation in the cell wall of diseased plants. Plant Physiology 70, 82-86.

Trione, E.S. (1960). The HCN content of flax in relation to flax wilt resistance. Phytopathology 50, 482-486.

Troughton, J.H. & Hall, D.M. (1967). Extracuticular wax and contact angle measurements on wheat (*Triticum vulgare* L.). Australian Journal of Biological Sciences 20, 509-525.

Turner, E.M.C. (1961). An enzymic basis for pathogenic specificity in *Ophiobolus graminis*. Journal of Experimental Botany 12, 169-175.

Turner, J.G. & Novacky, A. (1974). The quantitative relation between plants and bacterial cells involved in the hypersensitive reaction. Phytopathology 64, 885-890.

Vance, T., Kirk, K. & Sherwood, R.T. (1980). Lignification as a mechanism of disease resistance. Annual Review of Phytopathology 18, 259-288.

Van den Heuvel, J. & Glazener, J.A. (1975). Comparative abilities of fungi pathogenic and non-pathogenic to bean (*Phaseolus vulgaris*) to metabolize phaseollin. Netherlands Journal of Plant Pathology 81, 125-137.

Van der Molen, G.E., Beckman, C.H. & Rodehorst, E. (1977). Vascular gelation: a general response phenomenon following infection. Physiological Plant Pathology 11, 95-100.

Vanderplank, J.E. (1968). Disease resistance in plants. Academic Press, New York & London. 206 pp.

Vanderplank, J.E. (1978). Genetic and molecular basis of plant pathogenesis. Springer-Verlag, Berlin, Heidelberg & New York. 107 pp.

Vanderplank, J.E. (1982). Host pathogen interactions in plant disease. Academic Press, New York & London. 207 pp.

Van Dijkman, A., Dieleman, S.J. & Kaars Sijpesteijn, A. (1973). Differences in disc gel electrophoresis patterns of soluble proteins excreted by different physiological races and isolates of *Cladosporium fulvum*. Netherlands Journal of Plant Pathology 79, 70-80.

Van Dijkman, A. & Kaars Sijpesteijn, A. (1971). A biochemical mechanism for the gene-for-gene resistance of tomato to *Cladosporium fulvum*. Netherlands Journal of Plant Pathology 77, 14-24.

Van Dijkman, A. & Kaars Sijpesteijn, A. (1973). Leakage of pre-absorbed ^{32}P from tomato disks infiltrated with high molecular weight products of incompatible races of *Cladosporium fulvum*. Physiological Plant Pathology 3, 57-67.

Van Etten, H.D., Matthews, P.S., Tegtmeier, K.J., Dietert, M.F. & Stein, J.I. (1980). The association of pisatin tolerance and demethylation with virulence on pea in *Nectria*

Van Etten, H.D., Matthews, D.E. & Smith, D.A. (1982). Metabolism of phytoalexins. In 'Phytoalexins' (Bailey, J.A. & Mansfield, J.W., eds), pp. 181-217. Blackie, Glasgow & London.

Vessey, J.C. & Pegg, G.F. (1973). Autolysis and chitinase production in cultures of Verticillium albo-atrum. Transactions of the British Mycological Society 60, 133-143.

Volk, R.J., Kahn, R.P. & Weintraub, R.L. (1958). Silicon content of the rice plant as a factor influencing its resistance to infection by the blast fungus, Pyricularia oryzae. Phytopathology 48, 179-184.

Wade, M. & Albersheim, P. (1979). Race-specific molecules that protect soybeans from Phytophthora megasperma var. sojae. Proceedings of the National Academy of Sciences USA 76, 4433-4437.

Walker, J.C. & Stahmann, M.A. (1955). Chemical nature of disease resistance in plants. Annual Review of Plant Physiology 6, 351-366.

Waller, J.M. (1970). Sugarcane smut (Ustilago scitaminea) in Kenya. II Infection and resistance. Transactions of the British Mycological Society 54, 405-414.

Wang, S.C. & Pinckard, J.A. (1973). Cotton boll cuticle, a potential factor in boll rot resistance. Phytopathology 63, 315-311.

Ward, H.M. (1905). Recent researches on the parasitism of fungi. Annals of Botany 19, 1-54.

Ward, E.W.B. & Buzzell, R.I. (1983). Influence of light, temperature and wounding on the expression of soybean genes for resistance to Phytophthora megasperma f. sp. glycinea. Physiological Plant Pathology 23, 401-409.

Ward, E.W.B., Stoessl, A. & Stothers, J.B. (1977). Metabolism of the sesquiterpenoid phytoalexins capsidiol and rishitin to their 13-hydryoxy derivatives by plant cells. Phytochemistry 16, 2024-2025.

Ward, E.W.B., Stössel, P. & Lazarovits, G. (1981). Similarities between age-related and race-specific resistance of soybean hypocotyls to Phytophthora megasperma var. sojae. Phytopathology 71, 504-508.

Wargo, P.M. (1975). Lysis of the cell wall of Armillaria mellea by enzymes from forest trees. Physiological Plant Pathology 5, 99-105.

Watson, I.A. & Luig, N.H. (1968). Progressive increase of virulence in Puccinia graminis f. sp. tritici. Phytopathology 58, 70-73.

Wehtje, G.L., Littlefield, L.J. & Zimmer, D.E. (1979). Ultrastructure of compatibile and incompatible reactions of sunflower to Plasmopara halstedii. Canadian Journal of Botany 57, 315-323.

Whalley, W.M. & Taylor, G.S. (1973). Influence of pea-root exudates on germination of conidia and chlamydospores of physiologic races of Fusarium oxysporum f.sp. pisi. Annals of Applied Biology 73, 269-276.

Whatley, M.N., Hunter, N., Cantrell, M.A., Hendrik, C., Keegstra, K. & Sequeira, L. (1980). Lipopolysaccharide composition of the wild pathogen Pseudomonas solanacearum: correlation with the hypersensitive response in tobacco. Plant Physiology 65, 557-559.

Wheeler, H. (1969). The fate of victorin in susceptible and resistant oat coleoptiles. Phytopathology 59, 1093-1097.

Wheeler, H. (1975). Plant Pathogenesis. Springer-Verlag, Berlin, Heidelberg & New York. 106 pp.

Wilcoxon, R.D. (1958). Peduncle sclerenchyma tissue of Nugget wheat in relation to development of pustules of Puccinia graminis var. tritici. Phytopathology 48, 518-519.

Williams, B.J. & Boone, D.M. (1963). Venturia inaequalis (Cke) Wint. XVI Amino acids in relation to pathogenicity of two wild-type lines to two apple varieties. Phytopathology 53, 979-983.

Wilson, C.L. (1973). A lysosomal concept for plant pathology. Annual Review of Phytopathology 11, 247-272.

Wilson, A.R. & Jarvis, W.R. (1963). White blister on a glossy-leafed mutant of Brussels sprout. Plant Pathology 12, 91.

Wilson, U.E. & Coffey, M.D. (1980). Cytological evaluation of general resistance to Phytophthora infestans in potato foliage. Annals of Botany 45, 81-90.

Wolkow, P.M., Sisler, H.D. & Vigil, E.L. (1983). Effect of inhibitors of melanin biosynthesis on structure and function of appressoria of *Colletotrichum lindemuthianum*. Physiological Plant Pathology 22, 55-71.

Wood, R.K.S. (1967). Physiological Plant Pathology. Blackwell, Oxford & Edinburgh 570 pp.

Wood, R.K.S., Ballio, A. & Graniti, A. (1972) Phytotoxins in Plant Disease. Academic Press, London & New York. 530 pp.

Wyman, J.G. & Van Etten, H.D. (1978). Antibacterial activity of selected isoflavonoids. Phytopathology 68, 583-589.

Wyman, J.G. & Van Etten, H.D. (1982). Isoflavonoid phytoalexins and non-hypersensitive resistance of beans to *Xanthomonas campestris* pv. *phaseoli*. Phytopathology 72, 1419-1424.

Wynn, W.K. & Staples, R.C. (1981). Tropisms of fungi in host recognition. In 'Plant Disease Control, Resistance and Susceptibility' (Staples, R.C. & Toenniessen, G.H., eds), pp. 45-69. J. Wiley and Sons, New York.

Yang, S.L. & Ellingboe, A.H. (1972). Cuticle layer as a determining factor for the formation of mature appressoria of *Erysiphe graminis* on wheat and barley. Phytopathology 62, 708-714.

Yoder, O.C. (1973). A selective toxin produced by *Phyllosticta maydis*. Phytopathology 63, 1361-1366.

Yoder, O.C. (1976). Evaluation of the role of *Helminthosporium maydis*, race T toxin in Southern Corn Leaf Blight. In 'Biochemistry and Cytology of Plant-Parasite Interaction' (Tomiyama, K., Daly, J.M., Uritani, I., Oku, H. & Ouchi, S., eds), pp. 16-24. Kodansha Ltd, Tokyo; Elsevier, Amsterdam.

Yoder, O.C. (1980). Toxins in pathogenesis. Annual Review of Phytopathology 18, 103-129.

Yoder, O.C. (1981). Genetic analysis as a tool for determining the significance of host-specific toxins and other factors in disease. In 'Plant Disease Control, Resistance and Susceptibility' (Staples, R.C. & Toenniessen, G.H., eds), pp. 3-11. J. Wiley and Sons, New York.

Yoder, O.C. & Gracen, V.E. (1975). Segregation and pathogenicity types and host-specific toxin production in progenies of crosses between races T and O of *Helminthosporium maydis* (*Cochliobolus heterostrophus*). Phytopathology 65, 273-276.

Yoder, O.C. & Sheffer, R.P. (1969). Roles of toxin in early interactions of *Helminthosporium victoriae* with susceptible and resistant oat tissues. Phytopathology 59, 1954-1959.

Yoshikawa, M., Keen, N.T. & Wang, M.C. (1983). A receptor on soybean membranes for a fungal elicitor of phytoalexin accumulation. Plant Physiology 73, 497-506.

Yoshikawa, M., Yamauchi, K. & Masago, H. (1978). Glyceollin: its role in restricting fungal growth in resistant soybean hypocotyls infected with *Phytophthora megasperma* var. *sojae*. Physiological Plant Pathology 12, 73-82.

Young, D.H. & Pegg, G.F. (1982). The action of tomato and *Verticillium albo-atrum* glycosidases on the hyphal wall of *Verticillium albo-atrum*. Physiological Plant Pathology 21, 411-424.

Ziegler, E. & Pontzen, R. (1982). Specific inhibition of glucan-elicited glyceollin accumulation in soybeans by an extracellular mannan-glycoprotein of *Phytophthora megasperma* f. sp. *glycinea*. Physiological Plant Pathology 20, 321-331.

CHAPTER 8. RESISTANCE AND PATHOGENICITY: EPIDEMIOLOGICAL AND ECOLOGICAL MECHANISMS

M. J. JEGER and J. V. GROTH

8.1. INTRODUCTION

Resistance in host plants and pathogenicity in the organisms causing disease are inseparable and defining attributes of the disease interaction. These matching attributes have been intensively studied in two major contexts: to meet the often pragmatic needs of plant breeders; and to understand the host-pathogen relationship, perhaps the most subtle and intimately-linked area of inter-organismal biology. Advances in these two contexts do not always coincide, however, and many achievements of plant breeders have been made in the absence of detailed knowledge of the underlying physiological and genetic relationships. Shaner (1981) describes the accomplishments of plant breeders with regard to crops grown in the US. About 650 diseases of about 100 major crop plants are tabulated; some level of resistance to about 400 diseases is reported.

By contrast with plant breeding and host-pathogen relationships, there have been few intensive studies of disease in functioning populations. This is due, in part, to the origins of modern-day plant pathology in schools of botany. There has always been a communications barrier between plant pathology and most areas of population biology: especially the areas of ecology and population genetics. Accordingly, epidemiology, the study of disease in plant populations, has developed largely within the compass of plant pathology, and outside of the mainstream of thought in population biology. This has left a legacy of confused, and even contradictory concepts and terminologies. In recent years there has been some evidence that the situation is changing and, indeed, that the epidemiology of plant disease can make a unique contribution to the mainstream of thought in population biology.

At the population level, the details of the means by which a host population and a pathogen population reach a certain level of compatibility need be known only to a limited extent. If a particular example of an interaction between a host entity (a line or clone) and a pathogen entity (an isolate or culture) becomes fully understood in biochemical or histological terms, the information will probably not be of particular use in contributing to our understanding of inter-population interactions. A much broader view of population processes is needed. Functional definitions of resistance in the host and pathogenicity in the pathogen are strictly in terms of fitness. By this criterion, any mechanism of the host that curtails pathogen reproduction contributes to its resistance. Any mechanism of the pathogen that enhances reproduction on the host can be considered an increase in pathogenicity.

In practical terms, more detailed information about the nature of resistance or pathogenicity is often needed in order to account for their effects more accurately. For example, the phenomenon of adult plant resistance to rust fungi in cereals might have been overlooked, had not a discrepancy been noted between seedling reactions in the greenhouse and the level of compatibility of the same host-parasite combinations in field epidemics (Aamodt, 1927). Likewise, long latent periods were shown to correlate with, and perhaps account for, a large part of the field resistance of certain barley cultivars to leaf rust (Parleviet, 1976). Had the latent period been ignored, such resistance would have been incompletely understood.

Functional resistance can therefore take a variety of forms. The oversimplified division of resistance into constitutive (physical) and induced (physiological) mechanisms is a reasonable beginning, but combinations of mechanisms are common. To continue the oversimplification, the essential difference between the two mechanisms, in population genetical terms, is that the pathogen population is unable to adapt to physical resistance because variants that might allow this to occur do not exist in the short term. By contrast variants that allow adaptation to physiological resistance may occur. Their occurrence, however, does not ensure that they will predominate; there are other factors which influence whether selection will ultimately permit the variants to render the resistance worthless.

Because of the wide disparity in the corpus of knowledge between epidemiology and population biology, the treatment in this Chapter is necessarily rudimentary, fragmentary, and speculative. This, we believe, is a true reflection of the current situation; hence there is little of the well-rounded and balanced view. Much has to be achieved before we can catch up with the advances made in other areas of population biology. In the meantime, we have to convince ecologists, population geneticists, and modellers, of the unique and challenging problems of disease in populations.

The approach we shall take is from the standpoints of population dynamics and genetics. Our goal is a better understanding of disease epidemics and the use of this understanding for managing disease to man's best long-term advantage. Our bias will largely be with agricultural systems in which the host is controlled or manipulated by man, and where the evolution of plant pathogens is managed as a consequence (Knott, 1972). The evolution of crop plants and their pathogens will be emphasised much less. We will try to avoid the usual biases to be found in treatments of the subject: i.e. temperate zone crops, fungal diseases, biotrophic pathogens, annual cropping systems, and airborne dissemination of infective propagules. This combination of bias may be important in western agriculture, but is hardly representative of world agronomic crops or the diseases that afflict them. On this limited basis, however, generalizations have been made, with little regard for the totality of epidemiological and ecological processes that occur. We consider first the nature of resistance and pathogenicity in populations and the difficulties in definition and interpretation. Next we consider how a combined population dynamic and genetic approach to plant disease epidemics may develop in the future. We then review several topics that have received some attention recently: race frequency dynamics, heterogeneous crop populations, components of resistance, and fungicide insensitivity. The last two sections examine how an overtly ecological perspective may assist understanding, and the implications for disease in agricultural crops.

8.2. RESISTANCE AND PATHOGENICITY

There is now an almost overwhelming literature on proposed categories of resistance and pathogenicity. Many recent contributions have been made (Parlevliet, 1977; Parleviet & Zadoks, 1977; Nelson, 1978; Parlevliet, 1981; Johnson, 1981; Fleming & Pearson, 1982; Johnson, 1983; Parlevliet, 1983; Person et al., 1983; Robinson, 1983). Much of this literature can be considered as equivalent to the medieval theological debate as to the number of angels that can sit on the point of a needle. Where no concensus emerges, and definitions and qualification of definitions merely proliferate, then it can be argued that these categories are serving little purpose. We do not intend to add to the proliferation of definitions, nor to indulge in endless citations of the various protagonists, but shall extract the definitions we consider appropriate at the level of biological organization with which we are concerned.

8.2.1. <u>Definitions and levels of biological organization</u>. Zadoks & Schein (1977) have expressed a general view of processes at the various scales or levels of biological organization. They illustrate how processes are divided into subprocesses which are themselves further subdivided. By their definition, processes which are at the submicroscopic level (molecular to organelle) are not within the scope of epidemiology. Epidemiology and population genetics operate at similar levels, and this restriction can be extended to the latter discipline. This is not to say that there is no relevance of submicroscopic processes to population genetics; knowledge of the functional mechanisms of genetic variation is essential for the prediction of rate and direction of genetic change. We are, however, concerned with interacting host and pathogen populations, and must first define what we mean by population. A classical definition is that of any freely-reproducing and inter-breeding set of individuals. This hardly applies to a population of crop plants where the function of the plant, as imposed by man, has little to do with reproduction and survival. Nor need it apply to pathogens in which sexual reproduction either rarely occurs or is unknown. A simpler definition of a population is some local aggregation of individuals of a single species in time and space: for example, in the case of crops, a field of wheat or some other well-defined geographical entity.

Resistance in a host population is the ability to restrict disease development. Pathogenicity in a pathogen population is the ability to promote disease development. When the host and pathogen populations are homogeneous, and with no environmental heterogeneity, then resistance and pathogenicity at the population level are basically equivalent to the corresponding phenomena at the level of individual plants. Where there is heterogeneity in time and space then resistance and pathogenicity are not equivalent at the two levels of biological integration. A hierarchical view must be taken of the processes involved. In brief, we distinguish between explanation and description of biological processes. Explanation of processes at one level requires a description of subprocesses at a lower level. Thus an explanation of resistance and pathogenicity in populations requires a description of resistance and pathogenicity at the individual plant level. The basic premise we start with is that a description of resistance at the whole plant level is necessary but not sufficient for explanation of resistance at the population level.

Accordingly, as heterogeneity of at least one element of the host, pathogen and environment triangle is the norm, the populations have to be partitioned into more or less uniform classes of interaction. The disease interaction for each class must then be described and the consequences at the population level deduced. Thus, for example, resistance in populations may be characterized by a value in the range 0 to 1 with strict inequalities at either end of the scale (both 0 and 1 indicate homogeneity of host, pathogen and environment). It is not so easy, however, to characterize the disease interaction at the level of the individual plant. In many cases, genetic analysis is not possible in the pathogen, and the phenotypic expression of disease must suffice.

Resistance and pathogenicity at the level of individual plants is defined by the ability of a pathogen to infect, colonize and reproduce in an individual plant. In principle, there is no reason to consider any one of these components as qualitatively different from any other: they are equally important in explaining events at the population level, although the total reproduction of the pathogen in the host best integrates these components (Johnson & Taylor, 1976). By the same argument as before, a description of events at a lower (cellular) level is necessary but not sufficient to explain infection, colonization and reproduction at the individual plant level. These aspects are considered elsewhere in this book (Chapters 2, 3, 6 and 7).

8.2.2. Classification of resistance and pathogenicity. Structure in host and pathogen populations, and environmental heterogeneity, can be recognized or at least inferred from the disease phenotypes observed in an experiment or survey. The idea of partitioning leads naturally to an interactive classification of disease phenotypes and to the degree of specificity expressed in the interaction. Resistance and pathogenicity can be classified according to the level of specificity. In some cases this may effectively be a binary trait but only rarely. There are many ways, other than level of specificity, to classify resistance and pathogenicity: whether few or many loci are involved; whether the alleles at each locus are 'weak' or 'strong'; whether there are quantitative or qualitative effects; and whether different patterns of epidemic progress are observed in time and space. We are concerned with a classification that explains resistance and pathogenicity in populations. The epidemiological classification can be rejected as this only describes but does not explain resistance and pathogenicity in populations. Similarly, the classification of resistance as durable or ephemeral is merely a description at the population level. At lower levels of integration, resistance has been classified in terms of gene function and whether this is beyond the pathogen's capacity for micro-evolutionary change. As, in most cases, we have no idea what genes for resistance or pathogenicity do (or, indeed, what other function they serve in plants or pathogens), this classification can also be rejected. Indeed, the existence of different time scales in evolution (e.g. as in macro and micro evolution, sensu Robinson, 1983) can seriously be questioned, and the concept is clearly not testable in any conventional sense.

The various genetic classifications are often appropriate and these have implications for population processes that arise largely from biometric considerations. Several problems remain, however, in applying this information to populations. In particular, there is no easy correspondence between classifications in terms of specificity and those in terms of the number or quality of genes. The terms vertical and horizontal resistance were first introduced by Vanderplank (1963). They have since been modified and expanded (Robinson 1976) as largely abstract concepts to overcome the lack of correspondence between specificity and the other classifications. Similarly, virulence and aggressiveness are qualifying attributes of pathogenicity. In fact the additional information conveyed by these terms is largely superfluous to the task of explaining resistance and pathogenicity at the population level. The more important need in this task is a description of specificity, in particular whether of differential interaction or of constant ranking, although this is an over-simplification to be expanded on later. Where genetic analysis is possible in both host and pathogen then an explanation of the specificity can also be sought.

Differential interaction between host and pathogen. The basic procedure for describing specificity can be illustrated by considering a simple example: host and pathogen are partitioned into two classes (two cultivars and two isolates), and environmental constancy is ensured. The disease phenotypes that result are denoted by a value ranging from 0-1, and are measured with respect to one of the components of the disease cycle: infection, colonization, or reproduction. It should be stressed, of course, that results may differ according to the component measured. The possibilities that arise in an analysis of variance are (Figure 8.1):

a) no main effects with no interaction (constant ranking);

b) main effect due to host with no interaction (constant ranking);

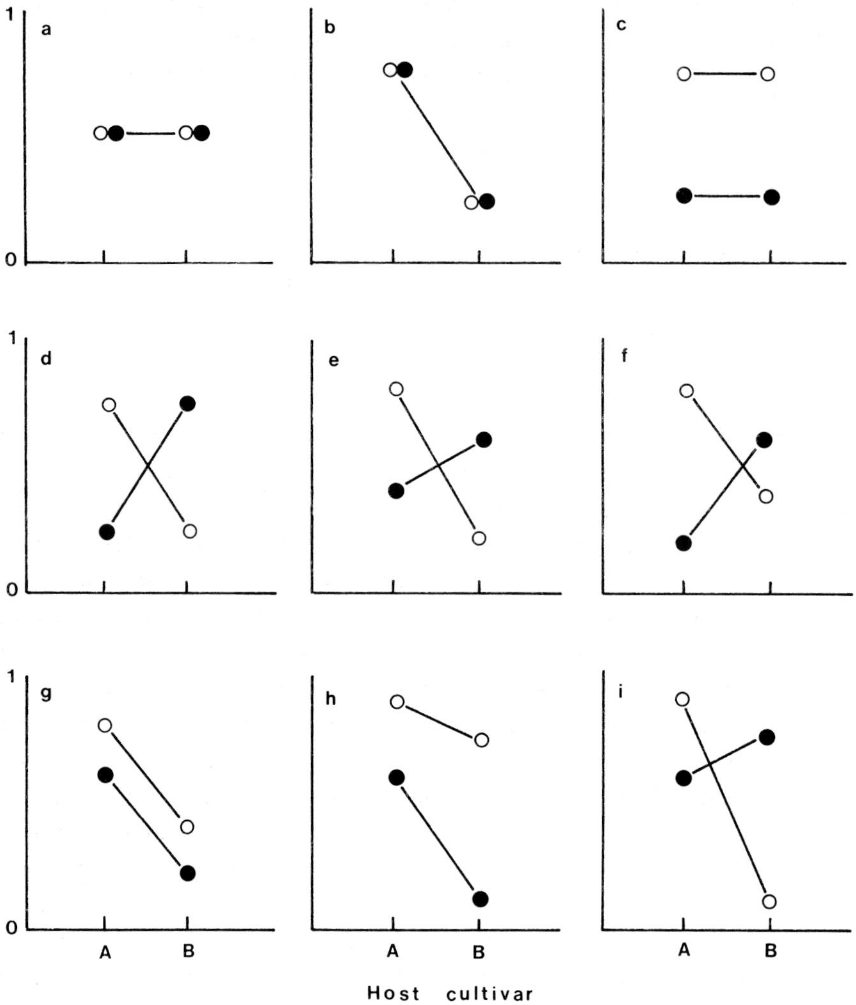

FIGURE 8.1. Analysis of disease phenotypes. Schematic representation of possible outcomes of a replicated experiment in which two cultivars (A and B) are inoculated with two isolates of a pathogen (shown as ○ and ●) and the outcomes are measured on a quantitative scale 0-1. See text for discussion of outcomes a) - i). The outcomes shown are not exhaustive: for example in e) the mean value for cultivar A is greater than that for B, and there is reversed ranking of pathogen isolate (with respect to A, isolate ● is lesser than isolate ○ ; with respect to B, isolate ● is greater than isolate ○ . However, the ranking of host cultivar can be reversed (as shown) or constant. Similarly, in f) the ranking of pathogen isolate can be reversed (as shown) or constant.

c) main effect due to pathogen with no interaction (constant ranking);
d) no main effects with interaction (reverse ranking);
e) main effect due to host with interaction (reverse ranking of pathogen);
f) main effect due to pathogen with interaction (reverse ranking of host);
g) main effect of host and pathogen with no interaction (constant ranking);
h) main effect of host and pathogen with interaction (constant ranking) and
i) main effect of host and pathogen with interaction (reverse ranking).

The outcome typified by h), in particular, shows the kind of methodological problem that can arise in analysis of even simple systems. An interaction can certainly be statistically significant at some accepted probability level, but is of a non-parallel nature rather than a reversal in ranking and constant ranking is maintained. The outcome observed also depends on the error variance of the experiment and the scale analyzed, particularly where a transformation is used. It can readily be shown that disease phenotypes which give constant ranking on an untransformed scale may give an interaction, even a reversed ranking, on a transformed scale (Scott et al., 1980). The scale analyzed should be appropriate statistically but should also, ideally, make sense epidemiologically. The procedure has at least one advantage, however, in that the null hypothesis of constant ranking can be rejected (except in outcome h) by the positive result of a significant interaction. This means, in effect, that the original hypothesis of a differential interaction, or 'gene-for-gene' system, can be accepted. In the literal sense, however, the latter demands further corroboration by genetic analysis to be acceptable. The problems in using infection type, reactions, or disease phenotypes to infer possible gene-for-gene relationships are well illustrated in the correspondence of Browder & Eversmeyer (1982) and Knott & Johnson (1983).

The main problem with the procedure described above is that, in the absence of genetic studies of host and pathogen, the classification of resistance and pathogenicity as specific or non-specific depends upon the error variance and hence the particular cultivars and isolates included in the test. Hence, if vertical resistance is taken as synonymous with differential 'gene-for-gene' resistance, then it cannot be an attribute of any individual cultivar. Similarly, vertical pathogenicity (or virulence) cannot be an attribute of an individual pathogen isolate. If a recent paper (Norelli et al., 1984) is examined, then it is possible to find host and isolate combinations in which several of the possibilities listed above can be found.

According to Robinson (1976), resistance and pathogenicity conditioned by a gene-for-gene relationship is equivalent to vertical resistance and pathogenicity, and conversely. Also, all (but not only) vertical resistance and pathogenicity is inherited oligogenically. It follows that all resistance and pathogenicity conditioned by a gene-for-gene relationship is inherited oligogenically. Parlevliet (1978) showed that polygenic resistance in the host can give gene-for-isolate relationships, and postulated a true gene-for-gene relationship. If this were so then one or both of Robinson's hypotheses must be rejected. We prefer to take the position that the concepts of vertical resistance and pathogenicity are serving no useful purpose. Resistance and pathogenicity may or may not be conditioned by a gene-for-gene relationship. It may also be conditioned by few or many genes. The former is of most importance in explaining resistance and pathogenicity at the population level; the latter is of most importance for plant breeding and in considering evolutionary change.

The gene-for-gene relationship has been a most useful benchmark against which hypotheses could be erected and tested. It has been used in a wide variety of ways, sometimes to indicate a locus-for-locus and hence an allelic system, and sometimes not. Unfortunately, it has been considered by many to be self-contained and sufficient, resulting in considerable over-simplification in assessing its role in host-parasite co-evolution and in modelling the stability of such systems.

Two points should be made about gene-for-gene recognition. Firstly, whether specific gene-for-gene recognition occurs in polygenic systems is intrinsically interesting, but Parlevliet & Zadoks (1977) have presented a model which suggests that it may be very difficult to establish that such recognition exists. In additive cases of resistance and pathogenicity, the importance of specific recognition, as compared with strongly interactive, monogenic examples, may be diminished. Additivity and variance in main effects will still be the predominant sources of variation whether specific recognition exists or not.

Secondly, it is tempting, but inappropriate, to consider only the well-known binary traits that have been so prevalent in our thinking about resistance and pathogenicity, when developing models of host-parasite co-evolution, or when considering the value of single-gene resistance in breeding programmes. In pathogens, one should never confine oneself to single-gene changes. Since the fitness of a pathogen isolate is a function of the entire organism, thus including its whole genome, the effects of a single gene change can never describe, sufficiently, how such a gene will behave in a population.

Constant ranking. According to Robinson (1976), constant ranking of cultivars with respect to pathogen isolates (races) and conversely, is the definition of horizontal resistance and pathogenicity. Clearly we have to dispose of the case in which there is constant ranking but with interaction due to non-parallel responses. Again constant ranking, and hence horizontal resistance or pathogenicity, is an attribute of the particular choice of cultivars and isolates and not an attribute of any one cultivar or isolate. Further, the test for constant ranking is not symmetric with that for differential interaction. The null hypothesis in this case is of a significant interaction (involving at least one reversed ranking) between cultivars and isolates. If, under the conditions of the experiment, a significant interaction results then we can only accept the null hypothesis. If there is not a significant interaction then we can reject the null hypothesis, but we cannot use this negative result to accept the original hypothesis. In other words, the hypotheses of gene-for-gene relationship and constant ranking are not of the same inferential status. Perfect knowledge of the system is required for the latter.

A further point that needs stressing is that constant ranking can occur at any level of disease. Accordingly we can envisage situations in which cumulative increases in pathogenicity and decreases in resistance occur over time, but with constant ranking maintained throughout.

8.2.3. Epidemiological characterization of resistance and pathogenicity. Given parameter values for the various components of the disease cycle (infection, colonization and reproduction), with host and pathogen populations partitioned into classes where necessary, then a disease progress curve follows as a result, provided some means to integrate the components is available (Shaner & Hess, 1978; Kulkarni et al., 1982). The disease progress curve is thus explained by the description of the components at the whole plant level. The curve can be considered as a single entity but can also be broken down into various attributes: the intrinsic rate of disease increase (Vanderplank's apparent infection rate); the asymptotic value of disease, and possibly some measure of the symmetry of the disease progress curve. The decision as to which attributes of the curve are explained by the various components is not yet clear; only the intrinsic rate of increase has received any attention to date (e.g. Louvel, 1980; Villarreal et al., 1980). Anderson

(1982) has considered various hypotheses concerning the effects of host and pathogen genotypes conditioning susceptible disease phenotypes. He suggests that the intrinsic rate of increase of disease could be used as a measure of the effect. Rigorous evaluation of these hypotheses has not yet been undertaken and the large amount of progeny testing required seems a formidable task.

Disease progress curves have been used many times as a description of resistance at the population level, usually in the context of cultivar evaluation (Thompson & Rees, 1979; Rees et al., 1979a; 1979b; Beale & Thurling, 1980a; 1980b; Politowski & Browning, 1978; Browning et al., 1978). In some cases (Griggs et al., 1978) polynomial functions have proved perfectly adequate for describing disease progress. At a higher level of integration, a description of the attributes of a disease progress curve can serve to explain resistance in communities of interacting populations, e.g. multiple hosts and diseases, weed species and pests. Note that just as a description at the whole plant level is necessary but not sufficient to explain resistance and pathogenicity in populations, so a description of disease progress curves is necessary but not sufficient to explain the response of whole communities to disease.

8.2.4. Ecological characteristics of disease in populations. What are resistance and pathogenicity in the context of ecological systems, whether agricultural or natural? Traditionally, there have been two contrasting approaches to ecology: autecology in which a single species is studied in great detail with respect to a whole range of physical and biological conditions, and synecology in which a single system such as a tropical rainforest or wheat agricultural system and the interaction between all components of the system is examined. From a modelling point of view, the former approach has led to the theoretical models of population dynamics, the latter approach has led to systems analysis, drawing upon the flow of energy and other thermodynamic analogies. A study of disease epidemics can obviously be incorporated within either of these approaches although, by contrast to microbial decomposition or nitrogen fixation, the role of pathogens in communities has more often been ignored or appended as an afterthought.

The autecological approach, in fact, differs little from a conventional epidemiological approach to disease in populations; although it is the more self-consciously ecological when natural systems are considered (Burdon, 1980; Burdon & Marshall, 1981; Burdon et al., 1983b; Oates et al., 1983). We have still to define the equivalent concepts of resistance and pathogenicity in communities. Recently Pimm (1984) used the term resistance to denote the degree to which some ecological variable is changed following a pertubation, but was not motivated by any consideration of disease. Clearly, a definition must be made in terms of inter-population interactions, and invoke the related concepts of stability, resilience and persistence, in time and space. Fleming (1982) discusses some of these concepts as they relate to plant pathosystems. Schmidt (1978) considers the mechanisms that limit epidemics in time and space in forest systems to be components of functional diversity, which he terms ecosystem resistance. Otherwise these concepts are rarely to be found in the epidemiological literature.

8.2.5. Tolerance to disease. The phenomenon of tolerance is a special case in resistance studies and deserving of a short section in its own right. We ask four simple questions: is tolerance, in principle, independent of resistance; can its expression be disentangled from resistance; is it under genetic control; and can it be manipulated in a breeding programme or management scheme?

Tolerance is a term used in a variety of contexts (Browning et al., 1978; Schmidt, 1978; Gaunt, 1981; Heijbroek et al., 1977). Clearly the possession of tolerance, loosely the ability to yield in the presence of disease (Mussel & Malone, 1979; Mussel, 1980; Mussel, 1981), is a sound evolutionary attribute for natural populations, although these will not be considered as a special case. With care, a definition of tolerance can be made at the population level that is independent of resistance (Figure 8.2,A). A cultivar may be considered tolerant if the slope of its absolute yield response with respect to some measure of disease is shallower than another reference cultivar. The measure of disease should integrate the effects of the disease during the plant's lifetime and the relationship with yield should, where possible, be put in linear form. Relating yield response, as a proportionate loss, to disease, measured as a proportion at a single point in time, are not appropriate for the reasons outlined by MacKenzie & King (1980), and Gaunt (1981). It is clear that this definition of tolerance is independent of any proposed for resistance and moreover there appears to be no matching phenomenon in pathogen populations. Epidemiological and genetic studies, however, have not progressed far enough to determine whether tolerance defined as above can be identified and used as a breeding criterion.

At the whole plant level, a definition of tolerance to the pathogen

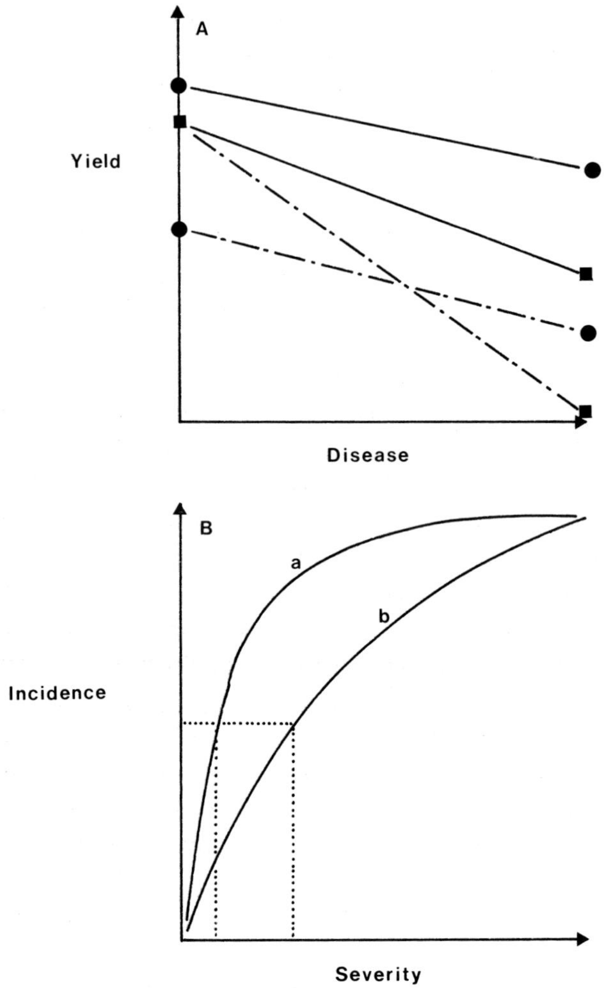

FIGURE 8.2. A). An example of tolerance to plant disease. Cultivar a (●) has different yield potentials in two different locations or years (shown as ——— and —·—-) but an equivalent yield response with disease. Cultivar b (■) has the same yield potential in the two locations or years but a consistently steeper yield response with disease. Cultivar a can be described as more tolerant to disease than cultivar b over the given range of disease.

B). Incidence-severity relationships showing tolerance to the pathogen. Cultivar a has a much lower severity of disease (e.g. lesions/leaf; proportion of leaf surface diseased) than cultivar b for any given incidence (proportion of leaves diseased). Cultivar a can be described as more tolerant to the pathogen over the given range of incidence.

(Shafer, 1971) can be made but does lead to some ambiguity. Cultivars are known to differ in incidence-severity relationship and this can possibly be related to tolerance (Figure 8.2,B). For example, comparing two cultivars at the same level of incidence: the cultivar with the lowest severity might be considered the more tolerant of the pathogen. Little consideration seems to have been given to mechanisms underlying different incidence-severity relationships until recently (Seem, 1984).

We conclude that although potentially important in agricultural and natural systems, the concepts of tolerance to the disease, or to the pathogen, are still too ill-defined to assist in either the understanding or management of disease epidemics and their consequences.

8.2.6. Disease escape. This subject was reviewed, comprehensively, by Agrios (1980). Escape has normally been considered as a distinct phenomenon to true resistance (Parlevliet, 1979). However, in terms of the definitions adopted in this Chapter, i.e. the ability to restrict disease development in a host population, this is hard to justify - especially where the host and pathogen characteristics conditioning disease escape are under direct genetic control. Thus earliness in maturation of the sexual stage of *Venturia inaequalis* in overwintering leaves of some cultivars and lateness in bud development may result in consistent escape from primary scab infection if these traits are combined in the same cultivar (Jeger *et al.*, 1982b).

There are many other examples of disease escape in a host, some of which do involve the lack of exposure to a pathogen, and some which are barely separable from resistance. There are those who would contend that mechanisms which are known to permit the individual plant to escape infection, and hence disease, should be categorized separately from resistance. There are perhaps three reasons why this is not useful or advisable. The first is the practical one that such mechanisms are already being used to protect plants from diseases and insects; plant breeders are able to transfer them in the same manner as they transfer active resistance, and would not regard these mechanisms as particularly distinctive, especially since crops possessing such mechanisms are as well protected as any. Secondly, from the population standpoint, a degree of escape can have identical effects on epidemiological parameters of disease progress as a level of 'active' resistance. The third reason is related to the second; usually our knowledge of the mechanism of resistance does not permit us to differentiate resistance from escape mechanisms. In some cases, such as that of the cereal smuts, the two may not be objectively separated, either in whole plants or in plant parts (Groth *et al.*, 1976). To go one step further, if resistance truly lacks specificity, it must curtail disease in a manner that is beyond the biological capacity of a pathogen population to adapt to or overcome - or at least our ability to measure this change. It is entirely possible that examples of apparently nonspecific resistance are mainly cases of cryptic escape, and that truly 'active' resistance is characterized by the capacity of the pathogen slowly or rapidly to overcome it. The important point here is that our collective knowledge of such matters is insufficient to reject such a possibility out of hand. We therefore caution against any strong attempts to isolate escape mechanisms, where understood as such, from other kinds of disease-curtailing mechanisms. The kinds of resistance to disease that operate in plants are numerous, and their numbers and limits are as yet unknown and unclassified.

8.3. POPULATION DYNAMICS AND THE GENETICS OF EPIDEMICS

This section forms a major part of this Chapter; an attempt will be made to reconcile two, traditionally different, approaches to population biology. This is a somewhat ambitious attempt in that neither approach is well developed in theoretical and experimental studies of plant disease epidemics. Our approach will be simplified somewhat by mainly considering agricultural systems in which crop evolution is prevented. The consequences of cultivar and fungicide (or other pesticide) usage in agricultural crops are, of course, reflected in evolution of the pathogen, and hence in the disease phenotypes that result. We first consider the development of population dynamic models in epidemiology and then show how some aspects of population genetics may be incorporated in dynamic models.

8.3.1. *Developments in population dynamic models*. The modelling of plant disease epidemics began with the pioneering, if idiosyncratic analyses of Vanderplank (1963). Idiosyncratic in the sense that virtually no reference was made to population dynamic models in ecology, or to epidemiological models of human diseases. The analogies made were with some ill-founded economic and financial models. The one common link between his analyses and ecological modelling was the logistic equation which, despite the disavowal by Vanderplank (1982), has always been considered as an extremely useful starting point in population models (Jeger, 1985a). Part of the problem with modelling plant disease epidemics has been the insistence of epidemiologists in measuring disease as a proportion rather than as absolute quantities (either dimensioned or as numbers). This insistence has led to some confusion in adequate definition of the populations being modelled. The essential feature of disease epidemics is the interaction of host and pathogen populations, and expressing one as a proportion of the other only obscures the underlying processes.

The logistic equation has been useful for obtaining insight into epidemic processes and, to some extent, resistance and pathogenicity in the interacting populations. When disease is measured as a proportion there are only two parameters available for characterizing resistance and pathogenicity. These parameters are the intrinsic rate of disease development, and the initial condition required for a particular solution of the equation - usually the initial amount of disease. The lumping of components such as infection and reproduction into the one rate parameter is quite valid for some purposes, but should be accepted only as a first attempt, or as a standard of comparison for subsequent modelling.

In particular, the weaknesses of the logistic equation are that the events characteristic of host-pathogen interactions, i.e. a discrete time lag before sporulation and a finite period of infectiousness, are not accounted for (Jeger, 1985a). Also, host growth and development and environmental influences on host and pathogen are not readily accounted for in the logistic equation. In general, the problem that has not adequately been defined in epidemiological models is to incorporate age or stage-structure into the equations. This can be achieved in models such as Vanderplank's differential-difference equation or linked differential equations (Jeger, 1985a; 1985b), but these solutions remained poorly developed in epidemiology. Some population dynamic models have been proposed which account for inter- and intra-specific interactions in disease epidemics by using forms of the Lotka-Volterra equations (Barrett, 1983a; Skylakakis, 1982a). Where a model is available for integrating the various components of the disease cycle, for each class of disease phenotype, then the dynamics of disease epidemics and the time trajectories of variation can be followed. Such models, however, are essentially static from the evolutionary point of view. The parameter values are usually constant, the various classes of interaction are known beforehand, if occurring at different frequencies, and there is no allowance for mutation and selection within the pathogen population. Hence population dynamic models, amended in the simple way described, are unlikely to predict, either qualitatively or quantitatively, the dynamics of pathogenicity genes or the associated disease phenotypes. Other aspects of population dynamic models of epidemics are discussed by Waggoner (1977), Rouse (1985), and Jeger (1985a; 1985b).

8.3.2. Genetic variation and population dynamic models. There have been several mathematical models that describe the dynamics of gene frequencies in interacting populations (Jayakar, 1970; Yu, 1972), but few attempts to incorporate genetic variation, either directly or indirectly, into population dynamic models. Basically there are five questions that can be posed:

a) how do population numbers affect population dynamics;

b) how does genetic variation affect disease dynamics;

c) how do disease dynamics affect genetic variation;

d) how do population numbers affect rate of genetic change, and

e) how does genetic variation affect rate of genetic change?

The first question is, of course, the basic problem of population dynamics as posed, in the case of interacting populations, by the Lotka-Volterra equations. The second and third are converse questions of the utmost importance and have been discussed to some extent. The fourth and fifth represent the special cases of density-dependent and frequency-dependent selection.

The influence of genetic variation in population ecology was considered in detail by Birch (1960), mainly in the context of animal populations. Natural selection tends to maximize the intrinsic rate of increase for the environment in which a species lives. The fitness of a genotype is defined by this criterion. This does not mean that birth-rates are maximized, indeed in some special cases they can be decreased, nor that mechanisms cannot evolve which ultimately limit population size irrespective of the rate of increase. Birch (1960) gives many examples of these mechanisms.

A rather different view was taken by Ginzburg (1981). If genetic variation in the population is not taken into account then the rate of change of a population responds immediately to any environmental change on the characteristic time scale being considered. One effect of genetic variation is that a population has 'inertia' and there is a delay in its response to change. The degree of inertia is related to the intensity of natural selection that occurs immediately after the environmental change. Hence natural selection has a direct effect on the population growth rate. The importance of genetic variation to populations experiencing changing environments, according to this view, is more for damping population fluctuations than as adaptations to environmental heterogeneity.

Ginsberg (1981) introduced a new variable into a population growth model to represent the inertial effect:

$$\left.\begin{aligned} \frac{dN}{dt} &= N\,(f\,(E) - x) \\ \frac{dx}{dt} &= g\,(x,\,E) \end{aligned}\right\} \qquad (1)$$

where $\underline{N}$ is population size, $\underline{E}$ represents the environmental parameters, $\underline{x}$ is the imbalance. The form of the function $\underline{g}$ is determined by the genetic model assumed. In a constant environment $\underline{x}$ tends to zero and the growth rate is a function of the environment ($f(\underline{E})$) alone. Hence many genetic variables are represented by the one extra variable. Ginzberg gives an example of the adaptation of a diploid population to a single environmental change. The adaptation is the change in average fitness of the population. It was found that the population can move from an initial to a new equilibrium structure at a rate determined by the relative values of the function parameters. Relationships are then derived, in the case of a one locus-two allelle polymorphism, for the average fitness of the population and the intensity of selection. Ginzberg defines all environments in which a population increases or decreases exponentially as stationary and, by so doing, excludes questions relating to initial conditions and concentrates more on the usually fewer factors causing change. Ginzberg's is one of the first attempts to look at the effect of genetic variability on population dynamics at the macroscopical level.

Antonovics (1976) discusses the development of theory in which ecological and genetic factors are considered in concert, and considers that the theory is applicable to plant populations. He also puts in intuitive terms the idea of population responses to stresses caused by new environments. Genetic change manifests itself through population dynamic (demographic) processes and, conversely, ecological change results in genotype-dependent life history shifts. Ecological and evolutionary change are thus inseparable. Intraspecific variation can be an important influence on population dynamics, even by comparison with interspecific influences.

Barrett (1983b) points out that even where models of host-parasite relationships have been proposed (e.g. Jayaker, 1970; Yu, 1972; Sorarrain et al., 1979), host models have considered only genetic changes and ignored population size changes. Again, he stresses that approaches fusing both population genetics and ecology are necessary: evolutionary change modifies the population dynamics, and conversely population dynamics affect evolution of host and parasite. Anderson & Arnold (1983) use the Lotka-Volterra equations to model a diploid population undergoing density-dependent and frequency-dependent selection. They claim that theirs is the first attempt to incorporate these effects, and obtain sufficient conditions for the existence of stable polymorphisms. Longini (1983) developed models of epidemics in genetically variable host populations and found necessary and sufficient conditions for disease to persist dependent upon both population dynamic and genetic processes.

8.3.3. Basic concepts and processes of population structure. Apart from the problem of developing a theoretical framework that will quantitatively encompass the critical aspects of the ecology and population genetics of plant pathogenic micro-organisms, we also lack sufficient descriptive tools to describe population diversity and pathogenicity dynamics. Initially, we can profitably borrow from what ecologists and geneticists have developed. Eventually we must apply this knowledge in some manner that will permit us to predict the likely genetic change and rate of occurrence when resistance is introduced into the host. Specifically we need to ask how genetic variation affects pathogenicity and disease dynamics? There are two parts to consider. First, pathogenicity dynamics is determined, in part, by disease dynamics. Secondly, the rapid rise in frequency of a new race is a function of its fitness relative to existing races but absolute fitness must be considered also. A poorly adapted parasite may not be able to predominate irrespective of its competition. Examples exist of isolates of cereal rusts being found in nature which possess the necessary pathogenicity genes to attack the host but which fail to become prevalent, presumably because they lack the appropriate complement of background genes necessary for their survival in the total environment (Roelfs & Rothman, 1976). In short, the variation in size and the absolute size of the pathogen population will influence the rate of change in frequency of pathogenicity.

Regular sexual recombination can exert dramatic influences on the amount and kind of genetic diversity of a pathogen population (Roelfs & Groth, 1980). Sexual recombination might not be expected to increase the mean fitness of a population in the short term. It is commonly accepted, however, that it can rapidly permit the assembly of highly fit genotypes. This is important to plant pathogenic micro-organisms because most of them have asexual reproduction, and hence the ability to conserve such individuals. Once these very fit individuals predominate then a highly pathogenic population - one of the necessary components of an epidemic - is obtained.

The alternative is where no effective sexual recombination is occurring. If pathogenicity is conditioned by a single gene, and it arises by mutation (which must be considered a recurring process), then the success of the individual possessing this gene depends on the adaptation to the environment, including the host, of the background genotype in which the mutation finds itself. This leads to some interesting, and as yet untested possibilities. It may generally be true that the durability of resistance in a new crop cultivar is in part determined by how different genetically that cultivar is from existing cultivars, exclusive of the presence in the new cultivar of one or more new resistance genes. If this was so, for at least a few cases, it would provide a further argument against maintaining too narrow a germplasm base over time. Similarly, durability of resistance might be improved where pathogen isolates are able to adapt very specifically to an environment, if a variety of cultural practices were used. For example, a widespread change in planting date might mean that a slightly different optimum temperature would be needed by the pathogen. This, in turn, might prevent an otherwise highly fit genotype from becoming predominant as quickly, and increase the durability of resistance.

8.3.2. Departures from Mendelian inheritance. Population genetics theory is based largely on two historically distinct areas of genetics: Mendelian genetics and biometrical genetics. Nearly all of the elementary theory begins with the assumption that traits are inherited in a simple Mendelian manner. Unfortunately, some of the premises that accompany this assumption are not or are only obliquely applicable to plant pathogens. To illustrate this point, several of the more important ways in which departures from this assumption can be seen are presented.

Functional details of the sexual cycle. Prokaryotes and eukaryotes are vastly different in the means by which chromosomes and genes are reassorted. Methods of genetic analysis for bacteria that recombine through conjugation, for example, are quite distinct from methods for higher eukaryotes. Even eukaryotic fungi exhibit a mixture of classical and unique means of genetic exchange. The parasexual cycle is the best documented of the alternative means available to the fungi. Modification of the existing structure of population genetics theory will be necessary if the parasexual cycle is to be covered. Generally speaking, alternative means of genetic recombination are not considered as efficient as the classical plasmogamy-karyogamy-meiosis that they replace. Apparently, in micro-organisms with large populations and rapid generation times such efficiency is rarely needed.

Ploidy. Classical population genetics was developed primarily with higher eukaryotes in mind. Diploidy is assumed. Plant pathogens, with some notable exceptions, are mostly not diploid. Haploidy and dikaryosis both have properties that require modifications of the theory before it can properly be applied to them. Likewise, higher levels of ploidy can strongly influence how phenomena such as gene dose and the expression of recessive traits are accounted for. Finally, fungi exhibit heteroploidy, which can be defined as the regular, although variable, occurrence of numbers of chromosomes per nucleus other than the normal euploid set (Tolmsoff, 1983).

Nuclear ratio. Since fungi have exploited the dikaryotic, as opposed to the diploid condition more than any other group of organisms, many of them are capable of phenotypic change through change in nuclear ratio. Consideration of heteroploidy above would suggest two important aspects. If there is no heteroploidy or simple chromosomal exchange between nuclei, selection can operate only in the relatively limited confines of changing the ratio of two fixed entities. When mechanisms of further variation, including parasexuality and heteroploidy are added, selection has a greater scope in which to operate. A nearly unlimited spectrum of genetically different nuclei can exist in a thallus and be selected among (Wright & Lennard, 1980).

Modification or synthesis of existing theory. Some of the above mechanisms can be accounted for theoretically by simple modifications of existing theory. In most cases, however, a new synthesis of theory may be desirable, beginning with descriptions of rather simple phenomena. A fungus thallus may be thought of as a population of nuclei or, perhaps, cells. Vegetative reproduction and competition among these nuclei can serve to change the thallus genetically, but perhaps not irreversibly. This can be manifested as a gradual change, or as in the case of colony sectoring following a large mutational event, as a sudden change. Loss of pathogenicity in imperfect fungi or in bacterial colonies maintained on agar is a good example of a notable change in a micro-organism that is sometimes gradual and sometimes sudden.

Burnett (1975) has provided perhaps the most lucid coverage of some aspects of the population genetics of fungi that primarily reproduce asexually. This development follows a notably different course than that for sexually reproducing organisms. For example, the number of generations $\underline{n}$ needed for recombination to produce an individual AB from a population of haploids composed of all Ab and aB individuals is a rather complex function if parasexual recombination or rare sexual recombination occurs. The relationship is:

$$n = \tfrac{1}{2}\left\{ \frac{\ln\left[1 + s\,\dfrac{(2+s)}{Nrx_o^2}\right]}{\ln(1+s)} - 1 \right\} \qquad (2)$$

where $\underline{s}$ is the selection coefficient against ab individuals, $\underline{x}_o$ is the initial frequency of Ab and aB, $\underline{N}$ is the population size, and $\underline{r}$ is the rate of recombination between A and B. This contrasts with the very simple statement that AB individuals will arise at the first sexual event in a sexual haploid population.

8.3.5. Fluctuating population size. One other characteristic of plant pathogenic micro-organisms must be mentioned because it severely limits the degree to which classical population genetics theory applies to such organisms. That characteristic is the striking dynamics of population size. Two aspects of the large, cyclic changes in population size will be noted. An important question is how such fluctuations in size might affect the likelihood that a gene for pathogenicity will be fixed after it arises by mutation. In cases where selection is not important, as is conceivable where specific pathogenicity is not matched by resistance in the host, Gale (1980) has shown that a population that fluctuates in size behaves more like one that is constantly small (closer to the fluctuating population at its low ebb) with respect to the likelihood of fixation due to random drift. Likewise, where selection is present and favours the rare mutant allele, as when at least a portion of the host population contains matching resistance, there is a definite relationship between probability of fixation and whether the population is increasing or decreasing. When the population is increasing rapidly, the probability of fixation is much higher for a give rare mutant allele than when it is in steady state or decreasing (Gale, 1980).

Associated with a strongly fluctuating population size in plant pathogenic micro-organisms is the low probability that a reproductive propagule is going to be successful. Most rust uredospores, for example, fail to result in a uredium. This complicates the definition and measurement of effective population size (roughly the size of an idealized population that should be substituted for the size of one whose members do not breed randomly in algebraic formulas that depend on idealized population assumptions for their validity). It also makes the definition of actual population size more subjective. If all propagules are considered as part of the population, it means that the disparity beween effective and actual population size is likely to be large. For example, if only 1 in 10^3 propagules is ever successful in infecting a plant, the base upon which a mutation rate should be considered in order to provide an estimate of _effective_ mutants per generation should be only 1/1000 of the actual number of propagules produced per generation.

The above two considerations are both germane to the accurate characterization of the population genetics of micro-organisms. They have not yet, however, become part of general population genetics theory. Special considerations are being developed regularly in the literature. It remains for those of us interested in population and ecological genetics of plant pathogens to take and adapt the theory to fit our own needs, and contribute to that theory where necessary.

8.3.6. _Sub-species terminology_. There have been several attempts to construct taxonomic criteria at ranks lower than that of the species. Races of various kinds (geographical, physiological) have been proposed which differ according to whether animals, plants or micro-organisms are being considered (Clairidge & Hollander, 1983; Conard & Radesevich, 1979; Jaenike, 1981). The 'deme' concept has been widely advocated and criticized (Briggs & Block, 1981) but seldom used in studies of plant pathogens. The most commonly accepted sub-species rank in plant pathology is _forma specialis_, which is determined solely on host species range, but is almost certainly an artefact of modern agriculture and plant breeding practices (Browning, 1981). Biotypes or strains that have not achieved the status of _forma specialis_ have commonly been described e.g. eyespot disease of cereals, couch, and ryegrass (Cunningham, 1981; Scott & Hollins, 1980). Within _forma specialis_ are physiological races defined by the reaction on known differential sets. In the wheat-stem rust system, differential sets are all cultivars of wheat, but in other systems such as apple scab, the differentials are different _Malus_ spp. Race means different things with different crops and diseases. The main question to be asked concerns the extent of variation within and between the so-called physiological races. Is the overall variation accounted for by race structure significantly greater than that within races? How do different measures of similarity, say by gel electrophoresis and by differential sets of cultivars, correspond (Burdon et al., 1983)? Are such comparisons even valid (Lewontin, 1984)?

Another trend has been to designate 'races' or strains in terms of morphology, growth and pigmentation on artificial media. Such classification does sometimes correspond to that in terms of pathogenicity in vivo (Hansen, 1983; Brazier et al., 1981) but equally there can be no correlation between cultural characteristics and pathogenicity (Asher, 1980; Blanch et al., 1981). A further use of the term 'race' arises with toxin producing pathogens and where a differential ability to produce host specific toxins is found in different isolates (Tegtmeier et al., 1982).

It should also be recognized that the terms used to describe sub-species rank differ according to the class of pathogen; particularly for nematodes and viruses. Luedders (1983) points out the vagueness of races or 'variants' with respect to nematodes. Conceptually, gene frequency is the correct way of looking at the population structure of nematodes; the problem is that genes for pathogenicity have yet to be identified. Although taxonomic criteria are somewhat different with viruses, there have at least been some estimates of the underlying commonality of genome. Van Der Walle & Siegel (1982) looked at strains of tobacco mosaic virus classified according to biological, physical and chemical attributes, as well as symptomatology, and found that typically about 15% of the genome was in common.

8.4. TOPICS IN RESISTANCE AND PATHOGENICITY

We have attempted a definition of resistance and pathogenicity at the population level and indicated how a truly population dynamic and genetic approach to epidemics may develop. We now consider several topics in resistance and pathogenicity that are currently receiving attention. Three topics are considered in detail.

8.4.1. Change and stability in race frequency. Quantitative methods of modelling changes in race frequency that permit one to account for at least some of the parameters that influence change, have been provided (Kiyosawa et al., 1975; Kiyosawa & Yabuki, 1976a; 1976b; Person et al., 1976; Groth &

Person, 1977; Leonard, 1977; Leonard & Czochor, 1980; Groth & Roelfs, 1982; Grant & Archer, 1983). A discussion of these and other parameters is found in Groth (1984). Approaches differ and comparisons are not always possible. Leonard (1977) and Kiyosawa & Yabuki (1976a; 1976b) use algebraic arguments; their models are deterministic and begin with parameters that best describe discrete generations. Kiyosawa et al. (1975), in an analysis of the breakdown of specific resistance to blast in rice in Japan use a statistical approach, bringing in stochastic methods. The objectives of these approaches differ somewhat. Deterministic models are valuable in exploring the effects of various parameters on the robustness (Fleming, 1980) of a particular model. If the model is reasonable, i.e. if it uses accurate initial assumptions (as some models do not; Yu, 1972) and relates parameters in such a way that artefactual behaviour is minimized if not obviated (May & Oster, 1976), then this process of selecting important parameters is a legitimate and valuable exercise. Sooner or later, however, stochastic methods must be used if a model is ever to have predictive value. Stochastic methods tacitly allow us to admit that we cannot account for all important parameters for whatever reason. Those parameters unaccounted for can then be at least recognized in the case of Kiyosawa's multiple regressions, as residuals. The objective of this approach is then to see whether, with an incomplete set of imperfectly measured parameters, more or better predictive ability can be realised than was possible previously. Such models also permit analysis of the relative importance of the various parameters; or, of the relative predictive accuracy of the various forms of the model. As Kiyosawa et al. (1975) state, an 'elegant combination' of stochastic analyses involving actual measurements and theoretical modelling will be necessary fully to understand the process of race frequency change.

A parameter that was shown to be important in determining the rate and extent of race frequency change was the percentage of the crop in which resistance occurred in the previous 5 years. The best estimator was always the most recent year: nevertheless, information from all 5 years contributed to the association of percent area and rate of frequency change of virulent races (Kiyosawa et al., 1975). This effect, having been shown to be important, was subsequently incorporated into the theoretical models of Kiyosawa & Yabuki (1976a; 1976b). The interaction of this parameter with the relative rate of multiplication of virulent and avirulent races on susceptible cultivars, on the assumption of some stabilizing selection or enhanced reproduction associated with avirulence alleles, is shown graphically. They conclude that stabilizing selection is the most poorly understood of the factors involved; implying that it deserves more attention. While this finding may not be too surprising, Kiyosawa and co-workers (see Kiyosawa, 1982) have shown how the complexities of population genetics may be approached in a systematic manner to identify the most critical parameters. The combination of obtaining relationships between measured variables using limited stochastic methods and then applying what is learned to more comprehensive theoretical model building is a logical step to take. Variations of the procedure will undoubtedly be used in the future. For example, once the influence of percent area in resistant cultivars is understood, and some of its interactions with other parameters worked out, the assumption that the resistant and susceptible cultivars occur at random can be relaxed. The qualitative implications of a non-random distribution of resistance have been discussed by Groth (1984): they are a function of the distribution (kind and degree of departure from non-randomness), and of the source and mobility of the pathogen, to name but two important parameters. A logical theoretical relationship can be developed, but eventually several different cases must be examined before the value of the relationship can be assessed. The scale of such events must be large, and it must be realized that examples will be hard to find. This makes the rather thorough analysis of Kiyosawa _et_ _al_. (1975) all the more necessary for those cases that are available.

8.4.2. _Heterogeneity in crop populations_. Most progress towards reconciling population genetic and dynamic approaches has been made in the area of disease in heterogeneous crop populations. This, in retrospect, is not incidental. Given the potential for variability in pathogen populations, and the ability of man to influence this variability by means of crop management and cultivar deployment, a wide range of possibilities for experimental and theoretical studies is opened up - although, as will be seen, these have not always progressed together.

The development of western agriculture has been dominated by a progressive trend to crop monoculture at all levels. The reasons for this trend, and the dangers, possibly disasters, in the widespread adoption of monoculture have been discussed many times (Harlan, 1972; Browning, 1974; Harlan, 1976; Frankel, 1977; Marshall, 1977; Trenbath, 1977; Zittelli & Vallega, 1977; Last, 1978; Barrett, 1981). Plant disease is a major component of crop vulnerability due to monoculture and the erosion in genetic base caused by its worldwide adoption.

Heterogeneity has many levels of meaning: in space and time, within and between taxa, and environmentally. Insofar as interactions with the environment are a characteristic feature of agricultural crops, then heterogeneity can be said to be the norm - even when the underlying genome is uniform - due to the spatial and temporal variation in environment. It is the quality of this heterogeneity rather than the quantity that is important although very few attempts have been made to characterize this property.

Traditionally, the use of mixed or multiple cropping in the tropics, or rotations in temperate-zone agriculture, has ensured some level of heterogeneity. There have been many agronomic studies on mixed cropping systems but very few have looked specifically at disease (Trenbath, 1974). By contrast there have been several studies on disease in heterogeneous natural populations (Browning, 1974; Dinoor, 1977; Schmidt, 1978; Burdon, 1978; 1982; Segal *et al.*, 1980; Browning, 1981) and agricultural crops (Browning & Frey, 1969; Wolfe & Barrett, 1977; Burdon & Shattock, 1982). There have not been a great number of experimental studies in heterogeneous crop populations other than cereals (Browning & Frey, 1969; Wolfe & Barrett, 1977). Most of the studies with cereal crops have mainly been concerned with air-borne diseases. The literature concerning disease in non-cereal crops, soil-borne diseases and non-fungal pathogens, was reviewed by Jeger (1979). Few studies have attempted to take the body of theory that has been developed and specifically test some of the hypotheses (Leonard, 1969a; Luthra & Rao, 1979).

Various mechanisms for the amelioration of disease caused by specialized pathogens have been considered: a reduction in the amount of inoculum produced within the mixture or multiline, an increase in the average distance between susceptible plants in the mixture or multiline, a spore-trapping effect of resistant plants in the mixture or multiline, the induced resistance of otherwise susceptible plants by spores with non-matching pathogenicity, and a reduced pathogenicity of spores capable of infecting both components of the mixture. These mechanisms have been discussed in detail, but only qualitatively (Burdon, 1978; Burdon & Shattock, 1982). They have yet to be posed in quantitative terms. The problem with this approach is that the hypothesized mechanisms can rarely be investigated in the field and there is a danger that they will remain convincing but untestable arguments. One modelling approach that should prove useful is to quantify the various mechanisms that are known to occur in mixtures and to examine whether or not these do result in amelioration of disease and, if so, the relative contribution of each mechanism.

The first theoretical prediction of the dynamics of disease in mixtures was proposed by Leonard (1969a)

$$r_m = r_s + c \ln m \qquad (3)$$

in which r_m is the rate of disease development in a mixture of resistant and susceptible plants, r_s is the rate in a uniformly susceptible crop, m is the proportion of susceptible plants, and c is the slope of the relationship. Kiyosawa (1976) incorporated a disease gradient parameter into this relationship, to account for a non uniform distribution of spores, and concluded that beneficial effects of mixing were less with less dispersing ability on the part of the pathogen. In particular, the numerical value of α/μ, where μ is the mean distance between plants and α is a gradient parameter, was found to be critical at a threshold value of 0.5. Burdon & Chilvers (1982) found curvilinear relationships between the rate of disease progress and the proportion of susceptible plants.

Most early studies were concerned not with the rate of disease progress, nor with the size of the diseased population, but with the frequency dynamics of racial components in the pathogen population. The development of these models is discussed by Leonard & Czochlor (1980) and Kiyosawa (1977; 1982). Here we will discuss the long term stability of multilines and mixtures. The question has been asked whether multilines and mixtures will quickly select for complex 'super races' resulting in a boom-and-bust cycle for the singly-used multiline and mixture. A number of different approaches have been taken in modelling this phenomenon. None of the models can be considered as definitive or of sufficient sophistication to have predictive value. The following sample of approaches should serve as a review.

The earliest theoretical contribution was that of Leonard (1969a; 1969b). The empirical work and resultant specific recommendations about how multilines should behave attracted most attention. A more generalized, deterministic mathematical model, that included essentially the same variables as used by Leonard, was developed by Groth (1976). He proposed two models describing the relative reproductive ability of a complex compared to a simple race. Parameters of this model included stabilizing selection (selection against 'excess' race specific pathogenicity), the number of components in the multiline, and race complexity in terms of the number of components that a given race could attack. The more realistic multiplicative version of the model was expressed as

$$R = m\ (1 - s)^{m-1} \qquad (4)$$

Where $\underline{m}$ = the complexity of a phenotype (= number of genes for pathogenicity, $\underline{s}$ = the selection against a phenotype by virtue of possession an additional gene for pathogenicity, $\underline{R}$ = the rate of reproduction of a race as a proportion of that of a simple race. If the number of unnecessary genes decrease the fitness of a complex race multiplicatively, then maximum reproduction occurs with $\underline{m} = 1/\ln(1 - \underline{s})$ races. The reproductive disadvantage, or selection coefficient, associated with complex races declined markedly in either the additive or multiplicative model. Marshall & Pryor (1978; 1979) formulated a very similar model to that proposed by Groth, but obtained results in which the number of component lines in the multiline did affect the selection of complex races. Groth's model can be further elaborated by introducing a variable mean fitness parameter dependent upon the number of lesions produced per parent per generation (Barrett & Wolfe, 1978; Groth, 1978). The outcome then is dependent in part on the dynamics of the epidemic. A response to this model was made by Barrett (1978) and expanded on by Barrett (1980).

Barrett (1978) was the first to propose a model which described both the frequency dynamics of genes and the population dynamics of disease. The model was based on a gene-for-gene relationship between the pathogen and host in a two locus system. The host was of constant density and a mixture of the two components. The pathogen was a haploid organism reproducing sexually by producing aerially disseminated spores. The model was also applicable to an asexually reproducing diploid organism where pathogenicity was fully dominant. A discrete model was proposed, using difference equations, giving the total number of infections of each pathogen genotype and their relative frequencies. The parameters of the model were the selection coefficients, the frequency of each host phenotype, the number of daughter lesions per parent per generation, and a parameter reflecting the proportion of dispersed spores that join an airborne pool and reinfect the crop. Thus two important elaborations were developed: one is the role of rate of increase (1 + $\underline{\alpha}$ per generation) and proportion of redistributed inoculum ($\underline{\phi}$) both of which, when low, can be expected to retard the role of increase in frequency of newly arisen pathogen components. For a single locus, this is expressed as:

$$N'_{ij} = (1 + \alpha)\, W_{ij} \left[N_{ij}\,(1 - \emptyset) + \emptyset g_j \sum_j N_{ij} \right] \qquad (5)$$

where g_j = the proportion of the jth host in the host population; N'_{ij} = the number of ith propagule units on the jth mixture component one generation beyond generation possessing number N_{ij}, W_{ij} = fitness of ith pathogen genotype on j^{th} host genotype. This relationship forms the basis for the further development of expressions of frequency change of pathogen components. The parameter α which describes 'old' or 'new' inoculum is important because it brings back into the picture some important epidemiological variables, namely the absolute rate of increase of the disease, which is dependent upon the number of host components n in the mixture or multiline. The model of Groth (1976) excluded n because under the assumptions of discrete generations and very rapid increase with random redistribution of inoculum, relative rate of reproduction of pathogen components is not a function of n. As was pointed out by Groth (1978), the parameter α is important in general, since a low rate of increase, by retarding change in frequency of pathogen phenotypes could change the equilibrium point (or limit cycle) of a given system. In the specific case of highly mobile rapidly-reproducing pathogens, however, the variable probably will not be a cardinal one. The other elaboration introduced by Barrett (1978), was the effect of incomplete resistance on the rate of change in the pathogen population; which is also manifested by virtue of reduced selection pressure against pathogenic phenotypes. A discussion of the importance of pathogen redistribution during disease increase, i.e. the percent of auto vs. alloinfection of an individual plant, is also included in Barrett (1980). Barrett's model has more recently been extended in a continuous time version by Ostergaard (1983a).

Fleming & Person (1978) presented a model which incorporated at least two further elaborations. They used a continuous growth model to account for differential reproduction and hence frequency change of each pathogen component:

$$Q_i(t) = Q_i(t-1)\, e^{r_i}\ ; \ r_i = \sum_j m_j R_{ij} \qquad (6)$$

Where Q_i (t) is the amount of pathogen genotype i at time t, R_{ij} is the rate of increase of inoculum of pathogen genotype i on host j, and m_j is the proportion of the host population made up of component j. Perhaps the most useful idea presented is that of breeding effort. Using the parameters of their model, they discuss the various strategies available to the plant breeder. In an attempt to bring cost/benefit analysis into theoretical models, they are the first workers to introduce a degree of reality into the discussion of how multilines or mixtures can best be used.

Other treatments of multilines have been published. A series of similar models that differ from one another by introducing some of the parameters discussed above, as well as others, have been developed by Marshall and co-workers (Marshall & Pryor, 1978; 1979). An interesting one is that of the multiline in which resistance genes are used in various combinations in disjoint sets instead of singly (Marshall & Burdon, 1981). As might be anticipated, their model shows that this approach increases the likelihood of selecting for complex pathogen phenotypes that can attack more than one component of the multiline or mixture.

Jeger et al. (1981) proposed models describing the disease progress of non-specialized pathogens in binary mixtures of crop cultivars compared to the pure stands grown alone. The models, one in discrete and one in continuous time, predicted that disease levels in the mixed stand would always be less than the arithmetic mean of the pure stands grown separately, during the early stages of epidemic, provided that the two cultivars were ranked the same with respect to two components of resistance: infection frequency and sporulation rate. Qualitative support for this prediction was provided in field studies of the splash-dispersed pathogens Septoria nodorum in wheat and Rhychosporium secalis in barley. The full range of possibilities predicted by the model, however, was not explored. The models were later extended to cover the entire time-course of an epidemic (Jeger et al., 1982). In this case it was predicted that the relative advantage, or disadvantage, of a mixture would increase to an optimum, but ultimately the asymptotic value of disease in a mixture would be the same as the arithmetic mean of asymptotic values in pure stands. Qualitative support for this prediction has been provided by Chin (1978) in work with barley powdery mildew where the components in the mixture did not discriminate against known pathogen races. The model of Ostergaard (1983a) also includes some elements of the effects of non-specialized pathogens and is thus the most general model yet presented of disease effects in heterogeneous crop populations.

A rather different simulation approach to the spatial dynamics of disease in mixtures or multilines was developed by Zadoks & Kampmeijer (1977). Their model was constructed by considering an epidemic to progress in various compartments with dissemination from compartment to compartment introduced by means of a Gaussian distribution. Their main conclusion was that mixtures or multilines worked because of focus suppression. Analytical work on this aspect of control has not been reported.

The theoretical treatment of how multilines or mixtures might influence pathogen dynamics and genetics is, of course, useful and needed. The various approaches discussed above have all contributed something to our understanding. Opinions vary as to how best to present the models. This situation is not likely to change. Empirical testing of such models has lagged and will probably continue to lag behind model development. Subsequent models, such as that of Ostergaard (1983a), are likely to be more comprehensive and mathematically sophisticated. They should, however, prove valuable in determining the relative importance of the various parameters and allow us to discount a few.

8.4.3. Components of resistance and pathogenicity. Resistance and pathogenicity are the phenomena we wish to explain at the population level using descriptions of the disease phenotype as measured on individual plants. Usually this description has been made on the basis of infection or lesion type. Although this description does incorporate other components in the life cycle (e.g. meagre or full sporulation), more realistic means of describing the disease phenotype by quantifying components of resistance or pathogenicity that are of epidemiological importance are available. This approach has often been recommended (Berger, 1977; Parlevliet, 1979) but has not been exploited fully in either theoretical or experimental studies. Many experimental studies in which components of resistance have been measured have been described recently (Eenink, 1981; Eenink et al., 1982; Groth & Urs, 1982; McDonald & Andrews, 1982; Neervoort & Parlevliet, 1978; Johnson & Wilcoxson, 1978; Nilsson, 1981; Sharma & Heather 1979; Wilcoxson, 1981).

In terms of the pathogen population the concepts of parasitic fitness and components of fitness have been proposed (Nelson, 1979); isolates of various fungi have been evaluated for these characteristics (Gregory et al., 1981; Royer & Nelson, 1981; Hamid et al., 1982a; 1982b; Hill & Nelson, 1982; 1983). There are many problems, however, in defining and measuring fitness in relation to more conventional usage (MacKenzie, 1978; Groth & Barrett, 1980; Barrett, 1983a; Kranz, 1983; Ostergaard, 1983b; Wolfe et al., 1983).

The components of resistance studied include infection frequency (efficiency, ratio), latent period, infectious period, sporulation rate, and colony growth rate. There have now been numerous studies of various combinations of these combinations in relation to host (uniform inoculum), pathogen isolate (uniform host) and environmental interactions. Apart from the technical problems involved in making these more detailed assessments, there are two major problems in using such assessments to determine resistance or pathogenicity in populations. These problems are: the extrapolation from monocyclic assessments usually done in a glasshouse and often on juvenile plant material, and the degree of association between components. The first problem has been approached in various ways. The most appropriate, theoretically, would be to use an epidemiological model, such as the differential-difference equation of Vanderplank (1963) to integrate the components into a disease progress curve and to use this to compare cultivars or pathogen isolates. Another approach is to correlate individual components with rates of disease progress observed in the field and make some judgement as to which component is most important (Parlevliet, 1979; Parlevliet et al., 1980). Components may be used to predict rates of development (Shaner & Hess, 1978; Leonard & Mundt, 1984) and the relative importance of components (Kulkarni et al., 1982) for comparison with observed rates. A more pragmatic approach that avoids modelling preconceptions altogether is to identify new variates by using standard multivariate techniques and correlate these with independent assessments of field performance (Jeger et al., 1983). Each of these techniques has its merits and uses and further developments may be expected.

A more interesting problem is the association between components of resistance. For example, do we find that high infection frequency, short latent period, long infectious period and high sporulation rate are always associated, or can they be independent? Despite a large number of studies, this question has never satisfactorily been addressed. Teleologically, there seems no good reason why a pathogen isolate must optimize all components of fitness in order to survive. Indeed the opposite argument, sometimes termed 'prudent parasitism', can be made with regard to quasi-natural systems. The 'prudent parasitism' argument need not involve altruistic or group selection; in the long term, the most fit isolate is the one which produces the most successful progeny. This might not correspond to high pathogenicity, which can shorten its period of reproduction through detrimental effects on the host. In agricultural crops, however, especially where there has been a high level of selection for resistance, then undoubtedly an association of components of resistance can result, although this is not always complete (Parlevliet, 1979). Such artificial circumstances, however, may not be typical. Where less selection has been applied, as with relatively unspecialized diseases such as Septoria diseases on wheat, then a degree of independence between components does emerge. In a study of 40 Triticum genotypes, Jeger et al. (1983) showed that although there was an underlying factor of general resistance involving an association of components, the patterns of necrosis (interpreted as symptomatic of toxin action) and spore production were independent of this general background, and were more important in explaining the total variation in the experiment. Tegtmeier et al. (1982) have shown that an allele associated with toxin production was essential for high pathogenicity, but not for fungal growth or survival.

8.4.4. Selected topics in resistance and pathogenicity. The three topics reviewed above are undoubtedly important in considering resistance and pathogenicity in populations. Other factors, however, may confound interpretation and in several cases be equally important in determining the expression of resistance and pathogenicity. Some of these factors are now considered.

Environmental interactions. The expression of resistance or some component of resistance to plant pathogens is often environmentally sensitive (Pochard & Daubeze, 1980; Hyde, 19182; Simons & McDaniel, 1983). Disease interactions and phenotypes vary with the environment, but the nature of this variation cannot by itself be used to draw conclusions as to the level of specificity in a particular host-pathogen combination. Recently it has been shown (e.g. Cirulli & Ciccarese, 1982; Heather & Chandrashekar, 1982; Chandrashekar & Heather, 1981a; 1981b; Kolkani & Chopra, 1982) that environmental interactions can cause apparent specificities in host-pathogen interactions that are largely spurious. Conversely, the absence of environmental interactions should not be used to infer that resistance is specific or non-specific, durable or ephemeral or any other dichotomy.

The problem with measuring or observing environmental interactions, as with location or year interactions, is that the real cause of the interaction (e.g. temperature) is obscured and seldom studied experimentally. This does not mean that information on environmental interactions is not useful; plant breeders and biometrical geneticists have used such information to great effect in analyzing the heritability and other quantitative aspects of many agronomic traits. Ball (1983) demonstrates how the variation in disease among host cultivars and pathogen isolates can be shown to be genetically and not environmentally determined when multi-location and hence multi-environment trials are undertaken.

Induced resistance and susceptibility. The resistance of a plant to a pathogen can be modified by the presence or absence of other micro-organisms, some of which may also be pathogens. Intraspecific interaction with non-pathogenic races of *Puccinia striiformis* and *Uromyces appendiculatus*, on wheat and beans respectively, modified subsequent disease phenotypes with pathogenic races (Johnson & Allen, 1975). Chin (1978) found that both induced resistance and susceptibility occurred in powdery mildew of barley, although the former was more common. The phenomena of induced resistance and susceptibility may be of some importance in heterogeneous plant populations, such as mixtures or multilines, but only when races are sufficiently frequent to ensure contact (Burdon, 1978).

At a higher level, interactions between pathogen (across species, genera and even kingdoms) which modify the disease phenotype have been reported. Matta (1978) and Chapters 9 and 10 review the general case of one pathogen inducing resistance to another. The interacting pair may be viruses, fungi, bacteria or any combination thereof. Of special interest is the case where the host is systemically infected by a virus and challenged by a pathogen causing a local lesion disease. The interaction can appear to go in either direction: resistance (Varughese & Griffiths, 1983; Gessler & Kuć, 1982), and susceptibility (Maszkiewiz *et al*., 1979; Stevens & Gudauskas, 1982; 1983). The effects of virus infection, often asymptomatic and not obviously debilitating, on the response to other pathogens is an intriguing area for future research and may be a further cause of spurious specificities in host-pathogen interactions. Notice that the modification of host resistance by an inducer is different from antagonism or competition between the inducer and the challenger.

Schonbeck *et al*. (1980; 1982) found that 10-15% of all bacteria and fungi were able to induce non-specific resistance in beans to Uromyces phaseoli. The concept of parasitic epistasis (Sidhu, 1983a) is a special case in which one pathogen inteferes with the disease expression of another; its importance is most likely to be felt in breeding for resistance to soil-borne pathogens (Sidhu, 1983b).

Relationships with other plant characteristics. A problem that has occasionally received some attention arises when resistance is associated with some other plant characteristic. For example, the apparent increase in susceptibility of modern semi-dwarf cultivars of wheat to Septoria diseases. According to Scott et al. (1982), the relationship of resistance to plant height with S. nodorum was possibly due to pleitropic effects. Danon et al. (1982) concluded that there was no evidence of linkage or pleiotropy of resistance to S. tritici with height, but possibly a linkage with date of heading. In either case it is probable that the epidemiological consequences of shorter plants (with respect to splash dispersal) will contribute to an apparent increase in susceptibility independent of the inherent susceptibility of plant tissue. With some diseases it has been found that host characteristics such as cuticle thickness, which might be supposed to correlate with resistance, are in fact a poor indication of host resistance to powdery mildew (Jarosz et al., 1982).

Ontogenetic susceptibility. It has long been recognized that the classification of a plant's resistance depends upon whether the plant is at the juvenile or adult stage of host growth. This dependence may occur in two ways and in either direction. For example, there appears to be a degree of specificity in wheat seedlings to S. nodorum that is not expressed in the adult plant (Frecha, 1963). Secondly, a plant classed as resistant or susceptible as a seedling may exhibit the opposite trait when an adult (Aust & Mau, 1983; Bennett, 1981a; 1981b; Populer, 1978; Hooker & Saxena, 1971; Huang & Heitefuss, 1982a; 1982b; Kinloch & Byler 1981).

Such phenomena undoubtedly occur, but dangers must be guarded against. Firstly, there is usually an organ age (leaf, shoot, etc.) effect on resistance which can be either increased or decreased (Populer, 1978; Groth & Urs, 1982) and is clearly different from adult plant resistance. Secondly, the environment, especially when reflected in the water relations of plants, can mimic the effects of adult plant resistance to plant pathogens; for example, the powdery mildews (Ayres & Woolacott, 1980). The last confounding detail to guard against is that the contrast is truly between juvenile and adult plants, and not between greenhouse and field growth plants where discrepancies in ratings can also occur (Rahe, 1981).

At the population level, ontogenetic susceptibility in agricultural crops can be handled quite readily as individual plants are normally at the same phenological stage, and proceed through each stage synchronously. Problems in accounting for resistance may arise when, for example, in a mixed planting of perennial tree crops, there is an intraseasonal variation in shoot growth, or interseasonal variation due to replanting within an existing planting. Such aspects appear not to have been considered at all in population studies.

Within plant variability. Although the whole plant has been defined as the unit for which a description of resistance is required, and although variation between plants in populations has an important effect on epidemics, there is a further class of variability that needs to be considered. This is within plant variability and is most likely to be important in perennial tree crops. This variability can arise in several ways. A pathogen may affect more than one plant part and these parts may differ widely in susceptibility. For example, *V. inaequalis* affects leaves, fruits and shoots of apple trees with little consistency of ranking on these different plant parts (Jeger, 1981). The relative contribution of disease on different plant parts to epidemic progress has been sorely neglected.

A further source of variability arises from the modular construction of plants and the role of somatic mutation in the development of mosaics of genetic variability in long-lived individuals (Whitham & Slobodchikoff, 1981). Individual plants are not then single islands but archipelagos of similar but distinct islands which may have a profound effect on herbivores and pathogen populations.

At a lower level of integration, but one that is important especially for epidemics of bacterial diseases, only some plant cells support growth of, for example, *Xanthamonas campestris* (Stall *et al.*, 1982). It is not clear how this relates to the normal aging of cells and hence the increasing rarity of susceptible sites in the host population.

8.5. TOPICS IN FUNGICIDE INSENSITIVITY

Examination of the phenomenon of fungicide insensitivity in pathogen populations makes an interesting comparative study to host resistance and pathogenicity. For clarity, the term insensitivity will be used throughout rather than the more common if misleading terms, fungicide resistance and tolerance. The link with the previous section is based on the premise that a fungicide-treated host is different from the same host untreated. There is the possibility that mutagenesis in the pathogen population may arise by direct action of the fungicide, but in many cases this appears not to be the case and the appearance of sensitive variants is essentially a random process that is occurring at different rates depending on the nature of the pathogen, chemical compound, and mode of action of the compound. Thus there is a similarity, superficially at least, to the appearance of highly pathogenic variants in the population.

8.5.1. Case studies. There have been many studies and reviews on the topic of insensitivity; for example, by class of compound (Georgopoulos, 1977), by pathogen (Lorbeer, 1982; Grindle, 1981), by class of compound x pathogen (Davis & Dennis, 1981; Pommer & Lorenz, 1982) and by mode of action, e.g. systemic fungicides (Van Tuyl, 1977). Many of these studies have been concerned with individual isolates and used *in vitro* techniques. The validity of these results for field populations has been questioned.

In most cases, whether insensitivity is likely to arise is linked to the class of compound. Some compounds show very durable characteristics in that insensitivity has never been observed despite their being used in a wide range of crops and against different pathogens (Skylakakis, 1982b). Insensitivity to other compounds has developed very rapidly. Some authors have attempted to make analogies with monogenically or polygenically-conditioned, and with ephemeral or permanent, host resistance but this is probably premature given the lack of field information and genetic analyses of insensitivity. Selection is able to operate among genetically heterogeneous populations and what appears as ephemeral might be an otherwise permanent trait that was not fixed, and which has a slight selective disadvantage. The reader is referred to the studies cited, especially the major reviews, for further details. In particular, competition between pathogen variants, as with pathogenicity (Brown, 1975; Falahatti & Manners, 1981; 1983), has rarely been examined (Horsten & Fehrmann, 1980; McGee & Zuck, 1981; Cohen *et al.*, 1983).

8.5.2. Mathematical modelling of fungicide insensitivity. From the theoretical point of view there are several questions of interest in fungicide insensitivity. These are: what is the mutation rate towards insensitivity and at what level are mutants maintained in the population in the absence of the fungicide; what is the magnitude of selection for insensitive mutants in a treated host population; what is the relation between epidemiological factors such as size and duration of epidemic, geographical area of crop, etc., on the frequency of insensitive strains; and how best to combat insensitivity by crop or disease management? These questions can be speculated on at length, but for speculation to form the basis for an experimental programme, it should at least be formalized into testable hypotheses. For this the language of mathematics is especially appropriate. There is not a great number of mathematical models of fungicide insensitivity, and all of these are of a reasonably tractable nature. The early models were reviewed by Skylakakis (1982a): these included the models of Delp (1980), Kable & Jeffrey (1980), and Skylakakis (1981). In addition, theoretical contributions have been made by Mackenzie (1978), Skylakakis (1982b), Groth & Barrett (1980), Jeffrey & Kable, (1982) and Barrett (1983a). Most recently Levy et al., (1983) have produced a more complex model based on Lotka-Volterra type equations. From this model it is rather difficult to draw conclusions on the qualitative behaviour of the system described although the behaviour seems highly dependent upon the time-course of fungicide weathering and efficacy. More work is needed on whether this model can be linked to experimental data; whether its conclusions agree with the earlier models, and over what range of conditions they hold.

Two elaborations, that intuitively seem important, have not been considered in any detail. The first is the effect of host growth which results in the rapid production of unprotected foliage; the second is the lack of attention given to the amount of foliage actually protected and the effect of this on the asympototic value of disease increase. Finally some consideration has now been given to the management of a pest or pathogen where the trade off between economic yield and the build up of insensitivity to a pesticide are coupled (Mangel & Plant, 1983).

8.5.3. Cross resistance, selectivity and relationship with pathogenicity. In some cases, pathogen isolates insensitive to one compound may be insensitive to another. This may occur with closely related compounds, or across widely different compounds with different biochemical modes of action. Additionally, negative cross resistance can occur with related or unrelated classes of compound (Leroux & Gredt, 1980).

Of special interest is the case where an initially broad-spectrum fungicide is applied and insensitivity develops in populations of some fungal species but not in others. The result is that an initially non-specific compound may become highly specific. There is some evidence that this has occurred and can be managed for pesticides. This concept would be of most value where insensitivity developed in a mycoparasitic population but not in the plant pathogen population. This phenomenon has been demonstrated with Ampelomyces quisqualis in powdery mildews (Sundheim, 1982; Sundheim & Amundson, 1982), and noted with respect to Trichoderma harzianum, a ubiquitous mycoparasite (Abt-El Moity et al., 1982; Hastie, 1981).

The relationship of insensitivity to pathogenicity has sometimes been examined. Are the two traits independent or can they be associated in some way? Hollomon (1978; 1981) could find no evidence for linkages between genes conditioning insensitivity to ethirimol in Erysiphe graminis and pathogenicity genes. Kilen & Barrentine (1983), however, found linkage relationships in soybean with respect to genes conditioning reactions to Phytophthora and metribuzin. Wolfe (1981) has shown that an association does occasionally occur but does not affect the soundness of control strategies in which host resistance and fungicide usage are combined.

8.6. ECOLOGICAL AND EVOLUTIONARY PERSPECTIVES

Epidemiological studies of disease in populations are essentially autoecological approaches to interacting host and pathogen populations. Studies of diseases in communities requires a synecological approach and as such should use the concepts and techniques of ecology.

8.8.1. Basic concepts. Concepts from three basic areas are considered: the related concepts of diversity, complexity and stability of ecosystems; density-dependent and independent influences on interacting populations; and the concept of niche as applied to plant pathogens.

Diversity, complexity and stability. There has been considerable debate in the ecological literature on how to characterize the complexity of ecosystems (Murdoch, 1975; King & Pimm, 1983; Pimm, 1984), the interrelationships between organisms (Petraitis, 1979), and the influence of these on the stability of population processes (Connell & Sousa, 1983; Hanski, 1983). Diversity is essentially concerned with the numbers and relative abundance of the different individuals (species or sub-species groupings) that comprise a population or community of interest. Clearly there are at least two components of diversity: the number of different classes of individuals and the distribution over these classes. Indices of diversity give more or less attention to one or other of these components. Complexity in ecosystems, however, reflects more than diversity as defined above. Some emphasis on the degree of connectance, i.e. interactions or inter-relationships between the individuals,

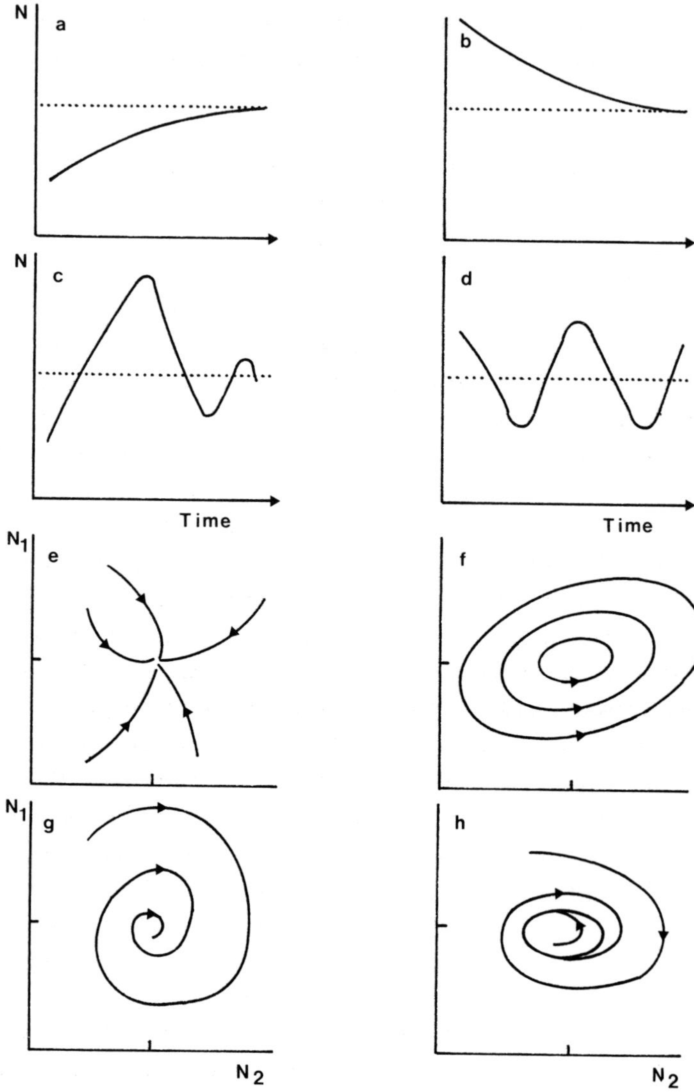

FIGURE 8.3. Concepts of stability. In a) and b), population size of an individual species (or sub-species unit) increases or decreases, respectively, to an equilibrium value. In c) and d), population sizes oscillate around a characteristic value; in the former case with damping, in the latter without. In e) and g), the trajectories of two species (or sub-species unit) approach an equilibrium value for each population, without or with a spiral. In f) trajectories take the form of closed loops around a value characteristic for each species, whereas in h) all trajectories converge eventually to a stable limit cycle.

is required to quantify complexity (Margalef & Gutierrez, 1983; Pimm, 1984). Also the number of trophic levels being considered (i.e. host-pathogen-mycoparasite) can make a difference to connectance. Thus interactions between individuals at different trophic levels are of a different nature from those between individuals at the same trophic level. Interestingly, some of the indices of diversity that have been used in ecological studies are now being used to assess genotypic diversity in pathogen populations (Groth & Roelfs, 1982; Lebeda, 1982).

It is on the relationship between diversity and complexity on the one hand and stability on the other that some debate occurs. It was originally thought that more complex systems were more stable in the sense that fluctuations in population size were less apparent (Pimm, 1984). This idea of stability is rather vague, however, and it becomes necessary further to define the concept in order to explore the relationship and quantify its effects (Figure 8.3). This clarification is also necessary with the related concepts of persistance and resiliance. May (1974) proposed, on the basis of mathematical models, that complex systems were not in fact more stable, and that as the number of connections increased, then the stability of the system in terms of its behaviour after pertubation, decreased. This view can be modified somewhat by considering biomass rather than population numbers (King & Pimm, 1983), but the hypothesis remains to be investigated rigorously in ecological and epidemiological studies. Certainly, the naive assumption that diversity per se can buffer against severe epidemics, certainly no measure of stability, can be challenged. However, it has been pointed out (Browning et al., 1978; Schmidt, 1978) that it is not so much the amount of diversity but rather its quality that is important. The complexities of disease in natural communities, in terms of interconnections and interactions, has always been appreciated in qualitative terms but not to our knowledge quantified. The nature of the different equilibria to be encountered in multi-species host/pathogen systems has only been investigated by Brittain (1983). These areas represent a fascinating and challenging problem for theoreticians; it is important, however, that detailed field assessments of disease in communities are made in order to evaluate the implications of theory.

Habitat and niche. Diversity and complexity are dependent, in part, upon the numbers and kinds of available habitats. A habitat is simply the place where an organism lives, its address and little more. The concept of niche is much wider and involves rate processes and the functioning of organisms as well as the physical space. For biotrophic pathogens, niche is defined by obligate parasitism. There has been much debate in the ecological literature on whether more than one organism can co-exist within the same niche (Petraitis, 1979). With plant pathogens there does seem to be some evidence for at least habitat overlap, in that pathogens which are similar with respect to plant part attacked, environmental conditions, and degrees of specialization do appear to overlap over part of their geographical ranges - e.g. the various rusts of maize. In other cases a habitat may be sufficiently altered, by fungicide application for example, to provide a complex change in occupancy. The more subtle question of niche differentiation in plant pathogens has barely been explored. Putter (1982) invoked the principle of competitive exclusion in theoretical and comparative studies of late and early blight epidemics. In a broader context there are few studies where the interactions between pathogenic fungi and quite different competitors for a shared food species have been reported (Inouye, 1981).

Density dependence and independence. The stability of a population is dependent, in part, upon whether density-dependent or -independent factors are determining population size. Where populations fluctuate randomly and independently of population size then the causal factors, if known, are considered to be density-independent. Alternatively where population numbers appear to stabilize around certain characteristic equilibrium values then the regulation of population size is said to be density-dependent. Most of the population dynamic models represented by autonomous equations are, of course, density-dependent, although density-independence can be readily incorporated. Density-dependence has become something of a dogma in ecological theory rather than a hypothesis or theory: often when a population does appear to stabilize, then it is assumed that some factor must be acting in a density-dependent manner, even though the factor(s) is unknown. In some cases, as in insect host-parasitoid relationships, the regulating function of parasitoids has been established and quite simple graphical techniques can be used to indicate such regulation (Varley *et al.*, 1973). Of special interest would be modelling studies to determine whether factors acting in a density independent manner can lead to population dynamics that are apparently density-dependent.

8.6.2. Applications to disease in communities. Disease, especially in natural systems, occurs in communities comprising many populations. Even in pure line, inbred or clonally propagated agricultural crops there is variation due to different plant parts, weed species, soil flora and fauna, and a range of immigrants and emigrants that lead to interactions between the component parts of the community. The ecological approach to studying such communities was outlined earlier. In general, there are few such studies in agricultural systems. In this section we shall consider a few of the ecological and evolutionary concepts that are now being employed in epidemiological studies.

Evolution of ecological parameters. The evolution of life characteristics has been discussed many times in evolutionary ecology (e.g. Levin, 1978; Maynard, 1977; Smith, 1978). The comparative advantages and disadvantages of life characteristics of fungal pathogens have rarely been considered in detail (but see Pugh, 1980): biotrophy and necrotrophy; whether obligate or facultative; dimorphism in fungi; ploidy and heterokaryosis; and the role of a sexual phase in the life cycle (Griffiths, 1958; Griffiths & Ao, 1980; Groth & Roelfs, 1982).

Life history traits - r and K selection. The concept of life history traits has become very popular since MacArthur & Wilson (1963) first introduced the concepts of *r* and *K* selection. Since then the concept has been expanded (Parry, 1981) and extended to plants (Grime, 1982) and micro-organisms (Pugh, 1980). With plant pathogens the concept has been most used by Thresh (1980a; 1980b) and Andrews & Rouse (1982). Sewell (1982) applies the concept to soil borne fungal pathogens. As with many theoretical constructs these concepts can be very useful indeed as metaphors, but if taken as actual phenomena can lead to problems (Harrison, 1981; Harper, 1982).

Additionally, there are dangers in taking concepts, such as *r* - *K* selection, developed in one discipline and applying them in another. For example, the cereal rusts and powdery mildews often cause explosive epidemics, with many of the features of *r* species (Conway, 1984); yet in one sense, their degree of physiological specialization, they are the very antithesis of *r* selected species. Physiologic specialization and biotrophy are hardly representative of *r* selected species and this is a major dimension that must be added if the concept is to be extended to plant pathogens. Andrews & Rouse (1982) point out the considerable variation in life history to be found in the rusts as a group.

Co-evolution of hosts and pathogens. In many cases remarkable specificity is shown between pairs of organisms that probably evolved in a stepwise and co-evolutionary manner (Barrett, 1983b; 1983c; Clarke, 1979; Hamilton, 1982; May & Anderson, 1983; Roughgarden, 1983; Person & Christ, 1983). It is one thing to come up with convincing *a posteriori* arguments as to why co-evolution as a mechanism can explain something observed at the present time. It is quite another matter to test the theory of co-evolution in any experimental sense, and one that has not been attempted in epidemiological studies. It cannot be achieved for example by manipulating the host or fungicide treatment, and determining evolutionary change in the pathogen population. Co-evolution at best is a stimulating and guiding paradigm in the study of disease in populations. But as with the gene-for-gene hypothesis and the concept of adaptation, it can be so glib and imprecise as to have virtually no operational meaning at all.

8.6.3. Influence of agricultural practices. Agricultural practices can have a marked effect on the expression of resistance and pathogenicity in populations (Apple, 1977). Good husbandry of growing crops, such as sanitation and appropriate crop rotation, can be used to supplement low levels of nonspecific resistance or, in the case of high levels of specific resistance, to prolong the useful life of a cultivar. On the other hand, the growing of crops under poor cultural conditions can give rise to a range of disease problems to which a cultivar may not normally be prone when not under stress. Several agricultural practices have been shown to exacerbate disease problems and lead to an apparent increase in cultivar susceptibility to a given disease. Some of these are irrigation, high density planting and semi-protected plantings. A comprehensive discussion of the effects of these and other cultural practices is given by Palti (1981).

Irrigation of crops, especially where the pathogen is dependent on wetness for much of its life cycle, can lead to disease problems. In many crops, resistance to such pathogens is directly related to the length of the period of leaf wetness required for a given level of infection to occur. If irrigation leads to longer periods of leaf wetness than is normal, under some conditions this will lead to a reduction in the apparent field resistance of cultivars, due largely to non-linear responses of disease increase to leaf wetness. High density plantings of perennial fruit crops such as apple are desirable for higher yields and better quality fruit, for smaller trees and precocity, and for improved spray coverage with respect to diseases such as powdery mildew (*Podosphaera leucotricha*). However, such high densities also give changed micro-environments and relationships with spore dispersal. Higher rates of disease incidence are usually associated with increased stand densities (Burdon & Chilvers, 1982) but the subject remains little studied. Growing crops, such as strawberries and flowers under semi-protected conditions, as in polythene tunnels, may again lead to quite different expressions of resistance to foliar pathogens, than those manifested under the previous cultural conditions, in which the cultivars were most likely evaluated for disease resistance. It is only by considering the ecological consequences of new agricultural practices that the effects on resistance and pathogenicity can be placed in proper context.

8.7. IMPLICATIONS FOR AGRICULTURE

The main emphasis of this Chapter has been on agricultural crops. Pathogenicity of parasite populations has largely been considered from the (often) deliberate manipulation of plant populations, and hence host resistance, by man. Throughout this Chapter we have indicated how aspects of resistance and pathogenicity have direct implications for agriculture; it is now time to bring these aspects together and draw the appropriate conclusions. This will be done by reference to three major topics: pathogenicity surveys; plant breeding; and the management of disease in plant populations.

8.7.1. Pathogenicity surveys. The structure of pathogen populations was always considered in terms of race, or some other sub-specific term. The problems inherent in 'race' with diseases such as stem rust, becomes transparently obvious; the numbers of races increases geometrically with each discovery of a new single race. Vanderplank (1983) has asked whether the concept of physiological race should die. In fact the usefulness of race in analyzing population structure has been questioned many times (Browder & Eversmeyer, 1977; Day 1974; 1978; Lebeda, 1982b; Wolfe & Schwarzback, 1978; Wolfe & Knott, 1982). Approaches placing less emphasis on the concept of race have developed from the studies of Wolfe (1973) and others, in which a phenotype-for-phenotype analysis is used to examine the frequencies of individual allelles, or genetic 'factors' (Dixon & Wright, 1978; Crute & Norwood, 1980; Welving & Crute, 1978), in populations. By examining conditions for linkage disequilibrium, the presence of complex combinations of individual alleles can be studied and conclusions drawn as to the dynamics of genetic change in response to some change in the host population or crop management practice. On the other hand, races of some pathogens do stay remarkably stable and few in number. In these cases the use of differential varieties with simple or complex combinations of resistance factors may continue to be useful. Roelfs & Groth (1980) provide a direct comparison of sexual with asexual populations of stem rust of wheat. In the former, the race concept is clearly much less useful than in the latter.

There are several ways in which plant pathogenic fungi have been surveyed for diversity in pathogenicity. The methods are largely determined by the objectives of the surveys and this should be considered when retrospective analyses (e.g. Nagarajan, 1983) are done. Some survey data have probably been misinterpreted, partly because more suitable data were not available to meet other needs. Two kinds of survey objectives can be discerned: 1) yearly detection of new and dangerous pathgenicities (Priestley, 1978), and 2) characterizing race composition of the pathogen population.

Objective 1) has been the driving force behind systemic surveys, such as those of the federal rust laboratories of the United States, Canada and Australia. Two kinds of collections are used. The first are collections from scattered localities and cultivars, which can be called representative collections. The second kind of collection is that taken from cultivars containing specific resistance, sometimes referred to as 'trap' cultivars; such collections fulfill objective 1) only. Unfortunately, they have been indiscriminately lumped together with representative collections in meeting objective 2). This results in difficulties in quantitatively describing race or pathogenicity frequency. A graphic example of the bias that can be introduced by non-representative sampling, because of certain methods of designating isolates and including collections from trap cultivars, was presented for stem rust of wheat in Kansas by Browder (1966), and paraphrased by Vanderplank (1968). Depending on the resistance present in trap cultivars, collections taken from trap cultivars were shown to inflate the estimated frequency of a race relative to what was found with the more representative methods. This occurred with race 15B in 1961 and 1962 (Table 8.1). Through frequency displacement, trap cultivars were also shown to decrease estimated frequencies of other races below their true values. This occurred in all four years presented with race 56.

Trap cultivars undoubtedly detect matching pathogenicity at much lower thresholds than do representative samples. Our ability to detect infrequent components of a sampled population is quite poor when we consider the practical limits of sampling; where samples generally do not exceed 1000 and quite often are as low as a few hundred. The more efficient trap methods, while capable of detecting components that occur at much lower frequencies, create new problems of quantitative interpretation. One is never certain what total inoculum falling on a plant is represented by the compatible events. In

Table 8.1. Percentages of race 56 of Puccinia graminis f.sp. tritici on wheat in Kansas for four years based on representative sampling versus sampling methods which included collections made on differentially resistant 'trap' cultivars. After Browder (1966).

Year	Representative collections	Collections including differential lines
1960	97-99[a]	80 (9)[b]
1961	99-100	74 (8)
1962	99-100	72 (14)
1963	90-99	56 (36)

[a] Range in percentages among separate composite collections of uredospores
[b] Parentheses enclose percentage of the second most prevalent race, race 15.

many cases, no attempt has been made to provide a denominator for an observed number of events, since absolute frequencies of pathogenicity were not needed. Where some indication of frequency was desired, a susceptible cultivar was included in the trap plot or mobile nursery (Schwarzbach, 1978). If differences in susceptible tissue surface area, plant architecture, tissue age (Groth & Urs, 1982), or tissue receptivity do not exist between the susceptible and trap cultivars, this should provide a good estimate of frequency. In trap plots, care must also be taken that polycyclic auto-infection is obviated, since the compounding of lesion numbers that results could seriously bias relative numbers to such an extent that no meaningful comparison can be made. The techniques for using trap plots should, of course, be as simple and 'portable' as is possible (Michelmore & Crute, 1982).

8.7.2. Plant breeding. Plant breeders have experienced many changing fashions in what is considered the best way to breed for disease resistance in crop plants. In the early days the selection of plants based on gross characteristics in the field was eminently respectable and practised. With the discovery of the monogenic nature of highly specific forms of resistance, the manipulation of one or a few resistance genes based on selection of easily recognized traits became universal in all but a few difficult crops. The limitations and dangers inherent in this approach, especially when combined with accompanying trends of increased acreage, of continuously grown and uniform crops, have since been recognized. The search has now turned to identifying more quantitative forms of resistance, based typically on polygenic inheritance; and the use of more complex selection procedures, based on the epidemiologically important components, although there is still an inherent tendency for the breeder to adopt procedures based upon infection characteristics alone. Reverting to simple criteria of field resistance, or even simply yield, has again become popular (Buddenhagen, 1983a; 1983b).

There is still controversy among plant pathologists and plant breeders as to whether resistance to some diseases is absolutely associated with a reduced yield potential. There are many examples of cases where incorporation of a degree of resistance to a pathogen has resulted in a cultivar with lowered yields compared with the susceptible but agronomically desirable original cultivar. While it has not been established conclusively to our knowledge, a logical and relatively simple explanation for this phenomenon lies in the high degree of very specific adaptation that major crop cultivars have attained in developed countries through narrow, intensive breeding. The likelihood that a source line of needed resistance will be poorly adapted can only increase as plant breeding has intensified. This means that the incorporation of the resistance, whatever its inheritance or the methods used, is likley to result in the incorporation of undesirable traits as well. In fact it is conceivable that the interpretation of what constitutes an undesirable trait has itself become more stringent because of rising expectations of how a cultivar should perform. The tendency is for resistance from unadapted germplasm, and more complexy-inherited resistance, to depress yields. Hence single locus resistance that is amendable to backcrossing and is already in an 'improved' background is less likely to cause yield depression.

Buddenhagen (1983b) has contrasted the needs of less-intensive but locally efficient agriculture of developing countries with those of developed countries; he points out some glaring but often unappreciated differences in effective breeding for resistance to biotic and abiotic diseases. Under suboptimal conditions, the importance of obtaining resistance from adapted sources, sometimes in the form of complexes of resistances that operate best in a unique local environment, becomes critical. Subsistence agriculture has relied on local material. More recently, exotic sources have been used, often with poor results. Under the unfavourable conditions often found in marginal locations, otherwise high yielding cultivars developed elsewhere will yield less than locally adapted cultivars, despite the higher efficiency of exotic cultivars in utilizing nitrogen under more ideal conditions. This is another aspect of elite germplasm: it is frequently associated with a strong genotype x environment interaction. Of course such an interaction may also be expected when locally adapted germplasm is grown elsewhere.

The main methodological problems facing plant breeders are not which procedures are best, but at which stage in the breeding programme the different procedures are appropriate; i.e. when to use simple criteria and when to use more complex procedures; when to subject plants to field evaluation. The answers must depend on the nature of the crop, the importance of crossing compared with clonal selection, the generation time, and the number of plants at each phase of the programme. At all stages two major sources of error must be guarded against and their relative importance assessed. By analogy with statistical inference, we can discern Type I errors in which plants initially selected as resistant turn out on a later screen to be susceptible, and Type II errors in which plants initially rejected as susceptible would have turned out to be resistant if selected. The Type I error may be costly but at least can be identified; but the effects of a Type II error cannot be quantified unless the appropriate preliminary and correlative studies have been done.

8.7.3. Disease management. Disease management covers a range of applications from the strategic decision of what proportion of land should be allocated to a particular crop enterprise to the tactical decision of what fungicide to use. The growers calculation of likely gross margins depends in part upon likely incidences of disease and the cost of chemical control. The deployment of resistant cultivars, if this choice is available, may have a profound influence on the decision. At the other extreme there is the practical problem of how to manage disease on a day-to-day basis. For example, should a cultivar with quantitatively more resistance to a disease be sprayed at lower rates and longer intervals, or should spraying be stopped at an earlier stage. There is now evidence that resistance and fungicide usage can be effectively manipulated and integrated (Jeger & Butt, 1983). The need is now strongly indicated to integrate the rational use of host resistance, and where possible manipulate the pathogen population via the host and control practice, into the whole of disease management (Wolfe, 1978; 1981; 1983). There is now some evidence that this integration is occurring and we are confident that it will continue. For it to be more than a transient and fashionable phase, it is essential that theoretical and experimental studies on functioning pathogen populations continue so that disease management can be placed on a sounder scientific basis. In most cases this will require the concerted and integrated efforts of population geneticists and epidemiologists.

REFERENCES

Aamodt, V.S. (1927). Breeding wheat for resistance to physiologic forms of stem rusts. Journal of the American Society for Agronomy 19, 218.

Abd-El Moity, T.H., Papavizas, G.C., & Shaha, M.N. (1982). Induction of new isolates of Trichoderma harzianum tolerant to fungicides and their experimental use for control of white rot of onion. Phytopathology 72, 396-400.

Agrios, G.N. (1980). Escape from disease. In 'Plant Disease, An Advanced Treatise. Vol. V. How Plants defend Themselves' (Horsfall, J.G. and Cowling, E.B., eds), pp. 18-37. Academic Press, New York.

Anderson, M.G. (1982). Interpreting residual effects of 'defeated' resistance genes. Phytopathology 72, 1383-1384.

Anderson, W.W. & Arnold, J. (1983). Density-regulated selection with genotypic interactions. American Naturalist 121, 649-655.

Andrews, J.H. & Rouse, D.I. (1982). Plant pathogens and the theory of r- and K-Selection. American Naturalist 120, 283-296.

Antonovics, J. (1976). The input from population genetics: 'The New Ecological Genetics'. Systematic Botany 1, 233-245.

Apple, J.L. (1977). The theory of disease management. In 'Plant Disease, an Advanced Treatise. Vol I. How Disease is Managed' (Horsfall, J.G. & Cowling, E.B., eds), pp. 79-101. Academic Press, New York.

Asher, M.J.C. (1980). Variation in pathogenicity and cultural characters in Gaeumannomyces graminis var tritici. Transactions of the British Mycological Society 75, 213-220.

Aust, H.J. & Hau, B. (1983). Die Latenzzeit von Septoria nodorum in Abhängigkeit von der ontogenetisch bedingten Anfälligkeit des Sommerweizens. Zeitschrift für Pflanzenkranheiten und Pflanzenschutz 90, 55-62.

Ayres, P.G. & Woolacott, B. (1980). Effects of soil water level on the development of adult plant resistance to powdery mildew in barley. Annals of Applied Biology 94, 255-263.

Ball, S.L. (1983). Pathogenic variability of downy mildew (Sclerospora graminicola) on pearl millet. I. Host cultivar reactions to infection by different pathogen isolates. Annals of Applied Biology 102, 257-264.

Barrett, J.A. (1978). A model of epidemic development in variety mixtures. In 'Plant Disease Epidemiology' (Scott, P.R. & Bainbridge, A., eds), pp. 129-137. Blackwell Scientific Publications, Oxford.

Barrett, J.A. (1980). Pathogen evolution in multilines and variety mixtures. Zeitschrift für Pflanzenkrankheiten und Pflanzenschutz 87, 383-396.

Barrett, J.A. (1981). The evolutionary consequences of monoculture. In: 'Genetic Consequences of Man Made Change' (Bishop, J.A. & Cook, L.M., eds), pp. 209-248. Academic Press, London.

Barrett, J.A. (1983a). Estimating relative fitness in plant parasites: some general problems. Phytopathology 73, 510-512.

Barrett, J.A. (1983b). The genetics of host-parasite interaction. In 'Evolutionary Ecology' (Shorrocks, B.R., ed.), pp. 275-294. Blackwell Scientific Publications, Oxford.

Barrett, J.A. (1983c). Plant-fungus symbioses. In: 'Co-evolution' (Futuyma, D.J. & Slatkin, eds), pp. 137-160. Sinauer Associates Inc., Sunderland, Mass.

Barrett, J.A. & Wolfe, M.S. (1978). Multilines and super races - a reply. Phytopathology 68, 1535-1537.

Beale, P.E. & Thurling, N. (1980a). Genetic control of resistance to Kabatiella caulivora in Trifolium subterraneum subspecies yanninicum. II. Variation in segregation populations. Australian Journal of Agricultural Research 31, 935-42.

Beale, P.E. & Thurling, N. (1980b). Genetic control of resistance to Kabatiella caulivora in Trifolium subterraneum subspecies yanninicum. I. Diallel Analysis of Variation in Disease Development Parameters. Australian Journal of Agricultural Research 31, 927-33.

Bennett, F.G.A. (1981a). The expression of resistance to powdery mildew infection in winter wheat cultivars. I. Seedling resistance. Annals of Applied Biology 98, 295-303.

Bennett, F.G.A. (1981b). The expression of resistance to powdery mildew infection in winter wheat cultivars. II. Adult plant resistance. Annals of Applied Biology 98, 305-317.

Berger, R.D. (1977). Application of epidemiological principles to achieve plant disease control. Annual Review of Phytopathology 15, 165-183.

Birch, L.C. (1960). The genetic factor in population ecology. American Naturalist 94, 5-24.

Blanch, P.A., Asher, M.J.C. & Burnett, J.H. (1981). Inheritance of pathogenicity and cultural characters in Gaeumannomyces graminis var. tritici. Transactions of the British Mycological Society 77, 391-399.

Brasier, C.M., Lea, J. & Rawlings, M.K. (1981). The aggressive and non-aggressive strains of Ceratocystis ulmi have different temperature optima for growth. Transactions of the British Mycological Society 76, 213-218.

Briggs, D. & Block, M. (1981). An investigation into the use of the '-deme' terminology. New Phytologist 89, 729-735.

Brittain, E.G. (1983). A model system of four species with a unique equilibrium point. Ecological Modelling 19, 199-211.

Browder, L.E. (1966). A rapid method of assaying pathogenic potential of populations of Puccinia graminis tritici. Plant Disease Reporter 50, 673-676.

Browder, L.E. & Eversmeyer, M.G. (1982). Sorting of infection-type data sets toward the gene-for-gene model: a reply. Phytopathology 72, 458-460.

Browder, L.E. & Eversmeyer, M.G. (1977). Pathogenicity associations in Puccinia recondita tritici. Phytopathology 67, 766-771.

Brown, J.F. (1975). Factors affecting the relative ability of strains of fungal pathogens to survive in populations. Journal of the Australian Institute of Agricultural Science 41, 3-11.

Browning, J.A. (1974). Relevance of knowledge about natural ecosystems to development of pest management programs for agro-ecosystems. Proceedings of the American Phytopathological Society 1, 191-199.

Browning, J.A. (1981). The agroecosystem-natural ecosystem dichotomy and its impact on phytopathological concepts. In: 'Pests, Pathogens and Vegetation' (Thresh, J.M. ed.), pp. 159-172. Pitmans, London.

Browning, J.A. & Frey, K.J. (1969). Multiline cultivars as a means of disease control. Annual Review of Phytopathology 7, 355-382.

Browning, J.A., Simons, M.D. & Torres, E. (1978). Managing host genes: epidemiologic and genetic concepts. In: 'Plant Disease, an Advanced Treatise, Vol II. How Disease develops in Populations' (Horsfall, J.G. & Cowling, E.B., eds), pp. 191-212. Academic Press, New York.

Buddenhagen, I.W. (1983a). Agroecosystems, disease resistance and crop improvement. In 'Challenging Problems in Plant Health' (Kommendahl, T. & Williams, P.H., eds), pp. 450-460. American Phytopathological Society, St. Paul.

Buddenhagen, I.W. (1983b). Breeding strategies for stress and disease resistance in developing countries. Annual Review of Phytopathology 21, 385-409.

Burdon, J.J. (1978). Mechanisms of disease control in heterogeneous plant populations - an ecologists view. In 'Plant Disease Epidemiology' (Scott, P.R. & Bainbridge, A., eds), pp. 193-200. Blackwell Scientific Publications, Oxford.

Burdon, J.J. (1980). Variation in disease resistance within a population of Trifolium repens. Journal of Ecology 68, 737-744.

Burdon, J.J. (1982). The effect of fungal pathogens on plant communities. In 'The Plant Community as a Working Mechanism' (Newman, E.I., ed.), pp. 99-112. Blackwell Scientific Publications, Oxford.

Burdon, J.J. & Chilvers, G.A. (1982). Host density as a factor in plant disease ecology. Annual Review of Phytopathology 20, 143-66.

Burdon, J.J. & Marshall, D.R. (1981). Inter- and intra-specific diversity in the disease-response of Glycine species to the leaf rust fungus Phakopsora pachyrhizi. Journal of Ecology 69, 381-390.

Burdon, J.J. & Shattock, R.C. (1982). Disease in plant communities. In 'Applied Biology, Vol. 5,' (Coaker, T.H., ed.), pp. 145-219. Academic Press, New York.

Burdon, J.J., Luig, N.H. & Marshall, D.R. (1983). Isozyme uniformity and virulence variation in Puccinia graminis f.sp. tritici and P. recondita f.sp. tritici in Australia. Australian Journal of Biological Sciences 36, 403-410.

Burdon, J.J., Oates, J.D. & Marshall, D.R. (1983). Interactions between Avena and Puccinia species. I. The wild hosts: Avena barbata Pott ex Link, A. fatua L., A. ludoviciana Durieu. Journal of Applied Ecology 20, 571-584.

Burnett, J.H. (1975). 'Mycogenetics.' John Wiley and Sons, New York, 375 pp.

Chandrashekar, M. & Heather, W.A. (1981a). Temperature sensitivity of reactions of Populus spp. to races of Melampsora larici-populina. Phytopathology 71, 421-424.

Chandrashekar, M. & Heather, W.A. (1981b). Influence of light intensity on reactions of certain poplar cultivars to races of Melampsora larici-populina. Transactions of the British Mycological Society 77, 375-380.

Chin, K.M. (1978). Aspects of the epidemiology and genetics of the foliar pathogen, Erysiphe graminis f. sp. hordei in relation to infection of homogeneous and heterogeneous populations of the barley host (Hordeum vulgare). Ph.D. Thesis, University of Cambridge, U.K.

Cirulli, M. & Ciccarese, F. (1982). Factors affecting early screening of tomatoes for monogenic and polygenic resistance to Fusarium wilt. Crop Protection 1, 341-348.

Clairidge, M.F. & Hollander, J. Den (1983). The biotype concept and its application to insect pests of agriculture. Crop Protection 2, 85-95.

Clarke, B.C. (1979). The evolution of genetic diversity. Proceedings of the Royal Society of London, Series B 205, 453-474.

Cohen, Y., Reuveni, M. & Samoucha, Y. (1983). Competition between metalaxyl-resistant and -sensitive strains of Pseudoperonospora cubensis on cucumber plants. Phytopathology 73, 1516-1520.

Conard, S.G. & Radesevich, S.R. (1979). Ecological fitness of Senecio vulgaris and Amaranthus retroflexus biotypes susceptible or resistant to atrazine. Journal of Applied Ecology 16, 171-177.

Connell, J.H. & Sousa, W.P. (1983). On the evidence needed to judge ecological stability or persistence. American Naturalist 121, 789-824.

Conway, G.R. (1984). Strategic models. In 'Pest and Pathogen Control: Strategic, Tactical and Policy Models' (Conway, G.R., ed.), pp. 15-28. International Institute for Applied Systems Analysis; John Wiley & Sons, Chichester.

Crute, I.R. & Norwood, J.M. (1980). Inter-isolate variation for virulence in Bremia lactucae. Annals of Applied Biology 94, 275-278.

Cunningham, P.C. (1981). Occurrence, role and pathogenic traits of a distinct pathotype of Pseudocercosporella herpotrichoides. Transactions of the British Mycological Society 76, 3-15.

Danon, T., Sacks, J.M. & Eyal, Z. (1982). The relationships among plant stature, maturity class and susceptibility to Septoria leaf blotch of wheat. Phytopathology, 72, 1037-1042.

Davis, R.P. & Dennis, C. (1981). Properties of dicarboximide-resistant strains of Botrytis cinerea. Pesticide Science 12, 521-535.

Day, P.R. (1974). 'Genetics of Host-Parasite Interaction.' W.H. Freeman and Company, San Francisco, 238 pp.

Day, P.R. (1978). The genetic basis of epidemics. In 'Plant Disease, an Advanced Treatise. Vol. II. How Disease develops in Populations' (Horsfall, J.G. & Cowling, E.B., eds), pp. 263-285. Academic Press, New York.

Delp, C.J. (1980). Coping with resistance to plant disease. Plant Disease 64, 652-657.

Dinoor, A. (1977). Genes for disease resistance from native populations. In 'Induced Mutations Against Plant Diseases' (International Atomic Energy Agency, ed.), pp. 15-27. IAEA, Vienna.

Dixon, G.R. & Wright, I.R. (1978). Frequency and geographical distribution of specific virulence factors in Bremia lactucae populations in England from 1973 to 1975. Annals of Applied Biology 88, 287-294.

Eenink, A.H. (1981). Partial resistance in lettuce to downy mildew (Bremia lactucae). 1. Search for partially resistant genotypes and the influence of certain plant characters and environments on the resistance level. Euphytica 30, 619-628.

Eenink, A.H., Bijker, W. & Den Toom, A. (1982). Partial resistance in lettuce to downy mildew (Bremia lactucae). 2. Differential interactions between plant genotypes and fungus races and the relationship of latent period, infection frequency and number of infected leaves with the level of partial resistance. Euphytica 31, 73-83.

Falahati-Rastegar, M., Manners, J.G. & Smartt, J. (1981). Effects of temperature and inoculum density on competition between races of Puccinia hordei. Transactions of the British Mycological Society 77, 359-368.

Falahati-Rastegar, M., Manners, J.G. & Smartt, J. (1983). Factors determining results of competition between physiologic races of Puccinia hordei. Transactions of the British Mycological Society 81, 233-239.

Fleming, R.A. (1980). Selection pressures and plant pathogens: robustness of the model. Phytopathology 70, 175-178.

Fleming, R.A. (1982). Evolutionarily persistent and evolutionarily stable strategies in plant pathosystems. Euphytica 31, 837-845.

Fleming, R.A. & Person, C.O. (1978). Disease control through use of multilines: A theoretical contribution. Phytopathology 68, 1230-1233.

Fleming, R.A. & Person, C.O. (1982). Consequences of polygenic determination of resistance and aggressiveness in nonspecific host: parasite relationships. Canadian Journal of Plant Pathology 4, 89-98.

Frankel, O.H. (1977). Genetic resources as the backbone of plant protection. In 'Induced Mutations Against Plant Diseases' (International Atomic Energy Agency, ed.), pp. 3-14. IAEA, Vienna.

Frecha, J.H. (1973). The inheritance of resistance to Septoria nodorum in wheat. Boletin Genetico del Instituto de Fitotecnia, Castelar 8, 29-30.

Gale, J.S. (1980). 'Population Genetics.' John Wiley and Sons, London. 189 pp.

Gaunt, R.E. (1981). Disease tolerance - an indicator of thresholds? Phytopathology 71, 915-916.

Georgopoulos, S.G. (1977). Pathogens become resistant to chemicals. In 'Plant Disease, an Advanced Treatise. Vol. I. How Disease is Managed' (Horsfall, J.G. & Cowling, E.B., eds), pp.327-345. Academic Press, New York.

Gessler, C. & Kuć, J. (1982). Induction of resistance to Fusarium wilt in cucumber by root and foliar pathogens. Phytopathology 72, 1439-1441.

Ginzburg, L.R. (1981). Genetic adaptation and models of population dynamics. In 'Quantitative Population Dynamics' (Chapman, D.G. & Gallucci, V.F. eds), pp. 1-10. Co-operative Publishing House, Fairland, Maryland.

Grant, M.W. & Archer, S.A. (1983). Calculation of selection coefficients against unnecessary genes for virulence from field data. Phytopathology 73, 547-551.

Gregory, L.V., Royer, M.H., Ayers, J.E. & Nelson, R.R. (1981). The evaluation of relative parasitic fitness of isolates of Helminthosporium maydis Race T. Phytopathology 71, 354-356.

Griffiths, E. (1958). Sexual reproduction and variation in Gloeotinia temulenta (Prill and Delacr.) Wilson and Gray. Transactions of the British Mycological Society 41, 461-482.

Griffiths, E. & Ao, H.C. (1980). Variation in Septoria nodorum. Annals of Applied Biology 94, 294-296.

Griggs, M.M., Nance, W.L. & Dinus, R.J. (1978). Analysis and comparison of fusiform rust disease progress curves for five slash pine families. Phytopathology 68, 1631-1636.

Grime, J.P. (1982). The concept of strategies: use and abuse. Journal of Ecology 70, 863-866.

Grindle, M. (1981). Variations among field isolates of Botrytis cinerea in their sensitivity to antifungal compounds. Pesticide Science 12, 305-312.

Groth, J.V. (1976). Multilines and 'super races' - a simple model. Phytopathology 66, 937-939.

Groth, J.V. (1978). Rebuttal to 'Multilines and super races - a reply'. Phytopathology 68, 1538-1539.

Groth, J.V. (1984). Virulence frequency dynamics of cereal rust fungi. In 'The Cereal Rusts Vol. I. Origins, Specificity, Structure and Physiology' (Bushnell, W.R.; & Roelfs, A.P., eds), pp. 231-252. Academic Press, New York.

Groth, J.V. & Barrett, J.A. (1980). Estimating parasitic fitness: A reply. Phytopathology 70, 840-842.

Groth, J.V. & Roelfs, A.P. (1982). Effect of sexual and asexual reproduction on race abundance in cereal rust fungus populations. Phytopathology 72, 1503-1507.

Groth, J.V. & Urs, N.V. (1982). Differences among bean cultivars in receptivity to Uromyces phaseoli var. typica. Phytopathology 72, 374-378.

Groth, J.V., Person, C.O. & Ebba, T. (1976). Relation between two measures of disease expression in barley - Ustilago hordei interactions. Phytopathology 66, 1342-1347.

Groth, J.V. & Person, C.O. (1977). Genetic interdependence of host and parasite in epidemics. Annals of the New York Academy of Science 287, 97-106.

Hamid, A.H., Ayers, J.E., Schein, R.D. & Hill, R.R. (1982a). Components of fitness attributes in Cochliobolus carbonum Race 3. Phytopathology 72, 1166-1169.

Hamid, A.H., Ayers, J.E. & Hill, R.R. (1982b). Host isolate interactions in corn inbreds inoculated with Cochliobolus carbonum Race 3. Phytopathology 72, 1169-1173.

Hamilton, W.D. (1982). Pathogens as causes of genetic diversity in their host populations. In 'Population Biology of Infectious Diseases' (Anderson, R.M. and May, R.M., eds), Dahlem Konferenzen. pp. 269-296. Springer-Verlag. Berlin.

Hansen, E.M. & Hamm, P.B. (1983). Morphological differentiation of host-specialized groups of Phytophthora megasperma. Phytopathology 73, 129-134.

Hanski, I. (1983). Coexistence of competitors in patchy environment. Ecology 64, 493-500.

Harlan, J.R. (1972). Genetics of disaster. Journal of Environmental Quality 1, 212-215.

Harlan, J.R. (1976). Diseases as a factor in plant evolution. Annual Review of Phytopathology 14, 31-51.

Harper, J.L. (1982). After description. In 'The Plant Community as a Working Mechanism' (Newman, E.I., ed.), pp. 11-25. Blackwell Scientific Publications, Oxford.

Harrison, B.D. (1981). Plant virus ecology: ingredients, interactions, and environmental influences. Annals of Applied Biology 99, 195-209.

Hastie, A.C. (1981). The genetics of conidial fungi. In 'Biology of Conidial Fungi, Vol. 2' (Cole, G.T. & Kendrick, B., eds), pp. 511-547. Academic Press, New York.

Heather, W.A. & Chandrashekar, M. (1982). Stability of disease resistance in natural- and agro-ecosystems. Transactions of the British Mycological Society 78, 381-383.

Heijbroek, W., McFarlane, J.S. & Doney, D.L. (1977). Breeding for tolerance to beet-cyst eelworm Heterodera schachtii Schm. in sugarbeet. Euphytica 26, 557-564.

Hill, J.P. & Nelson, R.R. (1982). The heritability of three parasitic fitness attributes of Helminthosporium maydis race T. Phytopathology 72, 525-528.

Hill, J.P. & Nelson, R.R. (1983). Genetic control of two parasitic fitness attributes of Helminthosporium maydis race T. Phytopathology 73, 455-457.

Hollomon, D.W. (1978). Competitive ability and ethirimol sensitivity in strains of barley powdery mildew. Annals of Applied Biology 90, 195-204.

Hollomon, D.W. (1981). Genetic control of Ethirimol resistance in a natural population of Erysiphe graminis f. sp. hordei. Phytopathology 71, 536-540.

Hooker, A.L. & Saxena, K.M.S. (1971). Genetics of disease resistance in plants. Annual Review of Genetics 5, 407-424.

Horsten, J. & Fehrmann, H. (1980). Fungicide resistance of Septoria nodorum and Pseudocercosporella herpotrichoides III. Survival ability of resistant strains in mixed

populations. Zeitschrift für Pflanzenkrankheiten und Pflanzenschutz 87, 577-586.
Huang, B.K. & Heitefuss, R. (1982a). Characterization of adult plant resistance of spring barley to powdery mildew (Erysiphe graminis f. sp. hordei) I. Race specificity and expression of resistance. Phytopathologische Zeitschrift 104, 168-178.
Huang, B.K. & Heitefuss, R. (1982b). Characterization of adult plant resistance of spring barley to powdery mildew (Erysiphe graminis f. sp. hordei II. Infection process at different leaf stages. Phytopathologische Zeitschrift 104, 179-190.
Hyde, P.M. (1982). Temperature-sensitive resistance of the wheat cultivar Maris Fundin to Puccinia recondita. Plant Pathology 31, 25-30.
Inouye, R.S. (1981). Interactions among unrelated species: Granivorous rodents; a parasitic fungus, and a shared prey species. Oecologia 49, 425-427.
Jaenike, J. (1981). Criteria for ascertaining the existence of host races. American Naturalist 117, 830-834.
Jarosz, A.M., Sheets, M. & Levy, M. (1982). Cuticle thickness in phlox and resistance to powdery mildew: an unreliable line of defence. American Journal of Botany 69, 824-828.
Jayakar, S.D. (1970). A mathematical model for interaction of gene frequencies in a parasite and its host. Theoretical Population Biology 1, 140-164.
Jeffery, H. & Kable, P.F. (1982). Effects of alternating and mixing pesticides on fungal resistance build up - a reply. Phytopathology 72, 274.
Jeger, M.J. (1979). Studies on disease spread in heterogeneous cereal populations. Ph.D. Thesis, University of Wales.
Jeger, M.J. (1981). Disease measurement in a study of apple scab epidemics. Annals of Applied Biology 99, 43-51.
Jeger, M.J. (1985a). Modelling the dynamics of pathogen populations. In 'Populations of Plant Pathogens: Their Dynamics and Genetics' (Wolfe, M.S. & Cater, C.E., eds), Blackwell Scientific Publications, Oxford, in press.
Jeger, M.J. (1985b). The potential of analytical compared with simulation approaches to modelling in plant disease epidemiology. In 'Plant Disease Epidemiology: Population Dynamics and Management' (Leonard, K.J. & Fry, W.E., eds), MacMillan, New York, in press.
Jeger, M.J. & Butt, D.J. (1983). Using partial resistance in the integrated control of apple powdery mildew. In 'Disease Resistance as a Component of Integrated Control in Orchards'. IOBC/WPRS Bulletin VI (4), 111-122.
Jeger, M.J., Griffiths, E. & Jones, D.G. (1981). Disease progress of non-specialized fungal pathogens in intraspecific mixed stands of cereal cultivars. I. Models. Annals of Applied Biology 98, 187-198.
Jeger, M.J., Griffiths, E. & Jones, D.G. (1982a). Asymptotes of diseases caused by non-specialized fungal pathogens in intraspecific mixed stands of cereal cultivars. Annals of Applied Biology 101, 459-464.
Jeger, M.J., Jones, D.G. & Griffiths, E. (1983). Components of partial resistance of wheat seedlings to Septoria nodorum. Euphytica 32, 575-584.
Jeger, M.J., Swait, A.A.J. & Butt, D.J. (1982b). Overwintering of Venturia inaequalis, the causal agent of apple scab, on different cultivars. Annals of Applied Biology 100, 459-464.
Johnson, D.A. & Wilcoxson, R.D. (1978). Components of slow-rusting in barley infected with Puccinia hordei. Phytopathology 68, 1470-1474.
Johnson, R. (1981). Durable resistance: definition of, genetic control, and attainment in plant breeding. Phytopathology 71, 567-568.
Johnson, R. (1983). Genetic background of durable resistance. In 'Durable resistance in crops' (Lamberti, F., Waller, J.M. & Van der Graaf, N.A., eds), pp. 5-26. Plenum Press, New York and London.
Johnson, R. & Allen, D.J. (1975). Induced resistance to rust diseases and its possible role in the resistance of multiline varieties. Annals of Applied Biology 80, 359-363.
Johnson, R. & Taylor, A.J. (1976). Spore yield of pathogens in investigations of the race-specificity of host resistance. Annual Review of Phytopathology 14, 97-119.
Kable, P.F. & Jeffery, H. (1980). Selection for tolerance in organisms exposed to sprays of biocide mixtures: a theoretical model. Phytopathology 70, 8-12.

Kilen, T.C. & Barrentine, W.L. (1983). Linkage relationships in soybean between genes controlling reactions to Phytophthora rot and metribuzin. Crop Science 23, 894-896.

King, A.W. & Pimm, S.L. (1983). Complexity, diversity and stability: a reconciliation of theoretical and empirical results. American Naturalist 122, 229-239.

Kinloch, B.B. & Byler, J.W. (1981). Relative effectiveness and stability of different resistance mechanisms to white pine blister rust in sugar pine. Phytopathology 71, 386-391.

Kiyosawa, S. (1976). A comparison by simulation of disease dispersal in pure and mixed stands of susceptible and resistant plants. Japanese Journal of Breeding 26, 135-145.

Kiyosawa, S. (1977). Development of methods for the comparison of utility values of varieties carrying various types of resistance. Annals of the New York Academy of Sciences 287, 107-123.

Kiyosawa, S. (1982). Genetics and epidemiological modelling of breakdown of plant disease resistance. Annual Review of Phytopathology 20, 93-117.

Kiyosawa, S. & Yabuki, S. (1976a). Modelling the race frequency in a host-pathogen system with genes for resistance and avirulence. Japanese Journal of Breeding 26, 247-255.

Kiyosawa, S. & Yabuki, S. (1976b). Simulation of race frequency change in a host-pathogen system with true resistance and avirulence genes. Japanese Journal of Breeding 26, 247-255.

Kiyosawa, S., Iwaka, K. & Asai, Y. (1975). Multiple regression analysis on the process of breakdown of Pi-K rice varieties by blast disease in Nigata Prefecture. Japanese Journal of Breeding 25, 109-120.

Knott, D.R. (1972). Using race-specific resistance to manage the evolution of plant pathogens. Journal of Environmental Quality 1, 227-231.

Knott, D.R. & Johnson, R. (1983). Some additional comments on sorting infection-type data sets. Phytopathology 73, 514-515.

Kranz, J. (1983). Epidemiological parameters of plant resistance. In 'Durable Resistance in Crops' (Lamberti, F., Waller, J.M. & Van der Graaf, N.A., eds), pp. 141-162. Plenum Press, New York and London.

Kulkarni, R.N. & Chopra, V.L. (1982). Environment as the cause of differential interaction between host cultivars and pathogenic races. Phytopathology 72, 1384-1386.

Kulkarni, R.N., Chopra, V.L. & Singh, D. (1982). Relative importance of components affecting the leaf rust progress curve in wheat. Theoretical and Applied Genetics 62, 205-207.

Last, F.T. (1978). Variety is the spice of life. Annals of Applied Biology 90, 303-322.

Lebeda, A. (1982a). Measurement of genetic diversity of virulence in populations of phytopathogenic fungi. Zeitschrift für Pflanzenkrankheiten und Pflanzenschutz 89, 88-95.

Lebeda, A. (1982b). Population genetic aspects in the study of phytopathogenic fungi. Acta Phytopathologica Academia Scientiarum Hungaricae 17, 215-219.

Leonard, K.J. (1969a). Factors affecting rates of stem rust increase in mixed plantings of susceptible and resistant oat varieties. Phytopathology 59, 1845-1850.

Leonard, K.J. (1969b). Selection in heterogeneous populations of Puccinia graminis f. sp. avenae. Phytopathology 59, 1851-1857.

Leonard, K.J. (1977). Selection pressures and plant pathogens. Annals of the New York Academy of Science 287, 207-222.

Leonard, K.J. & Czochor, R.J. (1980). Theory of genetic interactions among populations of plants and their pathogens. Annual Review of Phytopathology 18, 237-258.

Leonard, K.J. & Mundt, C.C. (1984). Methods for estimating epidemiological effects of quantitative resistance to plant diseases. Theoretical and Applied Genetics 67, 219-230.

Leroux, P. (1980). Effets de divers herbicides sur la croissance mycélienne de souches de Botrytis cinerea Pers et de Penicillium expansum Link, sensibles ou résistantes á certains fongicides. Weed Research 20, 249-254.

Levin, S.A. (1978). On the evolution of ecological parameters. In 'Proceedings in Life Sciences. Ecological Genetics: the Interface' (Brussard, P.F., ed.), pp. 1-26. Springer-Verlag, New York.

Levy, Y., Levi, R. & Cohen, Y. (1983). Build up of a pathogen subpopulation resistant to a systemic fungicide under various control strategies: a flexible simulation model. Phytopathology 73, 1475-1480.

Lewontin, R.C. (1984). Detecting population differences in quantitative characters as opposed to gene frequencies. American Naturalist 123, 115-124.
Longini, I.M. (1983). Models of epidemics and endemicity in genetically variable host populations. Journal of Mathematical Biology 17, 289-304.
Lorbeer, J.W. (1982). Variation in Botrytis and Botryotinia. In 'The Biology of Botrytis' (Coley-Smith, J.R., Verhoelf, K. & Jarvis, W.R. eds), pp. 19-39. Academic Press, New York.
Louvel, D. (1980). Comparisons des resultats d'utilisation du dispositif d'evaluation de la resistance horizontale du riz vis-a-vis de Pyricularia oryzae Cav. en Cote d'Ivoire et a Madagascar sous deux conditions epidemiologiques. I'Agronomie Tropicale 35, 259-283.
Luedders, V.D. (1983). Genetics of the cyst nematode-soybean symbiosis. Phytopathology 73, 944-948.
Luthra, J.K. & Rao, M.V. (1979). Multiline cultivars - how their resistance influences leaf rust diseases in wheat. Euphytica 28.
MacArthur, R.H. & Wilson, E.O. (1967). 'The Theory of Island Biogeography.' Princeton University Press, Princeton, New Jersey. 203 pp.
McDonald, G.I. & Andrews, D.S. (1982). Genetic variation of epidemiological fitness traits among single aecispore cultures of Cronartium ribicola. Phytopathology 72, 1391-1396.
McGee, D.C. & Zuck, M.G. (1981). Competition between benomyl-resistant and sensitive strains of Venturia inaequalis on apple seedlings. Phytopathologly 71, 529-532.
MacKenzie, D.R. (1978). Estimating parasitic fitness. Phytopathology 68, 9-13.
MacKenzie, D.R. & King, E. (1980). Developing realistic crop loss models for plant disease. In 'Crop Loss Assessment, Proceedings of the E.C. Stakman Commemorative Symposium', Minneapolis, August 20-23, pp. 85-89. University of Minnesota.
Mangel, M. & Plant, R.E. (1983). Multiseasonal management of an agricultural pest. I. Development of the theory. Ecological Modelling 20, 1-19.
Margalef, R. & Gutierrez, E. (1983). How to introduce connectance in the frame of an expression for diversity. American Naturalist 121, 601-607.
Marshall, D.R. (1977). The advantages and hazards of genetic homogeneity. Annals of the New York Academy of Sciences. 287, 1-20.
Marshall, D.R. & Burdon, J.J. (1981). Multiline varieties and disease control. III. Combined use of overlapping and disjoint gene sets. Australian Journal of Biological Sciences 34, 81-95.
Marshall, D.R. & Pryor, A.J. (1978). Multiline varieties and disease control I. The 'dirty crop' approach with each component carrying a unique single resistance gene. Theoretical and Applied Genetics 51, 177-84.
Marshall, D.R. & Pryor, A.J. (1979). Multiline varieties and disease control. II. The 'dirty crop' approach with components carrying two or more genes for resistance. Euphytica 28, 145-159.
Maszkiewicz, J., Blaszczak, W. & Millikan, D.F. (1979). Investigation on the apple proliferation disease I. Increased susceptibility of affected leaf tissue to Podosphaera leucotricha. Phytoprotection 60, 47-54.
Matta, A. (1980). Defences triggered by previous disease invaders. In 'Plant Disease, An Advanced Treatise. Vol V. How Plants defend Themselves' (Horsfall, J.G. & Cowling, E.B., eds), pp. 76-102. Academic Press, New York.
May, R.M. (1974). 'Stability and Complexity in Model Ecosystems.' Princeton University Press, Princeton, New Jersey. 265 pp.
May, R.M. & Anderson, R.M. (1983). Parasite-host coevolution. In 'Coevolution' (Futuyma, D.J. & Slatkin, M., eds), pp. 186-206. Sinaer Associates Inc., Sunderland, Massachusetts.
May, R.M. & Oster, G.F. (1976). Bifurcations and dynamic complexity in simple ecological models. American Naturalist 110, 573-599.
Maynard Smith, J. (1977). Mathematical models in population biology. In 'Mathematics and the Life Sciences' (Matthews, D.E., ed.), pp. 200-221. Springer-Verlag, Berlin.
Maynard Smith, J. (1978). Optimization theory in evolution. Annual Review of Ecology and Systematics 9, 31-56.
Michelmore, R.W. & Crute, I.R. (1982). A method for determining the virulence phenotype of

Bremia lactucae isolates. Transactions of the British Mycological Society 79, 542-546.

Murdoch, W.W. (1975). Diversity, complexity, stability and pest control. Journal of Applied Ecology 12, 795-807.

Mussell, H. (1980). Tolerance to Disease. In 'Plant Disease, An Advanced Treatise. Vol. V. How Plants defend Themselves' (Horsfall, J.G. & Cowling, E.B., eds), pp. 39-52. Academic Press, New York.

Mussell, H. (1981). Exploiting disease tolerance by modifying vulnerability. In 'Plant Disease Control' (Staples, R.C., ed.), pp. 273-284. John Wiley & Sons, New York.

Mussell, H.. & Malone, M.J. (1979). Disease tolerance: reducing the impact of disease-induced stress on crop yields. In 'Stress Physiology in Crop Plants' (Mussell, H. & Staples, R., eds), pp. 16-23. John Wiley & Sons, New York.

Nagarajan, S. (1983). Race dynamics of Puccinia graminis f. sp. tritici in the U.S.A. from 1973 to 1980. Phytopathology 73, 512-514.

Neervoort, W.J. & Parlevliet, J.E. (1978). Partial resistance of barley to leaf rust, Puccinia hordei. V. Analysis of the components of partial resistance in eight barley cultivars. Euphytica 27, 33-39.

Nelson, R.R. (1978). Genetics of horizontal resistance to plant diseases. Annual Review of Phytopathology 16, 359-378.

Nelson, R.R. (1979). The evolution of parasitic fitness. In: 'Plant Disease, An Advanced Treatise. Vol. IV. How Pathogens induce Disease' (Horsfall, J.G. & Cowling, E.B., eds), pp. 23-46. Academic Press, New York.

Nilsson, B.A. (1981). Component analysis of general resistance to Phytophthora infestans in clones from the Colombian potato collection. Potato Research 24, 239-244.

Norelli, J.L., Aldwinckle, H.S. & Beer, S.V. (1984). Differential host x pathogen interactions among cultivars of apple and strains of Erwinia amylovora. Phytopathology 74, 136-139.

Oates, J.D., Burdon, J.J. & Brouwer, J.B. (1983). Interactions between Avena and Puccinia species. II. The pathogens: Puccinia coronata Cda and P. graminis Pers. f. sp. avenae Eriks & Henn. Journal of Applied Ecology 20, 585-596.

Ostergaard, H. (1983a). Predicting development of epidemics on cultivar mixtures. Phytopathology 73, 166-172.

Ostergaard, H. (1983b). Problems in estimating parasitic fitness. In 'Durable Resistance in Crops' (Lamberti, F., Waller, J.M. & Van der Graaf, N.A., eds), pp. 101-104. Plenum Press, New York and London.

Palti, J. (1981). Cultural practices and infectious crop diseases. Springer-Verlag, Berlin, 243 pp.

Parlevliet, J.E. (1976). Partial resistance of barley to leaf rust, Puccinia hordei. III. The inheritance of the host plant effect on latent period in four cultivars. Euphytica 25, 241-248.

Parlevliet, J.E. (1977). Plant pathosystems: an attempt to elucidate horizontal resistance. Euphytica 26, 553-556.

Parlevliet, J.E. (1978b). Race-specific aspects of polygenic resistance of barley to leaf rust, Puccinia hordei. Netherlands Journal of Plant Pathology 84, 121-126.

Parlevliet, J.E. (1979). Components of resistance that reduce the rate of epidemic development. Annual Review of Phytopathology 17, 203-22.

Parlevliet, J.E. (1981). Stabilizing selection in crop pathosystems: An empty concept or a reality? Euphytica 30, 259-269.

Parlevliet, J.E. (1983). Models explaining the specificity and durability of host resistance derived from observations on the barley Puccinia hordei system. In: 'Durable Resistance in Crops' (Lamberti, F., Waller, J.M. & Van der Graaf, N.A., eds), pp. 57-80. Plenum Press, New York and London.

Parlevliet, J.E. & Zadoks, J.C. (1977). The integrated concept of disease resistance: A new view including horizontal and vertical resistance in plants. Euphytica 26, 5-21.

Parlevliet, J.E., Lindhout, W.H., Van Ommeren, A. & Kuiper, H.J. (1980). Level of partial resistance to leaf rust, Puccinia hordei, in West-European barley and how to select for it. Euphytica 29, 1-8.

Parry, G.D. (1981). The meaning of r- and K- selection. Oecologica 48, 260-264.

Person, C. & Christ, B. (1983). Population genetics and evolution of host-parasite interactions. In 'Challenging Problems in Plant Health' (Kommedahl, T. & Williams, P.H., eds), pp. 379-386. American Phytopathological Society, St. Paul.

Person, C., Groth, J.V. & Mylyk, O.M. (1976). Genetic change in host parasite populations. Annual Review of Phytopathology 14, 177-188.

Person, C., Fleming, R., Cargeeg, L. & Christ, B. (1983). Present knowledge and theories concerning durable resistance. In 'Durable Resistance in Crops' (Lamberti, F., Waller, J.M. & Van der Graaf, N.A., eds), pp. 27-40. Plenum Press, New York and London.

Petraitis, P.S. (1979). Likelihood measures of niche breadth and overlap. Ecology 60, 703-710.

Pimm, S.D. (1984). The complexity and stability of ecosystems. Nature 307, 321-326.

Pochard, E. & Daubeze, A. (1980). Recherche et evaluation des composantes d'une resistance polygenique: la resistance du Piment a *Phytophthora capsaci*. Annales Amelioration des Plantes. 30, 377-398.

Pommer, E.H. & Lorenz, G. (1982). Resistance of *Botrytis cinerea* Pers. to dicarboximide fungicides - a literature review. Crop Protection 1, 221-230.

Populer, C. (1978). Changes in host susceptibility with time. In 'Plant Disease, An Advanced Treatise. Vol. II. How Disease develops in Populations' (Horsfall, J.G. & Cowling, E.B., eds), pp. 239-262. Academic Press, New York.

Politowski, K. & Browning, J.A. (1978). Tolerance and resistance to plant disease: An epidemiological study. Phytopathology 68, 1177-1185.

Priestley, R.H. (1978). Detection of increased virulence in populations of wheat yellow rust. In 'Plant Disease Epidemiology' (Scott, P.R. & Bainbridge, A., eds), pp. 63-70. Blackwell Scientific Publications, Oxford.

Pugh, G.J.F. (1980). Presidential address: Strategies in fungal ecology. Transactions of the British Mycological Society 75, 1-14.

Putter, C.J. (1982). An epidemiological analysis of the *Phytophthora* and *Alternaria* blight pathosystem in the Natal midlands. Ph.D. Thesis, University of Natal, South Africa.

Rahe, J.E. (1981). Lack of correlation between field and laboratory tests for resistance with special reference to white rot of onions. In 'Plant Disease Control' (Staples, R.C. ed.), pp. 193-200. John Wiley & Sons, New York.

Rees, R.G., Thompson, J.P. & Mayer, R.J. (1979a). Slow rusting and tolerance to rusts in wheat I. The progress and effects of epidemics of *Puccinia graminis tritici* in selected wheat cultivars. Australian Journal of Agricultural Research 30, 403-19.

Rees, R.G., Thompson, J.P. & Goward, E.A. (1979). Slow rusting and tolerance to rusts in wheat, II. The progress and effects of epidemics in *Puccinia recondita tritici* in selected wheat cultivars. Australian Journal of Agricultural Research 30, 421-32.

Robinson, R.A. (1976). 'Plant Pathosystems.' Springer-Verlag, Berlin, 184 pp.

Robinson, R.A. (1983). Theoretical resistance models. In 'Durable Resistance in Crops' (Lamberti, F., Waller, J.M. & Van der Graaf, N.A., eds), pp. 45-56. Plenum Press, New York and London.

Roelfs, A.P. & Rothman (1976). Races of *Puccinia graminis* f. sp. *avenae* in the USA during 1975. Plant Disease Reporter 60, 703-706.

Roelfs, A.P. & Groth, J.V. (1980). A comparison of virulence phenotypes in wheat stem rust populations reproducing sexually and asexually. Phytopathology 70, 855-862.

Roughgarden, J. (1983). The theory of co-evolution. In 'Coevolution' (Futuyma, D.J. & Statkin, M. eds), pp. 33-64. Sinaer Associates Inc., Sunderland, Massachusetts.

Rouse, D.I. (1985). Construction of temporal models: I. Disease Progress of airborne pathogens. In 'Advances in Plant Pathology, Vol 4' (Gilligan, C.A., ed.), Academic Press, London. In press.

Royer, M.H. & Nelson, R.R. (1981). The effect of host resistance on relative parasitic fitness of *Helminthosporium maydis* Race T. Phytopathology 71, 351-354.

Schmidt, R.A. (1978). Diseases in forest ecosystems: the importance of functional diversity. In 'Plant Disease, an Advanced Treatise. Vol. II. How Disease develops in Populations' (Horsfall, J.G. & Cowling, E.B., eds), pp. 287-315, Academic Press, New York.

Schonbeck, F., Dehne, H.W. & Beicht, W. (1980). Activation of unspecific resistance mechanisms in plants. Zeitschrift für Pflanzenkrankheiten und Pflanzenschutz 87, 654-666.

Schonbeck, F., Dehne, H.W & Balder, H. (1982). The efficiency of induced resistance under practical culture conditions I. Powdery mildew of grape, cucumber and wheat. Zeitschrift für Pflanzenkrankheiten und Pflanzenschutz 89, 177-184.

Schwarzbach, E. (1978). Monitoring airborne populations of cereal mildew. In 'Plant Disease Epidemiology' (Scott, P.R. & Bainbridge, A., eds), pp. 55-62. Blackwell Scientific Publications, Oxford.

Scott, P.R. & Hollins, T.W. (1980). Pathogenic variation in Pseudocercosporella herpotrichoides. Annals of Applied Biology 94, 297-300.

Scott, P.R., Benedikz, P.W. & Cox, C.J. (1982). A genetic study of the relationship between height, time of emergence and resistance to Septoria nodorum in wheat. Plant Pathology 31, 45-60.

Scott, P.R., Johnson, R., Wolfe, M.S., Lowe, H. & Bennett, F. (1980). Host-specificity in cereal parasites in relation to their control. Applied Biology 5, 349-393.

Seem, R.C. (1984). Disease incidence and severity relationships. Annual Review of Phytopathology 22, 133-150.

Segal, A., Manisterski, G., Fischbeck, G. & Wahl, I. (1980). How plant populations defend themselves in natural ecosystems. In 'Plant Disease, An Advanced Treatise. Vol. V. How Plants defend Themselves' (Horsfall, J.G. & Cowling, E.B., eds), pp. 76-102. Academic Press, New York.

Sewell, G.W.F. (1981). Soil-borne fungal pathogens in natural vegetation and weeds of cultivation. In 'Pests, Pathogens and Vegetation' (Thresh, J.M., ed.), pp. 175-190. Pitmans, London.

Shafer, J.F. (1971). Tolerance to plant disease. Annual Review of Phytopathology 9, 235-252.

Shaner, G. (1981). Genetic resistance for control of plant disease. In 'CRC Handbook of Pest Management in Agriculture, Vol. I' (Pimental, D., ed.), pp. 395-444. CRC Press, Boca Raton.

Shaner, G. & Hess, F.D. (1978). Equations for integrating components of slow leaf-rusting resistance in wheat. Phytopathology 68, 1464-1469.

Sharma, J.K. & Heather, W.A. (1979). Comparison of disease parameters for quantitative assessment of Melampsora leaf rust in clones of Populus spp. Transactions of the British Mycological Society 72, 483-488.

Sidhu, G.S. (1983a). Horizontal resistance in tomato against the Meloidogyne-Fusarium complex: an artefact of parasitic epistasis. Crop Protection 2, 205-210.

Sidhu, G.S. (1983b). Introduction (Symposium on genetic aspects of breeding for resistance against soilborne plant pathogens) Phytopathology 73, 940.

Simons, M.D. & McDaniel, M.E. (1983). Gradual evolution of virulence of Puccinia coronata on oats. Phytopathology 73, 1203-1205.

Skylakakis, G. (1981). Effects of alternating and mixing pesticides on the build up of fungal resistance. Phytopathology 71, 1119-1121.

Skylakakis, G. (1982a). The development and use of models, describing outbreaks of resistance to fungicides. Crop Protection 1, 249-262.

Skylakakis, G. (1982b). Epidemiological factors affecting the rate of selection of biocide-resistant genotypes of plant pathogenic fungi. Phytopathology 72, 271-273.

Sorarrain, O.M., Boggio, R. & Favret, E.A. (1979). A mathematical model for the evolution of a host-pathogen system. Mathematical Biosciences 47, 1-13.

Stall, R.E., Marcó, G.M. & Canteros de Eechernique, B.I. (1982). Importance of mesophyll in mature-leaf resistance to cancrosis of citrus. Phytopathology 72, 1097-1100.

Stevens, C. & Gudauskas, R.T. (1982). Relation of maize dwarf mosaic virus infection to increased susceptibility of corn to Helminthosporium maydis Race O. Phytopathology 72, 1500-1502.

Stevens, C. & Gudauskas, R.T. (1983). Effects of maize dwarf mosaic virus infection of corn or inoculum potential of Helminthosporium maydis Race O. Phytopathology 73, 439-441.

Sundheim, L. (1982). Control of cucumber powdery mildew by the hyperparasite Ampelomyces quisqualis and fungicides. Plant Pathology 31, 209-214.

Sundheim, L. & Amundson, T. (1982). Fungicide tolerance in the hyperparasite Ampelomyces quisqualis and integrated control of cucumber powdery mildew. Acta Agriculturae Scandinavica 32, 349-355.

Tegtmeier, K.J., Daly, J.M. & Yoder, O.C. (1982). T-toxin production by near-isogenic isolates of Cochliobolus heterostrophus races T and O. Phytopathology 72, 1492-1495.

Thompson, J.P. & Rees, R.G. (1979). Pattern analysis in epidemiological evaluation of cultivar resistance. Phytopathology 69, 545-549.

Thresh, J.M. (1980a). The ecological approach to the epidemiology of plant virus diseases. In 'Comparative Epidemiology' (Palti, J. & Kranz, J., eds), pp. 57-70. Centre for Agricultural Publishing, Wageningen.

Thresh, J.M. (1980b). The origins and epidemiology of some important plant virus diseases. Applied Biology 5, 1-65.

Tolmsoff, W.J. (1983). Heteroploidy as a mechanism of variability among fungi. Annual Review of Phytopathology 21, 317-340.

Trenbath, B.R. (1974). Biomass productivity of mixtures. Advances in Agronomy 26, 177-210.

Trenbath, B.R. (1977). Interactions among diverse hosts and diverse parasites. Annals of the New York Academy of Sciences 287, 124-150.

Van de Walle, M.J. & Siegel, A. (1982). Relationships between strains of tobacco mosaic virus and other selected plant viruses. Phytopathology 72, 390-395.

Vanderplank, J.E. (1963). 'Plant Diseases: Epidemics and Control.' Academic Press, London. 349 pp.

Vanderplank, J.E. (1968). 'Disease Resistance in Plants.' Academic Press, London. 206 pp.

Vanderplank, J.E. (1982). 'Host-pathogen Interactions in Plant Disease.' Academic Press, London. 207 pp.

Vanderplank, J.E. (1983). Durable resistance in crops: should the concept of physiological races die? In 'Durable Resistance in Crops' (Lamberti, F., Waller, J.M. & Van der Graaf, N.A., eds), pp. 41-44. Plenum Press, New York and London.

Van Tuyl, J.M. (1977). Genetics of fungal resistance to systemic fungicides. Mededelingen Landbouewhogeschool, Wageningen 77-2, 121 pp.

Varley, G.C., Gradwell, G.R. & Hassell, M.P. (1973). 'Insect Population Ecology: an Analytical Approach.' Blackwell Scientific Publications, Oxford. 212 pp.

Varughese, J. & Griffiths, E. (1983). Effects of barley yellow dwarf virus on susceptibility of barley cultivars to net blotch (Pyrenophora teres) and leaf blotch (Rhynchosporium secalis). Plant Pathology 32, 435-440.

Villareal, R.L., MacKenzie, D.R., Nelson, R.R. & Coffman, W.R. (1980). Apparent infection rates of Pyricularia oryzae on different rice cultivars. Phytopathology 70, 1224-1226.

Waggoner, P.E. (1977). Contributions of mathematical models to epidemiology. Annals of the New York Academy of Sciences 287, 191-206.

Wellving, A. & Crute, I.R. (1978). The virulence characteristics of Bremia lactucae populations present in Sweden from 1971-1976. Annals of Applied Biology 89, 251-256.

Whitman, T.G. & Slobodchikoff, C.N. (1981). Evolution by individuals, plant-herbivore interactions, and mosaics of genetic variability: The adaptive significance of somatic mutations in plants. Oecologia 49, 287-292.

Wilcoxson, R.D. (1981). Genetics of slow rusting in cereals. Phytopathology 71, 989-993.

Wolfe, M.S. (1973). Changes and diversity in populations of fungal pathogens. Annals of Applied Biology 75, 132-6.

Wolfe, M.S. (1978). Some practical implications of the use of cereal variety mixtures. In 'Plant Disease Epidemiology' (Scott, P.R. & Bainbridge, A., eds), pp. 201-207. Blackwell Scientific Publications, Oxford.

Wolfe, M.S. (1981). Integrated use of fungicides and host resistance for stable disease control. Philosophical Transactions of the Royal Society of London, 295, 175-184.

Wolfe, M.S. (1983). Genetic strategies and their value in disease control. In 'Challenging Problems in Plant Health' (Kommedahl, T. & Williams, P.H., eds), pp. 461-473. American Phytopathological Society, St. Paul.

Wolfe, M.S. & Barrett, J.A. (1977). Population genetics of powdery mildew epidemics. Annals of the New York Academy of Sciences 287, 151-63.

Wolfe, M.S. & Knott, D.R. (1982). Populations of plant pathogens: some constraints of analysis of variation in pathogenicity. Plant Pathology 31, 79-90.

Wolfe, M.S. & Schwarzbach, E. (1978). Patterns of race changes in powdery mildews. Annual Review of Phytopathology 16, 159-180.

Wolfe, M.S., Barrett, J.A. & Slater, S.E. (1983). Pathogen fitness in cereal mildews. In 'Durable Resistance in Crops' (Lamberti, F., Waller, J.M. & Van der Graaf, N.A., eds), pp. 81-100. Plenum Press, New York and London.

Wright, R.G. & Lennard, J.H. (1980). Origin of a new race of Puccinia striiformis. Transactions of the British Mycological Society 74, 283-287.

Yu, P. (1972). Some host parasite genetic interaction models. Theoretical Population Biology 3, 347-357.

Zadoks, J.C. & Kampmeijer, P. (1977). The role of crop populations and their deployment, illustrated by means of a simulator, EPIMUL 76. Annals of the New York Academy of Science 287, 164-190.

Zadoks, J.C. & Schein, R.D. (1979). 'Epidemiology and Plant Disease Management.' Oxford University Press, Oxford. 427 pp.

Zitelli, G. & Vallega, V. (1977). Strategies to be used in the struggle between resistance and virulence genes. In 'Induced Mutations Against Plant Diseases' (International Atomic Energy Agency, ed.), pp. 97-107. IAEA, Vienna.

CHAPTER 9. MECHANISMS OF INDUCED RESISTANCE TO VIRUS DISEASE

R. S. S. FRASER

9.1. INTRODUCTION

Induced resistance to virus disease occurs when, as a result of a primary inoculation or chemical treatment, a plant becomes resistant to a second or challenge inoculation by a related or unrelated pathogen. The resistance induced may be localized or systemic. Such phenomena are also frequently referred to by the terms 'acquired systemic resistance' and 'localized acquired resistance'. The word 'induced' is preferred here purely because it implies some active process, rather than the passiveness which 'acquisition' tends to imply. However, the use of 'induced' in this connection should not be confused with its other use in the context of constitutive (gene-for-gene) resistance, i.e. the activation of an intrinsic but previously latent resistance mechanism during an early phase of the primary infection.

In animals, viral infection induces two defence mechanisms, the interferon and antibody systems, and amongst other effects these confer resistance to a second infection. In the case of interferon, the induced resistance is effective against unrelated viruses, and is transitory. The antibody system only confers resistance against immunologically related viruses, but can be very long lasting. Both the interferon and antibody systems can be activated by agents other than complete, multiplying virus. The underlying biochemistry of induced resistance in plants is much less well understood than is the case for these animal mechanisms. Parallels have certainly been drawn between plant and animal responses, but so far the evidence tends to suggest that the two groups of effect are separate, without shared features. In particular, there is no evidence from plants for effects akin to the immunological memory of non-self in the antibody system. Plants do have recognition systems equivalent to the self/non-self discrimination of the antibody system, for example that controlling the pollen-stigma interaction and compatibility. This is discussed as a model for host-pathogen recognition in Chapter 7. However, there is as yet no evidence for operation of such systems in induced resistance to virus diseases.

A commonly reported feature of plant resistance induced by prior infection is lack of specificity. Thus resistance induced by a virus infection may defend against subsequent attacks by unrelated viruses, as well as by fungi, bacteria and even insects (McIntyre et al., 1981). Conversely, resistance induced by fungal or bacterial pathogens, or by chemical or host developmental factors, may protect against subsequent virus infection. In practice, the concepts and experimental approaches to induced resistance to viruses on one hand, and to fungi and bacteria on the other, are quite different. It is therefore appropriate to consider them in separate Chapters. Bacteria and fungi are mentioned in this Chapter only in so far as they may induce resistance to viruses.

However, not all types of induced resistance display such broad spectrum effects: in the group of phenomena known as cross protection, prior infection by one virus may only confer resistance against subsequent infection by related viruses.

There appear therefore to be several types of mechanisms of induced resistance involving virus diseases. For convenience, they will be divided into those where the protecting virus is localized around the infection sites, and induces resistance in uninfected tissue; and those where the protecting virus is present systemically, and may be in direct contact with the challenge inoculum. Understanding how these mechanisms work is important, because the knowledge may eventually be exploited to develop novel methods of crop protection; progress to this end will be reviewed.

9.2. RESISTANCE INDUCED BY LOCALIZED INFECTIONS AND ANALOGOUS EFFECTS

9.2.1. Localized induced resistance. Fewer necrotic lesions are formed after challenge inoculation, in the immediate vicinity of previously formed lesions, than on tissue at some greater distance from the primary lesions. Most authors have interpreted localized induced resistance as that effective within a few mm from the primary lesion, i.e. within the limits of cell-to-cell spread of a stimulus, but not requiring its transport in the vascular system. Yarwood (1960) reported that TMV lesions gave TMV-specific localized resistance on bean leaves: the induced resistance was not effective against AMV or TNV. However, the majority of later reports suggested that localized induced resistance is non-specific. Ross (1961a) showed that TMV lesions on hypersensitive tobacco induced non-specific resistance to other viruses. Hecht and Bateman (1964) found that lesions of the fungus *Thielaviopsis basicola* prevented formation of TMV and TNV lesions, and that viral lesions inhibited the fungus.

In considering the mechanisms of localized induced resistance, it is difficult to distinguish between biochemical changes involved in preventing the secondary infection, and those that are involved in the localization of the primary inoculum. *A priori*, it might seem likely that the latter is a specific effect, a consequence of a unique gene-for-gene reaction between host and pathogen, while the former is generally non-specific. This does not help, however, to assign biochemical changes to either type of resistance. Because of this, the reader is referred to Chapter 6.3 for discussion of biochemical and ultrastructural changes around lesions, and their possible involvement in resistance.

Loebenstein *et al.* (1968) reported that actinomycin-D prevented induction of localized acquired resistance to TMV in tobacco and pinto bean. However, the same inhibitor caused an increase in size of lesions when applied to the primary infection (Loebenstein *et al.*, 1969): its effects on induction of localized resistance are therefore hard to assess.

9.2.2. *Induced systemic resistance: occurrence and effectiveness.* In the majority of cases, a primary necrotic infection induces systemic as well as localized resistance. The earliest report was by Gilpatrick & Weintraub (1952), who found that local lesion infection of a *Dianthus barbatus* clone with carnation mosaic virus resulted in formation of fewer lesions when upper, previously uninoculated leaves were challenged with the same virus. This property appeared to be a character of the particular clone used, as later experiments (Weintraub & Kemp, 1961) could not demonstrate it. The classical early work on acquired systemic resistance was by Ross (1961b) and his co-workers (Bozarth & Ross, 1964). These early experiments are reviewed in Ross (1966). He showed that in a number of host-virus combinations, primary lesion formation on lower leaves, or on half leaves, induced non-specific systemic resistance in the form of reduced number or size of lesions formed on subsequent challenge inoculation with the same or different viruses. Ross (1966) considered reduction in lesion size to be a better index of systemic induced resistance than reduction in number. He showed that systemic spread of resistance required living petiole tissue, and thus presumably involved phloem transport. Finally, comparison of biochemical and ultrastructural changes in the responses to primary and secondary inoculations (Simons & Ross, 1971; Ross

Table 9.1. Pathogenic agents, chemicals, and developmental treatments inducing systemic resistance to challenge virus infections, and / or pathogenesis-related proteins.

Host	Inducer	Challenge Virus	ISR	PR protein	References
	VIRUSES:				
tobacco Samsun-NN	SBMV	TMV	+	nt	Ross, 1961a
bean	TMV	SBMV			Ross, 1966
cowpea	CMV AMV	CMV AMV	+	nt	Bozarth & Ross, 1964
N. glutinosa	TMV protein	TMV	+	nt	Loebenstein, 1962
Datura stramonium	TMV	TMV	+	nt	Loebenstein, 1963
Gomphrena globosa	PVX	PVX	+	nt	
tobacco Xanthi-nc	PVY CMV PVX AMV	TMV	+	+	Kassanis *et al.*, 1974
tobacco Xanthi-nc	TMV	TMV	+	+	Gianinazzi *et al.*, 1970 Kassanis *et al.*, 1974
tobacco Sansum-NN	TMV	TMV	+	+	Van Loon & van Kammen, 1970 Van Loon & Dijkstra, 1976
tobacco Samsun-nn	TNV	TMV	+	+	Van Loon, 1975 Van Loon & Dijkstra, 1976
cucumber	TNV	TNV	+	+	Coutts & Wagih, 1983
bean	AMV PSV	-	nt	+	Abu-Jawdah, 1982
N. glutinosa	TMV	TMV	±[a]	+	Fraser *et al.*, 1979
tomato	TMV CMV	-	nt	+	Camacho-Henriquez & Sänger, 1982
	OTHER PATHOGENS:				
tobacco Samsun-NN	*Thielaviopsis basicola*	TMV, TNV	+	nt	Hecht & Bateman, 1964
tobacco Burley cvs	*Thielaviopsis basicola*	TMV	nt	+	Gianinazzi *et al.*, 1980
tobacco cvs	*Pseudomonas syringae*	TMV	+	+	Ahl *et al.*, 1981
Gynura aurantiaca	CEV	-	nt	+	Conejero & Semancik, 1977
tomato	PSTV *Cladosporium fulvum*	-	nt	+	Camacho-Henriquez & Sänger, 1982

ABBREVIATIONS: ISR: induced systemic resistance. PR: pathogenesis-related proteins. nt: not tested. BA: benzyladenine. IAA: indoleacetic acid. 2,4-D: 2,4-dichlorophenoxyacetic acid.

Host	Inducer	Challenge Virus	ISR	PR protein	References
	DEVELOPMENTAL FACTORS:				
tobacco Xanthi-nc	callus culture	-	nt	+	Antoniw *et al.*, 1981
N. glutinosa x *N. debneyi* hybrid	none	TMV	+[b]	+[b]	Ahl & Gianinazzi, 1982
tobacco cvs	grafting to hybrid rootstock	-	nt	+	Gianinazzi & Ahl, 1983
tobacco Xanthi-nc	flowering and senescence	TMV	+?[c]	+	Fraser, 1981
N. glutinosa	nutrient deficiency	-	nt	+	Ahl & Gianinazzi, 1982
	CHEMICALS				
tobacco Samsun NN	yeast nucleic acid	TMV	+	nt	Gicherman & Loebenstein, 1968
N. glutinosa	polycarboxyl- ates	TMV	+	nt	Stein & Loebenstein, 1972
tobacco Xanthi-nc Samsun-NN	polyacrylic acid	TMV	+ -	+ -	Gianinazzi & Kassanis, 1974
tobacco cvs	aspirin salicylic acid benzoic acid	TMV	+	±[d]	White 1979 Antoniw & White, 1980 Van Loon & Antoniw, 1982
Gomphrena globosa	aspirin	TBSV	+	-	Pennazio & Redolfi, 1980
tobacco Samsun-NN	$HgCl_2$	TMV	+	-	Van Loon, 1975
tobacco Xanthi-nc	$HgCl_2$	TMV	+	nt	Sziraki *et al.*, 1980
tobacco Samsun-NN	ethephon	TMV	+	+	Van Loon, 1977
tobacco Xanthi-nc	BA IAA 2,4-D	TMV	+	+	Antoniw *et al.*, 1981
tobacco Xanthi-nc	MBC ABA	TMV TMV	-[e] +	+ -	Fraser 1982
cucumber	mannitol	-	nt	+	Wagih & Coutts, 1982

[a]Prior inoculation caused a slight decrease in the size, but a large increase in numbers, of lesions formed on challenge inoculation.

[b]PR proteins were constitutive, and plants highly resistant to infection by TMV.

[c]Senescent, flowering plants are more resistant to infection by TMV than young plants.

[d]Untreated leaves on plants injected with salicylic acid often showed ISR but did not accumulate PRs.

[e]at low doses.

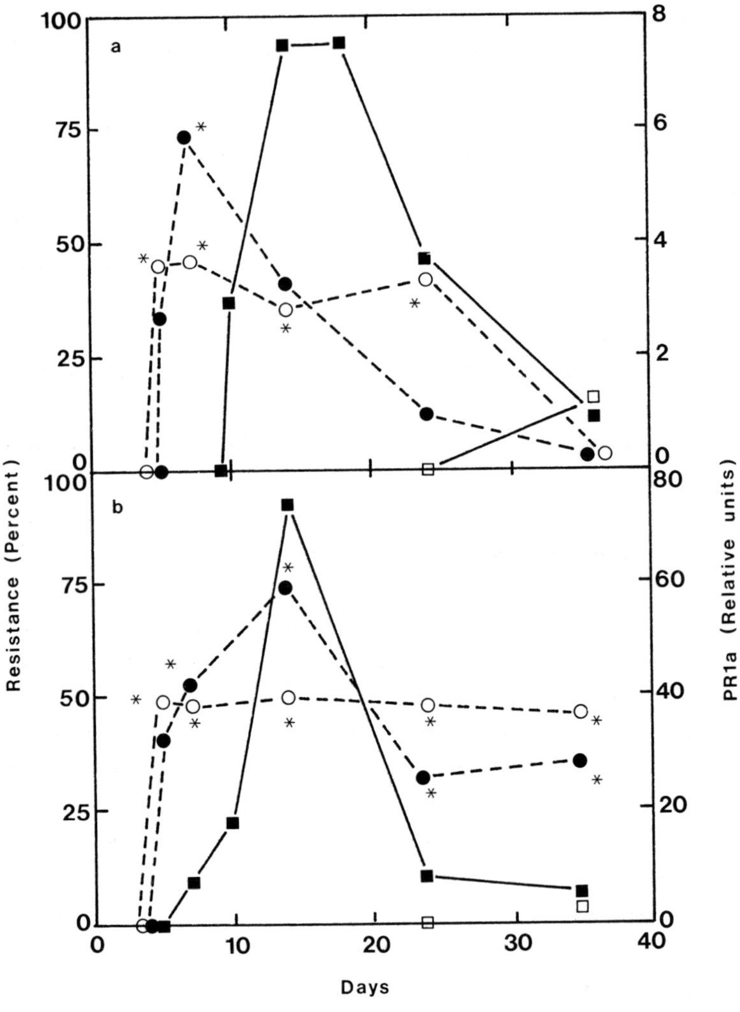

FIGURE 9.1. Changes in acquired systemic resistance and concentration of PR1a protein with time after inoculation with TMV. At day 0, Nicotiana tabacum cv. Xanthi-nc plants were inoculated on one half of each of two lower leaves. These formed 50 to 100 lesions per half leaf. Control plants were sham-inoculated. The concentration of PR1a protein in infected (■—■) or control (□—□) plants was measured in upper leaves (a) and in the uninoculated half of lower leaves (b). Acquired systemic resistance was measured in upper and lower half leaves by challenge inoculation at various times and expressed as the percentage reduction in lesion number (●--●) or diameter (○--○) caused by the primary inoculation. For points marked (*) the difference between control and primarily inoculated treatments was significant at P = 0.05. Reproduced from Fraser, 1982.

& Israel, 1970), led to the suggestion that the systemic effect was one of increased hypersensitivity, resulting from activation or enhancement of the normal localizing mechanism.

In the decades since the original discovery of induced systemic resistance, the phenomenon has been reported in numerous host-pathogen combinations. Several chemical and environmental treatments have also been found to induce systemic resistance. Some representative examples are given in Table 9.1, which is by no means exhaustive. In the great majority of cases, the challenging virus has been one which produces necrotic lesions. The effectiveness of induced systemic resistance may be as high as 97% for reduction in lesion numbers (Kassanis & White, 1975), and may reduce lesion size to the point of invisibility (Ross, 1966; Balazs et al., 1976; 1977). Induced resistance can be very long-lasting (Ross, 1961b) although some decline with time can be seen in some systems (Figure 9.1). Roberts (1983) showed that the progeny of hypersensitive tobacco plants with induced systemic resistance also showed some resistance, although it was necessary to inoculate the parent plants several times.

Surprisingly, the question of whether multiplication of the challenging virus is actually inhibited in leaves with induced systemic resistance is still much in doubt, and indeed has rarely been investigated. Ross (1966) measured lesion area and infectivity recoverable from challenge-inoculated leaves of TMV-infected Samsun NN tobacco. Leaves with induced resistance had smaller lesions, with lower levels of recoverable infectivity, than control leaves. These results could indicate that virus multiplication was inhibited, but no adjustment was made for any difference in lesion numbers between treatments, which could have influenced the outcome. In most experiments, moreover, the reduction was small. Balazs et al. (1977) and Fraser (1979) reported cases where lesion size was reduced in leaves with induced resistance, but virus multiplication per lesion was not. Balazs et al. (1977) suggested that induced resistance involved a decrease in necrotization rather than any effect on virus multiplication. Coutts & Wagih (1983) concluded that the resistance induced by TNV lesions on cucumber leaves was associated with a reduction of symptom expression during the challenge inoculation, rather than with inhibition of virus multiplication, and Pennazio et al. (1983) reached a similar conclusion for TNV on tobacco.

There have also been few investigations of whether the induced resistance is effective against viruses which cause systemic, non-necrotic infections.

Van Loon & Dijkstra (1976) showed that prior inoculation of Samsun _nn_ tobacco plants with TNV, and formation of necrotic lesions, reduced the systemic multiplication of TMV in a subsequent challenge infection. In contrast, Fraser (1979) found no reduction in multiplication of a systemic strain of TMV in White Burley tobacco plants with resistance induced by a necrotic strain.

9.2.3. Pathogenesis-related proteins and induced resistance. Van Loon & Van Kammen (1970) and Gianinazzi _et al_. (1970) first reported the occurrence of new, host-coded proteins in _N_-gene tobacco cultivars where systemic resistance had been induced by a local TMV infection. The proteins were named 'b'-proteins (Gianinazzi _et al_., 1970) or 'pathogenesis-related' (PR) proteins (Antoniw _et al_., 1980) (Figure 9.2). It has been suggested that they may be involved in the mechanism of induced resistance, perhaps in a way analogous to interferon (Kassanis _et al_., 1974; Gianinazzi, 1982). This, and the intrinsic interest of these proteins and their metabolism as model systems for study of control of protein synthesis in plants, has stimulated much research on them since their discovery. However, at the time of writing, their function remains unclear, and no definite role in resistance has been unequivocally demonstrated.

Occurrence. Proteins sharing some or many of the characteristics of the tobacco PRs and their mode of induction have been reported in several other species. Ahl _et al_. (1982) and Gianinazzi & Ahl (1983) have studied PR protein patterns in a large number of _Nicotiana_ species, as well as in many varieties of _N. tabacum_. They found evidence for intra- and interspecific variation, and at least 7 different PR proteins. Recent studies by Van Loon (1982) and Pierpoint (1983) have uncovered 7 further, potential PR proteins in tobacco, although they have not yet been fully characterized. (Figure 9.2).

Coutts (1978) and Coutts & Wagih (1981; 1983) reported two PR-like proteins in cowpea and one in cucumber. Pennazio & Redolfi (1980) and Redolfi _et al_. (1982) found up to 5 new proteins in _Gomphrena globosa_ leaves after hypersensitive virus infections. Further examples of species which accumulate PR-like proteins after infection or other treatments are listed in Table 9.1. It would appear that this type of response to infection is widespread in plants.

Properties. The extent to which PR-like proteins from different species have been characterized varies, but certain common properties emerge where data are available. The vast majority fall in the molecular weight range 10,000 to 20,000. They are soluble at comparatively low pH. Tobacco PRs are

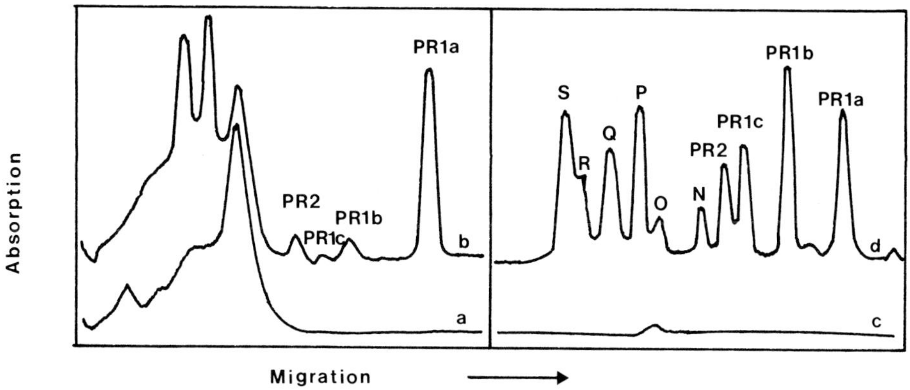

FIGURE 9.2. Densitometer tracings of electrophoretic separations of proteins extracted at pH 2.8 from N-gene containing tobacco leaves. a) From sham-inoculated healthy leaves. b) From leaves bearing an average of 50 TMV lesions, one week after inoculation. c) and d) are proteins from healthy and infected leaves respectively, after digestion with 12.5 μg trypsin / ml for 24 h at 2°C. Letters and numbers indicate particular PR proteins. a) and b) from Fraser, 1982; c) and d) from Van Loon, 1982, by permission of Dr L C Van Loon and Plenum Publishing Corporation.

commonly extracted at pH 2.8 (Gianinazzi et al., 1977; Antoniw & Pierpoint, 1978), while the tomato PR p14 can be extracted in acetic acid (Camacho-Henriquez & Sänger, 1982). Acid extraction provides a considerable enrichment of PRs, as most host proteins are insoluble under such conditions.

The first PRs to be studied showed particularly high electrophoretic mobility when fractionated on non-denaturing gels (Van Loon & Van Kammen, 1970; Gianinazzi et al., 1970); this ability to separate them from other host proteins undoubtedly facilitiated their discovery and early investigation. The high mobility of these particular proteins was a result of a high proportion of acidic amino acid residues (Antoniw & White, 1983). However, more recently-discovered PRs have lower electrophoretic mobilities (Pierpoint, 1983), and a high proportion of acidic residues may not turn out to be a universal characteristic. The tomato p14 protein is highly basic (Camacho-Henriquez et al., 1983).

Both the tomato p14 protein, and the 5 tobacco PRs for which amino acid compositions have been determined, have comparatively high proportions of apolar residues (Camacho-Henriquez et al., 1983; Antoniw & White, 1983). This would be consistent with an association with membranes (Benson & Jekela, 1976).

A striking feature of the tobacco PR proteins is very high resistance to degradation by proteolytic enzymes (Pierpoint et al., 1981; Van Loon, 1982; Figure 9.2). This property has been useful in the discovery of further potential PR proteins: those which are obscured on gel electrophoresis by healthy plant proteins with similar mobilities. Bean PRs have also been reported to be protease-resistant (Redolfi, 1983), but tomato p14 is protease-sensitive (Camacho-Henriquez et al., 1983).

Further consideration of the occurrence and properties of PRs can be found in recent reviews by Redolfi (1983) and Antoniw & White (1983).

Induction of PR proteins. Generally, PR proteins are absent from, or extremely difficult to detect, in normally growing, healthy plants (Van Loon, 1976a; 1983). Induction follows infection with necrotizing or non-necrotizing viruses, as well as by fungi, bacteria, viroids and a wide range of chemicals. Many examples of PR-inducers are given in Table 9.1, and further instances can be found in the review by Van Loon (1983). Necrosis itself does not appear to be necessary for induction, but the largest production of PRs seems to occur with necrotic infections (Van Loon, 1983), and concentrations of PRs are highest in tissue near the lesions (Rohloff & Lerch, 1977; Fraser, 1982).

Some insights into possible natural mechanisms of PR induction have come from experiments where induction is by developmental or plant-related factors. Fraser (1981) showed that healthy, untreated Xanthi-nc plants accumulated large concentrations of PRs in all leaves as the plant flowered and the lower leaves senesced. Removal of either the developing inflorescence or the lower leaves prior to yellowing reduced PR accumulation in the remaining leaves by around 90%. This suggests that signals from both ends of the shoot are necessary for induction. A possible model is that a signal molecule from one end induces production of a second signal from the other end, which then acts as the inducer.

Further evidence for a mobile inducer comes from grafting experiments by Gianinazzi & Ahl (1983). They showed that PR proteins could be induced in non-infected scions from n-gene tobacco cvs by grafting on to N-gene rootstocks with localized TMV infections. Virus was not transferred across the graft union, but some inducer of PRs clearly was. They were also able to exclude any transfer of PR proteins across the graft union: the pattern of PRs induced in the scion was characteristic of the scion genotype, and not that of the rootstock. This work also indicated that the mobile inducer is not PR protein gene-specific.

PR proteins can be induced in tobacco cvs lacking the N allele (Van Loon,

1975) and the PR and *N* genes are inherited independently (Ahl *et al.*, 1982). However, Gianinazzi & Ahl (1983) have suggested that the *N* gene may play a regulatory role in expression of the PR genes, possibly by controlling production of the 'mobile compound' responsible for PR induction.

A further finding was that the interspecific hybrid *N. glutinosa* (*NN*) x *N. debneyi* (*nn*) produced PRs constitutively without inducers (Ahl & Gianinazzi, 1982). Furthermore, the uninfected rootstock of this hybrid was able to induce PRs in healthy scions of Xanthi-nc (Gianinazzi & Ahl, 1983). Presumably the hybrid contains the mobile inducer as a constitutive component. It will be interesting to compare the constitutive inducer of the hybrid with the inducer(s) present in healthy plants during flowering (Fraser, 1981) and to determine whether the hybrid plants are altered in some way in hormonal control of development. The hybrids do tend to be smaller than the parents and may show early, senescence-like symptoms.

Turning to the more immediate mechanism of PR induction, Van Loon (1983) has drawn attention to a probable central role for ethylene. The ethylene-releasing compound ethephon (Van Loon, 1977), and the natural and immediate precursor of ethylene, 1-amino cyclopropane-1-carboxylic acid (ACC) (Van Loon, 1983) are both effective inducers of PRs. Incubation of TMV-infected leaf discs with aminoethoxyvinylglycine inhibited ACC formation and PR accumulation (Van Loon, 1983). Almost all other inducers of PR proteins so far studied appear to stimulate ethylene biosynthesis; some have not yet been examined.

The one group of inducers which does not stimulate ethylene biosynthesis comprises benzoic acid and derivatives such as salicylic acid (2-hydroxybenzoic acid) (Van Loon & Antoniw, 1982) and 2,6-dihydroxybenzoic acid (Van Loon, 1983).

Induction of PR proteins by ethylene is temperature sensitive: it does not occur above 30°C, whereas salicylic acid could induce PR proteins at 32°C (Van Loon & Antoniw, 1982). Van Loon (1983) proposed that ethylene induces a benzoic acid-type compound *in vivo* by a temperature-sensitive mechanism, and that this compound then induces PR proteins by a temperature-insensitive step. The nature of the putative, natural benzoic acid-type compound remains to be established: none of the exogenous inducers of the benzoic acid type appears to occur in tobacco plants, although they are found in other species.

A further, intriguing step towards understanding of the control of PR protein induction was made when Carr *et al.* (1982) showed that PR proteins were synthesized _in vitro_ by protein-synthesizing systems provided with messenger RNAs from TMV-infected or healthy Xanthi-nc leaves. The latter result contrasts with the fact that PR proteins are normally undetectable in healthy leaves, and suggests that healthy leaves do contain mRNA for PR proteins, but in a latent, untranslated form.

Carr (1983) reported that actinomycin-D, an inhibitor of transcription, would induce PR protein accumulation in healthy leaves. A possible model is that inhibition of translation of the latent PR mRNA is effected by an unstable transcription product. Similar mechanisms have been proposed for translational control in mammalian cells (Tomkins _et al._, 1969) and yeast (Fraser, 1975).

Earlier evidence indicated that actinomycin-D inhibited PR synthesis induced by TMV infection (Van Loon & Van Kammen, 1970) or polyacrylic acid (Kassanis & White, 1974). In contrast, actinomycin-D did not inhibit induction of PR proteins by ethephon or salicylic acid (Van Loon, 1983). These results might indicate that actinomycin-D can affect PR biosynthesis at two separate stages, depending on whether an unstable transcriptional product is required during PR induction or not.

It is clear that PR proteins offer a unique model system for study of the control of protein biosynthesis in higher plants.

9.2.4. _Are PR proteins involved in induced systemic resistance_? Table 9.1 shows that a wide range of treatments, both biotic and abiotic, induce both PR proteins and systemic resistance. The evidence for a link between the proteins and induced resistance is thus comprehensive, but entirely correlative. If PR proteins are involved in resistance, several predictions may be made:

a) the amount of resistance should show some correlation with the concentration of PR protein;

b) induced systemic resistance should not be detectable in the absence of PR proteins;

c) it should not be possible to induce PR proteins without also inducing resistance, and

d) the PR proteins should be shown to have some specific role in the mechanism of induced resistance.

Experimental demonstration that any of these predictions is invalid would not necessarily exclude a role for PR proteins in resistance, but would make involvement less likely. The available experimental evidence against each prediction will now be summarized.

a) Figure 9.2 shows time courses of amounts of induced resistance, measured by reduction in lesion size or number, and concentrations of the commonest PR protein, PR1a, in Xanthi-nc tobacco leaves. Resistance was measured in upper, previously uninoculated leaves, and in lower half leaves opposite halves which had received the primary inoculation. Upper leaves accumulated only about one tenth as much PR1a as opposite half leaves, but the amount of resistance induced in the two locations was similar. The concentration of PR1a in both types of leaf rose, then declined with time after the primary inoculation. The time course for induced resistance expressed as reduction in lesion diameter showed a more or less constant level of resistance once established. The time course of resistance measured as reduction in lesion number had a similar shape to that for PR1a concentration, but preceded it. In particular, resistance could be detected before PR1a was detectable.

In a separate type of experiment, Fraser (1981) altered the amount of PR proteins accumulating in flowering Xanthi-nc tobacco plants, by prior removal of either or both of the developing inflorescence and senescing lower leaves. When plants were inoculated with TMV, the size and numbers of lesions formed bore no relationship to the PR protein concentration, although this varied over a ten-fold range. What did emerge was that removal of the inflorescence in any treatment led to a significant reduction in lesion diameter, and removal of the lower leaves led to a significant increase in lesion number. This, together with the time course data, suggests that lesion size and number were influenced by separate factors. All the data argue against any close quantitative or temporal relationship between the concentrations of those PRs measured (principally PR1a) and amount of resistance induced.

b) If Xanthi-nc plants were sprayed with low concentrations of abscisic acid, the numbers and sizes of lesions formed on subsequent inoculation were less than on control plants, but no PR proteins were detectable (Fraser, 1982). Interpretation of this result as evidence for induction of resistance without PR proteins makes the assumption that the apparent resistance induced by abscisic acid was caused in the same way as authentic, virus-induced resistance. It should also be noted that many different compounds cause alterations in lesion size when applied to leaves.

Van Loon & Antoniw (1982) showed that injection of salicylic acid into leaves of Samsun-NN tobacco caused accumulation of PR proteins and induced resistance. Untreated leaves on injected plants did not accumulate PR proteins, but in several experiments did show induced resistance, measured as a reduction in lesion size.

These experiments therefore suggest that resistance-like effects can be induced in the absence of PR proteins.

c) Methyl benzimidazol-2-yl carbamate supplied at low dosages caused accumulation of PR protein, but did not induce resistance (Fraser, 1982). Very high, phytotoxic, concentrations of the compound, however, induced both resistance and PR proteins.

Nicotiana glutinosa plants accumulated large amounts of a PR protein in uninoculated leaves after infection with TMV. Lesions formed on challenge inoculation of these leaves were indeed slightly smaller than those on control plants, indicating resistance. However, lesion numbers were considerably increased as a consequence of the first inoculation, suggesting that the presence of PR protein was associated with an increase in sensitivity (Fraser *et al*., 1979).

These lines of evidence therefore suggest that PR proteins may be induced without concomitant induction of resistance.

d) Various workers have sought direct evidence for a role of PR proteins in resistance. Kassanis & White (1978) added purified PR protein to protoplasts, but could find no evidence for reduction in TMV multiplication. Experiments on the other obvious test, injection of purified PR proteins into leaves then assaying for enhanced resistance, do not seem to have been published. Rather more indirectly, Wieringa-Brants (1983) showed that injection of leaves with intracellular fluids obtained from leaves with local or systemic induced resistance caused a reduction in size and numbers of lesions formed on inoculation. The bulk of the PR proteins are known to be in the cell walls or intracellular spaces (L.C. van Loon, personal communication; Parent & Asselin, 1984). The possibility that some other factor in fluid from leaves with induced resistance is responsible for the reduction in lesion size and number remains to be excluded. However, these experiments form the first source of direct, as opposed to correlative, support for a role of PR proteins in induced resistance. Clearly, the system deserves further study.

At the gross level, the balance of the evidence tends to suggest that PR proteins do not play a central role in the mechanism of induced systemic resistance. Considered purely from a logical point of view, however, it is impossible to prove the negative statement, that PR proteins are not involved, as any hypothetical model of involvement could be modified to accommodate awkward experimental results.

If PR proteins are not involved in induced resistance, what do they do? A demonstration of a function in some process other than induced resistance would be a good indication, though by no means absolute proof, that they do not function in resistance. Van Loon (1982) reviewed evidence showing that the PR1 and PR2 groups did not show activity in tests for 30 enzymes known to increase in activity after infection. The high content of apolar residues of those PR proteins where amino acid compositions are known, and their extracellular location, might suggest an involvement with wall or membrane structure. The significance of their high protease resistance may also provide a clue, but available results from in vitro translation (Carr et al., 1982) and radiolabelling studies (Pierpoint, 1983) have given no suggestion that they are degradation end-products.

Fraser & Clay (1983) argued that induced resistance in the form of reduction of lesion size, and the probable wall location of PR proteins, might be reflected in ultrastructural changes in the cell wall and plasmodesmata. In a preliminary study by electron microscopy, they found no immediate differences in wall structure of control and induced resistant tissue. However, tissue with induced resistance did show many more myelinic bodies: multilamellar membrane structures located between the cell wall and plasma membrane. It would be premature, however, to associate these with PR proteins.

This section has concentrated on a possible role of PR proteins in the induced resistance to second inoculation. Their possible role in localization of virus around the site of the primary infection is considered in Chapter 6.3.4.

9.2.5. Other mechanisms possibly involved in induced resistance. The alterations in lesion size and number as a result of induced resistance might involve the same, or separate mechanisms. Ross (1966) and Balazs et al. (1976) suggested that reduced lesion number could occur as a consequence of reduced size: lesions on resistant leaves simply became too small to detect. In other cases, however, the two effects can be clearly separated. Thus Figure 9.1 shows that the time courses of resistance expressed as reduced size or number were clearly different, suggesting separate mechanisms. In N. glutinosa, reduced lesion size was accompanied by increased lesion number (Fraser et al., 1979), also suggesting separate controls.

Reduction in lesion number could represent a reduction in the number of successful infection events, i.e. this type of induced resistance could be expressed at a very early stage of pathogenesis. Reduction in lesion size might involve reduced necrosis, or rate of virus spread, and would operate after infection had established.

Resistance expressed as reduction in lesion size. Induced resistance might represent an increased level of activity of the primary resistance response, as suggested by Ross & Israel (1970), perhaps just advanced in time as a consequence of the first inoculation. Alternatively, it might represent some quite separate mechanism. In support of the former model, Simons & Ross (1971) reported that while the concentrations of phenolics, and activity of phenylalanine ammonia lyase (PAL) were unaltered during induction of systemic resistance, on challenge inoculation PAL activity increased more than in control leaves, and there was an earlier and greater reduction in concentration of phenolics. They interpreted this as indicating a faster rate of oxidation to quinones, giving more rapid virus inactivation. Fritig *et al.* (1973) also found no change in PAL activity during induction of systemic resistance. In contrast to Simons & Ross (1971), they suggested that changes in PAL activities after challenge inoculation of resistant or control leaves were of similar magnitude, and that differences in activity could not explain induced resistance. Van Loon (1982), however, has suggested that PAL activity per cell could be higher in resistant tissue.

Direct evidence supporting the idea that the mechanism of induced resistance is quite separate from the primary mechanism is scarcer. Chessin (1982) found that ultraviolet light increased the size of TMV lesions on pinto bean, suggesting that it inhibited the primary localization. But ultraviolet light had no effect on induced systemic resistance, suggesting a different mechanism.

Various authors have presented evidence suggesting that induction of systemic resistance does not involve increases in activities of other enzymes such as peroxidase, polyphenoloxidase or catalase (Batra & Kuhn, 1975; Stein & Loebenstein, 1976; Van Loon, 1976b; Faccioli, 1979). However, De Laat & Van Loon (1983) showed that the activity of the enzyme converting 1-aminocyclopropane-1-carboxylic acid to ethylene was higher in leaves with induced resistance, ensuring a more rapid and higher ethylene production on challenge inoculation than in previously uninoculated control plants.

Stein *et al.* (1979) induced resistance to TMV, expressed as reduction in both lesion size and number, by treatment of tobacco leaves with an ethylene maleic anhydride co-polymer. They suggested that this was associated with a 'specific ribosome fraction' which tended to aggregate during concentration *in vitro*. There was no indication of how this change of apparent 'stickiness' of ribosomes might be related to resistance.

There is evidence that induction of systemic resistance is accompanied by an increase in cytokinin concentration (Sziraki *et al.*, 1980). Application of exogenous cytokinin can also reduce lesion size (Balazs *et al.* 1976) although other effects have been reported (reviewed in Fraser & Whenham, 1982). Balazs *et al.* (1977) suggested that reduction of lesion size in induced resistance might represent reduction of necrosis caused by cytokinin effects. However, the changes in cytokinin concentration involved were small.

It is noteworthy that while all reports of resistance refer to reduction in lesion size, there are numerous treatments which cause increased lesion size (discussed in Chapter 6). It may be interesting to consider whether these represent separate mechanisms, or whether they merely interact in a different qualitative way with the mechanism modulating the control of lesion size in induced resistance.

Resistance expressed as reduction in lesion number. Reduction in lesion number as a result of failure to initiate infection could occur for mechanical reasons, because of thicker walls and cuticle, by structural changes in infection sites, or by changes in susceptibility caused by altered water relations. Surprisingly, there does not appear to have been extensive ultrastructural investigation of wall and cuticle thickness of leaves with induced resistance. Thomas and Fulton (1968a; 1968b) showed that the tobacco line T.I. 245 was constitutively more resistant to infection by several viruses, possibly because of a reduced number of ectodesmata, which they considered as potential infection sites. Cell-to-cell spread of viruses was also inhibited: the basis for this was not investigated, but deserves further study in view of the finding that systemic resistance could not be induced in this variety.

There is good evidence that altered lesion number could result from altered leaf water status. Turgid leaves are known to be more susceptible to mechanical wounding and hence infection, than flaccid ones (Yarwood, 1959). Cassells *et al.* (1978) induced resistance in tobacco leaves by polyacrylic acid treatment, then showed that the resistance could be reversed by treatment with an antitranspirant. The implication was that polyacrylic acid made leaves flaccid and less easy to wound. These experiments have been criticized because of the very high polyacrylic acid concentrations used (Kassanis, 1982) but this would not appear completely to invalidate the conclusion.

Fraser (1979); Fraser et al. (1979) and Whenham & Fraser (1981) have established an inverse correlation between alteration in lesion numbers, and changes in leaf concentration of abscisic acid (ABA), an indicator of leaf water stress. In N. glutinosa, where previous infection increases the number of lesions formed on challenge inoculation, ABA concentrations of leaves with induced susceptiblity were lower than in controls, indicating higher turgidity. In N. tabacum, where previous infection decreases lesion number formed on challenge inoculation, ABA concentration was higher than in control leaves. Whenham & Fraser (1981) showed that there was sufficient transport of ABA from the primarily-inoculated lower leaves to upper leaves, to explain their increased ABA concentration. The decreased lesion number on leaves with induced resistance may have been a direct effect of this increased ABA concentration, rather than involving any change in water status, as leaves of healthy plants treated with similar amounts of exogenous ABA also showed reduced lesion numbers on inoculation (Fraser, 1982).

9.3. RESISTANCE INDUCED BY SYSTEMIC INFECTION

9.3.1. Dark green islands. Systemic virus infection causes a light green or yellow/dark green mosaic pattern on leaves of numerous monocot and dicot species (e.g. Caldwell, 1934; Chamberlain et al., 1977). In some cases the dark green areas can have higher chlorophyll concentration than comparable healthy tissue. The dark green areas may be comparatively or completely virus-free, and variably resistant to challenge inoculation. It is important to distinguish these true areas of induced resistance, from mosaic regions of different colour formed due to infection with contaminating strains or spontaneous mutants (Chalcroft & Matthews, 1967).

Occurrence and resistance. Fulton (1951) inoculated Nicotiana sylvestris (N' gene) with a mosaic-forming strain of TMV, then challenge-inoculated with a local lesion strain. Lesions formed on the dark green areas, but were fewer, smaller, and slower to appear than those on previously uninoculated leaves. No lesions formed on the light green areas, indicating a greater level of interference associated with their higher virus concentration.

In CMV-infected tobacco, Loebenstein et al. (1977) could not detect infectious RNA or viral antigen in dark green areas. These areas were resistant to challenge infection by CMV, but not by TMV. At 30°C, however, the resistance broke down, and dark green areas were invaded. Infectivity subsequently reached the same level as in yellow areas.

Breakdown of resistance with time also occurred in some dark green areas of TMV-infected tobacco (Atkinson & Matthews, 1970) although in some cases they remained virus-free for longer. Murakishi & Carlson (1976) regenerated virus-free plants from protoplasts prepared from dark green islands. They showed that these plants had a transitory, residual level of resistance to challenge inoculation.

Origins. Clearly delimited light green/dark green mosaic develops on leaves which become infected by systemic spread of virus at an early stage of development: in Chinese cabbage leaves less than 1 cm long at infection by TYMV (Reid & Matthews, 1966) or tobacco leaves less than 2.4 cm long when infected by TMV (Atkinson & Matthews, 1970). The requirement for continued cell division during the development of green areas led Reid and Matthews (1966) to suggest that they might possibly have had a clonal origin, each from an individual resistant cell. However, later studies suggested a non-clonal origin (Atkinson & Matthews, 1970; Carlson & Murakishi, 1978), and led to the idea that an inducing substance diffused from cell to cell.

Structure and properties. Atkinson & Matthews (1970) showed that protoplasmic connections did exist between cells of yellow and dark green tissue. TMV concentration fell over a distance of several cells over the yellow-dark green boundary. Kluge (1976) found that both total and soluble protein concentrations were higher in dark green than in light green tobacco tissues. There were small differences in protein gel electrophoretic patterns. With TYMV-infected Chinese cabbage leaves, Reid & Matthews (1966) found that yellow areas contained lower concentrations of Fraction 1 protein (ribulose bisphosphate carboxylase) and no detectable 70 S (chloroplast) ribosomes, whereas the green areas had normal amounts of these two components. These differences were most likely secondary, and not related to the mechanism of resistance.

Sziraki & Balazs (1975) reported that dark green tissue contained a higher concentration of cytokinins than light green, although the differences were small. This topic deserves wider investigation in view of the effects of exogenous cytokinins in delaying senescence (e.g. Biddington & Thomas, 1978), and their inhibitory effects on some viruses (Fraser & Whenham, 1982). Fraser & Whenham (unpublished results) have shown that TMV infection also leads to higher abscisic acid concentrations in dark green areas. Healthy plants treated with exogenous ABA became darker green than controls, and to some extent resistant to TMV infection, in that fewer lesions formed after challenge inoculation (Fraser, 1982). This may suggest a role for ABA in green island formation and resistance.

9.3.2. Cross protection. McKinney (1929) showed that when tobacco plants were inoculated with a green mosaic strain of TMV, they did not develop further symptoms when inoculated by a strain causing a yellow mosaic. Salaman (1933, 1938) showed that mild strains of PVX and PVY protected plants against subsequent infection by severe strains of the same viruses, and suggested that deliberate inoculation with mild strains might be of use in crop protection. Since then, similar effects have been described for a large number of viruses, and some of these have indeed proved useful in control of virus diseases.

Perhaps inevitably, a complex terminology has evolved. Thus the terms 'interference', 'acquired immunity', 'antagonism', 'acquired tolerance', 'premunity', 'cross-immunization' and 'cross protection' have been used to describe a range of linked phenomema. Conversely, a single term such as 'cross protection' has been used to describe phenomena which probably involve different mechanisms. For example, 'cross protection' has been used to describe the green island type of induced resistance, where the protecting virus may not be present in the infecting tissue (Fulton, 1980), as well as for situations where presence of the initial strain in the tissue appears to be required for protection (e.g. Cassells & Herrick, 1977). In the discussion of mechanisms below, it should become apparent that there are many levels at which related viruses may interact. The term 'cross protection' is generally restricted to interaction between related viruses, and has historically been used as an experimental means of testing taxonomic relationships. It is recognized, however, that unrelated viruses may also interact in mixed or sequential infections, producing either inhibition of pathogenesis by one or both partners (e.g. Nitzany & Cohen, 1960; Nhu et al., 1982); synergistic effects leading to more severe disease than the sum of the two individually, or increased multiplication of one or both viruses (e.g. Damirdagh & Ross, 1967; Dodds & Hamilton, 1972). Whether some of these interactions between unrelated viruses might share any mechanism in common with cross-protection between related strains is an open question. Finally, the term 'superinfection' is frequently used to describe establishment of infection by a challenge virus in a plant already infected with another strain or virus. The term does not, however, imply achievement of levels of infection greater than in the normal virus-host relationship.

Occurrence. Cross protection is very widespread, and occurs with many different types of virus. However, it is by no means universal. Fulton (1978) studied cross protection between 8 isolates of TSV, giving 7! = 28 possible pairs of strains. Of these, 8 reciprocally protected tobacco against each other; 8 protected when inoculated in one order but not in the other, and 12 pairs showed no cross protection effect at all. Other viruses where absence of cross protection has been reported include beet curly top (Bennett, 1955) and ryegrass mosaic (Wilkins & Catherall, 1974). Paulsen & Sill (1970) found no cross protection between strains of MDMV in Sorghum, but other authors (e.g. Tosic, 1981) have found partial cross protection between strains, and also to the related virus SCMV. Clearly, any theory of the mechanisms of cross protection must also offer an explanation for these cases of variable occurrence.

Most experimental work on cross protection has been with mechanically inoculated plants; most use of cross protection in practical control of virus disease has involved either mechanical inoculation or grafting. However, cross protection can also be established in the plant against insect transmitted viruses, and may also operate within the insect (Harrison, 1958).

Effects which are probably related to cross protection have been reported in protoplast experiments. Prior inoculation with one strain can prevent subsequent establishment of infection by another strain or related virus, or inhibit multiplication of the challenge virus (Otsuki & Takebe, 1976; 1978; Watts & Dawson, 1980).

Use in crop protection. Cross protection has been widely used, but in only a small number of crops (Fletcher, 1970). In practice, it is necessary that the protecting strain is an attenuated isolate, causing the minimum of damage to the crop consistent with induction of resistance to natural, severe strains. Attenuated strains have been isolated from natural populations of virus (Posnette & Cropley, 1956; Vlasov, 1979; Costa & Müller, 1980), or prepared by nitrous acid mutagenesis (Rast, 1975; Tomlinson & Shepherd, 1978).

By 1980, some 8,000,000 citrus trees had been protected by mild isolates of citrus tristeza virus (Costa & Müller, 1980; Balaraman, 1981). The induced resistance is highly effective in preventing disease caused by natural severe strains, and persistent: breakdown of protection has occurred in only a small percentage of plants.

Cross protection of tomatoes by attenuated strains of TMV has been used worldwide (Rast, 1975; Fletcher & Rowe, 1975; Nagai, 1977; Vlasov, 1979;

Oshima, 1981). The attenuated strain itself causes a small loss of yield, normally around 5% (Channon *et al.*, 1978, Ahoonmanesh & Shalla, 1981), but this compares well with the losses of 25 - 50% occurring with severe strain infection of unprotected plants. Prior inoculation with attenuated strain has been reported to improve the yield of plants subsequently inoculated with severe strain by 14% (Channon *et al.*, 1978) to 50% (Ahoonmanesh & Shalla, 1981). The latter authors found that a naturally-occurring mild strain and an attenuated, nitrous acid mutant strain were equally effective in cross protection. Broadbent (1976) discussed evidence suggesting that cross protection in tomatoes is not effective in plants grown at high temperatures; it is thought that the multiplication of the attenuated strain may be temperature sensitive. Cassells & Herrick (1977a) studied the persistence of protection and ratios and spread of mild and severe strains through the plant. The time of appearance of severe strain symptoms on protected plants depended strongly on the concentration of the challenge inoculum, and varied from a minimum of 20 days to over 3 months. The development of tomato cultivars with constitutive resistance to TMV has recently diminished the use of cross protection in this crop.

Posnette & Cropley (1956) and Chamberlain *et al.* (1964) found that attenuated isolates of apple mosaic virus gave very durable control of the disease in apple trees. The former authors reported protection for at least four years, with little or no loss of yield. Other crops where cross protection has proved of potential benefit include cocoa infected with cocoa swollen shoot virus (Posnette & Todd, 1955) and passion fruit with passion fruit woodiness virus (Greber, 1966). Sequiera (1984) has suggested that many clonal lines of potato which are universally infected with PVX may have become passively cross protected, by the practice of rogueing those plants showing severe symptoms and thus probably infected with severe strains. The remaining plants are infected with predominantly mild strains.

Apart from the slight yield reduction, cross protection has several other potential disadvantages. Cross protected crops may act as sources of virus for infection of other areas or crops. Contaminants in the inoculum, or spontaneous mutants, may cause development of severe disease symptoms. There is also the danger of synergistic interaction between the protecting virus and other, unrelated viruses, leading to increased disease severity. Except in the case of clonal, vegetatively-propagated material, it is necessary to inoculate every plant; this requires highly efficient or laborious inoculation methods, and can introduce a temporary check to growth (Rast, 1975). Finally, selection of attenuated strains is difficult, and they must not be such poor multipliers that production of inoculum becomes inefficient and expensive.

Mechanisms. Cross protection has been detected and quantified in different ways in different systems. Use of virus strains which produce visibly different symptoms can provide evidence for protection, but no quantitative data on the effects of the interaction on multiplication of each strain. Bercks (1948) was able to show that the challenge strain of PVX in tobacco could often be detected serologically, even though it had not produced symptoms on previously-inoculated plants. In the TMV-*N. sylvestris* system, use of a local lesion strain as the challenge inoculum allows estimation of reduction in the number of infections established (De Zoeten & Fulton, 1975; Sherwood & Fulton, 1982). Where both strains are systemic, multiplication of each can be assessed by various methods including differential local lesion bioassays; strain-specific antisera (Cassells & Herrick, 1977a; 1977b) and electrophoretic separation of strains with different net charges (Dodds, 1982).

When two viruses or two isolates of the same virus are inoculated together, there may be interference between them, which may be a result of competition for a limited number of specific infection sites (Sadasivan, 1940). Wu (1964) showed that the VM strain of TMV reduced formation of strain U1 lesions on tobacco. Exposure of leaves to ultraviolet light immediately after inoculation allowed appearance of U1 lesions, suggesting that VM had blocked an early infection event. Cross protection by an already established strain could also operate by blocking infection sites, and this does seem a possibility in the light green areas of the TMV-*N. sylvestris* system. However, other types of cross protection clearly operate at different levels, and information on the nature of infection sites is in any case scant. Sherwood & Fulton (1983) suggested that there was both competition for infection sites and interference with multiplication of the challenging virus, in *N. sylvestris* doubly inoculated with TMV.

De Zoeten & Fulton (1975) proposed a theory for cross protection based on coat protein. With particular reference to the TMV-*N. sylvestris* system, they suggested that the RNA of the incoming challenge virus might be sequestered in coat protein of the protecting virus. They argued that it was unlikely that the incoming RNA would find an environment suitable for uncoating and replication, when that environment was already favouring coating of the progeny of the protecting virus. Independent evidence that sequestration is possible comes from the experiments of Dodds & Hamilton (1974), which showed that TMV RNA could be masked in BSMV protein in doubly-infected plants. However, multiplication of TMV was actually enhanced by BSMV (Dodds & Hamilton, 1972).

The sequestration theory was criticized by Zaitlin (1976), on the grounds that cross protection could be induced by the PM-1 mutant of TMV, which has a defective coat protein unable to assemble with the RNA. Sarkar & Smitamana (1981), using a proteinless mutant of TMV, and Niblett et al. (1978), working with viroids, also showed that cross protection could be induced by pathogens with no protein component. Shalla & Petersen (1978) prepared protoplasts from 'protected' leaves, and were unable to detect protecting virus antigen in many of them by fluorescent antibody staining. These diverse lines of evidence have been used to argue that coat protein and sequestration are not involved in cross protection. However, all that these experiments actually prove is that a recognizable cross protection effect can be induced independently of the coat protein mechanism. They do not exclude the possibility that viral coat protein might be active in some other form of cross protection.

In this context, the fact that the N. sylvestris system may indeed show two distinct types of mechanism, one in the dark green tissue, and another in the even more resistant, heavily infected tissue, may be important. Sherwood & Fulton (1982) compared the ability of challenge virus and its naked RNA to superinfect, and concluded that cross protection resulted from an inability of the challenge virus to uncoat in the light green areas. This might be mediated through the coat protein. Heterologous coat protein has been shown to be capable of stabilizing TMV particles to alkaline disassembly (Pelcher et al., 1980).

The other major theory of cross protection is the site competition hypothesis, where the protecting strain, or one of its components, saturates a component required for virus replication. Gibbs (1969) and Barker & Harrison (1978) have suggested the viral replicase as possible sites, while Ziemiecki & Wood (1976) suggested that the protecting strain might alter ribosome binding characteristics. In some of these theories, it is difficult to see how the observed levels of specificity might be explained. With the replicase model, an initial difficulty was the general lack of preference of available enzyme preparations for binding or transcription of different templates (Ikegami & Fraenkel-Conrat, 1980), although more recent reports indicate greater specificity (Miller & Hall, 1984). There is certainly a need to provide evidence for mechanisms which work at levels other than that of coat protein. Work with pseudorecombinants of multicomponent viruses (Murant et al., 1968; Fulton, 1978) has given evidence that the control of specificity and ability to cross protect can be located on genomic segments other than that containing the coat protein gene.

Two minor theories of cross protection have also been proposed. The suggestion that interference might operate by competition for metabolites (Köhler & Hauschild, 1947) cannot explain specificity. Moreover, Fulton (1978) showed that ability of different TSV isolates to cross protect was not related to differences in their concentration. Also, the combined amount of the two isolates accumulated in doubly-inoculated plants can be much less than the amount of virus synthesis in plants infected only with the severe isolate, suggesting that metabolite supply might actually be less limiting in protected plants.

Kavanau (1949) suggested that invading severe strain virus might be adsorbed on to crystals or aggregates of the protecting virus. This model could account for a certain level of specificity, but would not apply to experiments involving defective particles or viroids.

9.4. CONCLUSION

Induced mechanisms of resistance are clearly very diverse, and our understanding of how they work is limited. Paradoxically, the type which has been most researched at the mechanistic level has been least tested for potential in crop protection. Others with established usefulness in the field have been comparatively little studied.

REFERENCES

Abu-Jawdah, Y. (1982). Changes in the soluble protein patterns of bean leaves upon fungal or viral infections or after chemical injury. Phytopathologische Zeitschrift 103, 272-279.

Ahl, P., Benjama, A., Samson, R. & Gianinazzi, S. (1981). New host proteins (b-proteins) induced together with resistance to a secondary infection following a bacterial infection in tobacco. Phytopatholgische Zeitschrift 102, 201-212.

Ahl, P., Cornu, A. & Gianinazzi, S. (1982). Soluble proteins as genetic markers in studies of resistance and phylogeny in Nicotiana. Phytopathology 72, 80-85.

Ahl, P. & Gianinazzi, S. (1982). b-proteins as a constitutive component in highly (TMV) resistant interspecific hybrids of Nicotiana glutinosa x Nicotiana debneyi. Plant Science Letters 26, 173-181.

Ahoonmanesh, A. & Shalla, T.A. (1981). Feasibility of cross-protection for control of tomato mosaic virus in fresh market field-grown tomatoes. Plant Disease 65, 56-57.

Antoniw, J.F., Kueh, J.S.H., Walkey, D.G.A. & White, R.F. (1981). The presence of pathogenesis-related proteins in callus of Xanthi-nc tobacco. Phytopathologische Zeitschrift 101, 179-184.

Antoniw, J.F. & Pierpoint, W.S. (1978). The purification and properties of the 'b' proteins from virus-infected tobacco plants. Journal of General Virology 39, 343-350.

Antoniw, J.F., Ritter, C.E., Pierpoint, W.S. & Van Loon, L.C. (1980). Comparison of three pathogenesis-related proteins from plants of two cultivars of tobacco infected with TMV. Journal of General Virology 47, 79-87.

Antoniw, J.F. & White, R.F. (1980). The effects of aspirin and polyacrylic acid on soluble leaf proteins and resistance to virus infection in five cultivars of tobacco. Phytopathologische Zeitschrift 98, 331-341.

Antoniw, J.F. & White, R.F. (1983). Biochemical properties of the pathogenesis-related proteins from tobacco. Netherlands Journal of Plant Pathology 89, 255-264.

Atkinson, P.H. & Matthews, R.E.F. (1970). On the origin of dark green tissue in tobacco leaves infected with tobacco mosaic virus. Virology 40, 344-356.

Balaraman, K. (1981). Preimmunization for control of citrus tristeza virus on citron. Zeitschrift für Pflanzenkrankheiten und Pflanzenschutz 88, 218-220.

Balazs, E., Barna, B. & Kiraly, Z. (1976). Effect of kinetin on lesion development and infection sites in Xanthi nc tobacco infected by TMV: single cell lesions. Acta Phytopathologica Academiae Scientiarum Hungaricae 11, 1-9.

Balazs, E., Sziraki, I. & Kiraly, Z. (1977). The role of cytokinins in the systemic acquired resistance of tobacco hypersensitive to tobacco mosaic virus. Physiological Plant Pathology 11, 29-37.

Barker, H. & Harrison, B.D. (1978). Double infection, interference and superinfection in protoplasts exposed to two strains of raspberry ringspot virus. Journal of General Virology 40, 647-658.

Batra, G.K. & Kuhn, C.W. (1975). Polyphenoloxidase and peroxidase activities associated with acquired resistance and its inhibition by 2-thiouracil in virus-infected soybean. Physiological Plant Pathology 5, 239-248.

Bennett, C.W. (1955). Recovery of water pimpernel from curly top and the reaction of recovered plants to re-inoculation with different virus strains. Phytopathology 45, 531-536.

Benson, A.A. & Jekela, A.T. (1976). Cell Membranes. In 'Plant Biochemistry' (3rd Edition) (Bonner, J. & Varner, J.E., eds), pp. 65-89. Academic Press, New York.

Bercks, R. (1948). Serologische Beiträge zür Frage der Abwehr von Zweitinfektionen bei X-Viren. Phytopathologische Zeitschrift 25, 54-61.

Biddington, N.L. & Thomas, T.H. (1978). Influence of different cytokinins on the transpiration and senescence of excised oat leaves. Physiologia Plantarum 42, 369-374.

Bozarth, R.F. & Ross, A.F. (1964). Systemic resistance induced by localized virus infections: extent of changes in uninfected plant parts. Virology 24, 446-455.

Broadbent, L. (1976). Epidemiology and control of tomato mosaic virus. Annual Review of Phytopathology 14, 75-96.

Caldwell, J. (1934). The physiology of virus diseases in plants. V. The movement of the virus agent in tobacco and tomato. Annals of Applied Biology 21, 191-205.

Camacho-Henriquez, A. & Sänger, H.L. (1982). Analysis of acid-extractable tomato leaf proteins after infection with a viroid, two viruses and a fungus, and partial purification of the pathogenesis-related protein p14. Archives of Virology 74, 181-196.

Camacho-Henriquez, A., Lucas, J. & Sänger, H.L. (1983). Purification and biochemical properties of the 'pathogenesis-related' protein p14 from tomato leaves. Netherlands Journal of Plant Pathology 89, 308.

Carlson, P.S. & Murakishi, H.H. (1978). Evidence on the clonal versus non-clonal origin of dark green islands in virus-infected tobacco leaves. Plant Science Letters 13, 377-381.

Carr, J.P. (1983). Translational control of pathogenesis-related protein synthesis. Netherlands Journal of Plant Pathology 89, 311-312.

Carr, J.P., Antoniw, J.F., White, R.F. & Wilson, T.M.A. (1982). Latent messenger RNA in tobacco (Nicotiana tabacum). Biochemical Society Transactions 10, 353-354.

Cassells, A.C., Barnett, A. & Barlass, M. (1978). The effect of polyacrylic acid treatment on the susceptibility of Nicotiana tabacum cv. Xanthi-nc to tobacco mosaic virus. Physiological Plant Pathology 13, 13-21.

Cassells, A.C. & Herrick, C.C. (1977a). Cross protection between mild and severe strains of tobacco mosaic virus in doubly-inoculated tomato plants. Virology 78, 253-260.

Cassells, A.C. & Herrick, C.C. (1977b). The identification of mild and severe strains of tobacco mosaic virus in doubly inoculated tomato plants. Annals of Applied Biology 86, 37-46.

Chalcroft, J.P. & Matthews, R.E.F. (1967). Role of virus strain and leaf ontogeny in the production of mosaic patterns by turnip yellow mosaic virus. Virology 33, 659-673.

Chamberlain, E.E., Atkinson, J.D. & Hunter, J.A. (1964). Cross protection between strains of apple mosaic virus. New Zealand Journal of Agricultural Research 7, 480-490.

Chamberlain, J.A., Catherall, P.C. & Jennings, A.J. (1977). Symptoms and electron microscopy of ryegrass mosaic virus in different grass species. Journal of General Virology 36, 297-306.

Channon, A.G., Cheffins, N.J., Hitchon, G.M. & Barker, J. (1978). The effect of inoculation with an attenuated mutant strain of tobacco mosaic virus on the growth and yield of early glasshouse tomato crops. Annals of Applied Biology 88, 121-129.

Chessin, M. (1982). Interference in plant virus infection: ultraviolet light and systemic acquired resistance. Phytopathologische Zeitschrift 104, 279-283.

Conejero, V. & Semancik, J.S. (1977). Exocortis viroid: alteration in the proteins of Gynura aurantiaca accompanying viroid infection. Virology 77, 221-232.

Costa, A.S, & Müller, G.W. (1980). Tristeza control by cross protection: a U.S.-Brazil cooperative success. Plant Disease 64, 538-541.

Coutts, R.H.A. (1978). Alterations in the soluble protein patterns of tobacco and cowpea leaves following inoculation with tobacco necrosis virus. Plant Science Letters 12, 189-197.

Coutts, R.H.A. & Wagih, E.E. (1981). Alterations in RNA and protein metabolism in uninoculated half-leaves of cowpea adjacent to tobacco necrosis virus-infected halves. Plant Science Letters 21, 51-59.

Coutts, R.H.A. & Wagih, E.E. (1983). Induced resistance to viral infection and soluble protein alterations in cucumber and cowpea plants. Phytopathologische Zeitschrift 107, 57-69.

Damirdagh, I.S. & Ross, A.F. (1967). A marked synergistic reaction of potato viruses X and Y in inoculated leaves of tobacco. Virology 31, 286-307.

De Laat, A.M.M. & Van Loon, L.C. (1983). The relationship between stimulated ethylene production and symptom expression in virus-infected tobacco leaves. Physiological Plant Pathology 22, 261-273.

De Zoeten, G.A. & Fulton, R.W. (1975). Understanding generates possibilities. Phytopathology 65, 221-222.

Dodds, J.A. (1982). Cross-protection and interference between electrophoretically distinct strains of cucumber mosaic virus in tomato. Virology 118, 235-240.

Dodds, J.A. & Hamilton, R.I. (1972). The influence of barley stripe mosaic virus on the replication of tobacco mosaic virus in Hordeum vulgare L. Virology 50, 404-411.

Dodds, J.A. & Hamilton, R.I. (1974). Masking of the RNA genome of tobacco mosaic virus by the protein of barley stripe mosaic virus in doubly infected barley. Virology 59, 418-428.

Faccioli, G. (1979). Relation of peroxidase, catalase and polyphenoloxidase to acquired resistance in plants of Chenopodium amaranticolor locally infected by tobacco necrosis virus. Phytopathologische Zeitschrift 95, 237-249.

Fletcher, J.T. (1978). The use of avirulent virus strains to protect plants against the effects of virulent strains. Annals of Applied Biology 89, 110-114.

Fletcher, J.T. & Rowe, J.M. (1975). Observations and experiments on the use of an avirulent mutant strain of tobacco mosaic virus as a means of controlling tomato mosaic. Annals of Applied Biology 81, 171-179.

Fraser, R.S.S. (1975). Turnover of polyadenylated messenger RNA in fission yeast: evidence for the control of protein synthesis at the translational level. European Journal of Biochemistry 60, 477-486.

Fraser, R.S.S. (1979). Systemic consequences of the local lesion reaction to tobacco mosaic virus in a tobacco variety lacking the N gene for hypersensitivity. Physiological Plant Pathology

14, 383-394.
Fraser, R.S.S. (1981). Evidence for the occurrence of the "pathogenesis-related" proteins in leaves of healthy tobacco plants during flowering. Physiological Plant Pathology 19, 69-76.
Fraser, R.S.S. (1982). Are 'pathogenesis-related' proteins involved in acquired systemic resistance of tobacco plants to tobacco mosaic virus? Journal of General Virology 58, 305-313.
Fraser, R.S.S. & Clay, C.M. (1983). Pathogenesis-related proteins and acquired systemic resistance: causal relationship or separate effects? Netherlands Journal of Plant Pathology 89, 283-292.
Fraser, R.S.S., Loughlin, S.A.R. & Whenham, R.J. (1979). Acquired systemic susceptibility to infection by tobacco mosaic virus in Nicotiana glutinosa L. Journal of General Virology 43, 131-141.
Fraser, R.S.S. & Whenham, R.J. (1982). Plant growth regulators and virus infection: a critical review. Plant Growth Regulation 1, 37-59.
Fritig, B., Gosse, J., Legrand, M. & Hirth, L. (1973). Changes in phenylalanine ammonia lyase during the hypersensitive reaction of tobacco to TMV. Virology 55, 371-379.
Fulton, R.W. (1951). Superinfection by strains of tobacco mosaic virus. Phytopathology 41, 579-592.
Fulton, R.W. (1978). Superinfection by strains of tobacco streak virus. Virology 85, 1-8.
Fulton, R.W. (1980). The protective effects of systemic virus infection. In 'Active Defense Mechanisms in Plants' (Wood, R.K.S., ed.), pp. 231-246. Plenum Press, New York and London.
Gianinazzi, S. (1982). Antiviral agents and inducers of virus resistance: analogies with interferon. In 'Active defense mechanisms in plants' (Wood, R.K.S., ed.), pp. 275-298. Plenum Press, New York & London.
Gianinazzi, S. & Ahl, P. (1983). The genetic and molecular basis of b-proteins in the genus Nicotiana. Netherlands Journal of Plant Pathology 89, 275-281.
Gianinazzi, S., Ahl, P., Cornu, A., Scalla, R. & Cassini, R. (1980). First report of host b-protein appearance in response to a fungal infection in tobacco. Physiological Plant Pathology 16, 337-342.
Gianinazzi, S. & Kassanis, B. (1974). Virus resistance induced in plants by polyacrylic acid. Journal of General Virology 13, 1-9.
Gianinazzi, S., Martin, C. & Valee, J.-C. (1970). Hypersensibilite aux virus, temperature et proteines solubles chez le Nicotiana Xanthi n.c. Apparition de nouvelles macromolecules lors de la repression de la synthese virale. Compte rendu hebdomadaire des seances de l'Academie des Sciences de Paris 270D, 2383-2386.
Gianinazzi, S., Pratt, H.M., Shewry, P.R. & Miflin, B.J. (1977). Partial purification and preliminary characterisation of soluble leaf proteins specific to virus-infected tobacco plants. Journal of General Virology 34, 345-351.
Gibbs, A. (1969). Plant virus classification. Advances in Virus Research 14, 263-328.
Gicherman, G. & Loebenstein, G. (1968). Competitive inhibition by foreign nucleic acids and induced interference by yeast RNA with the infection of tobacco mosaic virus. Phytopathology 58, 405-409.
Gilpatrick, J.D. & Weintraub, M. (1952). An unusual type of protection with the carnation mosaic virus. Science 115, 701-702.
Greber, R.S. (1966). Passionfruit woodiness virus as the cause of passionvine tip blight disease. Queensland Journal of Agricultural and Animal Science 23, 533-536.
Harrison, B.D. (1958). Ability of single aphids to transmit both avirulent and virulent strains of potato leafroll virus. Virology 6, 278-286.
Hecht, E.I. & Bateman, D.F. (1964). Nonspecific acquired resistance to pathogens resulting from localized infections by Thielaviopsis basicola or viruses in tobacco leaves. Phytopathology 54, 523-530.
Ikegami, M. & Fraenkel-Conrat, H. (1980). Lack of specificity of virus-stimulated RNA-dependent RNA polymerases. Virology 100, 185-188.
Kassanis, B. (1982). Some speculations on the nature of the natural defence mechanism of plants against virus infection. Phytopathologische Zeitschrift 102, 277-291.

Kassanis, B., Gianinazzi, S. & White, R.F. (1974). A possible explanation of the resistance of virus-infected tobacco plants to second infection. Journal of General Virology 23, 11-16.
Kassanis, B. & White, R.F. (1974). Inhibition of acquired resistance to tobacco mosaic virus by actinomycin-D. Journal of General Virology 25, 323-324.
Kassanis, B. & White, R.F. (1975). Polyacrylic acid-induced resistance to tobacco mosaic virus in tobacco cv. Xanthi. Annals of Applied Biology 79, 215-220.
Kassanis, B. & White, R.F. (1978). Effect of polyacrylic acid and b-proteins on TMV multiplication in tobacco protoplasts. Phytopathologische Zeitschrift 91, 269-272.
Kavanau, J.L. (1949). On correlation of the phenomena associated with chromosomes, foreign proteins and viruses. III. Virus-associated phenomena, characteristics and reproduction. American Naturalist 83, 113-138.
Kluge, S. (1976). Proteingehalt in hell und dunkelgrünen Blattbezirken mosaikkranker Tabakpflanzen. Biochemie und Physiologie der Pflanzen 170, 91-95.
Köhler, E. & Hauschild, I. (1947). Betrachtungen und Versuche zum Problem der „erwarbenen Immunität" gegen Virusinfektionen bei Pflanzen. Zuchter 17/18, 97-105.
Loebenstein, G. (1962). Inducing partial protection in the host plant with native virus protein. Virology 17, 574-581.
Loebenstein, G. (1963). Further evidence on systemic resistance induced by localized necrotic virus infections in plants. Phytopathology 53, 306-308.
Loebenstein, G., Cohen, J., Shabtai, S., Coutts, R.H.A. & Wood, K.R. (1977). Distribution of cucumber mosaic virus in systemically-infected tobacco leaves. Virology 81, 117-125.
Loebenstein, G., Rabina, S. & van Praagh, T. (1968). Sensitivity of induced localized acquired resistance to actinomycin-D. Virology 34, 264-268.
Loebenstein, G., Sela, I & van Praagh, T. (1969). Increase of tobacco mosaic virus lesion size and virus multiplication in hypersensitive hosts in the presence of actinomycin-D. Virology 37, 42-48.
McIntyre, J.L., Dodds, J.A. & Hare, J.D. (1981). Effects of localized infections of Nicotiana tabacum by tobacco mosaic virus on systemic resistance against diverse pathogens and an insect. Phytopathology 71, 297-301.
Miller, W.A. & Hall, T.C. (1984). RNA-dependent RNA polymerase isolated from cowpea chlorotic mottle virus-infected cowpea is specific for bromovirus RNA. Virology 132, 53-60.
Murakishi, H.H. & Carlson, P.S. (1976). Regeneration of virus-free plants from dark green islands of tobacco mosaic virus-infected tobacco leaves. Phytopathology 66, 931-932.
Murant, A.F., Taylor, C.E. & Chambers, J. (1968). Properties, relationships and transmission of a strain of raspberry ringspot virus infecting raspberry cultivars immune to the common Scottish strain. Annals of Applied Biology 61, 175-186.
Nagai, Y. (1977). Control of the mosaic disease of tomato by seedling inoculation with the attenuated strain of TMV. Tropical Agricultural Research Series 10, 179-183.
Nhu, K.M., Mayee, C.D. & Sarkar, S. (1982). Reduced synthesis of PVY in PVX-infected tobacco plants. Naturwissenschaften 69, 183-184.
Niblett, C.L., Dickson, E. Fernow, K.H. Horst, R.K. & Zaitlin, M. (1978). Cross protection among four viroids. Virology 91, 198-203.
Nitzany, F.E. & Cohen, S. (1960). A case of interference between alfalfa mosaic virus and cucumber mosaic virus. Virology 11, 771-773.
Oshima, N. (1981). Control of tomato mosaic disease by attenuated virus. Japan Agricultural Research Quarterly 14, 222-228.
Otsuki, Y. & Takebe, I. (1976). Interaction of tobacco mosaic virus strains in doubly infected tobacco protoplasts. Annals de Microbiologie (Institut Pasteur) 127A, 21.
Otsuki, Y. & Takebe, I. (1978). Production of mixedly coated particles in tobacco mesophyll protoplasts doubly infected by strains of tobacco mosaic virus. Virology 84, 162-171.
Parent, J.G. & Asselin, A. (1984). Detection of pathogenesis-related (PR or b) and of other proteins in the intercellular fluid of hypersensitive plants infected with tobacco mosaic virus. Canadian Journal of Botany 62, 564-569.
Paulsen, A.Q. & Sill, W.H. (1970). Absence of cross protection between maize dwarf virus strains A and B in grain sorghums. Plant Disease Reporter 54, 627-629.

Pelcher, L.E., Walmsley, S.L. & Mackenzie, S.L. (1980). The effects of heterologous and homologous coat protein on alkaline disassembly of tobacco and tomato isolates of tobacco mosaic virus. Virology 105, 287-290.

Pennazio, S. & Redolfi, P. (1980). Resistance to tomato bushy stunt virus localized infection induced in Gomphrena globosa by acetylsalicylic acid. Microbiologica 3, 475-479.

Pennazio, S., Roggero, P. & Lenzi, R. (1983). Some characteristics of the hypersensitive response of White Burley tobacco to tobacco necrosis virus. Physiological Plant Pathology 22, 347-355.

Pierpoint, W.S. (1983). The major proteins in extracts of tobacco leaves that are responding hypersensitively to virus infection. Phytochemistry 22, 2691-2697.

Pierpoint, W.S., Robinson, N.P. & Leason, M.B. (1981). The pathogenesis-related proteins of tobacco: their induction by viruses in intact plants and their induction by chemicals in detached leaves. Physiological Plant Pathology 19, 85-97.

Posnette, A.F. & Cropley, R. (1956). Apple mosaic viruses: host reactions and strain interference. Journal of Horticultural Science 31, 119-133.

Posnette, A.F. & Todd, J.M. (1955). Virus diseases of cacao in west Africa. IX. Strain variation and interference in virus 1A. Annals of Applied Biology 43, 433-453.

Rast, A.T.B. (1975). Variability of tobacco mosaic virus in relation to control of tomato mosaic in glasshouse crops by resistance breeding and cross protection. Agricultural Research Reports, Wageningen (Netherlands) 834, 1-76.

Redolfi, P. (1983). Occurrence of pathogenesis-related (b) and similar proteins in different plant species. Netherlands Journal of Plant Pathology 89, 245-254.

Redolfi, P., Vecchiati, M. & Gianinazzi, S. (1982). Changes in the soluble leaf protein constitution of Gomphrena globosa during the hypersensitive reaction to different viruses. Phytopathologische Zeitschrift 103, 48-54.

Reid, M.S. & Matthews, R.E.F. (1966). On the origin of the mosaic induced by turnip yellow mosaic virus. Virology 28, 563-570.

Roberts, D.A. (1983). Acquired resistance to tobacco mosaic virus transmitted to the progeny of hypersensitive tobacco. Virology 124, 161-163.

Rohloff, H. & Lerch, B. (1977). Soluble leaf proteins in virus-infected plants and acquired resistance. 1. Investigations on Nicotiana tabacum cvs Xanthi-nc and Samsun. Phytopathologische Zeitschrift 89, 306-316.

Ross, A.F. (1961a). Localized acquired resistance to plant virus infection in hypersensitive hosts. Virology 14, 329-339.

Ross, A.F. (1961b). Systemic acquired resistance induced by localized virus infections in plants. Virology 14, 340-358.

Ross, A.F. (1966). Systemic effects of local lesion formation. In 'Viruses of Plants' (Beemster, A.B.R. & Dijkstra, J., eds), pp. 127-150. North-Holland Publishing Co., Amsterdam.

Ross, A. F. & Israel, H.W. (1970). Use of heat treatments in the study of acquired resistance to tobacco mosaic virus in hypersensitive tobacco. Phytopathology 60, 755-770.

Sadasivan, T.S. (1940). A quantitative study of the interaction of viruses in plants. Annals of Applied Biology 27, 359-367.

Salaman, R.N. (1933). Protective inoculation against a plant virus. Nature 131, 168.

Salaman, R.N. (1938). The potato virus "X": its strains and reactions. Philosophical Transactions of the Royal Society (B) 229, 137-217.

Sarkar, S. & Smitamana, P. (1981). A proteinless mutant of tobacco mosaic virus: evidence against the role of a viral coat protein for interference. Molecular and General Genetics 184, 158.

Sequeira, L. (1984). Cross protection and induced resistance: their potential for plant disease control. Trends in Biotechnology 2, 25-29.

Shalla, T.A. & Petersen, L.J. (1978). Studies on the mechanism of viral cross protection. Phytopathology 68, 1681-1683.

Sherwood, J.L & Fulton, R.W. (1982). The specific involvement of coat protein in tobacco mosaic virus cross protection. Virology 119, 150-158.

Sherwood, J.L. & Fulton, R.W. (1983). Competition for infection sites and multiplication of the competing strain in plant viral interference. Phytopathology 73, 1363-1365.

Simons, T.J. & Ross, A.F. (1971). Changes in phenol metabolism associated with induced systemic

resistance to tobacco mosaic virus in Samsun NN tobacco. Phytopathology 61, 1261-1265.
Stein, A. & Loebenstein, G. (1972). Induced interference by synthetic polyanions with the infection of tobacco mosaic virus. Phytopathology 62, 1461-1466.
Stein, A. & Loebenstein, G. (1976). Peroxidase activity in tobacco plants with polyanion-induced interference to tobacco mosaic virus. Phytopathology 66, 1192-1194.
Stein, A., Loebenstein, G. & Spiegel, S. (1979). Further studies of induced interference by a synthetic polyanion of infection by tobacco mosaic virus. Physiological Plant Pathology 15, 241-255.
Sziraki, I. & Balazs, E. (1975). The effect of infection by TMV on cytokinin level of tobacco plants, and cytokinins in TMV-RNA. In 'Current Topics in Plant Pathology' (Kiraly, Z. ed.), pp. 345-352. Akademiai Kiado, Budapest.
Sziraki, I., Balazs, E. & Kiraly, Z. (1980). Role of different stresses in inducing systemic acquired resistance to TMV and increasing cytokinin level in tobacco. Physiological Plant Pathology 16, 277-284.
Thomas, P. & Fulton, R.W. (1968a). Correlation of the ectodesmata number with nonspecific resistance to initial virus infection. Virology 34, 459-469.
Thomas, P.E. & Fulton, R.W. (1968b). Resistance to spread of virus from cell to cell in T.I. 245 tobacco. Virology 55, 108-111.
Tomkins, G.M., Gelehrter, T.D., Martin, D., Samuels, H.H. & Thompson, E.B. (1969). Control of specific gene expression in higher organisms. Science 166, 1474-1480.
Tomlinson, J.A. & Shepherd, R.J. (1978). Studies on mutagenesis and cross-protection of cauliflower mosaic virus. Annals of Applied Biology 90, 223-231.
Tosic, M. (1981). Cross protection among some strains of sugarcane mosaic virus and maize dwarf mosaic virus. Agronomie 1, 83-85.
Van Loon, L.C. (1975). Polyacrylamide disc electrophoresis of the soluble leaf proteins from Nicotiana tabacum var 'Samsun' and 'Samsun-NN'. IV. Similarity of qualitative changes of specific proteins after infection with different viruses and their relationship to acquired resistance. Virology 67, 566-575.
Van Loon, L.C. (1976a). Specific soluble leaf proteins in virus-infected tobacco plants are not normal constituents. Journal of General Virology 30, 375-379.
Van Loon, L.C. (1976b). Systemic acquired resistance, peroxidase activity and lesion size in tobacco reacting hypersensitively to TMV. Physiological Plant Pathology 8, 231-242.
Van Loon, L.C. (1977). Induction by 2-chloroethylphosphonic acid of viral-like lesions, associated proteins and systemic resistance in tobacco. Virology 80, 417-420.
Van Loon, L.C. (1982). Regulation of changes in proteins and enzymes associated with active defence against virus infection. In 'Active Defense Mechanisms in Plants' (Wood, R.K.S., ed.), pp. 247-274. Plenum Press, New York and London.
Van Loon, L.C. (1983). The induction of pathogenesis-related proteins by pathogens and specific chemicals. Netherlands Journal of Plant Pathology 89, 265-273.
Van Loon, L.C. & Antoniw, J.F. (1982). Comparison of the effects of salicylic acid and ethephon with virus induced hypersensitivity and acquired resistance in tobacco. Netherlands Journal of Plant Pathology 88, 237-256.
Van Loon, L.C. & Dijkstra, J. (1976). Virus-specific expression of systemic acquired resistance in tobacco mosaic virus- and tobacco necrosis virus-infected 'Samsun NN' and 'Samsun' tobacco. Netherlands Journal of Plant Pathology 82, 231-237.
Van Loon, L.C. & Van Kammen, A. (1970). Polyacrylamide disc electrophoresis of the soluble leaf proteins from N. tabacum var. Samsun and Samsun NN. II. changes in protein constitution after infection with TMV. Virology 40, 199-211.
Vlasov, Y.I. (1979). A method of vaccinating against viruses. Zashchita Rastenii (Moscow) 7, 20-22.
Wagih, E.E. & Coutts, R.H.A. (1982). Comparison of virus-elicited and other stresses on the soluble protein fraction of cucumber cotyledons. Phytopathologische Zeitschrift 104, 364-374.
Watts, J.W. & Dawson, J.R.O. (1980). Double infection of tobacco protoplasts with brome mosaic virus and cowpea chlorotic mottle virus. Virology 105, 501-507.
Weintraub, M. & Kemp, W.G. (1961). Protection with carnation mosaic virus in Dianthus barbatus.

Virology 13, 256-257.
Whenham, R.J. & Fraser, R.S.S. (1981). Effect of systemic and local-lesion-forming strains of tobacco mosaic virus on abscisic acid concentration in tobacco leaves: consequences for the control of leaf growth. Physiological Plant Pathology 18, 267-278.
White, R.F. (1979). Acetylsalicylic acid (aspirin) induces resistance to tobacco mosaic virus in tobacco. Virology 99, 410-412.
Wieringa-Brants, D.H. (1983). A model to simulate acquired resistance induced by localized virus infections in hypersensitive tobacco. Phytopathologische Zeitschrift 106, 369-372.
Wilkins, P.W. & Catherall. P.L. (1974). The effect of some isolates of ryegrass mosaic virus on different genotypes of Lolium multiflorium. Annals of Applied Biology 76, 209-216.
Wu, J.H. (1964). Release of inhibited tobacco mosaic virus infection by ultraviolet irradiation as a function of time and temperature after inoculation. Virology 24, 441-445.
Yarwood, C.E. (1959). Virus susceptibility increased by soaking bean leaves in water. Plant Disease Reporter 43, 841-844.
Yarwood, C.E. (1960). Localized acquired resistance to tobacco mosaic virus. Phytopathology 50, 741-744.
Zaitlin, M. (1976). Viral cross protection: more understanding is needed. Phytopathology 66, 382-383.
Ziemiecki, A. & Wood, K.R. (1976). Proteins synthesised by cucumber cotyledons infected with two strains of cucumber mosaic virus. Journal of General Virology 31, 373-381.

CHAPTER 10. INDUCED RESISTANCE TO FUNGAL AND BACTERIAL DISEASES

P. J. G. M. DE WIT

10.1. INTRODUCTION

Plant pathologists have long been intrigued by the idea that plants like animals could be immunized to potential pathogens. One of the first critical reviews was by Chester (1933). At that time he could not find strong evidence for what he called acquired physiological immunity in plants. Since then many reports and reviews have been written (Hamilton, 1980; Goodman, 1980; Kuc, 1982a; 1982b; Matta, 1980; McIntyre, 1982; Ouchi, 1983; Sequeira, 1983; Suzuki, 1980). During recent years the pesticide industry has also become interested: could induced resistance have potential as a new method for controlling plant diseases?

As Sequeira (1983) points out, the terminology in the field of induced resistance is confusing. In this Chapter we will restrict ourselves to the terms 'induced local resistance' and 'induced systemic resistance'. In relation to this, plants exhibit local or systemic protection, respectively, after challenge by a potential pathogen. The terms 'acquired resistance' and 'induced immunity' which are frequently used in the literature and which are more or less synonymous with induced resistance will not be used, to avoid confusion.

'Cross protection' and 'premunition' imply the presence of both inducing and challenging organisms in the protected tissue. These terms are frequently used in plant virology to indicate that (systemic) infection by one strain of a virus protects a plant against subsequent infection by a second strain. Cross protection between virulent and avirulent races or between species of various plant pathogenic fungi and bacteria in host tissues is also known. In a system developed by Rishbeth, cited in Bruehl (1975), spores of *Peniophora gigantea* were applied to pine stumps to protect them against invasion by *Fomes annosus*. Cross protection through the production of a bacteriocin by the protecting organism was reported by Kerr & Htay (1974). They showed that the pathogen *Agrobacterium radiobacter* var. *tumefaciens* could be controlled by a closely related non-pathogen A. radiobacter var. radiobacter which produced the bacteriocin.

The opposite of induced resistance viz. induced susceptibility or predisposition has also been observed. Ouchi et al. (1974) demonstrated that both barley and wheat leaves could be predisposed to non-pathogenic forma specialis of Erysiphe graminis by prior inoculation with a virulent race. Induced susceptibility was not only effective to avirulent races of the same pathogen, but also to other powdery mildew fungi such as Sphaerotheca fuliginea and even to some necrotrophic fungi (Ouchi, 1983). Induced susceptibility has been reported mainly for obligate fungus-host interactions. Induced susceptibility is mainly localized near the site of the primary infection and may not become systemic; induced resistance can be systemic as well as localized. Induced susceptibility will not be treated in this Chapter unless there are aspects in common with induced resistance which would clarify the latter phenomenon.

Irrespective of whether one studies induced resistance, or induced susceptibility, the assays for both phenomena are indirect. Thus there will always be uncertainty about whether the physiological changes induced by the first (inducer) inoculation are specifically directed to prevent or to give access to the challenge inoculation, respectively. For example, a compatible interaction often functions as a sink and as a result there is a surplus of food substances present, which could sustain growth of non-pathogens. Similarly, at the infection court of an incompatible interaction, many physiological changes occur which are deleterious to many pathogens.

Some aspects of induced local resistance have been discussed in detail in Chapter 7 (sections 3, 4 and 5), where active defence mechanisms and their induction are treated. The mechanisms involved are discussed mainly in relation to race-specific resistance, in other words for systems where gene-for-gene interactions are involved. However, induced local and induced systemic resistance can also be shown in plants having no known genes for resistance to a particular parasite. In this Chapter the emphasis will be on induced resistance phenomena which are not directed against one specific organism. The mechanisms are neither race nor species-specific. The discussion will concentrate on whether resistance responses, once they are induced in plants, protect these plants locally and/or systemically against subsequent invaders. The possible mechanism of the protective response will also be discussed.

10.1.1. Induced local resistance (phenomenon). Locally induced resistance to fungi was first described by Bernard (1909) working with orchids and mycorrhizae. He observed that a class of mycorrhizal fungi which penetrated the host tissue, but only partly colonized it, could protect the same host from becoming infected by another pathogenic fungus. These results suggested that the first inoculation (inducing inoculation) induced resistance to the second inoculation (challenging inoculation). Initially it was not known whether the phenomenon was caused by direct interaction between the inducing and challenging organisms (cross protection) or by physiological changes in the host plant (induced resistance) which were deleterious to the challenging agent. Müller & Börger (1941) studied the same phenomenon using virulent and avirulent races of the fungus Phytophthora infestans and potato tubers. They showed that prior treatment of cut surfaces of potato tuber with an avirulent race of P. infestans prevented colonization of the tuber by a virulent race. The protection was localized, but if a thin layer of tuber tissue which had been treated with the avirulent race was excised, the underlying tissue where the avirulent race was not present, was still protected against the virulent race. This indicated that protection was most likely caused by locally induced changes in the host and not by interaction or competition between the inducing and challenging agent; there was still the possibility that a diffusable factor produced by the avirulent race would inhibit the virulent race in the underlying tissue, but this would be very difficult to prove.

It was on the basis of the induced resistance experiments of Müller & Börger (1941) that the 'phytoalexin hypothesis' was introduced. Later the association of induced resistance with the accumulation of phytoalexins was found by others (Rahe et al., 1969; Svoboda & Paxton, 1972; Elliston et al., 1976a; 1976b; 1976c; 1977). However, there are other examples of local induced resistance where there is yet no evidence that phytoalexins are involved (Yarwood, 1956; Kochman & Brown, 1975; Doverall et al., 1979). Induced local resistance should be interpreted as being effective within a few mm from the primary site of infection. Generally, induced local resistance can be observed in many host pathogen interactions, while induced systemic resistance occurs more rarely. Most of the mechanisms involved in the local resistance response against the primary infecting pathogen are also effective against the challenger. If different responses to the two pathogens were involved, they would be very difficult to disentangle, as the infection court of inducer and challenger are more or less the same.

10.1.2. Induced systemic resistance (phenomenon). The phenomenon of induced systemic resistance was demonstrated much later than induced local resistance. Induced systemic resistance implies a spatial distribution of the protection against the challenging organism provided by prior inoculation with the inducing organism. The protection can spread over a great distance in the plant, and sometimes covers the whole plant including newly emerging plant parts. In contrast with induced local resistance direct interaction between the two organisms can generally be entirely excluded. Cruickshank & Mandryk (1960) were among the first to demonstrate a decrease of foliar infection by Peronospora tabacina caused by a prior stem infection with the same fungus. However, the more usual way to induce systemic resistance is to treat lower leaves of a plant with the inducing organism, and inoculate leaves at a distance with the challenging organism. Induced systemic resistance can also be demonstrated at shorter distances from the inducing inoculation i.e. on one half of the same leaf or on the opposite leaflet of a compound leaf. This raises the question whether differences between localized and systemically induced resistance are of a quantitative rather than a qualitative nature.

10.1.3. Quantification of induced resistance. The quantification of induced resistance varies significantly between the different host-pathogen interactions studied. Protection against local lesion-forming fungal pathogens is measured as a reduction in number and size of the local lesions formed by the challenger. A good example of such a type of interaction is that of Colletotrichum lagenarium on Cucurbitaceae (Caruso & Kuć, 1977). Protection against wilt causing pathogens is often qualitatively assessed and expressed as decrease in symptom development (Tjamos, 1979; Gessler & Kuć, 1982a). With obligate parasites like powdery mildews, protection is usually very local and is microscopically measured by determining decreased fungal growth around the inducing inoculation (Ouchi et al. 1974).

Measuring induced resistance involving bacterial pathogens is more difficult. Usually inducer and challenger bacterium are administered in precisely the same place. The protection is measured as a decrease of bacterial multiplication. Local induced resistance can be determined by this method. The occurrence of systemic induced resistance has also been claimed (Novacky et al., 1973; Sequeira, 1983) but this has never been substantiated.

10.2. INDUCERS OF LOCAL RESISTANCE

Different types of inducers of the local resistance response have been treated in detail in Chapter 7.3, 4 and 5). Non-pathogens, avirulent races of a pathogen and various types of biotic and abiotic elicitors can induce a local resistance response.

10.3. THE PROTECTIVE RESPONSE IN INDUCED LOCAL RESISTANCE

Induced local resistance and cross protection are difficult to disentangle especially if viable inducers and challengers are administered at the same place. The observed protection can be a result of a direct interaction between inducer and challenger or can be mediated by a host response. When the inducer is a non-viable elicitor the challenger interacts with the host and other types of interactions can be excluded. A number of host responses have been proposed to be involved in the protective response of induced local resistance.

10.3.1. Phytoalexins. Evidence that phytoalexins are involved in the protective response of induced local resistance dates back to the pioneering work of Müller & Börger (1941) with P. infestans on potato tuber. Later a number of workers have also shown that phytoalexins are involved. Paxton & Chamberlain (1967) and Svoboda & Paxton (1972) showed protection of soybean against Phytophthora megasperma var. sojae by prior inoculation with an avirulent race of the same fungus or a non-pathogen like Phytophthora cactorum. Protection was localized and limited to the wound area of the initial inoculation. They showed that phytoalexins accumulated after the inducer inoculation were sufficiently inhibitory to the challenger. Phytoalexins were also involved in induced local resistance in Phaseolus vulgaris to Colletotrichum lindemuthianum (Rahe et al., 1969). In cotton the phytoalexins gossypol and related compounds which accumulated in xylem exudates and vascular tissue after the inducer inoculation were proposed to explain protection against the pathogen Verticillium albo-atrum (Bell & Presley, 1969).

Sequeira & Hill (1974) and Rathmell & Sequeira (1975) suggested that accumulation of phytoalexin-like material induced by prior inoculation with heat-killed bacteria might explain protection against Pseudomonas solanacearum.

10.3.2. Other resistance mechanisms. Not all plants accumulate phytoalexins after the inducer inoculation. Most monocots do not accumulate phytoalexins after inoculation with non-pathogens or avirulent races of a pathogen (Bailey & Mansfield, 1982). Deverall et al. (1979) failed to detect significant accumulation of phytoalexins in wheat roots protected against *Gaeumannomyces graminis* var. *tritici* by prior inoculation with *G. graminis* var. *graminis*. They suggested that enhanced lignification might be involved in the protective response. In cucumber no phytoalexins have been detected so far, but enhanced deposition of lignin and cell wall hydroxyproline are associated with the primary resistance response (Hammerschmidt *et al.*, 1984). It is possible that deposition of these two compounds could explain local protection against secondary investing parasites also.

In potato tuber, in addition to the accumulation of the sesquiterpenoid phytoalexins, deposition of lignin was also found in the local resistance response to fungi (Hammerschmidt, 1984).

Despite circumstantial evidence that accumulated phytoalexins and other resistance responses such as deposition of lignin and hydroxyproline-rich glycoproteins are involved in local protection, one should be aware of the fact that the concentrations of phytoalexins were determined in extracts from infected host tissue and not *in situ*. To prove that the fungus really comes in close contact with the accumulated phytoalexins at the right time, more information about their localization is required. The localization of accumulated phytoalexins is also relevant in relation to the growth habit of a pathogen, i.e. intracellular accumulation of phytoalexins probably does not explain inhibition of growth of intercellularly growing fungi. Recently Moesta *et al.* (1983) reported an immuno-bioassay to detect accumulated glyceollin in infected host tissue. In the future such assays will be very useful for the localization of phytoalexins at the cellular level.

Most other determinants of resistance are also quantified indirectly in tissue extracts. Lignin and hydroxyproline-rich glycoproteins are quantified by measuring their breakdown products.

10.3.3. *Prospects for use of induced local resistance in disease control*. The local resistance response is an energy requiring process. Under natural infection pressure only a limited number of cells, which are invaded by avirulent strains of a pathogen or non-pathogens, will respond, so the host plant can afford this strategy. For a while it was thought that indirect control of plant diseases, achieved by activation of the natural defence mechanisms, would be promising. However, the inducers of the resistance response (elicitors) are rather unrealistic tools for use in plant disease control. Toxicity problems could occur as a result of the ubiquitous and continuous presence of phytoalexins and other secondary metabolites.

10.4. INDUCERS OF SYSTEMIC RESISTANCE

10.4.1. Fungi. In contrast to the inducers of local resistance (Chapter 7) the inducers of systemic resistance are less well characterized. Prior inoculation of plants with viable fungi can systemically protect these plants against challenging organisms including bacteria, fungi and viruses (Kuć, 1982a; 1982b). Heat-killed fungi or their cell walls or extracellular products did not induce systemic resistance (Kuć, 1982a; 1982b).

It is worth mentioning here that systemic resistance can only be induced in a certain number of plants among which green bean, cucumber and tobacco are the most important (Kuć, 1982b). In potato, tomato and many other plants induction of systemic resistance has remained unsuccessful, the reason for which remains unknown.

In green bean, local and systemic resistance can be induced simultaneously (Elliston et al., 1971, 1976a; 1976b; 1976c, 1977; Rahe & Kuć, 1970; Skipp & Deverall, 1973). Prior inoculation of green bean with an avirulent race of Colletotrichum lindemuthianum or the non-pathogen C. lagenarium protected locally as well as systemically against virulent races of C. lindemuthianum. Even a heat-attenuated virulent race could induce local and systemic resistance. The time required to achieve both types of resistance was 18 h.

Cucumber has probably been studied most intensively with respect to systemically induced resistance. Systemic protection has been observed 48-96 h after the inducing inoculation with C. lagenarium, or Pseudomonas lachrymans against a whole series of pathogens including C. lagenarium, Mycosphaerella melonis, Fusarium oxysporum f. sp. cucumerinum, Pseudoperonospora cubensis, P. lachrymans, Erwinia tracheiphila, CMV and TNV (Kuć, 1982b). Systemic resistance is graft transmissible and is not cultivar, genus or species-specific (Jenns & Kuć, 1979; Jenns et al., 1979). Roots can be protected against F. oxysporum f. sp. cucumerinum by an inducing inoculation of leaves (Gessler & Kuć, 1982a). Girdling the petiole (removal of the cambial zone) of the induced leaf prevented protection above and below this leaf (Guedes et al., 1980). Girdling the petioles of leaves, while leaving the petiole of the inducing leaf intact, prevented protection only in challenged leaves which had been girdled. These data suggest that a signal has to be transported from the induced leaf to the leaves to be challenged before protection becomes evident.

Systemically induced resistance of tobacco to blue mould (P. tabacina) was reported (Cruickshank & Mandryk, 1960). They injected conidia into the stub of a cut petiole near the base of tobacco stems and found that the upper foliage became completely resistant to sporulation of the blue mould sprayed onto the leaves four weeks after injection. However, they found that stem-injected controls became prematurely senescent compared with water-injected controls. Thus it is not clear whether the induced systemic resistance was caused indirectly by premature senescence. Induction of systemic resistance by stem-injection was also associated with dwarfing and a reduction of leaf area. Recently Tuzun & Kuć (1983a) developed a new technique for stem inoculation. When tobacco stems were injected externally to the cambium, plants were systemically protected by 95-99%, based on the area of necrosis and amount of sporulation. The plants protected by this method were approximately 40% taller, showed a 40% increase in dry weight, 30% increase in fresh weight and had 5-7 more leaves than untreated control plants. Systemically protected plants flowered 2-3 weeks earlier than control plants.

Also with tobacco it was found that plants girdled above the induced area in the first 6 days after induction remained susceptible to blue mould, though enhanced growth was evident already. Girdling plants below the induced area did not diminish induced resistance and enhanced growth. Plants were protected when they were girdled above the site of injection at 9 days or later (Tuzun & Kuć, 1983b).

Tobacco could also be protected against Erysiphe cichoracearum by prior inoculation with P. tabacina (Cohen, 1978). Prior inoculation of tobacco with avirulent races of Phytophthora parasitica var. nicotiana induced systemic resistance to virulent races of the same pathogen (McIntyre & Miller, 1977).

Systemic as well as local resistance to Thielaviopsis basicola could be induced by the same fungus (Hecht & Bateman, 1964) or Endogone mosseae (Baltruschat & Schonbeck, 1975).

10.4.2. Bacteria. In tobacco leaves the introduction of living or heat-killed pathogenic or saprophytic bacteria can induce non-specific resistance to bacteria and various other pathogens (Lovrekovich & Farkas, 1965; Lozano & Sequeira, 1970; Novacky et al., 1973). In cucumber, living but not heat-killed bacteria induced non-specific systemic resistance (Jenns et al., 1979; Caruso & Kuć, 1979). The inducers of systemic resistance to bacteria are not well-characterized and it is disputable whether one is dealing with inducers of local or systemic resistance, as in most studies the challenging inoculation was applied to the tissue which was previously treated with the inducer (Graham et al., 1977; Mazzuchi & Pupillo, 1976; Mazzuchi et al., 1979, 1982; McIntyre et al., 1975). Mazzuchi et al. (1976; 1979; 1982) characterized the inducer of the protection response as a protein-lipopolysaccharide (pr-LPS) while Graham et al. (1977) found lipopolysaccharide (LPS) as the inducer. The latter authors found that the lipid A portion but neither the O-side nor the core polysaccharide seemed to be important for activity. The chemical nature of bacterial inducers may vary in different host-pathogen systems, as McIntyre et al. (1975) found that DNA from virulent and avirulent *Erwinia amylovora* protected Bartlett pear against fire blight.

10.4.3. Abiotic substances. Fungal, bacterial and viral inducers of systemic resistance usually produce restricted necrotic lesions in the induced leaves (Richmond et al., 1979), although some examples where the inducing inoculation does not produce necrotic lesions are known (Staub & Kuć, 1980).

With the exception of some bacterial components (LPS, pr-LPS, DNA, section 10.4.2.) no cell-free components of biotic origin which induce systemic resistance have been found.

It is worth mentioning in this context that a wide range of synthetic chemical compounds can stimulate systemic resistance to fungi in plants (Rathmell, 1984). In screening tests designed to detect antifungal chemicals, he found that a high proportion of chemical compounds that caused visible damage to rice plants also gave some measure of systemic resistance to the fungus *Pyricularia oryzae*. Similar results were obtained with wheat and *Puccinia recondita*. These results suggest that a necrogenic reaction can condition a plant towards systemic resistance.

10.5. THE PROTECTIVE RESPONSE IN INDUCED SYSTEMIC RESISTANCE

10.5.1. The protective response to fungal pathogens.

Phytoalexins. In systems involving induction of local resistance a number of host defence responses which may be involved in protection have been suggested including the accumulation of phytoalexins, lignin, suberin, melanin and hydroxyproline-rich glycoproteins, and formation of papillae (Chapter 7).

In contrast, the nature of the protective response in systemically induced resistance has remained virtually unknown. Although the accumulation of phytoalexins in green bean is associated with race-specific resistance to *C. lindemuthianum* and *C. lagenarium* it cannot be directly involved in systemic resistance, as phytoalexins were not detectable in systemically protected unchallenged bean tissue (Elliston *et al.*, 1977). Accumulation of phytoalexins occurred only after challenge inoculation of systemically protected tissue. Protected tissue is possibly only conditioned to react with a resistance response after challenge by a potential pathogen.

In tobacco not much is known about the protective response. In primarily inoculated leaves there is a significant accumulation of sesquiterpenoid phytoalexins (Guedes *et al.*, 1982; Fuchs *et al.*, 1983; Uegaki *et al.*, 1981). In unchallenged systemically protected leaves, however, these phytoalexins could not be detected. Accumulation of phytoalexins in protected tissue also occurred only after challenge inoculation with a potential pathogen.

Inhibition of conidial germination was suggested to be responsible for systemic induced resistance in tobacco to *P. tabacina* (Leppik *et al.*, 1972). However, as systemically induced resistance to *P. tabacina* is not confined to reduced germination and penetration of the fungus (Cohen & Kuć, 1981), the germination inhibitors are not determinative factors in systemically induced resistance to this fungus.

Lignin. In cucumber, there is as yet no evidence for phytoalexins being involved in defence responses to fungi. Accumulation of lignin has been proposed as a defence response in cucumber resistant to *C. cucumerinum* (Hijwegen, 1963, Hammerschmidt *et al.*, 1984) and in systemically protected cucumber after challenge with *C. cucumerinum* or *C. lagenarium* (Hammerschmidt & Kuć, 1982). Richmond *et al.* (1979) found that penetration of cucumber leaves by *C. lagenarium* was reduced in plants systemically protected by prior inoculation with the same pathogen, while penetration of *C. cucumerinum* in systemically protected cucumber was not affected. Mechanisms for containment of the latter fungus in protected plants have features in common with those in naturally resistant plants (Hijwegen, 1963, Hammerschmidt *et al.*, 1984).

Peroxidase. It has long been known that increase in peroxidase activity is associated with the local resistance response (Benedict, 1972; Fehrman & Dimond, 1967; Kedar, 1959; Macko *et al.*, 1968). However, there are also reports questioning the role of peroxidase in disease resistance (Seevers & Daly, 1970; Nadolny & Sequeira, 1980).

In systemically protected unchallenged cucumbers a threefold increase in peroxidase has been found (Hammerschmidt *et al.*, 1982). A similar phenomenon has been observed in tobacco (Simons & Rose, 1970; Rathmell & Sequeira, 1975). However, increase in peroxidase activity could generate free radicals as precursors for lignin synthesis and therefore the localization and the time of appearance of increased peroxidase activity in systemically protected tissue could be important. The conditioned increase in peroxidase in systemically protected plants is probably only a manifestation of a general phenomenon which is not causally related to resistance, as artificially increasing peroxidase activity in plant tissue does not render these tissues resistant (Nadolny & Sequeira, 1980).

Pathogenesis-related (PR) or b-proteins. Pathogenesis-related (PR) or b-proteins are host-coded proteins which are detected systemically after inducing inoculation. They were first reported in virus-tobacco interactions (Gianinazzi *et al.*, 1970; Van Loon & Van Kammen, 1970) but later they were also found systemically after inducing inoculations with fungi (Gianinazzi *et al.*, 1980). In hosts other than tobacco, PR-proteins were also detected after inducing inoculation with fungi but no proteins or only small amounts were detected at a distance from the infection site (Abu-Jawdah, 1982; Andebrhan *et al.*, 1980; Camacho-Henriquez & Sänger, 1982; De Wit & Bakker, 1980; Gessler & Kuc, 1982b).

It has been suggested that PR-proteins are involved in induced resistance, but antibacterial, antifungal or antiviral activity of PR-proteins has not yet been shown. Although normally barely detectable in young healthy plants, PR-proteins have now also been found in plants at the onset of flowering (Fraser, 1981). With respect to systemically induced resistance to fungi, there is as yet no evidence that PR-proteins play a role.

10.5.2. The protective response to bacterial pathogens.

Envelopment. Since the protective response induced by living bacteria, heat-killed bacteria, pr-LPS or LPS is both light- and time-dependent there is presumptive evidence that the protective mechanism depends entirely on a host response.

Electron microscopic observations indicated that treatment of plants with pr-LPS from *Pseudomonas syringae* p.v. *tabaci* (Mazzuchi *et al.*, 1982) or with LPS from *P. solanacearum* (Graham *et al.*, 1977) caused the invagination of the plasmalemma and vesiculation in the area immediately opposite the place on the host cell wall where the inducer was present, and multiplication of virulent as well as avirulent bacteria was prevented.

Envelopment of avirulent bacteria in resistant host plants has long been thought to be one of the most important resistant responses. The specificity as well as the development of membrane vesicles and thick cell wall appositions at the bacterial attachment site led some authors to believe that attachment is specific for incompatible interactions and that envelopment of avirulent bacteria by fibrillar material is an essential part of the hypersensitive response (HR) (Goodman *et al*., 1976; Whatley & Sequeira, 1981). However, envelopment as a defence response has been seriously questioned. In bean and soybean, attachment of bacteria to the plant cell wall seems to be non-specific, as initially it occurs with virulent, avirulent and non-pathogens (Daub & Hagedorn, 1980; Fett & Jones, 1982; Hildebrand *et al*., 1980; Sigee & Epton, 1976). However, 12 h after inoculation of bean leaves, differences between compatible and incompatible interactions became evident (Hildebrand *et al*., 1980). Cells of virulent *P. syringae* p.v. *syringae* had multiplied more frequently, were less tightly packed and were surrounded by a more electron-lucent matrix than were four avirulent pathovars.

In cotton, as in tobacco, bacteria have been observed consistently to be enveloped in incompatible interactions and not in compatible ones (Al-Mausawi *et al*., 1982a; 1982b; 1983; Cason *et al*., 1978). Although envelopment may be necessary for elicitation of HR, envelopment alone is not sufficient to cause HR. Nor is it dependent on HR, since a saprophytic *Micrococcus lysodeikticus*, corn starch grains, and killed virulent *Xanthomonas campestris* pv. *malvacearum* could be enveloped in cotton in the absence of HR. So envelopment does not appear to require specific surface features (Al-Mousawi *et al*., 1983). Lack of envelopment of multiplying virulent bacteria might be caused by rapid synthesis of extracellular polysaccharide (EPS) which would prevent deposition of an enveloping film close to the bacteria (Sequeira *et al*., 1977).

Lectins. The presence of factors which are capable of agglutinating avirulent strains of *P. solanacearum in vitro* has been known for a long time (Main, 1968). Sequeira & Graham (1977) found that all virulent but not avirulent strains of *P. solanacearum* were agglutinated by the potato lectin which is similar to lectins of other Solanaceae (Kilpatrick *et al*., 1980). It appeared that all virulent strains produced extracellular polysaccharides (EPS) and it was the EPS which prevent virulent strains from being agglutinated. Therefore, the lectin receptors are present in virulent as well as avirulent strains. The bacterial receptors for the lectin appeared to contain N-acetylglucosamine since addition of chitin oligomers inhibited both the agglutination of the avirulent strain and the precipitation of its lipopolysaccharide (LPS). The receptors for the lectin, however, seemed to be present in both EPS and LPS of *P. solanacearum*, a finding which does not support the hypothesis that EPS inhibits the attachment of virulent strains.

A closer look at the LPS showed that avirulent HR-inducing strains of P. solanacearum produce LPS which behaved differently from LPS of non-HR-inducing virulent and avirulent strains on SDS-polyacrylamide gels (Whatley et al., 1980). In addition, LPS from HR-inducing strains had lower xylose : glucose and rhamnose : glucose ratios than non-HR-inducing strains. Moreover, the LPS of the avirulent HR-inducing strains bound more strongly to the potato lectin than those not inducing HR. Most probably, the structure of the O-side chain in HR and non-HR-inducing strains is different. However, protection in tobacco could be induced by both LPS-types, indicating that tobacco leaves do not differentiate between the two types of LPS (Whatley et al., 1980).

The concentration of lectin-like material increased following infiltration of tobacco leaves with heat-killed bacteria. Granular material could be seen surrounding the bacterium at the point of attachment. The granular material surrounding the bacterium was contained by a membrane-like structure and did not diffuse throughout the intercellular spaces.

Agglutinating granules related to induced resistance to bacterial pathogens have also been described by Goodman (1980). However, challenge inoculation of protected leaves did not show extensive agglutination or envelopment of the challenging bacteria although their populations were much reduced (Sequeira & Hill, 1974; Sequeira et al., 1977). This suggests that lectins themselves may only be involved in the initial recognition process leading to a protective response but not in the protective response itself.

Phytoalexin-like compounds. Rathmell & Sequeira (1975) reported the presence of antibacterial substances in the intercellular fluid of tobacco leaves protected by prior inoculation with heat-killed cells of P. solanacearum. The compounds were only secreted under light conditions that resulted in the development of resistance. P. solanacearum cultured in the intercellular fluids from protected leaves grew differently, and slower than bacteria grown in fluids from unprotected leaves. The intercellular fluids contained low-molecular weight inhibitors which are probably terpenoid phytoalexins.

Peroxidases. In unchallenged tobacco plants, systemically protected by heat-killed cells of P. solanacearum, peroxidase increased substantially (Rathmell & Sequeira, 1975). The appearance of peroxidase in the intercellular fluid was light-dependent. Peroxidase may be involved in the production of antibacterial lignin-like compounds. These compounds themselves are probably not antibacterial but their precursor coniferyl alcohol, as well as peroxidase-generated free radicals may be.

Pathogenesis-related (PR) or b-proteins. Pathogenesis-related (PR) or b-proteins were found in tobacco and cucumber after inducing inoculations with bacteria (Ahl et al., 1981; Gessler & Kuć, 1982b; Leach et al., 1983). Whether the PR-proteins play a role in systemically induced resistance to bacteria is not known.

10.5.3. Factors conditioning systemic resistance. From girdling experiments carried out by Kuć & coworkers it was suggested that as a result of the inducing inoculation a movable signal is released which induces systemic resistance in the host to a wide variety of organisms. This signal would condition the host plant to a multimechanism resistance response (Kuć, 1982b). Efforts to protect plants against micro-organisms by inducing inoculation with cell wall preparations and extracellular material from fungi have been unsuccessful. Garas & Kuć (unpublished data cited in Kuć, 1982b) extracted a 'signal' from cucumber leaves inoculated with C. lagenarium or P. lachrymans. Injection of this active preparation into cotyledons and first true leaf of cucumber induced protection in the leaves above to disease caused by C. lagenarium. However, a lag of 4-5 days was required between injection of the active preparation and expression of protection, indicating that the preparation probably contained the inducer itself but not the alleged release signal conditioning protection.

The inducer of systemic accumulation of proteinase inhibitor (proteinase inhibitor inducing factor, PIIF) in tomato plants and other Solanaceous plant species is present in extracts of many higher and lower plants (McFarland & Ryan, 1974). Recently it has been found that PIIF is associated with a pectic polysaccharide which can be released by polygalacturonases of plant origin (released after wounding) or fungal origin (released around the infection site) (Ryan et al., 1981).

Pectic fragments of plant cell wall are also known to induce HR and the accumulation of phytoalexins (Bruce & West, 1982; Hahn et al., 1981; Yamazaki et al., 1983), phenomena always associated with the induction of local resistance. Sequeira (1983) suggested a role for PIIF in the induction of systemic resistance. However, wounding of tomato plants alone releases PIIF and the accumulation of proteinase inhibitors, but does not induce systemic resistance. This indicates that the induction of proteinase inhibitor and the induction of systemic resistance are two separable phenomena. Besides, in tomato plants induced systemic resistance has never been reported.

A somewhat different explanation for conditioning of induced resistance is based on changes in hormone balance caused by the inducing inoculation. Haberlach et al. (1978) found that a shift in kinetin : IAA ratio could modify the resistance of tobacco callus to P. parasitica var. nicotianae. Kinetin suppressed the hypersensitive response, while IAA could reverse the kinetin effects. Tobacco showing systemic resistance to P. tabacina by stem injection showed stunting and suckering, two features indicative of an hormonal imbalance (Cruickshank & Mandryk, 1960). Recently, Tuzun & Kuć (1983a; 1983b) reported that tobacco plants stem-inoculated with conidial suspensions of P. tabacina external to the cambium showed growth increase instead of stunting and suckering. Girdling experiments showed, however, that the phenomena of increase in growth and induced resistance were separable.

Systemic virus infection can cause a dark green mosaic pattern on host leaves (dark green islands). These dark green islands contained a higher concentration of cytokinin than light green leaf tissue (Sziraki & Balazs, 1975) and showed resistance to challenge inoculations (Chapter 9). Green islands are also associated with the infection sites of several microbial obligate parasites (Dekhuijzen, 1976). They appear as a green ring with a diameter of about 3 mm around the chlorotic primary infection site. Increase in cytokinin levels are involved in green island formation. However, whether increased cytokinin levels could be involved in induced local or systemic resistance to bacteria and fungi is not known.

As will be clear from this section, factors involved in conditioning systemic resistance are still much a matter of speculation.

10.5.4. Prospects for use of induced systemic resistance in disease control. Induced systemic resistance has not been used extensively in disease control, although many reports of growth room and greenhouse tests look promising and an example of protection in the field has been reported (Kuć, 1982a). Induced systemic resistance rather than induced local resistance could offer an alternative to the use of pesticides and plant breeding. The problems which remain to be solved include development of suitable inducers without damaging side effects, and possible phytotoxic effects of the induced resistance itself (Chapter 7.7).

REFERENCES

Abu-Jawdah, Y. (1982). Changes in the soluble protein patterns of bean leaves upon fungal or viral infections or after chemical injury. Phytopathologische Zeitschrift 103, 272-279.

Ahl, P., Benjama, A., Samson, R. & Gianinazzi, S. (1981). Induction chez le tabac par Pseudomonas syringae de nouvelle proteines (proteines 'b') associees au developement d'une resistance non-specifique a une deuxieme infection. Phytopathologische Zeitschrift 102, 202-212.

Al-Mousawi, A.H., Richardson, P.E., Essenberg, M. & Johnson, W.M. (1982a). Ultrastructural studies of a compatible interaction between Xanthomonas malvacearum and cotton. Phytopathology 72, 1222-1230.

Al-Mousawi, A.H., Richardson, P.E., Essenberg, M. & Johnson, W.M. (1982b). Cotyledon and leaf ultrastructure of a bacterial blight-immune cotton line inoculated with a low level of Xanthomonas campestris pv. malvacearum. Phytopathology 72, 1230-1234.

Al-Mousawi, A.H., Richardson, P.E., Essenberg, M. & Johnson, W.M. (1983). Specificity of the envelopment of bacteria and other particles in cotton cotyledons. Phytopathology 73, 484-489.

Andebrhan, T., Coutts, R.H.A., Wagih, E.E. & Wood, R.K.S. (1980). Induced resistance and changes in the soluble protein fraction of cucumber leaves locally infected with Colletotrichum lagenarium or tobacco necrosis virus. Phytopathologische Zeitschrift 98, 47-52.

Bailey, J.A. & Mansfield, J.W. (1982). 'Phytoalexins.' Blackie & Sons, Glasgow. 334 pp.

Baltruschat, H. & Schonbeck, F. (1975). Untersuchungen über den Einfluss der endotropen Mycorrhiza auf den Befall von Tabak mit Thielaviopsis basicola. Phytopathologische Zeitschrift 84, 172-188.

Bell, A.A. & Presley, J.T. (1969). Heat-inhibited or heat-killed conidia of Verticillium albo-atrum induce disease resistance and phytoalexin synthesis in cotton. Phytopathology 59, 1147-1151.

Benedict, W.G. (1972). Influence of light on peroxidase activity associated with resistance of tomato cultivars to Septoria lycopersici. Canadian Journal of Botany 50, 1931-1936.

Bernard, N. (1909). L'evolution dans la symbiose. Les orchidees et leurs champignons commenseaux. Annales des Sciences Naturelles (Botanique) 9, 1-196.

Bruce, R.J. & West, C.A. (1982). Elicitation of casbene synthetase activity in caster bean. The role of pectic fragments of the plant cell wall in elicitation by a fungal endopolygalacturonase. Plant Physiology 69, 1181-1188.

Bruehl, G.W. (1975). Biology and control of soil borne plant pathogens. American Phytopathological Society, St. Paul, Minnesota.

Camacho-Henriquez, A. & Sänger, H.L. (1982). Analysis of acid-extractable tomato leaf proteins after infection with a viroid, two viruses and a fungus and partial purification of the "pathogenesis-related" protein p14. Archives of Virology 74, 181-196.

Caruso, F. & Kuć, J. (1977). Protection of watermelon and muskmelon against Colletotrichum lagenarium by Colletotrichum lagenarium. Phytopathology 67, 1285-1289.

Caruso, F. & Kuć, J. (1979). Induced resistance of cucumber to anthracnose and angular leaf spot by Pseudomonas lachrymans and Colletotrichum lagenarium. Physiological Plant Pathology 14, 191-201.

Cason, E.T., Richardson, P.E., Essenberg, M.K., Brinkerhoff, L.A., Johnson, W.M. & Venere, R.J. (1978). Ultrastructural cell wall alterations in immume cotton leaves inoculated with Xanthomonas malvacearum. Phytopathology 68, 1015-1021.

Chester, K.S. (1933). The problem of acquired physiological immunity in plants. Quarterly Review of Biology 8, 129-154, 275-324.

Cohen, Y. (1978). Protection of tobacco against Erysiphe cichoracearum by Peronospora tabacina. Annals of Applied Biology 89, 317-321.

Cohen, Y & Kuć, J. (1981). Evaluation of the systemic resistance to blue mold induced in tobacco leaves by prior stem inoculations with Peronospora tabacina. Phytopathology 71, 783-789.

Cruickshank, I.A.M. & Mandryk, M. (1960). The effect of stem infestation of tobacco with Peronospora tabacina Adam on foliage infection to blue mold. Journal of Australian Institute of Agricultural Sciences 26, 369-372.

Daub, M.E. & Hagedorn, D.J. (1980). Growth kinetics and interactions of Pseudomonas syringae with susceptible and resistant bean tissue. Phytopathology 70, 429-436.

Dekhuijzen, H.M. (1976). Endogenous cytokinins in healthy and diseased plants. In 'Physiological Plant Pathology' (Heitefuss, R. & Williams, P.H., eds), pp. 526-559, Springer Verlag, Berlin.

Deverall, B.J., Wong, P.T.W. & McLeod, S. (1979). Failure to implicate antifungal substances in cross-protection of wheat against take-all. Transactions of the British Mycological Society 72, 233-236.

De Wit, P.J.G.M. & Bakker, J. (1980). Differential changes in soluble tomato leaf proteins after inoculation with virulent and avirulent races of Cladosporium fulvum (syn. Fulvia fulva). Physiological Plant Pathology 17, 121-130.

Elliston, J., Kuć, J. & Williams, E. (1971). Induced resistance to anthracnose at a distance from the site of the inducing interaction. Phytopathology 61, 1110-1112.

Elliston, J., Kuć, J. & Williams, E. (1976a). Protection of bean against anthracnose by Colletotrichum species non-pathogenic on bean. Phytopathologische Zeitschrift 86, 117-126.

Elliston, J., Kuć, J. & Williams, E. (1976b). A comparative study of the development of compatible, incompatible and induced incompatible interactions between Colletotrichum species and Phaseolus vulgaris. Phytopathologische Zeitschrift 87, 289-303.

Elliston, J., Kuć, J. & Williams, E. (1976c). Effect of heat treatment on the resistance of Phaseolus vulgaris to Colletotrichum lindemuthianum and Colletotrichum lagenarium. Phytopathologische Zeitschrift 88, 43-52.

Elliston, J., Kuć, J. Williams, E. & Rahe, J. (1977). Relation of phytoalexin accumulation to local and systemic protection of bean against anthracnose. Phytopathologische Zeitschrift 88, 114-130.

Fehrmann, H. & Dimond, A.E. (1967). Peroxidase activity and Phytophthora resistance in different organs of the potato plant. Phytopathology 57, 68-72.

Fett, W.F. & Jones, S.B. (1982). Role of bacterial immobilization in race-specific resistance of soybean to Pseudomonas syringae pv. glycinea. Phytopathology 72, 488-492.

Fraser, R.S.S. (1981). Evidence for the occurrence of the "pathogenesis-related" proteins in leaves of healthy tobacco plants during flowering. Physiological Plant Pathology 19, 69-76.

Fuchs, A., Slobbe, W., Mol, P.C. & Posthumus, M.A. (1983). GS/MS analysis of fungitoxic terpenoids from tobacco. Phytochemistry 22, 1197-1199.

Gessler, C. & Kuć, J. (1982a). Induced resistance to Fusarium wilt in cucumber by soil and foliar pathogens. Phytopathology 72, 1439-1441.

Gessler, C. & Kuć, J. (1982b). Appearance of a host protein in cucumber plants infected with viruses, bacteria and fungi. Journal of Experimental Botany 33, 58-66.

Gianinazzi, S., Ahl, P., Cornu, A. & Scalla, R. (1980). First report of host b-protein appearance in response to a fungal infection in tobacco. Physiological Plant Pathology 16, 337-342.

Gianinazzi, S., Martin, C. & Vallee, J.C. (1970). Hypersensibilite aux virus, temperature et proteines solubles chez le Nicotiana Xanthi nc. Apparition de nouvelle macromolecules lors de la repression de la synthese virale. Compte rendu hebdomadaire des Seances de l'Academie des Sciences de Paris 270D, 2383-2386.

Goodman, R.N. (1980). Defenses triggered by previous invaders: Bacteria. In: 'Plant Disease, An Advanced Treatise' (Horsfall, J.G. & Cowling, E.B., eds), Vol. 5, pp. 305-317. Academic Press, New York.

Goodman, R.N., Huang, P.Y. & White, J.A. (1976). Ultrastructural evidence for immobilization of an incompatible bacterium, Pseudomonas pisi, in tobacco leaf tissue. Phytopathology 66, 754-764.

Graham, T.L., Sequeira, L. & Huang, T.R. (1977). Bacterial lipopolysaccharides as inducers of disease resistance in tobacco. Applied and Environmental Microbiology 34, 424-432.

Guedes, M., Kuć, J., Hammerschmidt, R. & Bostock, R. (1982). Accumulation of sesquiterpenoid phytoalexins in tobacco leaves infected with Pseudomonas lachrymans. Phytochemistry 21, 2987-2988.

Guedes, M.E., Richmond, S. & Kuć, J. (1980). Induced systemic resistance to anthracnose in cucumber as influenced by the location of the inducer inoculation with Colletotrichum lagenarium and onset of flowering and fruiting. Physiological Plant Pathology 17, 229-233.

Haberlach, G.T., Budde, A.D., Sequeira, L. & Helgeson, J.P. (1978). Modification of disease resistance of tobacco callus tissues by cytokinins. Plant Physiology 62, 522-525.

Hahn, M.G., Darvill, A.G. & Albersheim, P. (1981). Host-pathogen interactions. XIX. The endogenous elicitor, a fragment of plant cell wall polysaccharide that elicits phytoalexin accumulation in soybeans. Plant Physiology 68, 1161-1169.

Hamilton, R.I. (1980). Defenses triggered by previous invaders: Viruses. In: 'Plant Disease, An Advanced Treatise' (Horsfall, J.G. & Cowling, E.B., eds), Vol. 5, pp. 279-303. Academic Press, New York.

Hammerschmidt, R. (1984). Rapid deposition of lignin in potato tuber tissue as a response to fungi non-pathogenic on potato. Physiological Plant Pathology 24, 33-42.

Hammerschmidt, R. & Kuc, J. (1982). Lignification as a mechanism for induced systemic resistance in cucumber. Physiological Plant Pathology 20, 61-71

Hammerschmidt, R., Lamport, D.T.A. & Muldoon, E.P. (1984). Cell wall hydroxyproline enhancement and lignin deposition as an early event in the resistance of cucumber to Cladosporium cucumerinum. Physiological Plant Pathology 24, 43-47.

Hammerschmidt, R., Nuckles, E. & Kuc, J. (1982). Association of enhanced peroxidase activity with induced systemic resistance of cucumber to Colletotrichum lagenarium. Physiological Plant Pathology 20, 73-82.

Hecht, E.I. & Bateman, D.F. (1964). Non-specific acquired resistance to pathogens resulting from localized infection by Thielaviopsis basicola or viruses in tobacco leaves. Phytopathology 54, 523-530.

Hijwegen, T. (1963). Lignification, a possible mechanism of active disease resistance against pathogens. Netherlands Journal of Plant Pathology 69, 314-317.

Hildebrand, D.C., Alosi, M.C. & Schroth, M.N. (1980). Physical entrapment of pseudomonads in bean leaves by films formed at air-water interfaces. Phytopathology 70, 98-109.

Jenns, A., Caruso, F. & Kuć, J. (1979). Non-specific resistance to pathogens induced systemically by local infection of cucumber with tobacco necrosis virus, Colletotrichum lagenarium (Pass) Ell et Halst or Pseudomonas lachrymans (Smet Bryan) Carsner. Phytopathologia Mediterranea 18, 129-134.

Jenns, A. & Kuc, J. (1977). Localized infection with tobacco necrosis virus protects cucumber against Colletotrichum lagenarium. Physiological Plant Pathology 11, 207-212.

Kedar, N. (1959). The peroxidase test as a tool in the selection of potato varieties resistant to blight. American Potato Journal 36, 315-324.

Kerr, A. & Htay, K. (1974). Biological control of crown gall through bacteriocin production. Physiological Plant Pathology 4, 37-44.

Kilpatrick, D.C., Jeffree, C.E., Lockhart, C.M. & Yeoman, M.M. (1980). Immunological evidence for structural similarity among lectins from species of the Solanaceae. FEBS Letters 113, 129-133.

Kochman, J.K. & Brown, J.F. (1975). Studies on the mechanism of cross-protection in cereal rusts. Physiological Plant Pathology 6, 19-27.

Kuć, J. (1982a). Plant immunization - mechanisms and practical implications. In: 'Active Defence Mechanisms in Plants' (Wood, R.K.S., ed.), pp. 157-178. Plenum Press, New York and London.

Kuć, J. (1982b). Induced immunity to plant disease. Bioscience 32, 854-860.

Leach, J.E., Sherwood, J., Fulton, R.W. & Sequeira, L. (1983). Comparison of soluble proteins associated with disease resistance induced by bacterial lipopolysaccharide and by viral necrosis. Physiological Plant Pathology 23, 377-387.

Leppik, R.A., Hollomon, D.W. & Bottomley, W. (1972). Quiesone: an inhibitor of germination of Peronospora tabacina conidia. Phytochemistry 11, 2055-2063.

Lovrekovich, L. & Farkas, G.L. (1965). Induced protection against wildfire disease in tobacco leaves treated with heat-killed bacteria. Nature 205, 823-824.

Lozano, J.C. & Sequeira, L. (1970). Prevention of the hypersensitive reaction in tobacco leaves by heat-killed bacterial cells. Phytopathology 60, 875-879.

McFarland, D. & Ryan, C.A. (1974). Proteinase inhibitor-inducing factor in plant leaves. A phylogenetic survey. Plant Physiology 54, 706-708.

McIntyre, J.L. (1982). Induced resistance. In 'Phytopathogenic Prokaryotes' (Mount, M.S. & Lacy, G.H., eds), Vol. 2, pp. 137-147. Academic Press, New York.

McIntyre, J.L., Kuc, J. & Williams, E.B. (1975). Protection of Bartlett pear against fire blight with deoxyribonucleic acid from virulent and avirulent *Erwinia amylovora*. Physiological Plant Pathology 7, 153-170.

McIntyre, J.L. & Miller, P.M. (1977). Protection of tobacco against *Phytophthora parasitica* var. *nicotianae* by cultivar non-pathogenic races, cell free sonicates, and *Pratylenchus penetrans*. Phytopathology 68, 235-239.

Macko, F., Woodbury, W. & Stahmann, M.A. (1968). The effect of peroxidase on germination and growth of mycelium of *Puccinia graminis* f. sp. *tritici*. Phytopathology 58, 1250-1254.

Main, C.E. (1968). Induced resistance to bacterial wilt in susceptible tobacco cuttings pretreated with avirulent mutants of *Pseudomonas solanacaearum*. Phytopathology 58, 1058-1059.

Matta, A. (1980). Defenses triggered by previous diverse invaders. In: 'Plant Disease, An Advanced Treatise' (Horsfall, J.G. & Cowling, E.B., eds), Vol. 5, pp. 345-361. Academic Press, New York.

Mazzuchi, U., Bazzi, C. & Medeghini Bonatti, P. (1982). Encapsulation of *Pseudomonas syringae* pv. *tabaci* in relation to its growth in tobacco leaves both pretreated and not pretreated with protein-lipopolysaccharide complexes. Physiological Plant Pathology 21, 105-112.

Mazzuchi, U., Bazzi, C. & Pupillo, P. (1979). The inhibition of susceptible and hypersensitive reactions by protein-lipopolysaccharide complex from phytopathogenic pseudomonads: relationship to polysaccharide antigenic determinants. Physiological Plant Pathology 14, 19-30.

Mazzuchi, U. & Pupillo, P. (1976). Prevention of hypersensitive necrosis in tobacco leaves by a bacterial protein-polysaccharide complex. Physiological Plant Pathology 9, 101-112.

Moesta, P., Hahn, M.G. & Grisebach, H. (1983). Development of a radioimmunoassay for the soybean phytoalexin glyceollin I. Plant Physiology 73, 233-237.

Müller, K.O. & Börger, H. (1941). Experimentelle Untersuchungen über die *Phytophthora* Resistenz der Kartoffel. Arbeiten aus dem Biologischen Reichsanstalt (Reichsaust.) Berlin 23, 189-231.

Nadolny, L. & Sequeira, L. (1980). Increases in peroxidase activities are not directly involved in induced resistance in tobacco. Physiological Plant Pathology 16, 1-8.

Novacky, A., Acedo, G. & Goodman, R.N. (1973). Prevention of bacterially induced hypersensitive reaction by living bacteria. Physiological Plant Pathology 3, 133-136.

Ouchi, S. (1983). Induction of resistance or susceptibility. Annual Review of Phytopathology 21, 209-315.

Ouchi, S., Oku, H., Hibino, C. & Akiyama, I. (1974). Induction of accessibility and resistance in leaves of barley by some races of *Erysiphe graminis*. Phytopathologische Zeitschrift 79, 24-34.

Paxton, J.D. & Chamberlain, D.W. (1967). Acquired local resistance of soybean plants to *Phytophthora* spp. Phytopathology 57, 352-353.

Rahe, J.E. & Kuć, J. (1970). Metabolic nature of the infection-limiting effect of heat on bean anthracnose. Phytopathology 60, 1005-1009.

Rahe, J.E., Kuć, J., Chuang, C. & Williams, E.B. (1969). Induced resistance in *Phaseolus vulgaris* to bean anthracnose. Phytopathology 59, 1641-1645.

Rathmell, W.G. (1984). The discovery of new methods of chemical disease control, current developments, future prospects and the role of biochemical and physiological research. Advances in Plant Pathology 2, 261-288.

Rathmell, W.G. & Sequeira, L. (1975). Induced resistance in tobacco leaves: the role of inhibitors of bacterial growth in the intercellular fluid. Physiological Plant Pathology 5, 65-73.

Richmond, S., Kuć, J. & Elliston, J. (1979). Penetration of cucumber leaves by *Colletotrichum lagenarium* is reduced in plants systemically protected by previous infection with the pathogen. Physiological Plant Pathology 14, 329-338.

Ryan, C.A., Bishop, P., Pearce, G., Darvill, A.G., McNeil, M. & Albersheim, P. (1981). A sycamore cell wall polysaccharide and a chemically related tomato leaf polysaccharide possess similar proteinase inhibitor-inducing activities. Plant Physiology 68, 616-618.

Seevers, P.M. & Daly, J.M. (1970). Studies on wheat stem rust resistance controlled at the Sr6 locus. II. Peroxidase activities. Phytopathology 60, 1642-1647.

Sequeira, L. (1983). Mechanisms of induced resistance in plants. Annual Review of Microbiology 37, 51-79.

Sequeira, L., Gaard, G. & De Zoeten, G.A. (1977). Interaction of bacteria and host cell walls: its relation to mechanisms of induced resistance. Physiological Plant Pathology 10, 43-50.

Sequeira, L. & Graham, T.L. (1977). Agglutination of avirulent strains of Pseudomonas solanacearum by potato lectin. Physiological Plant Pathology 11, 43-54.

Sequeira, L. & Hill, L.M. (1974). Induced resistance in tobacco leaves: the growth of Pseudomonas solanacearum in protected tissues. Physiological Plant Pathology 4, 447-455.

Sigee, D.C. & Epton, H.A.S. (1976). Ultrastructural changes in resistant and susceptible varieties of Phaseolus vulgaris following artificial inoculation with Pseudomonas phaseolicola. Physiological Plant Pathology 9, 1-8.

Simons, T.J. & Ross, A.F. (1970). Enhanced peroxidase activity associated with induction of resistance to tobacco mosaic virus in hypersensitive tobacco. Phytopathology 60, 383-384.

Skipp, R. & Deverall, B. (1973). Studies on cross-protection in the anthracnose disease of bean. Physiological Plant Pathology 3, 299-313.

Staub, T. & Kuc, J. (1980). Systemic protection of cucumber plants against disease caused by Cladosporium cucumerinum and Colletotrichum lagenarium by prior localized infection with either fungus. Physiological Plant Pathology 17, 389-393.

Suzuki, H. (1980). Defenses triggered by previous invaders: Fungi. In 'Plant Disease, An Advanced Treatise' (Horsfall, J.G. & Cowling, E.B., eds), Vol. 5, pp. 319-332. Academic Press, New York.

Svoboda, W.E. & Paxton, J.D. (1972). Phytoalexin production in locally cross-protected Harosoy and Harosoy 63 soybeans. Phytopathology 62, 1457-1460.

Sziraki, I. & Balazs, E. (1975). The effect of infection by TMV on cytokinin level of tobacco plants, and cytokinins in TMV-RNA. In 'Current Topics in Plant Pathology' (Kiraly, Z. ed.), pp. 345-352. Akademiai Kiado, Budapest.

Tjamos, E.C. (1979). Induction of resistance to Verticillium wilt in cucumber (Cucumis sativus). Physiological Plant Pathology 15, 223-227.

Tuzun, S. & Kuć, J. (1983a). A new technique which immunizes tobacco against blue mold (Peronospora hyoscyami f. sp. tabacina) and increases the growth of tobacco. Phytopathology 73, 823.

Tuzun, S. & Kuć, J. (1983b). Movement of a factor in tobacco which protects against blue mold (Peronospora hyoscyami f. sp. tabacina). Plant Physiology 72 Supplement, 158.

Uegaki, R., Fujimori, T., Kubo, S. & Kato, K. (1981). Sesquiterpeniod stress compounds from Nicotiana species. Phytochemistry 20, 1567-1568.

Van Loon, L.C. & Van Kammen, A. (1970). Polyacrylamide disc electrophoresis of the soluble leaf proteins from Nicotiana tabacum var. "Samsun" and "Samsun NN". II. Changes in protein constitution after infection with tobacco mosaic virus. Virology 40, 199-211.

Whatley, M.H., Hunter, N., Cantrell, M.A., Hendrick, C., Keegstra, K. & Sequeira, L. (1980). Lipopolysaccharide composition of the wild pathogen Pseudomonas solanacearum. Plant Physiology 65, 557-559.

Whatley, M.H. & Sequeira, L. (1981). Bacterial attachment to plant cell walls. In 'The Phytochemistry of Cell Recognition and Cell Surface Interactions' (Loewus, A. & Ryan, C.A., eds), pp. 213-240. Plenum Press, New York.

Yamazaki, N., Fry, S.C., Darvill, A.G. & Albersheim, P. (1983). Host pathogen interactions. XXIV. Fragments isolated from suspension-cultured sycamore cell walls inhibit the ability of the cells to incorporate (^{14}C)leucine into proteins. Plant Physiology 72, 864-869.

Yarwood, C.E. (1956). Cross protection with two rust fungi. Phytopathology 46, 540-544.

CHAPTER 11. PRESENT AND FUTURE PROSPECTS FOR EXPLOITATION OF RESISTANCE IN CROP PROTECTION BY NOVEL MEANS

A. G. HEPBURN, M. WADE and R. S. S. FRASER

11.1 INTRODUCTION

11.1.1. Conventional and novel approaches. Historically, plant breeders have placed much emphasis on simply-inherited resistance which has often proved to be vertical. Some of the resistances conferred by single genes have been spectacularly successful in agricultural terms. However, a single mutation in the pathogen can be sufficient to overcome such resistance, and the apparent advantage of improved plant varieties has often been quickly lost. The reliance on major gene resistance becomes an especial problem when the selection pressures exerted on the pathogen population increase with the extensive use of any one variety. Possibilities of epidemics are enhanced by the demands of modern high technology farming for uniformity in crop performance. This, coupled with the narrowing of the genetic base of crops indicates the need for an alternative approach.

One possibility, using either conventional or novel breeding techniques, is to switch emphasis from major gene resistance to characteristics typical of a more general or horizontal resistance. Accordingly, interest is increasing in screening new plant crosses for features such as increased capacity to synthesize phytoalexins and lignin; increased passive defences such as thicker cuticles and more pre-formed inhibitors; greater sensitivity to elicitors, and increased levels of enzymes capable of degrading fungal and bacterial cell walls.

The other approach is to seek to make greater use of the available genetic resources, whether single gene or oligogenic. A variety of techniques such as artifical mutation, somaclonal variation and use of haploids attempt to expose existing variation, increase it, or offer fresh combinations of resistance with other characteristics.

Further techniques aim to widen the genetic basis of resistance by performing transfers of resistance genes between species where hybridization by conventional techniques is impossible. These may operate at the cell level, as in protoplast fusion, or at the molecular level, using recombinant DNA technology.

Chemical control of disease has been largely based on fortuitous discoveries combined with routine screening of large numbers of chemicals on a purely empirical basis. While this approach has given highly effective agents for control of fungal diseases, chemicals of practical use against bacterial and viral infections are respectively few or completely lacking. There is also the problem that many chemical controls are relatively unspecific in their action, and may have undesirable environmental side effects. Biochemical plant pathology is now beginning to provide the insights into mechanisms of host-pathogen interaction which should allow disease control chemicals to be designed on a rational basis. These should include not only chemicals with a direct action against the pathogen, but also those which may act indirectly by stimulating host defence mechanisms.

The aim of this Chapter is to review current progress in these developing technologies, and to identify the problems which remain.

11.2. MUTAGENESIS AND VARIATION: TECHNIQUES AT THE CELL AND TISSUE LEVEL

11.2.1. Artifical mutation techniques. Chemical and physical mutagens such as gamma, X-ray and ultraviolet radiation have been been widely used in three types of approach:

a) with cultivars or species where no natural resistance to a pathogen is available, in an attempt to induce resistance. An example is the apparent resistance to MYMV induced in mungbean by gamma irradiation (Shakoor *et al.*, 1977). Such induced resistances may represent deletion or inactivation of a host susceptibility function normally required for pathogenesis, and are usually recessive (Flor, 1960). They also tend to be associated with other, possibly pleiotropic effects on the host. For example, several of the mung bean mutants were dwarves compared to the parent.

b) with cultivars or species where natural resistance is available, mutagenesis has been used to modify existing genes, in attempts to widen the basis of resistance and to combat gene or allele erosion occurring as a consequence of modern agricultural practices. Thus at the *ml-o* locus for resistance to powdery mildew in barley, there are at least ten alleles induced by mutagenesis (Jørgensen, 1977).

c) use of mutants of the host or pathogen which alter the pattern of their interaction, as a means to study this interaction. This aspect is considered in Chapters 4 - 7.

Despite the fact that resistance of varying effectiveness can be induced in a wide variety of crop species, against many viral, bacterial and fungal pathogens (compiled in International Atomic Energy Agency, 1977), there appears to have been little or no introduction of resistance genes or alleles so made into commercial cultivars. This may indicate the frequency of deleterious side effects.

A different approach to the use of irradiation in resistance breeding was suggested by experiments with pollen given massive doses of gamma rays. Progeny were found to resemble the maternal parent very closely, but to contain a few features characteristic of the paternal genotype (Pandey, 1975; Jinks et al., 1981). It was suggested that the irradiated pollen might induce a form of parthenogenesis in the female, but that some of the genetic material of the pollen could be stably incorporated in the offspring. Davies (1981) noted the attractive possibility of using this as a rapid means of transfer of resistance genes between cultivars or compatible species, without having to go through a series of back-cross generations. Since then, transfer of paternal genes, including one for resistance to mildew, has been reported in other species using the irradiated pollen technique (Powell et al., 1983; Snape et al., 1983). However, Snape et al. (1983) and Werner et al. (1984) argued from cytological evidence that seed produced using irradiated pollen was in fact the result of a conventional fertilization, and that the maternal appearance of the offspring was due to radiation damage, although post-meiotic selection might also be involved. The evidence for a gene transfer mechanism as suggested by Pandey (1975; 1980) is therefore strongly challenged.

11.2.2. Somaclonal variation and haploidy. Clones derived via callus culture may show a greater range of variation in response to pathogens than the parent plant, as well as in other characteristics of agronomic importance. Somatic variation between cells of crops which are normally propagated vegetatively was thought to account for some of this variation. However, it is now clear that the process of callus culture itself is mutagenic. This type of variation is normally referred to as 'somaclonal' (Larkin & Scowcroft, 1981; Chaleff, 1983; Cassells et al., 1983). It is interesting that cases of variation induced by callus culture have been shown to be under monogenic or polygenic control, raising the possibility that this may provide a method for handling polygenic resistance which is free from the difficulties associated with conventional hybridization.

As examples of somaclonal variation, Shepard et al. (1980) showed that several clones regenerated from single leaf cell protoplasts of potato had enhanced resistance to Phytophthora infestans and Alternaria solani. Sugar cane clones have been produced with improved resistance to FDV (Scowcroft & Larkin, 1982).

For commercial production of F_1 hybrid cultivars, it is normally desirable that the parent lines are highly inbred. A possible short cut to this is to use anther culture to produce haploid plants, which by natural or chemically-induced chromosome doubling give rise to completely homozygous dihaploids. These may expose genetic variation present in the parent, by preventing dominance and hypostasis. They can also offer a useful route to combinations of resistance genes with other agronomically useful characters (Burk & Chaplin, 1980). An example is the rapid production of barley lines containing BaYMV resistance (Foroughi-Wehr & Friedt, 1984). Furthermore, the process of haploidization may also be mutagenic and create new disease resistance (Burk et al., 1979).

11.2.3. Somatic hybridization. Many pathogens, especially viruses, can affect a wide range of crops, but resistance present in one species can generally not be exploited in another, because of incompatibility barriers (Stavely et al., 1973). This may be especially relevant to wild species related to crop species, which can be a useful source of resistance genes. Transfer of such genes may be possible via an intermediate host but this is protracted. This led to attempts to fuse protoplasts. Such 'somatic hybrids' or 'cybrids' have now been produced between several pairs of sexually incompatible species (Power et al., 1980) and have been used to transfer disease resistance genes (Evans et al., 1981; Stavely et al., 1973).

However, several problems remain before the method reaches full application in crop protection. Generally, somatic hybridization is easier the closer the two species are related. Regeneration of intergeneric hybrid plants has proved most difficult. Somatic hybridization between Solanum and Lycopersicon species has been achieved (Melchers et al., 1978; Schiller et al., 1982), although the significance of this as an intergeneric hybrid should be tempered by the taxonomic difficulties of separating the two genera (Melchers, 1984).

There may also be problems of genetic instability. Thus in callus cultures derived from fusion of Glycine max and Nicotiana glauca, Kao (1977) recorded a gradual elimination of chromosomes, and several other authors have reported aneuploidy and other variations. It is clear that any attempt to transfer resistance genes by these techniques would require an extensive programme of backcrossing to the crop species parent, to remove unwanted characteristics of the donor species, and to obtain a stable karyotype. Dudits _et al._ (1980) irradiated the protoplasts of one species in an attempt to obtain selective gene transfers, by radiation damage to the remainder of the genome. This might accelerate production of commercially useful species, but would require a selection procedure for suitably transformed clones at an early stage.

11.3. RECOMBINANT DNA TECHNIQUES

Application of recombinant DNA techniques to manipulation of resistance genes involves several main steps:

a) isolation of useful genes from the complete plant genome;

b) cloning the resistance gene as an insert in a plasmid which is multiplied in a bacterium;

c) transfer of the gene to a suitable vector which will carry it to the cells of a recipient plant;

d) control of stability of the transferred genetic information in the new plant, either as an episome or by chromosomal integration;

e) selection of transformed host cells, and regeneration of whole plants from thom;

f) control of the level of expression of the transferred gene, and

g) control of inheritance of the transferred gene.

Most of these steps and their attendant problems apply to all proposed gene transfers, not only those involving resistance genes. However, some steps involve particular problems for work with resistance genes, and these will be discussed in more detail.

11.3.1. _Isolation of useful genes._ For most of the resistances described in earlier Chapters, the scientific background is largely descriptive or symptomatic, with no clear understanding of how the pathogenicity is controlled. In these cases, a molecular approach requires more background information. To decide whether a molecular transformation would be worth attempting, it is necessary to determine, at the genetic level, how many loci are involved. A single gene resistance offers potential at the molecular level, while success with polygenic resistance is more likely if grosser techniques are used.

It is also important to know whether conferring resistance on the host will involve adding or removing a character; that is to say whether the resistance mechanism is positive, involving active or passive defence mechanisms, or negative, implying absence of a susceptibility function or recognition event. Some authors (e.g. Russell, 1978) have suggested that it might be possible to transfer genes for 'non-host' resistance to host species. However, if such 'genes' represent lack of a recognition or susceptibility function found in hosts, there is nothing useful in the non-host which can be transferred, and what is required is a deletion in the host genome.

All these criteria provide a framework within which transfer of genes between species can be assessed. Assuming that it is decided that there is a useful gene to transfer, how might this be isolated at the molecular level? There are several techniques, of varying degrees of difficulty, and the protocol chosen will depend on the extent of knowledge of the gene in question and its mode of action. The currently available techniques for isolating and manipulating genes from eukaryotes, and for constructing clones and genomic libraries, are well documented (Maniatis _et al._, 1982).

Isolation of genes with known products. If the gene produces a defined polypeptide product which can be purified in moderate quantity, then using antibodies raised against that product, the mRNA from which it was translated can be partially purified, by using the antibody to select polyribosomes synthesizing the gene product. Some enrichment may be obtained if the polyribosomes are fractionated by size prior to selection with the antibody. If the synthesis of the product is a response to the presence of the pathogen, then further enrichment of the cell mRNA population for the resistance gene mRNA may be obtained.

A population of cDNA clones is then prepared from the population of enriched mRNAs, and the resistance gene clone is identified by hybrid arrest (Paterson _et al._, 1979) or hybrid release (Goldberg _et al._, 1979) translation. In the first case, the cDNAs are used individually to remove mRNAs from a total preparation and the resistance gene clone identified as that which eliminates translation of the protein of interest. In hybrid release, each clone is used to 'fish' for the mRNA of interest in the total mRNA population; each mRNA is then released from the mRNA-cDNA hybrid, translated and identified by its translation product. Once the cDNA has been isolated and identified (and it need not be a cDNA for the whole gene, as long as it provides a specific probe), it can be used to isolate the chromosomal gene from a library of the DNA from the resistant species. The assumption is that the chromosomal gene with its controlling sequences intact will be a better unit to transfer than a cDNA with an alternative promoter. However, there are some situations where this assumption would not be valid, and these are considered below.

Isolation by 'genome walking'. In some cases, although the resistance gene product may be unknown, the gene position may be well documented on the genetic map, and may be located close to genes with known products. If one of the latter genes can be isolated from the library by the techniques detailed above, then the possibility of isolating the resistance gene by 'genomic walking' arises (Bossy et al., 1984). In principle, this involves use of sequence overlaps between clones in a genomic library, to move hand-over-hand outwards from the starting gene. If cosmids are used as the cloning vehicles, then the clones can be of 30-40,000 base pairs in length, and it should be possible to scan large regions of linear DNA.

In practice, the technique is fraught with problems. In pea, for example, one map unit equals approximately 1,000,000 bases, and thus even genes which map close together may be widely separated at the molecular level. The structure of the genome can also interfere. If members of repeated sequence families occur between the two genes, the repeat sequence would act as a branch point to a large number of genetic locations, and the specificity of the system would be degraded. It is important to note, however, that occurrence of a large amount of repeated DNA in a species does not necessarily indicate that the structural genes are embedded in these repeats. Nor does the presence of large amounts of what appears to be single-copy DNA imply that this represents the bulk of the genic DNA, and that the organization of the bulk of the single-copy sequences reflects the organization of the genes. Murray & Thompson (1982), also working with pea, have shown that the bulk of the apparent single-copy DNA which shows normal short period interspersion is in fact ancient repeated sequences which have diverged significantly, and that the genes are in a longer period interspersion pattern.

The next problem is that of identifying the resistance gene as it is 'walked over'. Either each cosmid or phage clone would have to be used to select mRNAs for test, or each segment would have to be sequenced. From this it is clear that a map location and genomic walking without a means of assaying for the gene product are of very little use.

Isolation using the pathogen. The ultimate method for isolation of resistance genes would be to use the specific interaction between the resistance gene product and the pathogen as the screen. Randomly cloned DNA fragments from the genomic library of the resistant plant would be transferred to the susceptible host and tested for resistance by challenge inoculation with the pathogen. This assay might be done at the cell, callus or regenerated plant level, and would obviously require massive resources, given the size of the higher plant genome. It would probably be necessary to screen in the region of 1,000,000 clones, and with transformation efficiencies of around 10% using *Agrobacterium tumefaciens* or 0.0001% using naked DNA, it is clear that some persistance and a rather large amount of greenhouse space would be required. It is possible that combining 'genome walking' with such a functional assay could reduce the numbers to reasonable levels.

Isolation of susceptibility or recognition genes. In cases where resistance is the lack of something required for normal pathogenesis, there is a further level of complexity in genetic manipulation of the susceptible host. It would be necessary to know what the pathogen is recognising in the host; to isolate the gene for that feature and either remove it from the host; alter its function and replace it, or substitute it by a suitably different gene from a resistant plant. The aim would be to eliminate the interaction with the pathogen, while maintaining any function necessary to the host. Here again the limiting molecular constraint is having to recognize the gene product.

Isolation by transposon mutagenesis. A method which might offer the ability to isolate both resistance genes and susceptibility genes may soon be available. Transposon mutagenesis has already been used extensively in bacteria to identify functional genes, and is currently being used to identify pathogenicity characters in *Xanthomonas* (Turner *et al*., 1984). Plant transposable elements show some of the characteristics of their bacterial counterparts, in that they are capable of moving around the genome and affecting the expression of the plant genes that they lie next to. Robertson's mutator (Strommer *et al*., 1982), when inserted into tobacco cells using an *Agrobacterium tumifaciens* delivery system, increases its copy number above that of its delivery vehicle, and appears to alter its location (Lillis *et al*., 1984). If this is a general phenomenon, then it may be possible to use transposon mutagenesis to identify resistance genes by loss of resistance function.

First, a transformed plant or cell line is selected in which the resistance function has been lost. If the transposable element has been appropriately constructed, for example with the addition of an *Escherichia coli* replicon and a phage lambda *cos* site, then the total DNA from the selected plant could be digested with restriction enzymes; circularized by ligation; packaged in phage lambda heads (using the *cos* site to select only clones containing transposon and linked DNA), and thus transfected to *E. coli*, where clones carrying the *E. coli* replicon could be selectively multiplied. Some clones carrying the transposable element would, it is to be hoped, carry chromosomal DNA containing the resistance gene; further constructs lacking the tranposable element could then be derived and used to transform susceptible plants. These kinds of technique for gene isolation are still in the future, but information is accumulating that the future they inhabit is close.

11.3.2. *Transfer of genes to new plants*. In contrast to the difficulties of isolating resistance genes, potential methods for transferring them to plants are well advanced, and in recent years genes for specific proteins have been sucessfully transferred between families which are taxonomically distant (Murai *et al*., 1983; Broglie *et al*., 1984; Herrera-Estrella *et al*., 1983; Matzke *et al*., 1984). Two basic approaches are currently in use; one is totally artificial while the other relies on adaptations of a natural delivery system.

Direct insertion of DNA. Purified DNA can be inserted into plant protoplasts by a number of techniques. The molecules are either taken up by a process of endocytosis (Cocking, 1972), or are encapsulated in natural or artificial liposomes (Lurquin, 1979; Tanaka *et al*, 1984) with which the protoplasts are encouraged to fuse. Delivery, however, is only the first stage in transformation; for the new character to be of use it must be maintained in a stable form. There is very recent evidence that some DNA is directly integrated into the host chromosome (Paszkowski *et al*., 1984), and such mechanisms clearly deserve further study for plants such as most monocots, where alternative delivery mechanisms are less well developed.

Adapted natural delivery systems. There are no good episomal vectors available yet for higher plants, although much research is currently being done on virus-based, cytoplasmically maintained vectors (Gronenborn *et al*., 1981). The best current technique is to deliver the gene in such a way that it will become integrated into the host genome. The major emphasis here is on vector systems based on the plant pathogenic bacterium *Agrobacterium tumefaciens*. This is of interest both for its potential in genetic manipulation of plants, and as an example of a pathogen where recombinant DNA techniques have given great insights into the mechanisms of pathogenesis. It

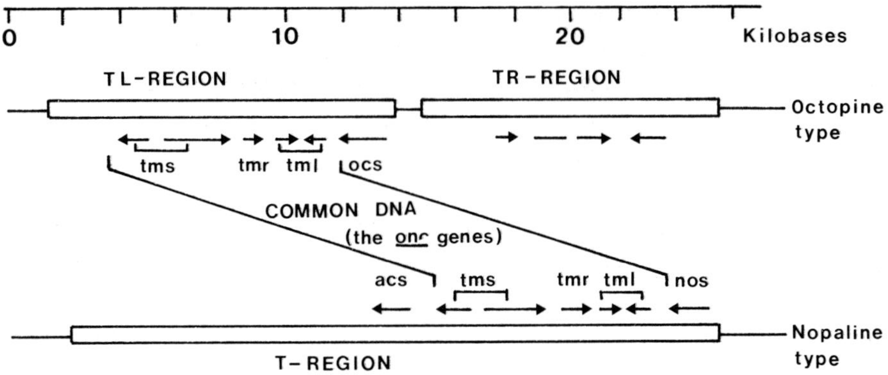

FIGURE 11.1. Comparison of the structure of the T-regions of the two major groups of Ti-plasmids. acs = agrocinopine synthase; nos = nopaline synthase; ocs = octopine synthase. The other named transcripts are described in Garfinkel *et al*. (1981) and on this page. The transcripts of the TR-region are involved in the synthesis of agropine and its relatives (see Barker *et al*. (1983) for detailed analysis). The left half of the nopaline T-region encodes transcripts that are as yet only crudely mapped and have no known function.

therefore deserves description in some detail.

The molecular basis of Agrobacterium pathogenicity. This is associated with a family of large plasmids, the Ti-plasmids. During pathogenic transformation, a precisely-defined segment of the plasmid, the T-region, becomes integrated into the host genome, where it is maintained through somatic cell divisions as the T-DNA (Hille *et al*., 1984) (Figure 11.1).

The T-region carries the *onc* genes which are responsible for the tumorigenic phenotype. There are three major loci: *tmr* (*roi*), *tms* (*shi*) and *tml* (Garfinkel *et al*., 1981). The *tms* locus comprises two genes which encode auxin synthesis (Inze *et al*., 1984), while the *tmr* locus is a single gene apparently involved in cytokinin synthesis (Barry *et al*., 1984). The precise nature of the *tml* locus is not known, but it may comprise one or two genes which produce similarly sized transcripts (Bevan & Chilton, 1982; Willmitzer *et al*., 1983). This *onc* region contains sequences common to all Ti-plasmids.

The remainder of the T-region contains genes coding for enzymes involved in opine synthesis. Opines are relatively simple molecules, the results of condensations of amino acids, sugars, nucleotides and ketoacids, and are used by the inducing bacterium as a food source. Occurrence of opines, which are not normal plant constituents, is taken as an assay for transformation.

The T-region is delimited by 25 base-pair repeats which, although they define the limits of DNA to be transferred and integrated, are not usually fully integrated themselves (Zambryski et al., 1982; Yadav et al., 1982). Insertion has been shown to include approximately half of the repeat at the left end of the molecule and none of the repeat at the right end (Holsters et al., 1980; Thomashow et al., 1980). These terminal repeats are not exact, but conform to a concensus (Barker et al., 1983).

Different plasmid families, defined by their opine fingerprints, have T-regions which differ in structure (Figure 11.1). The nopaline family have one block of 23 kb flanked by the repeats, with the onc region in the right half. The octopine family in contrast have two T-regions side-by-side, each bracketed by copies of the repeats with approximately 2 kb of DNA between them. TL (13 kb) comprises the onc region and the major opine synthesis gene (octopine synthase), while TR (approximately 8 kb) contains the remaining opine synthases (Barker et al., 1983; Leemans et al., 1982; Ellis et al., 1984). These structures mean that while the nopaline plasmid T-region is inserted as one large piece (Hepburn et al., 1983a; Lemmers et al., 1980), those of the octopine plasmids are inserted in two pieces. TR is not normally found by itself in tumours, but since TR does not contain any onc functions, this is hardly surprising! Since in tumours TL by definition must be present, and since the insertion of TL and TR is independent, the copy number of the two T-DNAs is variable. TL is usually in fairly low copy number, while the number of TR copies per cell varies from none to 20 - 30 (Merlo et al., 1980; Ooms et al., 1982).

A further set of genes, the vir region, comprises seven loci. These are required for pathogencity, but only in trans with the T-region (Hille et al., 1984) and are not integrated into the host DNA. These genes have several functions, but only a few such as auxin synthesis have been identified (Kao et al., 1982). Recent evidence (Koukolokova-Nicola et al., 1984) suggests that one role of the vir genes is the excision of the T-region between the two repeats to generate minicircles containing one intact copy of the T-region and one chimaeric copy of the repeat, probably resulting from recombination.

Use of Ti-plasmids as delivery vehicles. Two major methods of using the Agrobacterium system as a gene delivery method are currently employed. The first is based on a method originally described for use with Rhizobium (Ruvkun

& Ausubel, 1981) and involves three stages.

In the first stage, a cloned piece of Ti-plasmid from the T-region is modified to carry the foreign DNA of choice, which is inserted into the clone so as to have at least 600 bases of T-region on either side of it. The inserted DNA should be close, ideally next to, a bacterial antibiotic resistance gene capable of operating in Agrobacterium. The second stage involves the transfer of this chimaeric construct to Agrobacterium, and there are as many methods as there are groups working on the system. The aim is to allow the modified clone to coexist with the Ti-plasmid for long enough to allow recombination to take place between the modified DNA in the clone, and its unmodified homologous counterpart in the resident Ti-plasmid. This is the third stage.

This temporary co-existence can be achieved in several ways. The replicon used in Escherichia coli is frequently one which will not replicate in Agrobacterium, but which can be transferred to Agrobacterium either by trans-mobilization (Matzke & Chilton, 1981) or by direct plasmid transformation (Holsters et al., 1978). This can be followed by selection for the antibiotic resistance marker next to the foreign gene, which will only succeed if the gene has become stably integrated into the Ti-plasmid. Assuming that the foreign gene and antibiotic resistance marker are close together, they will most likely have been integrated together.

This system tends to lead to marker rescue following a single recombination event, which puts the whole clone into the Ti-plasmid and leads to T-region duplication. This is still an acceptable construct for transforming plants. There is only a problem when the T-region clone used in the first part contains one of the copies of the terminal T-region repeats. In this case, a single integration event will lead to the insertion of a third border sequence and produce an A-B-C form of structure, where A, B and C represent border sequences, and (-) the intervening genes. Then the onc genes will lie between A and B, and a duplication containing a modified segment of DNA between B and C. In practical terms, integration into the plant DNA will result in three possible constructions: A-B, B-C and A-B-C. Only A-B-C will satisfy the requirement of being selectable by the onc genes and also carry the modification. (A double insertion A-B/B-C in one cell would of course also be acceptable). In practice, this occurs often enough for the system to be usable.

A simplified variant has recently been described where a resident

Ti-plasmid was constructed which contains a copy of pBR322 located in the T-region (Zambryski *et al.*, 1983). pBR322 is the *E. coli* multi-copy vector (Sutcliffe, 1978). Using this as recipient, plant genes cloned in pBR322 can be used directly in association with a suitable marker and recombination obtained in *Agrobacterium* between the pBR322 copies. The result is the insertion of the foreign gene of interest, bracketed between two copies of pBR322, the whole contained within the Ti-region of the resident plasmid.

Other variations on this general theme have been developed to allow the modified clone to persist in *Agrobacterium* for long enough to allow the double reciprocal exchange to take place. In this case, incP- or incW- based 'intermediate vectors' are commonly used (Matzke & Chilton, 1981; Leemans *et al.*, 1981). They may be present initially as the *E. coli* replicon, or they may be added after all modification has been completed either by *in vitro* cloning or by *in vivo* co-integration and transferred to *Agrobacterium* either by self- or trans-mobilization (Matzke & Chilton, 1981; Hepburn, unpublished observations). They can then be displaced from *Agrobacterium* by plasmid incompatibility where, following conjugation with a suitable donor strain, selection can be applied for the presence of an incompatible plasmid and hence for loss of the recombinant clone. Persistence of the selection marker can only be achieved by integration into the Ti-plasmid by recombination. If a test for loss of the *E. coli* replicon is included, the single recombination events can be eliminated and the double reciprocal events identified.

The second system uses the fact that the *vir* region genes need only be in trans with the T-region. This has resulted in the development of a binary vector system (Hoekema *et al.*, 1983; Framond *et al.*, 1983) comprising an *Agrobacterium* strain containing a helper or driver plasmid which is a derivative of a Ti-plasmid where the T-region has been deleted, and a clone containing the T-region on a separate plasmid capable of replicating both in *Agrobacterium* and *E. coli*. The T region is then modified and transferred to an *Agrobacterium* strain containing the helper. The new strain containing both plasmids can then be used to infect plants.

Whichever system is used, it is obvious that the tumour phenotype is a highly undesirable character. It eliminates the possibility of regeneration into whole plants because of the hormonal disruption it imposes. Some strains of *Agrobacterium* produce tumours where a certain amount of plant regeneration is possible, but in many cases these arise from the outgrowth of untransformed cells carried in the tumour mass (Sacristan & Melchers, 1977). Where such regenerated plants do indeed arise from transformed cells, they can be expected to show some tumour character, for example opine synthesis, and may be unstable, producing tumours either spontaneously or on wounding (Gresshoff et al., 1979). Only rarely is the T-DNA or a deleted portion of it inherited through the sexual cycle (Yang et al., 1980).

The reason for using the onc region which is responsible for this phenotype is to provide a transformation marker. Provided an alternative marker can be used, then the onc genes can be removed without any effect on the transfer and integration of the T-region. In practice, rather than removing the onc region, the rationale has been to reassemble those portions of the T-region required for the transfer and integration process. Thus the right and left ends of the T-region containing the repeats can be joined to eliminate all the genes of the T-region. However, it is useful to retain an opine synthesis gene as a means of unequivocally demonstrating transformation, and since those for nopaline and octopine lie adjacent to the right border repeat of the appropriate T-region, they are customarily included with the right border repeat.

Screening for transformants. There are several methods for applying the altered bacteria to the recipient plant material. In co-cultivation, bacteria are added to protoplasts regenerating a cell wall. If a high enough transformation efficiency can be obtained, then all viable calli produced can reasonably be screened for opine production. It the efficiency drops below 1%, then the numbers to be screened become excessive. Under these circumstances is is useful to have a positive selection marker within the reduced T-region. A bacterial gene for kanamycin resistance is the most widely used example and will serve as a model.

The resistance gene is the neomycin phosphotransferase II structural gene carried by the bacterial transposable element Tn5. Being of bacterial origin, this gene will not operate in eukaryotes unless under the control of a eukaryotic promoter. Both opine and viral promoters have been used (Herrera-Estrella et al., 1983; Fraley et al., 1983). In all cases, the ATG initiation codon is removed from the transcription start signals, or, since these may not always be well defined, the regions where such signals are known to be located from mutagenesis experiments. The promotor is then joined to the structural gene, from which the prokaryotic signals have been removed. It is important to note that any ATG between the plant promoter and the protein start will reduce the efficiency of expression, by providing a false start for translation. Eukaryotic ribosomes appear in general to start translation at the first AUG they encounter (Kozac, 1981).

In order for the mRNA to accumulate, the gene must also be provided with a eukaryotic transcription termination signal and poly(A) addition site. These again are often provided as the 3' ends of opine genes. The advantage of using T-region or viral control signals is that they usually derive from constitutively expressed genes and are not subject to developmental controls, as some host plant gene promotors may be.

With an antibiotic resistance marker replacing the onc function, transformants can be selected on the basis of antibiotic resistance. The assay for opine synthesis provides a check that the resistance observed is not spontaneous. Chloramphenicol acetyl transferase is also a useful marker, not for positive selection but for ease of assay (Herrera-Estrella et al., 1983).

Systems based on these general models and combinations of them have been used to transfer several genes between plant species (Broglie et al., 1984; Murai et al., 1983; Herrera-Estrella et al., 1983; Matzke et al., 1984). In some cases, the gene can be shown to be under the influence of control systems comparable to those operating in its original environment (Broglie et al., 1984). In others, the controls are less restrictive then normal, so that expression of the gene, at least in terms of mRNA, can take place in unusual tissues (Matzke et al., 1984). Thus although consensus promotor sequences similar to those of animal genes have been identified in plant genes, they do not necessarily give the tissue-specific expression observed when the gene is in its normal environment. It is not yet known whether specific expression results from other controlling sequences around the gene or from the position of the gene in the genome.

The plant itself can influence the expression of the inserted gene. Methylation of the cytosine residues in the DNA acquired at or soon after transformation is capable of suppressing the activity of opine genes (Hepburn et al, 1983b). Incubation of the tissue with 5-azacytidine which inhibits methylation results in the expression of the genes being restored.

Inhibition of expression of transformed genes could in theory happen to any gene. Thus it is clear that DNA delivery is only one aspect of the problem of stable genetic manipulation. The gene, once inserted, must be under reasonably normal control, and must maintain expression at a useful level.

Alternatives to co-cultivation. Although co-cultivation is probably the most efficient transformation system, it is not necessarily the most desirable, relying as it does on the production and regeneration of protoplasts. However, the use of non-tumorigenic vectors allows regenerating plant material to be infected. Normally either callus (Muller et al., 1984), suspension cultures (Smith, 1979) or leaf discs can be infected with the disabled bacteria. Callus is then allowed, or in the case of leaf discs, encouraged to grow. Kanamycin to select for transformed cells can be added at the same time as the antibiotics used to kill the Agrobacterium. After growth has been observed, small growing calli can be removed and cultured on medium which will promote regeneration. In the case of tobacco, this involves removing the auxin while retaining a reasonable level of cytokinin (Skoog & Miller, 1957). The protocol can be modified for other species. It still requires a degree of regenerative capacity from callus tissue, and thus is not likely to be universally applicable at present.

At the other extreme of transformation methods, the recombinant molecules with their selection marker might be used to transform protoplasts directly, without involving the bacterium. The Ti-plasmid is too large for effective manipulation at this level. Its supercoiled structure of 200 kbases is very fragile and breaks readily during manipulation. It is also difficult to purify to high enough concentration. The smaller pieces used in the construction work in E. coli, however, are eminently suitable. Since the T-region border repeats appear to be used by vir gene products, then in the absence of the vir region it should be unnecessary to include then in the constructs.

To date, there have been few well-documented cases of in vitro transformation of plant protoplasts with either native or encapsulated DNA. Where Ti-plasmid derivatives have been used, the integration has lacked the specificity of the natural system, and has shown fragmentary integration (Draper et al., 1982; Paszkowski et al., 1984). The time scale of these experiments, however, is fairly long and suitable non-oncogenic, antibiotic resistance-marked DNAs have only recently become freely available. It is likely that reports of direct transformations using small kanamycin-resistance marked DNA segments will appear in the near future. In the longer term future, vectors are likely to be based on, or at least contain, eukaryotic transposable elements.

Inheritance of transferred genes. Early fears that there was a meiotic block to the sexual transmission of the T-DNA proved unfounded. The inhibition found was a function of the onc genes, and once these were removed or lost by rearrangement, the T-DNA acted as a normal locus and showed Mendelian segregation (Barton *et al.*, 1983; Norton & Towers, 1984; De Greeve *et al.*, 1982).

11.3.3. Choice of approach to novel gene transfers. Resistance to disease is an obvious candidate for transfer between species. Where natural breeding techniques cannot be used, artificial genetic manipulation techniques may provide a method. However, as with all projects where the driving force originates from an agricultural need rather than scientific altruism, the question is where to start: whether it is better to attempt the transfer using techniques at the cell and tissue level, or to operate on isolated resistance genes. The molecular basis of many diseases is unknown; often it is even difficult to make a decision on what information is required to allow a sensible choice of route.

There are two primary considerations which affect the choice of technique: the genetic complexity of the characteristic to be transferred, and the susceptibility of the designated host species to manipulation. Both considerations are functional, in that they define current knowledge, and are fluid in that they vary with the state of basic research in relevant areas. At one extreme, they will indicate the procedures to be used to transfer a character; at the other they will specify areas of basic research required to establish transfer procedures.

As with any natural breeding programme, the preliminary product of any modification is likely to be one plant which can then be put into a breeding or bulking programme. Thus the primary requirement of novel breeding techniques is to regenerate, either immediately or following sexual or asexual propagation, fully transformed plants where the new genetic information is part of the heritable genetic information. It is at this point that a serious stumbling block arises. Technologies tend to be developed using 'model' species and genetic manipulation is no exception. Tobacco, tomato, petunia, sunflower, *Kalanchoe* and flax have all been used to test aspects of the new technology (Ooms *et al.*, 1982b; Draper *et al.*, 1982; Hepburn *et al.*, 1983b). The Solanaceae in general provide excellent models, and can normally be regenerated from protoplasts all the way to intact plants (Shepard *et al.*, 1980). Where the final regeneration results in shoots only, they can usually be grafted and whole plants obtained from a second generation. As one moves out of the Solanaceae, however, examples of regeneration become rarer and it becomes necessary to distinguish different levels of regenerative capacity. Protoplasts derived for purposes of manipulation either from whole plant tissue or from callus must first be regenerated into callus. Callus, derived from whole plant tissue or from protoplasts, can be regenerated into whole plants. Neither regenerative capacity is necessarily available outside the model species.

Several dicotyledonous crop species are manipulable at the tissue culture level. Apart from the Solanaceae, some members of the Brassicae and Cruciferae have well defined culture conditions (Lazzeri & Dunwell, 1984; Langridge, 1957). The legumes, on the other hand, are particularly intractable and rarely regenerate into plants. Where such regeneration has been observed, it usually has derived from chimaeric callus culture, for example from embryonic axes, where regeneration is likely to have arisen from persistent meristematic cells (Hussey & Gunn, 1984). Such chimaeric callus does not normally arise when the callus is obtained from protoplasts. The worst of all possible worlds is found with monocotyledonous plants, in that not only is regeneration rare, but development of delivery systems has lagged behind those for dicotyledonous hosts, due to the apparent inability of Agrobacterium to infect monocotyledonous plants (De Cleene & De Ley, 1976). However, this has now been shown to be an artifact of the assay procedure for at least some species. (Hooykaas-Van Slogteren et al., 1984). Opine assays showed that although tissue proliferation following inoculation was negligible, opine synthesis could be detected in the wound site. Thus, as was predicted (Hepburn, 1981), obtaining evidence for the transformation of at least some monocots depends on the ability to select and assay the transferred and presumably integrated functions.

If protoplasting eliminates the regenerative capacity of the species, then many of the potential manipulative procedures which demand the absence of the cell wall are correspondingly made impossible. Under such circumstances, the choice lies between working with the limited number of techniques which can be applied to callus tissue, and investing resources in basic research to overcome the regeneration problem. With a crop species of high commercial value, such as soybean, basic research and cruder transformation assays can proceed side by side. For less valuable crops, the choice between the two approaches may well be dictated by the genetic complexity of the characteristic to be transferred. It is possible that a single isolated gene could be transferred directly to regenerating tissue, as well as to protoplasts. If, however, the character is polygenic or of unknown complexity, then the cruder genetic techniques of protoplast fusion or total DNA transfer might be more appropriate. In some of these cases, the ability to regenerate from protoplasts is essential.

11.4. THE FUTURE FOR CHEMICAL CONTROLS

11.4.1. Perspective. The history of chemical control of plant diseases is as long as that of plant pathology: Homer described the use of burning sulphur as a fumigant, and the earliest known writings on plant pathology are those of Theophrastus (370-286 BC). In the nineteenth century, chance observations followed by deduction led to the discovery of copper-containing antifungal agents such as Bordeaux mixture. Commercial pesticide research programmes began in the 1890's, with the development of the chemical industry. Antinonnin, for instance, was based on the dyestuff dinitro-*o*-cresol. Mercuric chlorophenols were introduced in the early 1900's as seed dressings. Experiments on vulcanizing agents in the rubber industry led to the discovery of dithiocarbamates such as zineb and thiram and other organosulphurs. These continue to be used as broad spectrum foliar fungicides.

The success of systemic organo-chlorine and organo-phosphate insecticides led to the development of similarly systemic compounds for fungus control. Kitazin, introduced in the mid 1960's, was an early example, used for control of *Pyricularia oryzae* on rice. There then followed a succession of remarkably effective systemic and therapeutic products, from benomyl in 1967 to the sterol-inhibiting fungicides of today (Langcake *et al.*, 1983). Fungicides are now well established as a major cost-effective component of modern agriculture. Today, there are essentially two types of fungicide:

a) acting against the growth of fungi directly. Such compounds may or may not require conversion to the actual fungitoxic agent after application to the plant or fungus. They affect metabolic processes that are common to pathogenic and non-pathogenic fungi alike, and their activity can often though not always be demonstrated against the growth of certain fungi in culture. The majority of commercial fungicides belong to this group.

b) acting against the host-parasite association, either through effects on the pathogenic processes of fungi or on resistance mechanisms of the host. Such compounds are not inherently fungitoxic; they therefore have an indirect mode of action.

Plant pathology has had little impact on the discovery of fungicides: chance and inspirational observation have been more involved. Modern agrochemical screening methods have become more systematic but the process is still largely empirical. Only now is biochemical plant pathology beginning to provide the fundamental insights necessary for rational design of 'directly acting' compounds. Likewise, study of the biochemistry of host resistance and host-pathogen interactions may provide a basis for design of antimicrobial agents with an indirect mode of action.

11.4.2. Design of directly- and indirectly-acting fungicides. For disease to occur, host and potential parasite must meet several requirements. The parasite must have a means of entering the plant, and must be able to avoid or counteract host defence mechanisms. It must then multiply, and often causes the biochemical changes that are recognised as disease. Finally, parasite propagules must spread the disease further. The host must provide a suitable environment for parasite growth, including essential nutrients. In theory at least, chemical disruption of any of these host or parasite processes could lead to disease control. A number of possibilites present themselves.

Clearly, preventing the entry of a parasite into the plant would prevent subsequent disease development. Inhibition of cutinesterases and other lytic enzymes required by some pathogens would be one approach. Inhibition or disruption of the appressorial infection peg function is another. There are precedents for the validity of both approaches, particularly the latter. The commercial fungicide tricyclazole, used for the control of rice blast disease, appears to act by inhibiting melanin biosynthesis (Tokousbalides & Sisler, 1979). Without melanin, the appresorial peg does not have sufficient tensile strength to penetrate the cuticle (see Chapter 7.2.1). Note that in this example, tricyclazole is acting indirectly. It affects fungal metabolism directly enough, but this is of little consequence to the fungus growing on agar; it is only of consequence to the fungus when attempting to grow as a pathogen.

Once inside the plant, possibilities exist for reversing pathogen mechanisms which counter host defences. Thus suppressors of defence reactions or phytoalexin metabolizing enzymes might be antagonized to achieve disease control. Since the primary object of plant protection is to control disease rather than to kill the pathogen, control could also be brought about by inhibiting production or action of toxins which cause disease symptoms, for example by preventing them from binding to receptors. Some limited success in controlling disease in this way has been reported (Horsfall & Zentmeyer, 1942; Strobel, 1973). Besides low molecular weight toxins, many pathogens also utilize hydrolytic enzymes to aid their progress through host tissue. The possibility of inhibiting such enzymes to control disease has received considerable attention, but so far only with limited success.

There are two ways in which host resistance mechanisms might be exploited for disease control. The first is the activation of resistance mechanisms themselves. The second is the use of host defensive compounds directly.

Many chemicals, both biotic and abiotic, elicit the production of phytoalexins in the absence of the parasite (see Chapters 7 and 10). While this might suggest useful employment in disease control, a frequent problem is that phytoalexin production is accompanied by cell death and the hypersensitive response. The elegance of the natural resistance response lies in the fact that the hypersensitivity occurs only at the time and place where it is needed to counteract penetration by the pathogen. The sacrifice by the plant is minimal. Clearly, however, it would be a bad idea to trigger such a reaction indiscriminately throughout the plant.

Ideally, what is required is a compound that would alter the plant in a subtle way, such that on challenge by the pathogen, it would behave in the same way as a genetically resistant plant, and resistance mechanisms would be activated only where and when required. Fundamental knowledge on how such mechanisms might be accomplished is fragementary, but it is established that they can be achieved. Once class of molecules, the dichlorocyclopropanes, was discovered during screening of compounds for activity against rice blast disease. The most active compound, WL28325 (Langcake *et al.*, 1981) is systemically prophylactic, and highly specific for this disease. In this respect, it resembles tricyclazole. WL28325 is not sufficiently fungitoxic to account for disease control, and does not give rise to fungitoxic metabolites. Plants treated with the compound, however, responded to infection by *Pyricularia oryzae* in a manner similar to that of genetically-resistant varieties. The response includes hypersensitive cell death and accumulation of phytoalexins near the invading fungus.

Another example of a compound that appears to act through host defence mechanisms in the systemic fungicide tris-*o*-ethyl phosphonate (TEPA; Aliette). It stimulates defence reactions including the synthesis of phytoalexins in grape vines and tomato, when these plants are challenged by certain fungi. TEPA is active *in vivo* against members of the Peronosporales. However, it has very low antifungal activity *in vitro*. Its potent antifungal activity *in vivo* can be nullified by glyphosate and aminooxyacetic acid, inhibitors of the shikimic acid pathway which is the source of phenolics required for phytoalexin synthesis (Bompeix et al., 1981). There are precedents for chemical activation of other defence responses such as lignification for disease control (Kaars Sijpesteijn, 1969).

Direct use of phytoalexins in disease control has been tried several times but with limited success. There are several problems. Firstly, the compounds themselves are not readily manufactured, and in some cases cannot yet be synthesized. Secondly, they are not transported systemically in plants. Thirdly, they tend to be metabolically or sometimes intrinsically unstable. Finally, they are generally of low fungitoxicity, especially in comparison with systemic fungicides. Nevertheless, it is possible that some of these problems could be overcome by analogue synthesis.

For TMV infections of tobacco, Fraser & Whenham (1978a; 1978b) showed that virus multiplication could be inhibited by up to 90% by treatment with methyl benzimidazol-2yl-carbamate (MBC), the systemic antifungal constituent of the benomyl fungicides. No direct, specific antiviral activity could be demonstrated for MBC, and it was only effective if present in leaves from a very early stage of development. It was suggested that MBC might be maintaining leaves in some juvenile, antiviral stage, perhaps through its known cytokinin activity (Thomas, 1974). Interestingly, this occurred without other deleterious effects on plant growth (Fraser & Whenham, 1978b). Parallels might be drawn with the low virus content and resistance to infection of dark green islands on systemically-infected leaves (see Chapter 9).

Other substituted benzimidazoles and triazoles have also been reported to inhibit virus multiplication or eliminate infection, but the effects are complex and do not suggest a single mode of action. Virazole (1-β-D-ribofuranosyl-1,2,4-triazole-3-carboxamide (synonym ribavirin) suppressed multiplication of CMV and AMV in tissue culture, apparently showing an interaction with kinetin (Simpkins et al., 1981). 2-α-hydroxybenzylbenzimidazole (HBB) prevented TMV accumulation in tomato and tobacco leaf discs (Cassells & Cocker, 1982; Fraser & Gerwitz, 1984). HBB did not, however, display activity in cytokinin bioassays (Cassells & Cocker, 1982); did not show any selectivity for inhibition of TMV RNA synthesis in leaf discs, and was ineffective against virus multiplication in whole plants (Fraser & Gerwitz, 1984).

11.5. CONCLUSION

The novel strategies for disease control discussed in this Chapter range from those which are already capable of making a contribution, to those whose usefulness is hypothetical or questionable. This Chapter has therefore discussed the possibilities in the widest sense, taking a very long term view, but also giving prominence to the problems and difficulties which remain.

Deciding which strategy to pursue requires value judgements, balancing the costs of the research; the eventual benefits in crop protection, and the exploitable basic knowledge gained _en route_. These value judgements are heavily influenced by economic and commercial inputs: can a new disease control measure be protected in some way, such as by patent or plant breeder's rights, so that the investment in research can be recouped? Is the target crop of sufficiently high value, and are there economically viable ways of conferring novel types of resistance on minor crops? Will the novel type of resistance be nullified by evolution of the pathogen to overcome it, or can resistance be designed specifically to be durable?

The judgements are also influenced by environmental and safety considerations. Manipulating genes and metabolic pathways may produce human and animal toxicity problems which will require careful screening; for example in control of the levels of alkaloid production in the Solanaceae. However, the optimistic outlook is that by exploiting natural resistance mechanisms, there is the possibility for a greater degree of specificity in disease control, with less likelihood of side effects. The non-lethal but specific effects of indirectly-acting antifungal agents offer possibilites of disease control with the least impact on non-target organisms.

REFERENCES

Barker, R.F., Idler, K.B., Thompson, D.V. & Kemp, J.D. (1983). Nucleotide sequence of the T-DNA region from the _Agrobacterium tumefaciens_ octopine Ti plasmid pTi15955. Plant Molecular Biology 2, 335-350.

Barry, G.F., Rogers, S.G., Fraley, R.T. & Brand, L. (1984). Identification of a cloned cytokinin biosynthesis gene. Proceedings of the National Academy of Sciences, U.S.A. 81, 4776-4780.

Barton, K.A., Binns, A.N., Matzke, A.J.M. & Chilton, M.-D. (1983). Regeneration of intact tobacco plants containing full-length copies of genetically engineered T-DNA and transmission of T-DNA to R1 progeny. Cell 32, 1033-1043.

Bevan, M.W. & Chilton, M.-D. (1982). Multiple transcripts of T-DNA detected in nopaline crown gall tumours. Journal of Molecular and Applied Genetics 1, 539-546.

Bompeix, G., Fettouche, F. & Saindrenan, P. (1981). Mode d'action du Phosethyl Al. Phytiatrie-Phytopharmacie 30, 257-272.

Bossy, B., Hall, L.M.C. & Spierer, P. (1984). Genetic activity along 315 kb of the _Drosophila_ chromosome. EMBO Journal 11, 2537-2541.

Broglie, R., Coruzzi, G., Fraley, R.T., Rogers, S.G., Horsch, R.B., Niedermeyer, J.G., Fink, C.L., Flick, J.S. & Chua, N.-H. (1984). Light-regulated expression of a pea ribulose-1,5-bisphosphate carboxylase small subunit gene in transformed plant cells. Science 224, 838-843.

Burk, L.G. & Chaplin, J.F. (1980). Variation among anther-derived haploids from a multiple disease-resistant tobacco hybrid. Crop Science 20, 334-338.

Burk, L.G., Chaplin, J.F., Gooding, G.V. & Powell, N.Y. (1979). Quantity production of anther-derived haploids from a multiple disease-resistant tobacco hybrid. 1. Frequency of plants with resistance or susceptibility to tobacco mosaic virus (TMV), potato virus Y (PVY) and root knot (RK). Euphytica 28, 201-208.

Cassells, A.C. & Cocker, F.M. (1982). Antiviral activity of 2-(α-hydroxybenzyl)-benzimidazole and other 2-substituted benzimidazoles against TMV in tomato leaf discs. Zeitschrift für Naturforschung 37c, 390-393.

Cassells, A.C., Goetz, E.M. & Austin, S. (1983). Phenotypic variation in plants produced from lateral buds, stem explants and single-cell-derived callus of potato. Potato Research 26, 367-372.

Chaleff, R.S. (1983). Isolation of agronomically useful mutants from plant cell cultures. Science 219, 676-682.

Cocking, E.C. (1972). Virus uptake, cell wall regeneration and virus multiplication in isolated plant protoplasts. International Review of Cytology 28, 89-124.

Davies, D.R. (1981). Gene transfer in plants. Nature 291, 531-532.

De Cleene, M. & De Ley, J. (1976). The host range of crown gall. Botanical Review 42, 389-466.

De Greve, H., Leemans, J., Hernalsteens, J.-P., Thia-Toong, L., De Beuckeleer, M., Willmitzer, L., Otten, L., Van Montagu, M. & Schell, J. (1982). Regeneration of normal and fertile plants that express octopine synthase, from tobacco crown galls after deletion of tumour-controlling functions. Nature 300, 752-755.

Draper, J., Davey, M.R., Freeman, J.P., Cocking, E.C. & Cox, B.J. (1982). Ti-plasmid homologous sequences present in tissues from Agrobacterium plasmid transformed Petunia protoplasts. Plant and Cell Physiology 23, 451-458.

Dudits, D., Fejer, O., Hadlaczky, G., Koncz, C., Lazar, G.B. & Horvath, G. (1980). Intergeneric gene transfer mediated by plant protoplast fusion. Molecular and General Genetics 179, 283-288.

Ellis, J.G., Ryder, M.H. & Tait, M.E. (1984). Agrobacterium tumefaciens encodes a pathway for agropine biosynthesis. Molecular and General Genetics 195, 466-473.

Evans, D.A., Flick, C.E. & Jensen, R.A. (1981). Disease resistance: incorporation into sexually incompatible somatic hybrids of the genus Nicotiana. Science 213, 907-909.

Flor, H.H. (1960). The inheritance of X-ray induced mutations to virulence in a ureidospore culture of race 1 of Melampsora lini. Phytopathology 50, 603-605.

Foroughi-Wehr, B. & Friedt, W. (1984). Rapid production of recombinant barley yellow mosaic virus resistant Hordeum vulgare lines by anther culture. Theoretical and Applied Genetics 67, 377-382.

Fraley, R.T., Rogers, S.G., Horsch, R.B., Sanders, P.R., Flick. J.S., Adams, S.P., Bittner, M.L., Brand, L.A., Fink, C.L., Fry., J.S., Galluppi, G.R., Goldberg, S.B., Hoffmann, N.L., & Woo, S.C. (1983). Expression of bacterial genes in plant cells. Proceedings of the National Academy of Sciences, U.S.A. 80, 4803-4807.

Framond, A.J., Barton, K.A. & Chilton, M.-D. (1983). Mini-Ti: a new vector strategy for plant genetic engineering. Biotechnology 1, 262-269.

Fraser, R.S.S. & Gerwitz, A. (1984). Effects of 2-α-hydroxybenzylbenzimidazole on tobacco mosaic virus and host RNA synthesis in tobacco leaf discs and plants. Plant Science Letters 34, 111-117.

Fraser, R.S.S. & Whenham, R.J. (1978a). Inhibition of the multiplication of tobacco mosaic virus by methyl benzimidazol-2yl-carbamate. Journal of General Virology 39, 191-194.

Fraser, R.S.S. & Whenham, R.J. (1978b). Chemotherapy of plant virus disease with methyl benzimidazol-2yl-carbamate: effects on plant growth and multiplication of tobacco mosaic virus. Physiological Plant Pathology 13, 51-64.

Garfinkel, D.J., Simpson, R.B., Ream, L.W., White, F.F., Gordon, M.P. & Nester, E.W. (1981). Genetic analysis of crown gall: fine structure map of the T-DNA by site-directed mutagenesis. Cell 27, 143-153.

Goldberg, M.L., Lefton, R.P., Stark., G.R. & Williams, J.G. (1979). Isolation of specific RNAs using DNA covalently linked to diazobenzyloxymethylcellulose paper. In 'Methods in Enzymology', Vol. 68 (Wu, R., ed.), pp. 206-208. Academic Press, New York.

Gresshoff, P.M., Skotnicki, M.L., & Rolfe, B.G. (1979). Crown gall teratoma formation is plastid and plant controlled. Journal of Bacteriology 137, 1020-1021.

Gronenborn, B., Gardner, R.C., Schaefer, S. & Shepherd, R.J. (1981). Propagation of foreign DNA in plants using cauliflower mosaic virus as vector. Nature 294, 773-776.

Hepburn, A.G. (1981). The development of vectors for the genetic manipulation of higher plants based on the tumour-inducing plasmid of Agrobacterium tumefaciens. In 'Barley Genetics IV: Proceedings of the 4th International Barley Genetics Symposium' pp. 940-946. Edinburgh University Press, Edinburgh.

Hepburn, A.G., Clarke, L.E., Blundy, K.S. & White, J. (1983a). Nopaline Ti-plasmid, pTiT37, T-DNA insertions into a flax genome. Journal of Molecular and Applied Genetics 2, 211-224.

Hepburn, A.G., Clarke, L.E., Pearson, L. & White, J. (1983b). The role of cytosine methylation in the control of nopaline synthase gene expression in a plant tumour. Journal of Molecular and Applied Genetics 2, 315-329.

Herrerá-Estrellá, L., De Block, M., Messens, E., Hernalsteens, J.-P., Van Montagu, M. & Schell, J. (1983). Chimaeric genes as dominant selectable markers in plant cells. EMBO Journal 3, 1525-1531.

Hille, J., Hoekema, A., Hooykaas, P. & Schilperoort, R.A. (1984). Gene organization of the Ti-plasmid. In 'Genes Involved in Microbe-Plant Interactions' (Verma, D.P.S. & Hohn, T., eds) pp. 287-309. Springer Verlag, New York and Vienna.

Hoekema, A., Hirsch, P.R., Hooykaas, P.J.J. & Schilperoort, R.A. (1983). A binary vector strategy based on separation of vir- and T-region of the Agrobacterium tumefaciens Ti-plasmid. Nature 303, 179-180.

Holsters, M., de Waele, D., Depicker, A., Messens, E., Van Montagu, M. & Schell, J. (1978). Transfection and transformation of Agrobacterium tumefaciens. Molecular and General Genetics 163, 181-187.

Holsters, M., Silva, B., Van Fliet, F., Genetello, C., De Block, M., Dhaese, P., Depicker, A., Inze, D., Engler, G., Villarroel, R., Van Montagu, M. & Schell, J. (1980). The functional map of the nopaline A. tumefaciens Ti-plasmid pTiC58. Plasmid 3, 212-230.

Hooykaas-Van Slogtern, G.M.S., Hooykaas, P.J.J. & Schilperoort, R.A. (1984). Expression of Ti-plasmid genes in monocotyledonous plants infected with Agrobacterium tumefaciens. Nature 311, 763-763.

Horsfall, J.G. & Zentmeyer, G.A. (1942). Antidoting of toxins of plant diseases. Phytopathology 32, 22-23.

Hussey, G. & Gunn, H.V. (1984). Plant production in pea (Pisum sativum L. cvs Puget and Upton) from long-term callus with superficial meristems. Plant Science Letters, in press.

International Atomic Energy Agency (1977). Induced Mutations Against Plant Diseases. IACA, Vienna, 581 pp.

Inze, D., Follin, A., Van Lijsebettens, M., Simoens, C., Genetello, C., Van Montagu, M. & Schell, J. (1984). Genetic analysis of the individual T-DNA genes of Agrobacterium tumefaciens; further evidence that two genes are involved in indole-3-acetic acid synthesis. Molecular and General Genetics 194. 265-274.

Jinks, J.L., Caligari, P.D.S. & Ingram, N.R. (1981). Gene transfer in Nicotiana rustica using irradiated pollen. Nature 291, 586-588.

Jørgensen, J.H. (1977). Location of the ml-o locus on barley chromosome 4. In 'Induced Mutations Against Plant Diseases' (International Atomic Energy Agency, ed.), pp. 533-549. IEAE, Vienna.

Kaars Sijpesteijn, A. (1969). Mode of action of phenylthiourea, a therapeutic agent for cucumber scab. Journal of the Science of Food and Agriculture 20, 403-405.

Kao, J.C., Perry, K.L. & Kado, C.I. (1982). Indoleacetic acid complementation and its relation to host range specifying genes on the Ti plasmid of Agrobacterium tumefaciens. Molecular and

General Genetics 188, 425-432.
Kao, K.N. (1977). Chromosome behaviour in somatic hybrids of soybean and Nicotiana glauca. Molecular and General Genetics 150, 225-230.
Koukolikova-Nicola, Z., Shillito, R.D., Hohn, B., Wang, K., Van Montagu, M. & Zambryski, P. (1984). T-DNA transfer from Agrobacterium to the plant cell involves induction of circular intermediates. Nature, in press.
Kozac, M. (1981). Possible role of flanking nucleotides in recognition of the AUG initiator codon by eukaryotic ribosomes. Nucleic Acids Research 9, 5233-5252.
Langcake, P., Cartwright, D. & Ride, J.P. (1981). The dichlorocyclopropanes and other fungicides with indirect mode of action. In 'Systemic Fungicides and other Antifungal Compounds' (Lyr, H. & Polter, C., eds), pp. 199-210. Akademie-Verlag, Berlin, DDR.
Langcake, P., Kuhn, P.J. & Wade, M. (1983). The mode of action of systemic fungicides. In 'Progress in Pesticide Biochemistry and Toxicology', Vol. 3. (Hutson, D.H. & Roberts, T.R., eds), pp. 1-109. John Wiley & Sons, Chichester.
Langridge, J. (1957). The aseptic culture of Arabidopsis thaliana. Australian Journal of Biological Science 10, 243-252.
Larkin, P.J. & Scowcroft, W.R. (1981). Somaclonal variation, a novel source of variability from cell cultures for plant improvement. Theoretical and Applied Genetics 60, 197-214.
Lazzeri, P.A. & Dunwell, J.M. (1984). Establishment of isolated root cultures of Brassica species and regeneration from cultured root segments of Brasssica oleraceae var. italica. Annals of Botany 54, 351-361.
Leemans, J., De Blaere, R., Willmitzer, L., De Greve, H., Hernalsteens, J.P., Van Montagu, M. & Schell, J. (1982). Genetic identification of functions of TL-DNA transcripts in octopine crown galls. EMBO Journal 1, 147-152.
Leemans, J., Shaw, C., De Blaere, R.; De Greve, H., Hernalsteens, J.P., Maes,, M., Van Montagu, M. & Schell, J. (1981). Site-specific mutagenesis of Agrobacterium Ti plasmids and transfer of genes to plant cells. Journal of Molecular and Applied Genetics 1, 149-164.
Lemmers, M., De Beuckeleer, M., Hollsters, M., Zambryski, P., Depicker, A., Hernalsteens, J.P., Van Montagu, M. & Schell, J. (1980). Internal organization, boundaries and integration of Ti-plasmid DNA in nopaline crown gall tumours. Journal of Molecular Biology 144, 353-376.
Lillis, M., Spielman, A. & Simpson, R.B. (1984). Unusual activity of a maize transposable element in tobacco cells. Nature, in press.
Lurquin, P.F. (1979). Entrapment of plasmid DNA by liposomes and their interactions with plant protoplasts. Nucleic Acids Research 6, 3773-3784.
Maniatis, T., Fritsch, E.F. & Sambrook, J. (1982). 'Molecular Cloning, a Laboratory Manual'. Cold Spring Harbor Laboratory, U.S.A. 545 pp.
Matzke, A.J.M. & Chilton, M.-D. (1981). Site specific insertion of genes into T-DNA of the Agrobacterium tumor-inducing plasmid: an approach to genetic engineering of higher plant cells. Journal of Molecular and Applied Genetics 1, 39-49.
Matzke, M.A., Susani, M., Binns, A.N., Lewis, E.D., Rubinstein, I. & Matzke, A.J.M. (1984). Transcription of a zein gene introduced into sunflower using a Ti-plasmid vector. EMBO Journal 3, 1525-1531.
Melchers, G. (1984). Topatoes and pomatoes, somatic hybrids between tomatoes and potatoes. In 'Tissue Culture' (Röhlich, P. & Bacsy, E., eds), pp. 499-513. Akademiai Kiado, Budapest.
Melchers, G., Sacristan, M.-D. & Holder, A.A. (1978). Somatic hybrid plants of potato and tomato regenerated from fused protoplasts. Carlsberg Research Communications 43, 203-218.
Merlo, D.J., Nutter, R.C., Montoya, A.L., Garfinkel, D.J., Drummond, M.H., Chilton, M.-D., Gordon, M.P. & Nester, E.W. (1980). The boundaries and copy numbers of Ti-plasmid T-DNA vary in crown gall tumors. Molecular and General Genetics 177, 637-645.
Muller, A., Manzara, T. & Lurquin, P.F. (1984). Crown gall transformation of tobacco callus cells by co-cultivation with Agrobacterium tumefaciens. Biochemical and Biophysical Research Communications 123, 458-462.
Murai, N., Sutton, D.W., Murray, M.G., Slighton, J.L., Merlo, D.J., Reichert, N.A., Sengupta-Gopalan, C., Stock, C.A., Braker, R.F., Kemp, J.D. & Hall, T.C. (1983). Phaseolin gene from bean is expressed after transfer to sunflower via tumor-inducing plasmid vectors.

Science 222, 476-482.

Murray, M.G. & Thompson, W.F. (1982). Repeat sequence interspersion in coding DNA of peas does not reflect that in total pea DNA. Plant Molecular Biology 1, 143-153.

Norton, R.A. & Towers, G.H.N. (1984). Transmission of nopaline crown gall tumour markers through meiosis in regenerated whole plants of *Bidens alba*. Canadian Journal of Botany 62, 408-413.

Ooms, G., Bakker, A., Molendijk, L., Wullems, G.J., Gordon, M.P., Nester, E.W. & Schilperoort, R.A. (1982a). T-DNA organization in homogeneous and heterogeneous octopine-type crown gall tissues of *Nicotiana tabacum*. Cell 30, 589-597.

Ooms, G., Hooykaas, P.J.J., Van Veen, R.J.M, Van Beelen, P., Regensburg-Tuink, A.J.G. & Schilperoort, R.A. (1982b). Octopine Ti-plasmid deletion mutants of *Agrobacterium tumefaciens* with emphasis on the right side of the T-region. Plasmid 7, 15-29.

Pandey, K.K. (1975). Sexual transfer of specific genes without gametic fusion. Nature 256, 310-313.

Pandey, K.K. (1980). Gametic gene transfer in *Nicotiana* by means of irradiated pollen. Genetica 49, 56-69.

Paszkowski, J., Shillito, R.D., Saul, M., Mandak, V., Hohn, T., Hohn, B. & Potrykus, I. (1984). Direct gene transfer to plants. EMBO Journal 3, 2717-2722.

Paterson, B.M., Roberts, B.E. & Kuff, E.L. (1979). Structural gene identification and mapping by DNA-mRNA hybrid-arrested cell-free translation. Proceedings of the National Academy of Sciences, U.S.A. 74, 4370-4374.

Powell, W., Caligari, P.D.S. & Hayter, A.M. (1983). The use of pollen irradiation in barley breeding. Theoretical and Applied Genetics 65, 73-76.

Power, J.B., Berry, S.F., Chapman, J.V. & Cocking, E.C. (1980). Somatic hybridization of sexually incompatible Petunias: *Petunia parodii*, *Petunia parviflora*. Theoretical and Applied Genetics 57, 1-4.

Russell, G.E. (1978). 'Plant Breeding for Pest and Disease Resistance.' Butterworths, London, 485 pp.

Ruvkun, G.B. & Ausubel, F.M. (1981). A general method for site-directed mutagenesis in prokaryotes. Nature 289, 85-88.

Sacristan, M.-D. & Melchers, G. (1977). Regeneration of plants from 'habituated' and '*Agrobacterium*-transformed' single cell clones of tobacco. Molecular and General Genetics 152, 111-117.

Schiller, B., Herrmann, R.G. & Melchers, G. (1982). Restriction endonuclease analysis of plastid DNA from tomato, potato and some of their somatic hybrids. Molecular and General Genetics 186, 453-459.

Scowcroft, W.R. & Larkin, P.J. (1982). Somaclonal variation; a new option for plant improvement. In 'Plant Improvement and Somatic Cell Genetics' (Vasil, I.K., Scowcroft, W.R. & Frey, K.J., eds) pp. 159-178. Academic Press, New York.

Shakoor, A., Ahsamml-Haq, M., Sadiq, M. & Sarwar, G. (1977). Induction of resistance to yellow mosaic virus in mungbean through induced mutations. In 'Induced Mutations Against Plant Disease' (International Atomic Energy Agency, ed.) pp. 293-302. IEAE, Vienna.

Shepard, J.F., Bidney, M.R., & Shahin, E. (1980). Potato protoplasts in crop improvement. Science 208, 17-24.

Simpkins, I., Walkey, D.G.A. & Neely, H.A. (1981). Chemical suppression of virus in cultured plant tissues. Annals of Applied Biology 99, 161-169.

Skoog, F. & Miller, C.D. (1957). Chemical regulation of growth and organ formation in plant tissues cultured in vitro. Symposia of the Society for Experimental Biology 11, 851-855.

Smith, V.A. (1979). A study of Agrocin 84-mediated suppression of crown gall tumour induction. Ph. D. Thesis, University of Bristol. U.K.

Snape, J.W., Parker, B.B., Simpson, E., Ainsworth, C.C., Payne, P.I. & Law, C.N. (1983). The use of irradiated pollen for differential gene transfer in wheat (*Triticum aestivum*). Theoretical and Applied Genetics 65, 103-111.

Stavely, J.R., Pittarelli, G.W. & Burk. L.G. (1973). *Nicotiana rependa* as a potential source of disease resistance in *N. tabacum*. Journal of Heredity 64, 265-271.

Strobel, G.A. (1973). The helminthosporoside-binding protein of sugar cane. Journal of Biological

Chemistry 248, 1321-1328.
Strommer, J.N., Hake, S., Bennentzen, J., Taylor, W.C. & Freeling, M. (1982). Regulatory mutants of the maize Adh1 gene caused by DNA insertions. Nature 300, 542-544.
Sutcliffe, J.G. (1978). Complete nucleotide sequence of the Escherichia coli plasmid pBR322. Cold Spring Harbor Symposium 43, 77-90.
Tanaka, N., Ikegami, M., Hohn, T., Matsui, C. & Wantanabe, I. (1984). E. coli spheroplast-mediated transfer of cloned cauliflower mosaic virus DNA into plant protoplasts. Molecular and General Genetics 195, 378-380.
Thomas, T.H. (1974). Investigations into the cytokinin-like properties of benzimidazole-derived fungicides. Annals of Applied Biology 76, 237-241.
Thomashow, M.F., Nutter, R., Montoya, A.L., Gordon, M.P. & Nester, E.W. (1980). Integration and organisation of Ti-plasmid sequences in crown gall tumors. Cell 19, 729-739.
Tokousbalides, M.G. & Sisler, H.D. (1979). Site of inhibition by tricyclazole in the melanin biosynthetic pathway of Verticillium dahliae. Pesticide Biochemistry and Physiology 8, 26-32.
Turner, P., Barber, C. & Daniels, M. (1984). Behaviour of the transposons Tn5 and Tn7 in Xanthomonas campestris pv. campestris. Molecular and General Genetics 195, 101-107.
Werner, C.P., Dunkin, I.M., Cornish, M.A. & Jones, G.H. (1984). Gene transfer in Nicotiana rustica by means of irradiated pollen. II. Cytogenetical consequences. Heredity 52, 113-119.
Willmitzer, L., Dhaese, P., Schreier, P.H., Schmalenback, W., Van Montagu, M. & Schell, J. (1983). Size, location and polarity of T-DNA encoded transcripts in nopaline crown gall tumors; common transcripts in octopine and nopaline tumors. Cell 32, 1045-1056.
Yadav, N.S., Vanderleyden, J., Bennett, D.R., Barnes, W.M. & Chilton, M.-D. (1982). Short direct repeats flank the T-DNA on a nopaline Ti-plasmid. Proceedings of the National Academy of Sciences, U.S.A. 79, 6322-6326.
Yang, F., Montoya, A.L., Merlo, D.J., Drummond, M.H., Chilton, M.-D., Nester, E.W. & Gordon, M.P. (1980). Foreign DNA sequences in crown gall teratoma and their fate during the loss of the tumorous traits. Molecular and General Genetics 177, 707-714.
Zambryski, P., Depicker, A., Kruger, K. & Goodman, H.M. (1982). Tumor induction by Agrobacterium tumifaciens: analysis of the boundaries of the T-DNA. Journal of Molecular and Applied Genetics 1, 361-370.
Zambryski, P., Joos, H., Genetello, C., Leemans, J., Van Montagu, M. & Schell, J. (1983). Ti plasmid vector for the introduction of DNA into plant cells without alteration of their normal regeneration capacity. EMBO Journal 2, 2143-2150.

INDEX

Hosts and pathogens are listed only where there is major treatment of the species in the text. Viruses are indexed by abbreviated names; fungi and bacteria by binomials, and hosts by common names.